D1226923

Low-Speed Wind Tunnel Testing

Model of the Space Shuttle in the NASA Ames 40 ft × 80 ft low-speed wind tunnel. (Photograph courtesy of NASA Ames.)

Low-Speed Wind Tunnel Testing

SECOND EDITION

WILLIAM H. RAE, JR.
Associate Professor and Associate Director
University of Washington Aeronautical Laboratory
F. K. Kirsten Wind Tunnel

ALAN POPE
Formerly Director of Aerospace Projects,
Sandia National Laboratories

A WILEY-INTERSCIENCE PUBLICATION
JOHN WILEY & SONS
New York · Chichester · Brisbane · Toronto · Singapore

Library of Congress Cataloging in Publication Data:

Pope, Alan, 1913–
Low-speed wind tunnel testing.

"A Wiley-Interscience publication."
Includes index.
1. Wind tunnels. I. Rae, William H. II. Title.
TL567.W5P694 1984 629.134'52 84-3700
ISBN 0-471-87402-7

Printed in the United States of America

10 9 8 7 6 5

Preface

Over the last 15 years, the use of low-speed wind tunnels to solve engineering problems has been very active. During this period the wind tunnel has been used more frequently to solve nonaeronautical problems in the fields of ground transportation, architecture, and environmental studies.

The advent of solid-state electronics and the micro/mini digital computer has, and is having, a large impact on instrumentation, tunnel operation, data acquisition, and display. The computer now enables a tunnel to obtain, process, and display the data at rates and in amounts that were only dreamed of a short time ago.

Despite these shifts to automated systems, the basic methods and theory have not changed. In fact, because of the speed of the automated systems, it is now more necessary than before to clearly and thoroughly understand the basic test techniques and objectives so that tests can be intelligently planned and carried out.

As in the previous edition, the scope of the book remains the same: to help students taking a wind tunnel course and to furnish a reference source to wind tunnel engineers and those who use a wind tunnel to solve a problem. Because of this wide scope the reader will still find sections on tunnel design, calibration, and simple instrumentation that often prove most useful in checking the more sophisticated systems. The subject of high-speed wind tunnel testing is covered in *High-Speed Wind Tunnel Testing* by Pope and Goin (John Wiley & Sons, New York, 1965).

We hope that this revised edition will prove as useful to the engineering community as the previous edition has.

William H. Rae, Jr.
Alan Pope

Seattle, Washington
Albuquerque, New Mexico
May 1984

Acknowledgments

The senior author sincerely appreciates the cooperation received from members of the Subsonic Aerodynamic Testing Association (SATA) in providing photographs and ideas, especially Dick Day (Boeing), Myron Morrison (McDonnell–Douglas, St. Louis), Andy Morse (U.S. Army, Ames), and Joe Alford (Dynamic Engineering, Inc.). Special thanks go to Erik Hansen of UWAL for his darkroom work. For their help and suggestions, I am grateful to Harry Heyson (NASA, Langley), Jim Hackett (Lockheed, Georgia), Mal Holcomb (Beech Aircraft), Bob Joppa and my tireless assistant Luke Shindo (University of Washington), and the student crew members of the UWAL tunnel. Very special thanks to Nina Seco who typed and typed and retyped, and to my wife and son for patiently enduring their disrupted weekends and evenings.

W.H.R.

Contents

Abbreviations

In view of the large number of aeronautical research centers being set up, a list such as this must be considered incomplete. However, it may be of help in identifying the source of particular publications.

Abbreviation	Complete Meaning
ACA	Australian Council for Aeronautics, CSIR
AEDC	Arnold Engineering Development Center (Air Force), Tullahoma, Tennessee
AFCRC	Air Force Cambridge Research Center, Cambridge, Massachusetts
AFSWC	Air Force Special Weapons Center, Albuquerque, New Mexico
AFWL	Air Force Weapons Laboratory, Kirtland Air Force Base, Albuquerque, New Mexico
AGARD	Advisory Group for Aeronautical Research and Development
AIAA	American Institute of Aeronautics and Astronautics
APL	Applied Physics Laboratory, Johns Hopkins University, Silver Spring, Maryland
ARC	Air Research Committee (British)
ARC	Ames Research Center of the NASA, Moffett Field, California
ARIS	Aeronautical Research Institute of Sweden, Ulsvunda, Sweden
ARI, TIU	Air Research Institute, University of Tokyo (Japanese)
AVA	Aerodynamische Versuchsanstalt (Göttingen Institute for Aerodynamics), Göttingen, Germany

Abbreviation	Complete Meaning
BRL	Ballistic Research Laboratory, Aberdeen Proving Ground, Maryland
CAI	Central Aerohydrodynamic Institute, Moscow, U.S.S.R.
CSIR	Council for Scientific and Industrial Research, Australia
CNRC	Canadian National Research Council, Ottawa, Canada
DNW	German–Dutch Windtunnel, Noordoostpolder, The Netherlands
DTMB	David Taylor Model Basin (Navy), Carderock, Maryland
DVL	Deutsche Versuchsanstalt für Luftfahrtforschung (German Institute for Aeronautical and Space Research), Berlin and Göttingen, Germany
ETH	Eidgenossische Technische Hochschule (Swiss Institute of Technology)
GALCIT	Guggenheim Aeronautical Laboratory of the California Institute of Technology, Pasadena, California
JAM	*Journal of Applied Mechanics*
JAS	*Journal of the Aeronautical Sciences* (United States)
JPL	Jet Propulsion Laboratory, California Institute of Technology, Pasadena, California
JRAS	*Journal of the Royal Aeronautical Society* (British)
LFA	Luftfahrtforschungsanstalt Hermann Göring (Hermann Göring Institute for Aeronautics), Braunschweig, Germany
LRC	Langley Research Center of the NASA, Langley Field, Virgina
LRC	Lewis Research Center of the NASA, Cleveland, Ohio
MIT	Massachusetts Institute of Technology, Cambridge, Massachusetts
MSC	Manned Spacecraft Center of the NASA, Houston, Texas
MSFC	Marshall Space Flight Center of the NASA, Huntsville, Alabama
NACA	National Advisory Committee for Aeronautics, now NASA
NAMTC	Naval Air Missile Test Center, Point Mugu, California
NASA	National Aeronautics and Space Administration, Washington, D.C.
NOL	Naval Ordnance Laboratory, White Oaks, Maryland
NPL	National Physical Laboratory, Teddington, Middlesex, England
NRTS	National Reactor Testing Station, Arco, Idaho
ONERA	Office National d'Études et de Recherches Aeronautiques (National Bureau of Aeronautical Research), Paris, France
ONR	Office of Naval Research, Washington, D.C.
ORNL	Oak Ridge National Laboratory, Oak Ridge, Tennessee
PRS	*Proceedings of the Royal Society of London* (British)
QAM	*Quarterly of Applied Mechanics*

Abbreviation	Complete Meaning
R&M	*Reports and Memoranda* (of the Air Research Committee)
RAE	Royal Aeronautical Establishment, Farnborough, Hants, England
RM	*Research Memorandum of the NASA*
SAE	Society of Automotive Engineers (United States)
TCEA	Training Center for Experimental Aerodynamics, Rhode-Saint-Genèse, Belgium
TM	*Technical Memorandum of the NASA*
TN	*Technical Note of the NASA*
TR	*Technical Report of the NASA*
WADC	Wright Air Development Center, Wright-Patterson Air Force Base, Ohio

Low-Speed Wind Tunnel Testing

Introduction

The earliest attempts by humans to design heavier-than-air machines, or airplanes, was based on the observation of birds in flight. Most of these machines used flapping wings (ornithopters) powered by humans through various mechanisms. In the 15th century Leonardo da Vinci used this approach, among others, and he left a legacy of over 500 sketches and 35,000 words dealing with the problem of flight. All of the attempts at flight by human-powered ornithopters were failures. By the 18th and 19th centuries it was realized that our knowledge of what we now call aerodynamics was minuscule. This led to the concept of building instrumented facilities to measure aerodynamic forces and moments.

The early experimenters recognized that aerodynamic forces were a function of the relative velocity between the air and the model. Thus one could either move the model through still air or let the air move past a stationary model. Initially this led to the use of nature's natural winds as a source of moving air, but the perversity of nature quickly led to the building of mechanical devices to move the model through the air. The simplest method of accomplishing this was to mount the model on a whirling arm that could be powered by falling weights and pulleys.

In the 18th century Robins, one of the early experimental aerodynamicists, used a whirling arm to study various shapes. He reached the conclusion that the relationships among a body's shape, orientation to the relative wind, and aerodynamic forces was much more complex than indicated by the then current theory.

In the late 18th and early 19th centuries Cayley began a systematic investigation of the lift and drag of airfoils with a whirling arm. His studies resulted in a small unmanned glider that is believed to be the first heavier-than-air machine. Cayley also espoused the new design philosophy that the wings should support the vehicle and power should be used to provide forward motion, rather than having the wings produce both lift and thrust, as in an ornithopter.

Near the end of the 19th century the major fault of a whirling arm was apparent. This fault was that the wing was forced to fly in its own disturbed wake (still a problem in helicopter rotors). This led to the wind tunnel where the model was held stationary and the air was moved past the model.

Although wind tunnels have been built in many different configurations, they all have four basic parts, which are:

1. A contoured duct to control the passage of the working fluid through the test section where the model is mounted.
2. A drive system to move the working fluid through the duct.
3. A model of the test object that is either full size or, more often, a reduced-scale model.
4. Instrumentation that may either be quite simple, such as a spring scale to measure force, or extremely complex, such as a modern balance feeding its output to relatively large digital computers.

The Wright brothers used the natural wind and a biplane kite in 1899 to solve the problem of roll control (by warping the wings). From their results they built their No. 1 glider in 1900. This glider proved that their pitch and roll controls worked, but the lift was lower and the drag higher than was expected (this is still a problem). The Wrights next used the natural wind and a balance to measure lift to compare cambered and noncambered airfoils. From these results they began to suspect that the accepted aerodynamic design data were in error. The No. 2 glider in 1901 used their results, but the pitching moment due to large airfoil camber made the glider difficult to control. This forced them to reduce the camber. By now the Wrights were convinced that the available aerodynamic data were in error and they decided to rely on their own data.

The Wrights then used a bicycle to provide wind to compare two test models mounted on a horizontal wheel ahead of the bicycle. The crude data from these tests reinforced their opinion to reject all published data.

The Wrights then built a wind tunnel; first a simple tunnel and then a larger and more sophisticated tunnel with a 16 × 16 sq. in. test section to obtain data for their No. 3 glider in 1902, which was successful with the addition of a rudder to counteract the adverse yaw from the warped wing roll control. The famous 1903 Wright flyer closely duplicated the No. 3 glider with the addition of a 12-hp engine and two counterrotating propellers.

Curiously, the only mistake the Wrights made in their large wind tunnel was the placement of the fan upstream of the test section. The use of honeycombs and screens reduced, but did not remove, the fan's swirl and turbulence. Although it was the Wright brothers that conclusively demonstrated the value of wind tunnels in aerodynamic design in 1903, it was the Europeans who capitalized on this fact by using about a dozen major wind tunnels built in government-funded aeronautical laboratories to achieve technical leadership in aviation between 1903 and 1914 (Ref. 1.1).

This spurt in wind tunnel construction by the Europeans has led to the two basic types of wind tunnels being known by European names. The open circuit or nonreturn tunnel, where the surrounding room acts as an air return, is known as the Eiffel or NPL type, named for the French engineer and the British Laboratory (Fig. 1.1). The tunnel with a closed return or a continuous circuit using vanes at the corners to turn the air is known as the Prandtl or Göttingen type, named for the German aerodynamicist and the university at which it was built (Fig. 1.2).

Before a detailed discussion of the various tunnels that have been built, a brief discussion is in order of the testing parameters that must be considered in using scale-model test results to predict full-scale behavior.

CHAPTER 1

The Wind Tunnel

Experimental information useful for solving aerodynamic problems may be obtained in a number of ways: from flight tests; drop tests; rocket sleds; water tunnels; whirling arms; shock tubes; water tables; rocket flights; flying scale models; ballistic ranges; and subsonic, nearsonic, transonic, supersonic, and hypersonic wind tunnels. Each device has its own sphere of superiority, and no one device can be called "best."

This book considers only the design and use of low-speed wind tunnels, where low speed means below 300 mph or so. An alternative definition would be "where compressibility is negligible."

Because they make it possible to use models and because they are always available, wind tunnels offer a rapid, economical, and accurate means for aerodynamic research. Their use saves both dollars and lives.

The nations of the world support aeronautical research, of which wind tunnel testing is a major item, according to their abilities and desires. Usually each nation sets up a separate organization that augments the activities of the armed services, and further work is farmed out to universities and industry. In the United States this central agency is the National Aeronautics and Space Administration, with offices in Washington, D.C. and whose laboratories are at the Goddard Space Flight Center in Maryland, the Langley Research Center in Virginia, the John F. Kennedy Space Center in Florida, the Marshall Space Flight Center in Alabama, the Mississippi Test Facility in Mississippi, the Manned Spacecraft Center in Texas, the Lewis Research Center in Ohio, the Flight Research Center in California, the Ames Research Center in California, and the Jet Propulsion Laboratory in California.

In addition the armed services have tunnels of their own. The Air Force has several at Wright-Patterson AFB, Ohio, and at Arnold Engineering Development Center, Tennessee. The Navy has tunnels at the David Taylor Naval Ship R&D Center in Carderock, Maryland and the Naval Ordnance Laboratory at White Oaks, Maryland. The Army has tunnels at the Aberdeen Proving Grounds, Maryland and Ames Research Center, California. In addition, nearly every aircraft corporation has at least one wind tunnel.

Since the wind tunnel is a device intended primarily for scale-model testing, it is proper that we pause now and consider how scale tests can best be conducted so that the results may be most effectively applied to full-scale craft.

1.1. IMPORTANT TESTING PARAMETERS

When a body moves through a medium, forces arise that are due to the viscosity of the medium, its inertia, its elasticity, and gravity. The inertia force is proportional to the mass of air affected and the acceleration given that mass. Thus, while it is true that a very large amount of air is affected by a moving body (and each particle of air a different amount), we may logically say that the inertia force is the result of giving a constant acceleration to some "effective" volume of air. Let this effective volume of air be kl^3, where l is a characteristic length of the body and k is a constant for the particular body shape. Then we may write

$$\text{Inertia force} \sim \rho l^3 V/t$$

where ρ = the air density, slug/ft^3; V = velocity of the body; ft/sec; t = time, sec.

Substituting l/V for t, we get

$$\text{Inertia force} \sim \frac{\rho l^3 V}{l/V} \sim \rho l^2 V^2 \tag{1.1}$$

The viscous force, according to its definition, may be written

$$\text{Viscous force} \sim \mu V l \tag{1.2}$$

where μ = coefficient of viscosity, slug/ft-sec.

The gravity force is simply

$$\text{Gravity force} = \rho l^3 g \tag{1.3}$$

where g = acceleration of gravity.

Be definition, the bulk modulus of elasticity of a gas is the stress needed to develop a unit change in volume. It is given the symbol E and has the units of pounds per square foot. We have then

$$\text{Elastic force} \sim El^2 \tag{1.4}$$

The speed of sound in air a is related to its elasticity according to

$$E = \rho a^2$$

so that we may write

$$\text{Elastic force} \sim \rho a^2 l^2$$

The important force ratios (as identified with the men who first drew attention to their importance) then become

$$\text{Reynolds number} = \frac{\text{Inertia force}}{\text{Viscous force}} = \frac{\rho}{\mu} Vl \tag{1.5}$$

$$\text{Mach number} = \frac{\text{Inertia force}}{\text{Elastic force}} = \frac{V}{a} \tag{1.6}$$

$$\text{Froude number} = \frac{\text{Inertia force}}{\text{Gravity force}} = \sqrt{\frac{V^2}{lg}} \tag{1.7}$$

Many wind tunnel tests are seriously sensitive to Reynolds number effects and no test should be attempted without reading Sections 7.1 and 7.2, and following the study with a discussion with the experienced operators of the tunnel to be used.

The last equation, it will be noted, uses the square root of the ratio rather than the ratio itself.

If a model test has the same Reynolds and Mach numbers as the full scale vehicle then the flow about the model and the full scale vehicle will be identical. Under these conditions, the forces and moments developed by the model can be directly scaled to full scale. Furthermore, for a free flight model (a spin or dynamic model) the Froude number must be matched.

The largest portion of wind tunnel tests are made with rigid models held in a fixed attitude within the tunnel thus it is not necessary to match the Froude number. The matching of Mach number usually applies only to flight vehicles in the high speed flight region as Mach number effects predominate and the matching of Reynolds number effects is not as critical. In the low speed flight region Reynolds number effects predominate and matching of Mach number is not as critical. However, for any test a careful evaluation of the effect of Reynolds and Mach numbers should be made to insure that the results are valid.

Despite the fact that it is difficult, if not impossible, to match both Reynolds and Mach numbers in most wind tunnel tests, the wind tunnel still is one of the most useful tools an aerodynamics engineer has available to him or her. The wind tunnel is, of course, an analog computer, and in it the aerodynamics engineer can quickly and efficiently optimize his or her design. In the last few years the use of dedicated mini digital computers has greatly decreased the time required to present the final corrected data in tabulated form, and in most cases large facilities can provide the results in plotted format in real time as the data are acquired. This enables the aerodynamics engineer both to check his or her predicted results and, based on the results of one tunnel run, to decide on the next run. Furthermore, properly conducted wind tunnel tests, with results extrapolated for full scale, can drastically reduce the amount of flight testing required and thus pay for both the model and wind tunnel testing.

1.2. TYPES OF WIND TUNNELS

There are two basic types of wind tunnels and two basic test-section configurations.

The first basic tunnel type is an open circuit tunnel (Fig. 1.1). In this type of tunnel the air follows a straight path from the entrance through a contraction to the test section, followed by a diffuser, a fan section, and an exhaust of the air. The tunnel may have a test section with no solid boundaries (open jet or Eiffel type) or solid boundaries (closed jet or NPL type).

The second basic type is a closed return wind tunnel (Prandtl or Göttingen type). This tunnel has a continuous path for the air. The great majority of the closed circuit tunnels have a single return (Fig. 1.2), although tunnels with both double and annular returns have been built. Again, the closed circuit

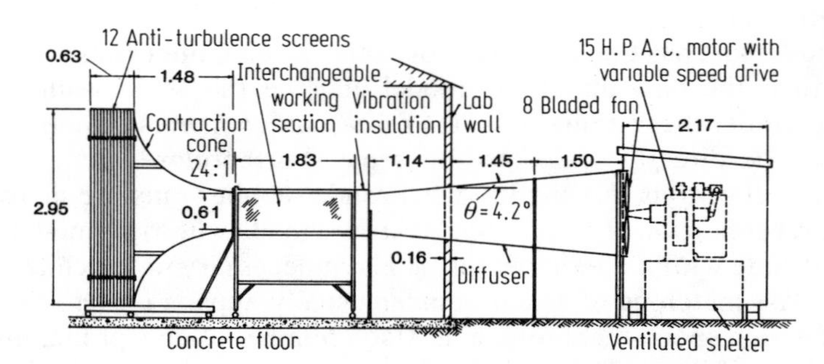

FIGURE 1.1. An NPL-type tunnel, or open return tunnel. Many small tunnels are of this type and use a surrounding room to return the air to the inlet. The tunnel shown is a smoke tunnel. (Reprinted from *Aeronautics & Astronautics,* Vol. 20, No. 1, Jan. 1982, p. 52. Copyright American Institute of Aeronautics and Astronautics.)

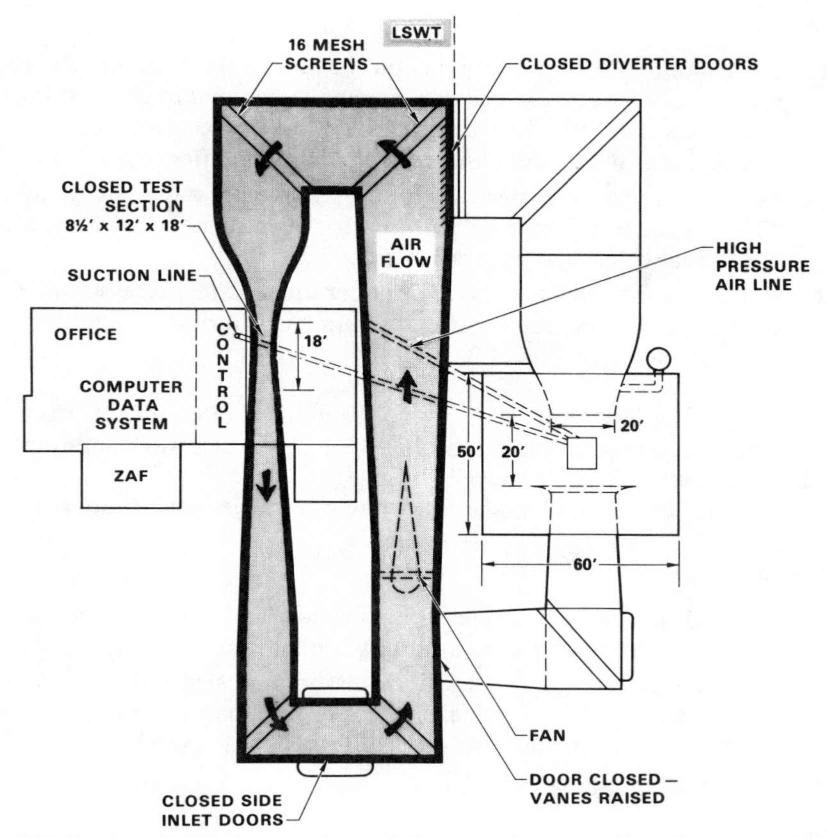

FIGURE 1.2. A Göttingen-type or closed return tunnel. This tunnel has a single return duct but tunnels have been built with annular and double returns. (Courtesy of McDonnell Aircraft Company.)

tunnel may have either a closed or open test section, and some have been built that can be run with either an open or closed test section, as desired.

As with any engineering design, there are advantages and disadvantages with both the open and closed circuit type tunnels. In general, the type of tunnel decided upon depends on funds available and purpose.

An open circuit tunnel has the following advantages and disadvantages:

Advantages

1. Construction cost is less.
2. If one intends to run internal combustion engines or do much flow visualization via smoke, there is no purging problem if both inlet and exhaust are open to the atmosphere.

Disadvantages

1. If located in a room, depending on the size of the tunnel to the room size, it may require extensive screening at the inlet to get high-quality flow. The same may be true if the inlet and/or exhaust is open to the atmosphere, when wind and cold weather can affect operation.
2. For a given size and speed the tunnel will require more energy to run. This is usually a factor only if used for developmental testing where the tunnel has a high utilization rate.
3. In general, a tunnel is noisy. For larger tunnels (test sections of 70 sq. ft and more) noise may cause environmental problems and limits on hours of operations.

Because of the low initial cost, an open circuit is often ideal for schools and universities where a tunnel is required for classroom work and research and high utilization is not required.

A closed return tunnel has the following advantages and disadvantages:

Advantages

1. Through the use of corner turning vanes and possibly screens, the quality of the flow can be easily controlled.
2. Less energy is required for a given test-section size and velocity. This can be important for a tunnel used for developmental testing with high utilization (two or three shifts, five to six days a week).
3. Less noise when operating.

Disadvantages

1. Higher initial cost due to return ducts and corner vanes.
2. If used extensively for smoke tests or running of internal combustion engines, there must be a way to purge tunnel.
3. If tunnel has high utilization, it may have to have an air exchanger or some other method of cooling during hot summer months.

Open or Closed Test Section

An open test section in conjunction with an open circuit tunnel will require an enclosure around the test section to prevent air being drawn into the tunnel from the test section rather than the inlet.

For closed return tunnels of large size with an external balance, the open test section tends to have one solid boundary, since the balance must be shielded from the wind.

The mounting of ground planes is more difficult in an open test section. In larger size tunnels, access to the model for changes is also more difficult.

Thus, in general, most tunnels tend to have closed test sections. It also should be noted that in larger size tunnels a rectangular test section is preferable because it is easier to change a model when working off a flat surface.

Test-Section Size

The test section should have as large a cross-sectional area as possible. Ideally, a tunnel should be large enough to handle a full scale airplane. In fact, several tunnels were built in the 1920s through the 1940s to achieve this goal. However, in the late 1940s through the 1970s, and presumably in the future, the size of aircraft has grown. If one uses the rule of thumb that the model span should be less than 0.8 of tunnel width, then Howard Hughes' *Hercules,* or as more popularly known, the "Spruce Goose," with a 320 ft span would require a test section 400 ft wide. The cost of building and operating a tunnel of this size is staggering. The cost of building a model, transporting it, and erecting it in the tunnel, as well as making changes during a test program, would also be an interesting, albeit expensive task. Thus, it is apparent that for large modern aircraft, the concept of a full size tunnel is out of the question based on costs. Thus, most aircraft testing is done in tunnels with widths from 10 to 20 ft.

High-Reynolds-Number Tunnels

It is not practicable to obtain full-scale Reynolds numbers by use of a full-scale aircraft; however, there are methods of increasing the Reynolds number with smaller tunnels and models.

One of the oldest methods is to build a tunnel that can be pressurized. In fact, some of the earliest work in Reynolds number effects was done in the old (no longer in service) variable density tunnel at NASA Langley. This tunnel could be pressurized to 20 atm. The reason this approach was taken can be seen by examination of the equation of state for a perfect gas and the equation for Reynolds numbers:

$$\text{Equation of state} = P = \rho RT \tag{1.8}$$

$$\text{Reynolds number} = \frac{\rho}{\mu} Vl \tag{1.5}$$

Thus, if one increases the pressure by a factor of 20, the density, and hence the Reynolds number, for a given size and speed is increased by a factor of 20. There are some basic drawbacks to this method:

1. Cost of a tunnel shell to withstand high internal pressures.
2. Time to pump the tunnel up to pressure, and cost of pumps, etc.

3. High densities yield large dynamic pressures and large—and possibly excessive—model loads.
4. For larger tunnels there must be some way to seal the test section and to allow it to be depressurized for access to the model.

Despite these problems, there have been many tunnels built that can be pressurized to obtain higher Reynolds numbers.

A second approach is to change the working fluid. For a given power input the use of Freon 12 can increase the Mach number by a factor of 2.5 and the Reynolds number by a factor of 3.6. Again, many of the problems of a pressure tunnel will exist, such as initial cost, cost of pumps, cost of the gas, and a method of making the test section habitable for model changes.

A third approach is that of a cryogenic tunnel. NASA has built such a facility at its Langley Research Center called the National Transonic Facility (Ref. 1.2). Although this tunnel is for transonic testing, the same concept should work for a low-speed tunnel. This tunnel has the ability to change pressure up to 9 atm. The working fluid is nitrogen, and by injecting liquid nitrogen upstream of the fan, the gas is cooled. By this technique it is possible to operate over a range of dynamic pressures and Reynolds numbers at a constant temperature to the tunnel's stagnation pressure limit, similar to any pressure tunnel. Or, the tunnel can be run at constant dynamic pressure and by changing temperature, the Reynolds number can be changed. The range of Reynolds number for a given model size and Mach number is impressive, varying from 1×10^6 to over 100×10^6. This sort of facility is very expensive both to build and to operate, but it does show what can be achieved in a wind tunnel.

V/STOL Wind Tunnels

These tunnels require a much larger test section for a given size model owing to large downwash angles generated by powered lift systems in the transition flight region. Flight velocities in the transition region are low, thus tunnels with large test sections do not need high velocities, the maximum being in the 60–100 mph range, compared to the 200–300 mph range for a conventional low-speed wind tunnel. Since power varies with the cube of velocity, this reduces the installed power requirement. There will be, however, a demand to run the tunnel at higher speeds with conventional models, thus the tunnel will be powered for the higher speeds. This was the solution in both the Boeing Vertol tunnel (test section area is 400 sq. ft) and the NASA Langley tunnel (test section area about 300 sq. ft).

Another solution to the problem of building V/STOL tunnels is that taken by Lockheed Georgia. Lockheed built a tunnel with tandem test sections with two contractions. The first test section is for V/STOL or powered lift models and has a cross-sectional area of 780 sq. ft with speeds from 23 to 115 mph. The second test section has a cross section of 378 sq. ft and speeds from 58 to 253 mph. This design avoids the high installed power required to drive the larger test section at high speeds. The length of the tunnel is increased by this solution, increasing the shell cost.

Another approach to a large test section at low speeds for V/STOL and a smaller test section at higher speeds would be to design inserts for the large test section to reduce its area and increase its speed capability. This is attractive, but the time required to install and remove the large inserts may reduce the utility of the tunnel facility. The most economical approach to providing V/STOL tunnels has been to modify an existing wind tunnel.

One of the least expensive methods of obtaining V/STOL capability is to use some portion of an existing tunnel return circuit for a V/STOL test section. This could be the settling area ahead of the contraction cone as done by NASA Langley or, possibly, the end of the diffuser. These test sections

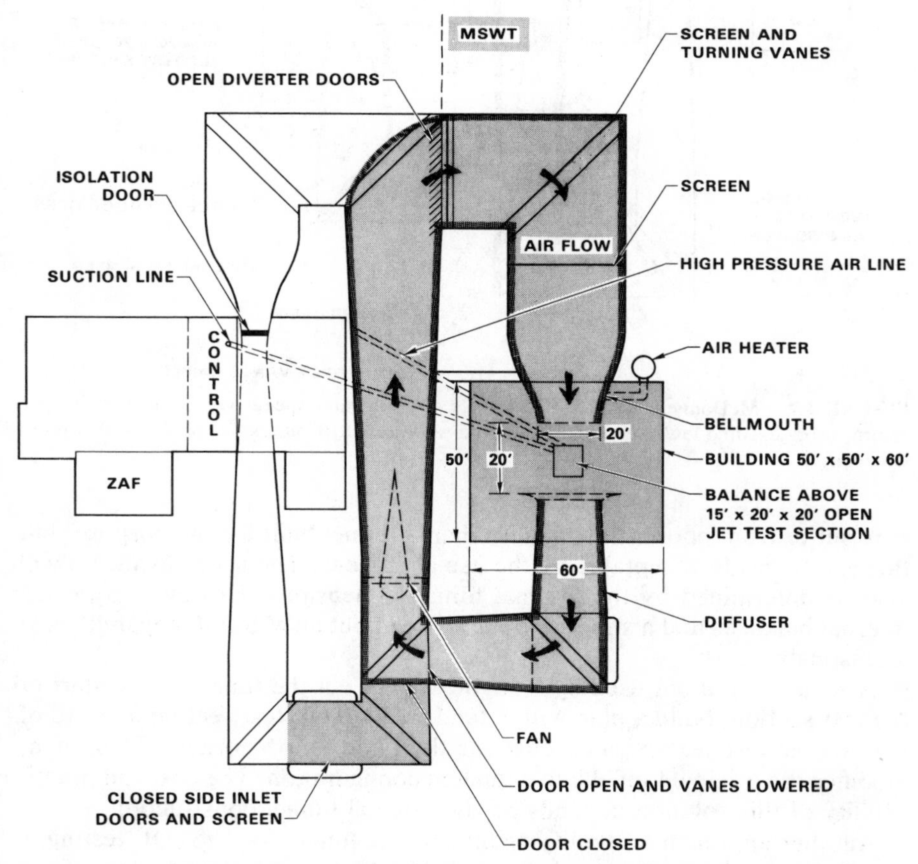

FIGURE 1.3. McDonnell Aircraft Company's 80 mph mini speed closed return tunnel using existing tunnel fan and return duct. (Courtesy of McDonnell Aircraft Company.)

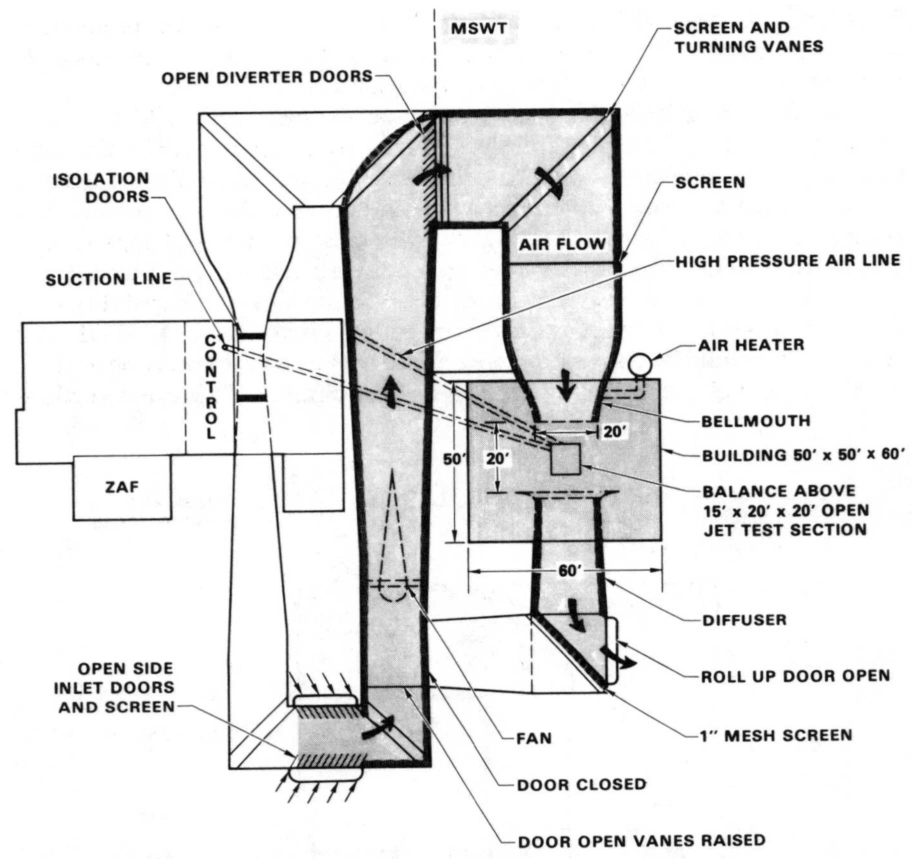

FIGURE 1.4. McDonnell Aircraft Company's 60 mph mini speed wind tunnel with open return, using existing low-speed tunnel fan and return duct. (Courtesy of McDonnell Aircraft Company.)

may suffer from poorer flow quality than a tunnel built for the purpose, but this can be made acceptable by the use of screens. The speed available will also be determined by the original tunnel dimensions. However, by using internal balances and a sting support, one can obtain V/STOL capabilities at a reasonable cost.

A second solution, used by Vought, was to cut the tunnel at the start of the test section, build a plug with a tandem V/STOL test section forward of the original test section, and reuse the third and fourth corners. This, then, modifies the original tunnel into a tandem configuration. The cost and practicability of this solution depends on the original tunnel construction.

Another approach to modifying an existing tunnel for V/STOL testing is to add another leg or legs to the tunnel. The McDonnell-Douglas low-speed tunnel actually has three legs in their tunnel (Fig. 1.3). With this arrange-

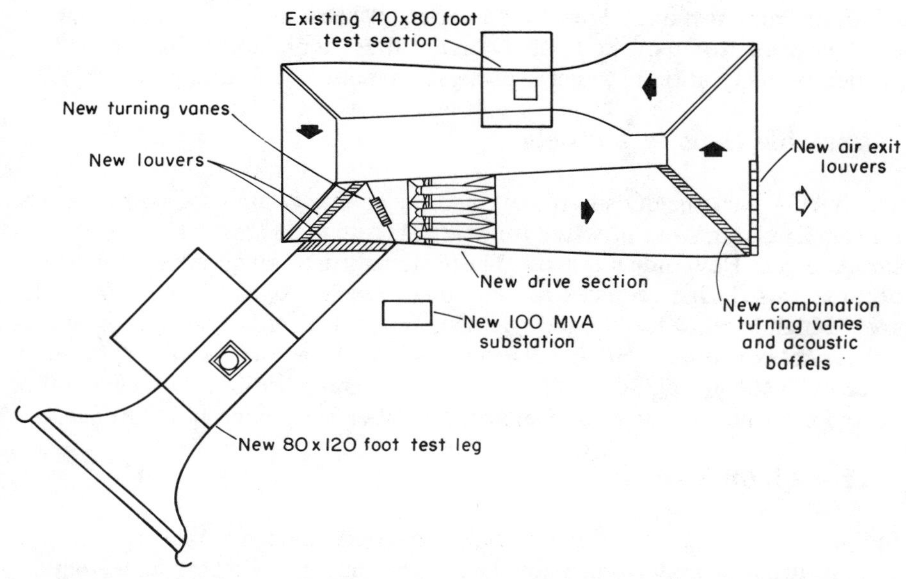

FIGURE 1.5. Modification to NASA Ames Research Laboratories 40 × 80 ft tunnel to add an 80 × 120 ft closed V/STOL test section. (Courtesy NASA Ames.)

ment, they can operate with an $8\frac{1}{2} \times 12$ ft closed test section up to 200 mph, or as a closed tunnel with an open throat V/STOL test section 15 × 20 ft up to 80 mph. The V/STOL 15 × 20 ft open throat test section can also be operated as an open return tunnel up to 60 mph (Fig. 1.4).

A second approach to adding another leg is the NASA Ames modification to the 40 by 80 ft closed throat tunnel (Fig. 1.5). The new leg forms basically an open circuit tunnel with a closed throat test section 80 × 120 ft. In this modification the tunnel was also repowered and new fans were built to increase the speed in both test sections (Ref. 1.3).

Another approach to an inexpensive V/STOL tunnel was the modification of an engine test cell by British Aerospace Aircraft. Again, this is an open circuit tunnel with a closed test section 18 × 18 ft. In the design and construction of the tunnel, many problems associated with large open circuit tunnels, such as the effect of gusts on the test section flow and the effect of weather, were addressed and solved.

These few examples show that there are many ingenious and practical solutions to adapting an existing facility for V/STOL testing.

1.3. SPECIAL-PURPOSE TUNNELS

Over the years there have been many tunnels built to meet either specific research or testing requirements. Many of these tunnels have been decommissioned and dismantled for a large variety of reasons or have been adapted to other uses. Some of the testing done in these tunnels can also be accomplished by adaptation of general-purpose tunnels, which are quite versatile.

Variable Density Tunnels

The NACA variable density tunnel, and the similar compressed-air tunnel at NPL in England, were pressure tunnels (VDT up to 20 atm) and were used to simulate high Reynolds numbers. These tunnels used an annular return duct, because this design required the minimum amount of steel. The VDT, despite a high level of turbulence, yielded a good deal of insight into the effect of Reynolds number on the characteristics of 78 airfoils as reported in NACA TR460 in 1933. The VDT is no longer in operation. However, other pressure tunnels to raise the Reynolds number have been built (Ref. 1.4).

Free-Flight Tunnels

In the 1930s several free-flight tunnels were built. These tunnels were of the open return type and a dimensional and dynamically scaled model flew in the tunnel under the influence of gravity. The tunnel could be tilted to match the glide path of the model. The dynamic behavior of the model could be studied in these tunnels, and often control surfaces could be deflected by command through a trailing wire. At present, none of these tunnels are in operation as a free-flight tunnel. NASA Langley still does free-flight testing in the 30 × 60 ft tunnel (Ref. 1.5) with powered models. This facility is used to study V/STOL transition, stalls, and loss of control of aircraft models (Fig. 1.6).

Spin Tunnels

The tendency of some aircraft to enter a spin after a stall and the spin recovery have been perennial problems of the aircraft designer. The recovery from a spin is studied in a spin tunnel. This is a vertical wind tunnel with the air drawn upward by a propeller near the top of the tunnel. The spin tunnel uses an annular return with turning vanes. The dynamically similar model is inserted into the tunnel by an operator in a spinning attitude. The tunnel airspeed is adjusted to hold the model at a constant height, and the model's motion is recorded by both movies and TV for later analysis. Spin tunnels have been built in several countries. See Section 5.17 for spin testing techniques (Refs. 1.6 and 1.7).

Stability Tunnels

In 1941 a stability tunnel was built at NASA Langley. This tunnel had two interchangeable test sections about 6 ft in size. One of the test sections had a set of rotating vanes that created a swirl in the airstream. The second test section was curved to simulate turning flight. This tunnel was decommis-

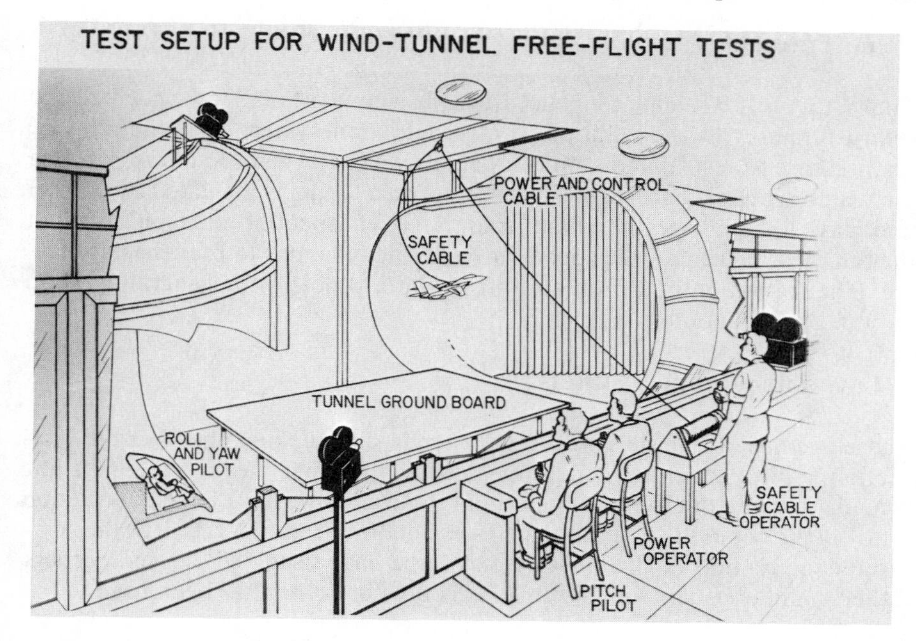

FIGURE 1.6. The test set up for free-flight tunnel tests in the NASA 30 × 60 ft tunnel. (Courtesy NASA Langley.)

sioned for a time in the late 1950s. Similar results can also be obtained by using oscillating model techniques in conventional tunnels or free-flight tests in the 30 × 60 ft tunnel (Ref. 1.8).

Propeller Tunnels

Propeller tunnels are similar to conventional tunnels with the exception that they usually have an open test section and a round cross section (see Sections 5.15 and 6.32). One of the earliest propeller tunnels was built at Stanford University in 1917 with a 5.5-ft-diameter test section. The propeller tunnel at NASA Langley with a 20-ft test section went into operation in 1922. Besides propeller tests, this tunnel gave insight into the location of engine nacelles relative to the wing and the design of cowls (NASA cowl) for radial engines to reduce drag and increase cooling (Ref. 1.9).

Propulsion Tunnels

Testing of aircraft engines, either reciprocating or jet, requires simulation of both flight velocity and the variation of atmospheric pressure and temperature. Since the engine must be operated in the tunnel, the exhaust gases must be removed from a closed circuit tunnel, or an open circuit tunnel must be used. The altitude requirement necessitates pumps to provide the low density, and the lower temperature at altitudes requires a refrigeration system.

Ice Tunnels

The NASA 6 × 9 ft icing tunnel at Lewis is a conventional low-speed closed return tunnel with the addition of a refrigeration system to reduce the air temperature to −40°F and atomizers upstream of the test section to produce the water droplets that freeze. The shell of this tunnel is insulated by 3 in. of fiberglass to help keep the tunnel cold. A novel aspect of this facility is that the fan must be run at idle speed during model changes to prevent its freezing. The formation of ice on aircraft is still a problem for general aviation aircraft and helicopters (Fig. 1.7).

Low-Turbulence Tunnels

These are usually tunnels with a wide angle diffuser just ahead of the test-section settling chamber. The large settling chamber has honeycombs and screens to damp out the turbulence and its size allows for a large contraction ratio to further reduce turbulence (see Sections 2.2 and 2.11). Some low-turbulence tunnels of the closed return type have used 180° curved corners rather than the usual two 90° turns. This type of corner has been used in the

FIGURE 1.7. Ice formation on rear of turbofan nacelle. (Photograph courtesy of NASA Lewis.)

NASA Langley low-turbulence pressure tunnel (LTPT), a two-dimensional tunnel, and the Ames 12-ft high-speed pressure tunnel. The Langley LTPT uses a wide angle diffuser ahead of the settling chamber. This tunnel has a 17.6 to 1 contraction ratio and 11 screens.

Two-Dimensional Tunnels

Two-dimensional tunnels are used primarily for testing of airfoil sections. They have been built both as open circuit and closed return types. These tunnels have tall narrow test sections with height-to-width ratios of 2 or greater. The tunnels are usually of the low-turbulence type and may be pressurized to increase the Reynolds number.

Smoke Tunnels

Smoke tunnels are used primarily for flow visualization. Usually these tunnels are of the nonreturn type, as photographs are the primary method of data recording (Fig. 8.4 and Refs. 1.10, 1.11, and 1.12). Both two- and three-dimensional smoke tunnels have been built. For smoke sources current practice seems to favor vaporized light oils (see Section 3.10). Smoke tunnels used for research rather than demonstration purposes tend to have large contraction ratios (up to 24 : 1) and a large number of antiturbulence screens at the inlet to obtain smooth laminar flow. In general, the speed of smoke tunnels tends to be low—around 30–60 fps. A supersonic smoke tunnel was built at the University of Notre Dame in 1959. The use of a modified Schlieren permitted simultaneous smoke and shock wave patterns to be photographed.

Lippisch developed a periodic smoke-injection system and was able to obtain pressure distribution about the model using high-speed movies. Smoke has been injected both just before the model and at the inlet. For research tunnels, injecting ahead of the inlet and antiturbulence screens reduces the turbulence from the smoke-injector rake.

Meteorological–Environmental Tunnels

These tunnels are designed to specifically simulate the earth's natural boundary layer, which can be as high as 1000–2000 feet. These tunnels are used to determine wind loads on buildings and their surrounding area, air pollution, soil erosion, snow drifts, and so on. See Section 9.2 for these tunnels.

Automobile Tunnels

Tests for aerodynamic parameters that affect an automobile's performance and stability are made with either scale models or full-scale cars in large tunnels (see Section 9.13).

Following is a tabular list of low-speed wind tunnels grouped by country.

TABLE 1.1. List of Low-Speed Wind Tunnels

Tunnel Name	Location	Type	Test Section (Open or Closed)	V_{max} (mph)	Turbulence Factor	Test-Section Shape	Test-Section Length	Contraction Ratio	Power, Remarks, etc.
United States: National Aeronautics and Space Administration									
ARC 7 × 10 #1	Moffett Field, CA	Single return	Both	280	0.3%	7 × 10 ft	1.4 B	14	1600 hp
ARC 7 × 10 #2	Moffett Field, CA	Single return	Both	280	0.3%	7 × 10 ft	1.4 B	14	1600 hp
Full Scale 40 × 80	Moffett Field, CA	Single return	Closed	300	.5%	40 × 80 ft, oval	1.0 W	8.4	135,000 hp
Full scale 80 × 120	Moffett Field, CA	Open circuit	Closed	100	0.5%	80 × 120 ft	1.0 W	5	135,000 hp
LAL 4 × 7 Tunnel	Langley Field, VA	Single return	Both	230		14.5 × 21.75 ft	2.3 B	9	8,000 hp
LAL 30 × 60 ft	Langley Field, VA	Double return	Open	110	1.1	60 × 30 ft, elliptical	0.9 D	4.9	8000 hp
LAL 20-ft Spin Tunnel	Langley Field, VA	Annular return	Closed	60		12-sided polygon	1.25 D	2.9	400 hp
LAL 12-ft Low-speed tunnel	Langley Field, VA	Open circuit	Closed	60		12-sided polygon	1.25 D		280 hp
LERC Ice Tunnel	Cleveland, OH	Single return	Closed	350		6 × 9 ft	2.2 W	14.0	4160 hp, −40°F
LERC AWT	Cleveland, OH	Single return	Closed	500 60		20 ft round; 45 ft round	2.0 D		60,000 hp, proposed 1985
LERC 9 × 15	Cleveland, OH	Single return	Closed	150		9 × 15 ft	1.9 W		87,000 hp

				United States: Other Federal Services					
DTNSRDC #1	Carderock, MD	Single return	Closed	190	1.01	8 × 10 ft	1.4 W	4.5	1000 hp
DTNSRDC #2	Carderock, MD	Single return	Closed	165	1.01	8 × 10 ft	1.4 W	4.5	700 hp
DTNSRDC Anechoic	Carderock, MD	Single return	Open or closed	135		8 × 8 ft	2.4 W	10	1350 hp, sound studies
DTNSRDC V/STOL	Carderock, MD	Single return	Closed	150		12 × 15 ft	2.0 W	3	
WADC 12-ft vertical	WPAFB, OH	Annular return	Open	80		12-ft, round	1.2 D	4.0	
W-P 10 × 7 low turbulence	WPAFB, OH	Open circuit	Closed	400		10 × 7 ft	1.5 H	35.0	Flow visualization, 20,000 hp
U.S. Naval Academy, Blue Tunnel	Annapolis, MD	Single return	Closed	180	1.1	3 × 4.5 ft	2.0 W	6.2	
				United States: Colleges and Industry					
Boeing Research Wind Tunnel	Seattle, WA	Single return	Closed	150	0.1%	5 W × 8 H, ft	4.0 W	7.2	650 hp, low turbulence
Boeing VSTOL BVWT	Philadelphia, PA	Single return	Open, closed, or slotted	250	1.04	20 × 20, ft	2.2 W	6.0	
California Institute of Technology	Pasadena, CA	Single return	Closed	170	1.05	10 ft, round	1.0 D	4.0	600 hp
Douglas Aircraft Co.	Long Beach, CA	Single return	Closed	195	1.02	3.2 × 4.5 ft	2.2 W	6.0	100 hp
General Dynamics, Convair	San Diego, CA	Single return	Closed	300 70		8 × 12 ft 16 × 20 ft	1.25 W	6.25	2250 hp
Georgia Institute of Technology	Atlanta, GA	Single return	Closed	170	1.2	7 × 9 ft	1.2 W	5.8	600 hp
University of Illinois	Urbana, IL	Single return	Closed	95	<1%	3 × 4 ft	1.0 W	4.0	50 hp
University of Kansas	Lawrence, KA	Single return	Closed	200	1.1	3 × 4.2 ft	1.4 W	9.0	300 hp

TABLE 1.1. (*Continued*)

Tunnel Name	Location	Type	Test Section (Open or Closed)	V_{max} (mph)	Turbulence Factor	Test-Section Shape	Test-Section Length	Contraction Ratio	Power, Remarks, etc.
Lockheed Aircraft Corp.	Burbank, CA	Single return	Closed	220		8 × 12 ft	1.25 W	6.56	1250 hp
Lockheed Georgia #1	Marietta, GA	Single return	Closed	250 125	1.04 1.13	16 × 23 26 × 30	1.9 W 2.2 W	7.0 3.5	9000 hp
Lockheed Georgia MTF	Marietta, GA	Single return	Closed	150		2.5 × 3.5 ft	1.5 W	4	400 hp
Glenn L. Martin Wind Tunnel	College Park, MD	Single return	Closed	220	1.05	7.75 × 11 ft	1.27 B	7.31	2200 hp
McDonnell-Douglas Low Speed	St. Louis, MO	Single return	Closed	200	1.13	8.5 × 12 ft	1.5 W	7.0	
McDonnell-Douglas Mini-speed	St. Louis, MO	Single return	Closed Open	80 60	1.18 1.18	15 × 20 ft 15 × 20 ft	1.25 W 1.25 W	3.0 3.0	
MIT Structures	Cambridge, MA	Single return	Closed	90		7.5 × 5 ft			
MIT Wright Brothers	Cambridge, MA	Single return	Closed	140	1.08	10 × 7.5 ft, elliptic	1.5 W		2000 hp
University of Michigan	Ann Arbor, MI	Single return	Closed	200	$<0.35\%$	5 × 7 ft	5 W	15.0	1200 hp
Northrop Aviation	Hawthorne, CA	Single return	Closed	300	1.02	7 × 10 ft	2 W	12.2	
University of Notre Dame #1	Notre Dame, IN	Open circuit	Closed	61	$<0.1\%$	2 × 2 ft	3.0 W	24.0	15 hp, smoke tunnel
University of Notre Dame #2	Notre Dame, IN	Open circuit	Closed	61	$<0.1\%$	2 × 2 ft	6.0 W	24.0	15 hp, smoke tunnel
Pennsylvania State University	University Park, PA	Single return	Both	100		4 × 5 ft	1.2 W	7.0	500 hp
Princeton University	Princeton, NJ		Closed	100	1.03	4 × 5 ft	1.6 W	5.0	Student tunnel

Princeton University	Princeton, NJ		Closed	100		3 × 4 ft 1 × 6 ft	1.0 W 4.0 W	4	Two alternate circuits
Rensselaer Polytechnic Institute	Troy, NY	Single return	Both	150	1.06	4 × 6 ft	1.3 W		
Rockwell–Los Angeles	El Segundo, CA	Single return	Closed	200	1.1	7.75 × 11 ft	1.1 W	7.5	1250 hp
Rockwell–Columbus	Columbus, OH	Single return	Closed	300		7 × 10 ft or 14 × 16 ft			
Stanford University	Stanford, CA	Open circuit	Closed	100	2.4	7.5 round		6.7	125 hp
Texas A&M University	College Station, TX	Single return	Closed	200	1.1	7 × 10 ft	1.2 W	10.4	
United Aircraft	East Hartford, CT	Single return	Closed	650 200	1.02 1.14	8 ft, octagonal 18 ft, octagonal	2.2 D	24.0 4.7	9000 hp
Vought Aircraft	Dallas, TX	Single return	Closed	240		7 × 10 ft	1.6 W	7.0	1500 hp
Vought Aircraft VSTOL	Dallas, TX	Single return	Closed	52		15 × 20 ft	1.9 W	1.7	1500 hp
University of Texas	Austin, TX	Open circuit	Closed	200	0.2%	5 × 7 ft	7.0 W	10	800 hp
University of Washington Aeronautical Lab.	Seattle, WA	Double return	Closed	250	1.1	8 × 12 ft	0.8 W	6.0	1500 hp
Wichita State University	Wichita, KS	Single return	Closed	180	1.09	7 × 10 ft	1.2 W	6.0	1000 hp
United States: Colleges and Industry—Nonaeronautical									
Calspan Atmospheric Simulation	Buffalo, NY	Open return	Closed	55		8 × 8 ft	60 ft		
Cermak/Peterka	Ft. Collins, CO	Single return	Closed	45		7.5 × 10 ft	7 W	2.4	75 hp, wind engineering
CSU Meteorological	Ft. Collins, CO	Single return	Closed	82	0.1%	6 × 6	16 H	9.0	400 hp temperature controlled, humidity controlled

TABLE 1.1. (*Continued*)

Tunnel Name	Location	Type	Test Section (Open or Closed)	V_{max} (mph)	Turbulence Factor	Test-Section Shape	Test-Section Length	Contraction Ratio	Power, Remarks, etc.
CSU Industrial Aero	Ft. Collins, CO	Single return	Closed	55	0.5%	6 × 6	10 H	4.0	75 hp
CSU Environmental	Ft. Collins, CO	Open circuit	Closed	26	1.0%	8 × 12	7.5 H	3.35	50 hp
Chrysler	Chelsea, MI	Single return	Open	120		6 × 8.5 ft	1.5 W	6.0	700 hp, 1972 Full-scale automobiles
Ford Motor Company #2	Dearborn, MI	Single return	Closed	151	0.6%	12.5 × 20	1.5 W	3.8	2000 hp, Environmental automobile testing
General Motors	Detroit, MI	Single return	Closed	155		18 × 34 ft	2.0 W	5.0	4000 hp, full-scale automobiles
University of Illinois	Urbana IL	Open circuit	Closed	35	1%	5 × 5 ft	5.0 W	1.0	100 hp
			Closed	170	<1%	1 × 6 ft	1.0 W	6.0	
Northwest Hydraulic Consultants	Kent, WA	Open circuit	Closed	30		7 × 12 ft	4.5 W		20 hp
				Australia					
ARL 9 × 7	Melbourne	Single return	Closed	215	1.2	9 × 7 ft, octagonal	1.2 W	4.0	890 hp (1962)
RMIT (Aero)	Melbourne	Single return	Closed	150	1.2	$4\frac{1}{2} \times 3\frac{1}{2}$ ft, octagonal	1.5 W	4.0	134 hp (1953)
RMIT (ME)	Melbourne	Single return	Closed	112		$9\frac{3}{4} \times 6\frac{1}{2}$ ft, rectangular	3.0 W	2.0	300 hp (1981)
Monasch University		Single return	Closed	226		13 × 10 ft	1.5 W		600 hp
University of NSW		Single return	Closed	20		10 × 10 ft	0.9 W		125 hp, industrial aerodynamics
				66		5 × 5 ft	1.2 W		
				139		3 × 5 ft	2.2 W		
University of Sydney	Sydney	Single return	Closed			5 × 7 ft			80 hp

University of Sydney	Sydney	Single return	Closed			8 × 6 ft			100 hp, building loads
University of Sydney	Sydney	Single return	Closed			3 × 2 ft	2.0 W		150 hp, wind engineering
					Belgium				
TCEA L-1	Rhode-St.-Genèse	Single return	Open	120	1.4	9.3 ft, round	1.5 D	4.0	750 hp
TCEA, 2 meter	Rhode-St.-Genèse	Single return	Closed	200	1.4	6.2 ft, round	1.25 D	9.0	750 hp
TCEA Vertical	Rhode-St.-Genèse	Open circuit	Open	60		9.3 ft, round		4.0	750 hp
				Canada: Government Establishments					
NAE 15 ft	Ottawa	Annular return	Open	50	1.92	15 ft, round	0.7 D	4.0	275 hp, bridges, smoke dispersal
NAE 6 × 9 ft	Ottawa	Single return	Closed		1.1	6.3 × 9.0 ft	1.6 W	9.0	2000 hp
Atmospheric Environment Service	Downsville, Ontario	Single return	Closed		1.16	30 × 30 ft	2.5 W	6.0	9200 hp, full-scale automobiles
NAE Meteorological	Ottawa	Single return	Closed			6 × 8 ft	10.0 W		75 hp
NAE 3 ft	Ottawa	Single return	Closed	230		1.4 × 1.4 ft	2.5 W	28.0	100 hp
				125		3 × 3 ft	2.5 W	6.0	
NAE Propulsion	Ottawa	Open circuit	Closed	90		20 × 10 ft	4.0 B	6.0	1000 hp
				Canada: Universities and Industry					
University of British Columbia	Vancouver	Single return	Closed	55		8 × 6 ft	10 W		125 hp, atmospheric testing
Laval University	Quebec	Single return	Closed	85		3 ft, round			50 hp
University of Toronto	Toronto	Single return	Closed	15		4 × 8 ft	4.5 W		7 hp, pollution dispersal
University of Toronto	Toronto	Single return	Closed	40		4 × 8 ft	4.5 W		60 hp, aeroelastic building studies

TABLE 1.1. (*Continued*)

Tunnel Name	Location	Type	Test Section (Open or Closed)	V_{max} (mph)	Turbulence Factor	Test-Section Shape	Test-Section Length	Contraction Ratio	Power, Remarks, etc.
University of Toronto	Downsville, Ontario	Single return	Closed	75		4 × 5 ft			60 hp fan & 75 hp blower
University of Western Ontario	London	Single return	Closed			8 × 7 ft	10.0 W		40 hp, wind engineering
MH TR Boundary layer	Guelphs Ontario					4 × 8 ft	3.2 W		7 hp, wind engineering
University of Western Ontario #1	London	Open circuit	Open	30		6 × 8 ft	10 W	3.0	36 hp, variable pitich, adjustable roof
University of Western Ontario #2	London	Single return	Closed	55 22		7 × 11 ft 12 × 16 ft	16 W 11.2 W	1.0	250 hp, dc motor, wave tank
					France				
Alger No. 1	Alger	Open circuit	Both	227		5.9 × 7.2 ft	1.5 W	4.0	860 hp
Hispano-Suiza	Paris	Open circuit	Closed	205		16.4 ft, round	1.6 D		4000 hp
Lille No. 1	Lille	Single return	Closed	192		11.1 ft, round			200 hp
Lille No. 2	Lille	Single return	Closed	78		6.6 ft, round			52 hp
MI Ice Tunnel	Mont-Lachat	Open circuit	Closed	200		4.6 × 10.5 ft	1.4 H	6.1	1260 hp
ONERA S1 Ma	Modane	Open circuit	Open	111		26 × 52 ft, elliptical	1.4 W	3.5	6000 hp
ONERA Spin tunnel	Modane	Open circuit	Open			16.4 ft, round			475 hp
ONERA S2 Ma	Modane			270		10 ft, round			2400 hp

ONERA S1 Ca	Cannes	Open circuit	Closed	111		9.8 ft, round	1.5 D	6.1	160 hp
TO No. 4	Toulouse	Open circuit	Open	94		6.6 × 9.8, elliptical			240 hp
TO No. 5	Toulouse	Open circuit	Open	94		13.9 ft, round			430 hp
St. Cyr	St. Cyr	Single return	Slotted	90				5.0	Full-scale automobiles
					India				
Hindustan Aeronautics Ltd.	Bangalore	Closed circuit	Closed	204	1.2	9 × 6 ft, octagonal	1.3 W	6.35	750 hp
Indian Institute of Science	Bangalore	Open circuit	Closed	235	1.12	9 × 14 ft, octagonal	1.4 W	14.0	2 × 640 hp
Indian Institute of Science	Bangalore	Closed circuit	Closed	214	1.4	7 × 5 ft, elliptical	1.4 W	4.1	530 hp
Indian Institute of Science	Bangalore	Open circuit		30		15 ft, round			150 hp
Indian Institute of Technology, spin tunnel	Madras	Closed circuit	Open	200		5.0 ft, round	1.7 W	6.1	350 hp
Space Science Center	Trivandum			310		2 × 2 ft			300 hp
Wind Tunnel Center	Bangalore			65		3 ft, round			100 hp
					Italy				
Aeritalia	Turin	Eiffel	Closed	197	1.15	23 × 23 ft, square	9.8 ft	6.0	410 hp
Breda	Milan	Single return	Open	205		6.6 ft, round	1.6 D		800 hp
Caproni	Milan	Double return	Open	121	1.9	4.9 ft, round	1.2 D		190 hp
Fiat, Aerodynamical Wind Tunnel	Turin	Single return	Semiopen	131	1.08	23 × 15 ft	1.5 W	4.0	2500 hp

TABLE 1.1. (*Continued*)

Tunnel Name	Location	Type	Test Section (Open or Closed)	V_{max} (mph)	Turbulence Factor	Test-Section Shape	Test-Section Length	Contraction Ratio	Power, Remarks, etc.
Fiat, Hot Wind Tunnel	Turin	Single return	Semiopen	105	1.52	13.8 × 9.8 ft	2.8 W	4.0	760 hp
Fiat, Cold Wind Tunnel	Turin	Single return	Semiopen	105	1.52	13.8 × 9.8 ft	2.8 W	4.0	760 hp
Piaggio	Finale-Ligure	Open circuit	Closed	90	2.2	6.6 ft, round	2.0 D		100 hp
Pininfarina	Grugliasco (Turin)	Open but housed	Semiopen	94	1.2	13 ft diam	2.5 D	6.5	1972, 850 hp, full-scale automobiles
Politecnico of Turin	Turin	Single return	Closed	229	1.3	9.8 ft, round	1.8 W	5.4	1500 hp
					Japan				
Defense Agency	Tachikawa, Tokyo	Single return	Open	160	1.21	6.4 × 6.4 ft	13 ft		260 hp, 1972
Defense Agency	Tachikawa, Tokyo	Single return	Open	160		11 × 11 ft	15 ft		260 hp, 1970
Defense Agency	Tachikawa, Tokyo	Single return	Open octagonal	75		13 × 13 ft	13 ft		260 hp, 1970
Fuja	Nishihara, Utsunomiya	Single return	Open	135	1.23	6.4 × 6.4 ft	9 ft		1969, remodeled
Kawasaki 3.5 meter	Kagamihara, Gifu	Single return	Open	145 78		8 × 8 ft 11 × 11 ft	1.9 W 1.2 W		500 hp, 1968 remodeled
University of Kyushu	Higashi, Fukuoka	Single return	Closed	60	164	6.4 ft, round			50 hp, 1959 remodeled
Mitsubishi	Minato, Nagoya	Single return	Closed	230		6.4 × 6.4 ft	8 ft		600 hp, 1956
University of Nagoya	Chidane, Nagoya	Open circuit	Open	90	1.54	6.4 ft, octagonal			300 hp, 1959

National Aerospace Laboratory	Chofu, Tokyo	Single return	Closed	135		21 × 18 ft	30	4000 hp, 1965
National Aerospace Laboratory	Mitaka	Single return	Open	160	1.21	8.2 ft, round	11	600 hp, 1960
Nippon Airplane	Kanazawa, Yokohama	Single return	Open	90		6.4 ft, round	8	100 hp, 1958
University of Tokyo	Bunkyo, Tokyo	Open circuit	Open	34		13 × 20		120 hp, 1962, free flight
University of Tokyo	Meguro, Tokyo	Open circuit	Open	112	1.08	10 ft, round	10 ft	500 hp, 1931
University of Tokyo	Meguro, Tokyo	Open circuit	Open	145	1.25	6 ft, round	8.5 ft	250 hp, 1929
Japan: Nonaeronautical Wind Tunnels								
Agricultural Research Lab.	Kita, Tokyo	Open Circuit	Open	54		7.5 × 7.5 ft		
Asahi Glass	Kanagawa, Yokohama	Open circuit	Closed	34		7 × 8 ft	4.8 W	70 hp, 1972, boundary layer studies
Central Utilities Research	Komae, Tokyo	Open circuit	Closed	34		5 × 10 ft	3.3 W	67 hp, 1965, atmospheric diffusion
Central Utilities Research	Komae, Tokyo	Open circuit	Closed	34		5 × 10 ft	9.8 W	20 hp, 1973, atmospheric diffusion
Ishikawaji Maharimo	Eto, Tokyo	Single return	Closed	27		4 × 26 ft	3.3 W	3 × 40 hp, 1959, suspension bridges
University of Kyushu	Munakata, Fukuoka	Single return	Closed	134		6.3 × 13 ft	1.5 W	800 hp, 1970
Ministry of Construction	Chiba	Single return	Closed	56		10 × 6 ft	6.4 H	134 hp, 1965
Ministry of International Trade	Kita, Tokyo	Open circuit	Closed	34		5 × 10 ft	5.2 W	67 hp, 1969, pollution and environment
Mitsubishi	Nagasaki	Single return	Closed	63		33 × 10 or 10 × 33	0.6 W	1540 hp, 1973, rotatable test section
Mitsubishi	Nagasaki	Open circuit	Closed	34		6.4 × 10 ft	8.2 W	270 hp, 1966, atmospheric diffusion

TABLE 1.1. (*Continued*)

Tunnel Name	Location	Type	Test Section (Open or Closed)	V_{max} (mph)	Turbulence Factor	Test-Section Shape	Test-Section Length	Contraction Ratio	Power, Remarks, etc.
Mitsubishi	Nagasaki	Open circuit	Closed	34		10 × 6.4 ft	8.2 H		270 hp, 1969, atmospheric diffusion
Nippon Glass	Ichihara, Chiba	Open circuit	Closed	45		6.3 × 6.3 ft	10.5 W		100 hp, 1972, buildings
Nissan Motor	Yokosuka, Kanagawa	Open circuit	Closed	74		11.5 × 20 ft	1.5 W	2.9	370 hp, 1968, full-scale automobiles
Toyota Motor	Toyota, Aichi	Single return	Closed	135		11.5 × 16 ft	1.2 W		2000 hp, 1969, full-scale automobiles
Toyota Motor	Toyota, Aichi	Single return	Closed	123		6.3 × 10 ft			1973, heated flow
University of Tokyo	Bunkyo, Tokyo	Single return	Open	38		6 × 52 ft			100 hp, 1964, bridges and structures
University of Tokyo	Bunkyo, Tokyo	Single return	Closed	27		6.3 × 13 ft	1.5 W		54 hp, 1969, buildings and structures
				Netherlands					
Delft University	Delft	Single return	Closed	250	0.02 to 0.09%	5.8 × 4.1 ft	1.4 W	18.0	730 hp, 10 test sections
DNW	Northeast Polder	Single return	Closed	250	0.1%	26 × 20 ft	2.5 W	9.0	17,000 hp, 1983, 4 test sections
				300	0.08%	20 × 20 ft	2.5 W	12.0	
				135	0.08%	31 × 31 ft	2.5 W	4.8	
			Open	185	0.15%	26 × 20 ft	2.5 W	9.0	
NLR LST-NOP	Northeast Polder	Single return	Open or closed	200	0.1%	9.8 × 7.4 ft	2.9 W	9.0	1983
NLR 0.6 × 0.8 m	Northeast Polder	Single return	Open or closed	325	0.07%	2 × 2.6 ft	2.5 W	90	

				New Zealand					
Ministry of Works	Gracefield	Single return	Open	60		9 × 4 ft	2.2 W		47 hp, buildings
University of Canterbury	Christchurch	Single return	Closed	123		3 × 4 ft	2.0 W		60 hp
University of Canterbury	Christchurch	Single return	Closed	50		4 × 4 ft			66 hp
				Spain					
Cuatro Vientos	Cuatro Vientos	Single return	Open			9.8 ft round			
				Sweden					
FFA	Stockholm	Single return	Closed	180		12 ft, round	2 D		1300 hp
KTH	Stockholm	Single return	Closed	80		5 × 7 ft			125 hp
Volvo Climatic Tunnel	Göteborg	Single return	Semiopen	60	1.4	10.5 × 11.5 ft	4.0 W	2.4	trucks
				122	1.2	7.9 × 5.9 ft	5.8 W	6.6	cars
				124	1.2	7.9 × 5.9 ft	2.6 W	6.6	models
				Switzerland					
ETH	Zurich	Single return	Both	185		7 × 10 ft			550 hp
Swiss Fed. Aircraft	Emmen	Single return	Open	134	1.24	8 × 6 ft	1.4 W	4.7	300 hp
Swiss Fed. Aircraft	Emmen	Single return	Open Closed	150 180	1.08	16 × 23 ft	1.0 W	4.0	3850 hp
Swiss Fed. Aircraft	Emmen	Open circuit	Closed	65		10 ft, round	0.8 D	2.7	300 hp, spin tunnel
Swiss Fed. Aircraft	Emmen	Single return	Closed	310	1.03	7.5 × 10 ft	1.3 W	19.0	3850 hp, flutter with quick stop

TABLE 1.1. *(Continued)*

Tunnel Name	Location	Type	Test Section (Open or Closed)	V_{max} (mph)	Turbulence Factor	Test-Section Shape	Test-Section Length	Contraction Ratio	Power, Remarks, etc.
United Kingdom: Government and Associations—Aeronautical									
BAe 9 × 7 ft	Wharton	Single return	Closed	140	0.25%	9 × 7 ft	2.0 W	5.0	500 hp
BAe V/STOL	Wharton	Open circuit	Closed	50		18 × 16.5 ft	1.8 W	2.2	300 hp diesel
BAe 12 × 10 foot	Bristol	Single return	Closed	280	0.25%	12 × 10 ft	2.1 W	6.6	2 × 550 plus 1 × 850 hp
BAe 9 × 7 foot	Hatfield	Single return	Closed	170	0.14%	9 × 7 ft	2.0 W	5.0	500 hp
BAe 15 foot	Hatfield	Open circuit	Closed	95		15 × 15 ft	2.6 W	6.8	7 × 100 hp
BAe 9 × 7 ft	Woodford	Single return	Closed	150		9 × 7 ft	2.0 W	5.0	500 hp
BAe	Brough	Single return	Closed	205		7 × 5 ft	1.7 W	5.0	400 hp
BAe 13-foot	Weybridge	Single return	Closed	245	1.07	13 × 9 ft	2.0 W	10.7	2000 hp
British Hovercraft #2	Cowes, Isle of Wight	Single return	Closed	68		10.2 × 5.5 ft	1.4 W		250 hp
British Hovercraft #1	Cowes, Isle of Wight	Single return	Open	89		6 × 4 ft, elliptical	1.3 W		250 hp
RAE 5 ft	Farnborough	Single return	Open	135	1.5	5 ft, round	1.8 D	2.78	Acoustic studies
RAE 11 ft	Farnborough	Single return	Closed	275		11.5 × 8.5 ft	1.75 D	6.00	4000 hp
RAE 24 ft	Farnborough	Single return	Open	115		24 ft, round	1.83 D	3.53	2000 hp
RAE 9 × 13	Bedford	Closed circuit	Closed	200		9 × 13 ft	3.0 W	16.0	1500 hp
RAE 5-meter	Farnborough	Single return	Closed	260		16 × 14 ft	2.4 W	7.6	17,000 hp, 209 mph at 3 atm

United Kingdom: Universities and Industry—Aeronautical									
University of Bath	Claverton Down, Bath	Single return	Closed Closed	137 24		7 × 5 ft 15 × 10 ft	1.8 W 2.0 W		170 hp, general aeronautical
University of Cambridge	Cambridge	Single return	Closed	136		4 × 5.5 ft	2.0 W	7.0	100 hp
Cranfield Institute of Technology	Cranfield	Single return	Closed	150		8 × 6 ft	2.1 W		500 hp
University of Glascow	Glasgow	Single return	Closed	137		7 × 5 ft	1.2 W		220 hp
University of Newcastle	Newcastle	Single return	Closed Open	60 20		5 ft, round			140 hp
United Kingdom: Government and Associations—Nonaeronautical									
British Gas	London	Open circuit	Closed	33		6.7 × 3.2 ft			26 hp
Building Research Station	Watford	Single return	Open	45		6.7 × 4 ft	3.0 W		147 hp, wind engineering
Central Electricity Board	Marchwood, Hampshire	Single return	Closed	36 157		30 × 9 ft 7 × 9 ft	2.8 W 2.1 W		1200 hp, wind loads, dispersal
Central Electricity Board	Leatherhead, Surrey	Single return	Closed	40		14 × 5 ft	2.6 W		145 hp, power-station components, heating
Fire Research Station	Boreham Wood, Herts		Open			15 × 10 ft	1.0 W		200 hp, fire studies
MIRA Fullscale	Nuneaton, Warwickshire	Open, in building	Semiopen	80	1.7	14 × 26 ft	1.9 W	1.45	1340 hp, full-scale automobiles
MIRA Model Tunnel	Nuneaton, Warwickshire	Open, in building	Closed	100	1.6	3 × 7 ft	1.9 W	2.1	50 hp
United Kingdom: Universities and Industry—Nonaeronautical									
University of Bristol	Bristol	Single return	Closed Closed	164 31		7 × 5 ft 18 × 8	1.6 W 2.5 W		270 hp, dispersion, wind engineering
City University	London	Single return	Closed	58		5 × 10 ft	2.6 W		77 hp, wind engineering
Clarke Chapman	Gateshead	Single return	Closed			1.5 × 1.8 ft	6.1 W		470 hp, tube tests for nuclear power plants

TABLE 1.1. ***(Continued)***

Tunnel Name	Location	Type	Test Section (Open or Closed)	V_{max} (mph)	Turbulence Factor	Test-Section Shape	Test-Section Length	Contraction Ratio	Power, Remarks, etc.
Cranfield Institute of Technology	Cranfield	Single return	Closed	60		8 × 4 ft	3.5 W		200 hp
University of Edinburgh	Edinburgh	Open return	Open	27		3 × 5 ft	1.0 W		
Heriot-Watt University	Edinburgh	Single return	Closed	40		10 × 5 ft	3.5 W		150 hp, buildings
University of Leicester	Leicester	Single return	Open	29		8 × 6 ft	1.9 W		125 hp, buildings
University of London	London	Single return	Closed	112	0.06%	4.5 × 4 ft	4.0 W	4.9	37 hp, buildings
National Engineering Lab.	East Kilbride	Single return	Closed	82		1.5 × 1.5 ft	4.8 W		115 hp, icing, heat transfer
National Institute of Agricultural Engineering	Silsoe, Bedford		Closed			5 × 7 ft	2.6 W		35 hp, agricultural spraying
NMI #5 North Tunnel	Teddington	Single return	Closed	150		9 × 7 ft	1.3 W		250 hp, general industrial work
NMI #6 South Tunnel	Teddington	Single return	Closed	105		9 × 7 ft	2.7 W		250 hp, building oscillations
NMI #9 3 × 3 ft	Teddington	Single return	Closed	100		3 × 3 ft	5.0 W		35 hp, fundamental research
NMI #8 wind/wave	Teddington	Single return	Closed	45		4 × 2 ft	3.5 W		30 hp, wind/wave studies
NMI #4 7 foot	Teddington		Closed	93		7 × 7 ft	3.0 W		110 hp
NMI #7	Teddington	Single return	Closed	100		15.7 × 7.8 ft	3.2 W	3.8	750 hp, wind loads
NMI Open Jet #1	Teddington	Double return	Open	15		7 × 9 ft, ellipitical	1.1 W	3.9	300 hp, smoke dispersal, landscapes

NMI Open Jet #2	Teddington	Double return	Open	75		7 × 9 ft, elliptical	1.1 W	3.9	300 hp, buildings
University of Nottingham	Nottingham		Open	36		4 × 8 ft	2.5 W		36 hp, building loads and pressures
Oxford University	Oxford	Single return	Closed	67		13 × 6.7 ft	3.5 N		470 hp, wind engineering
Redland Technology	Horsham, Sussex	Single return	Closed	60		7 × 6.8 ft	4.8 W		500 hp, building roofs and cladding
Royal Military College of Science	Shrivenham	Single return	Closed	90		3.5 × 4.5 ft	1.5 W		75 hp
			Closed	27		6.5 × 7.5 ft	1.9 W		
Royal Military College of Science	Schrivenham	Single return	Open	100		3.7 × 5 ft, elliptical	1.3 W		75 hp
University of Southampton	Southampton	Single return	Closed	136		7 × 5.5 ft	2.0 W		200 hp
			Closed	28		15 × 12 ft	1.3 W		
			Closed	70		7 × 5.5 ft			
University of Strathclyde	Glasgow	Single return	Open	90		5 ft, round	1.6 D		50 hp
				West Germany					
BMW	Munich	Closed	Slotted	110		23 × 14 ft	2.1 W	3.7	1150 hp, automobile tests
Daimler-Benz 20 foot	Stuttgart	Single return	Open	155	0.3%	27 × 19 ft	1.5 W	3.6	1944, 5000 hp, full-scale automobiles
Dornier Wind Tunnel	Friedrich-Shafen	Open	Open	146	1.3	10.5 × 7 ft	1.5 W		670 hp
DFVLR 3-meter	Göttingen	Single return	Open	145		10 × 10 ft		5.5	1959, 1600 hp
DFVLR	Göttingen	Single return	Open or closed	134		11.8 × 9.2 ft		5.6	1960, 1100 hp
DFVLR	Göttingen	Single return	Open	190		10.5 × 7.5 ft		10.0	1962, 1340 hp
Ford of Europe #1	Cologne	Single return	Open		1.2	9.5 × 14 ft	1.5 W		Climatic or environmental
Ford of Europe #2	Cologne	Single return	Open		1.2	24 × 16 ft	1.7 W	4.0	Automobile tests
VW Climatic Wind Tunnel	Wolfsburg (Göttingen)	Single return	Open	112	1.6	30 × 20 ft	1.3 W	4.0	1967, 3500 hp, full-scale automobiles

REFERENCES

1.1. D. D. Beals and W. R. Corliss, Wind Tunnels of NASA, *NASA SP* 440, 1981.

1.2. O. W. Nicks and L. W. McKinney, Status and Operational Characteristics of the National Transonic Facility, Paper 78-770, AIAA 10th Aerodynamic Testing Conference, 1978.

1.3. K. W. Mort, D. F. Engelbert, and J. C. Dusterberry, Status and Capabilities of the National Full Scale Facility, Paper 82-0607, AIAA 12th Aerodynamic Testing Conference, 1982.

1.4. E. N. Jacobs, The Variable-Density Tunnel, *TR* 416, 1932.

1.5. S. J. DeFrance, The NACA Full-Scale Wind Tunnel, *TR* 459, 1933.

1.6. C. H. Zimmerman, Preliminary Tests in the NACA Free-Spinning Tunnel, *TR* 557, 1936.

1.7. A. I. Neihouse, Design and Operating Techniques of Vertical Spin Tunnels, AGARD Memorandum AG17/P7, 1954.

1.8. R. J. Chambers et al., Stall/Spin Test Techniques Used by NASA. AGARD Flight Mechanics Panel Specialists Meeting on Stall/Spin Problems on Military Aircraft, Brussels, 1975.

1.9. J. Stack, The NACA High Speed Wind Tunnel and Tests of Six Propeller Sections, *TR* 463, 1933.

1.10. T. J. Muller, Smoke Visualization in Wind Tunnels, *Aeronautics and Astronautics*, **21**, 50–54, 1983.

1.11. T. J. Muller, On the Historical Development of Apparatus and Techniques for Smoke Visualization of Subsonic and Supersonic Flows, Paper 80-0420, AIAA 11th Aerodynamic Testing Conference, 1980.

1.12. D. C. Hazen and R. F. Lehnert, Smoke Flow Studies at Princeton University, Report 290, Princeton Univ. Dept. of Aeronautical Engineering, 1955.

CHAPTER 2

Wind Tunnel Design

No single wind tunnel is adequate for all possible aerodynamic tests. In general, wind tunnels can be divided into four broad categories by their speed ranges: subsonic with a maximum Mach number of up to 0.4; transonic with a maximum Mach number to 1.3; supersonic with a maximum Mach number up to 4.0 to 5.0; and hypersonic with a Mach number to 5.0 or higher. In this book only the subsonic tunnels are considered.

There are special-purpose subsonic tunnels, such as spin tunnels, icing tunnels, meteorological tunnels, and tunnels designed for the testing of buildings. The rest of the low-speed tunnels can be classified as general-purpose tunnels. These tunnels can be adapted for many types of special tests, such as buildings, flutter, store separations, windmills, skiers, skydivers, or dynamic tests for spins. Basically they are designed to test vehicles and their component parts. These tunnels will often have large test sections for V/STOL, full scale of smaller aircraft, and automobiles. In the medium size tunnels with a test-section area around 100 sq. ft large amounts of both research and aircraft developmental testing is accomplished. With smaller size test sections, the tunnels are used for research and instructional purposes. Ideally, the size of the tunnel is determined by its purpose.

The first step in the design of a tunnel is to determine the size and shape of the test section. In the following discussion it will be assumed that the primary use of the tunnel will be aircraft testing, since most tunnels are built for this purpose. Furthermore, most of the testing will be force model testing, where information is sought for performance, stability, and control of an aircraft. It also will be assumed that the (large) tunnel will be used for a minimum of 1000–2000 hours per year.

The cross-sectional area of the tunnel test section basically determines the overall size of the facility. The test-section size, speed, and design will determine the required power. The size of the facility will determine the structural or shell costs, and the power and operating hours will determine the energy portion of the operational cost. Although the major cost of operation is the salaries of the tunnel personnel, the electrical energy costs to run the tunnel and its auxiliaries is not an insignificant cost, and it is doubtful that this cost will decrease in the long term. Thus, in the design there is a balance between initial costs and operating costs. In the past many tunnels have been built with short-length diffusers, and so on, and hence short circuit length to hold down initial costs, while accepting higher energy costs of operation. This trade off should be carefully examined owing to escalating energy costs.

Test-Section Size. This is the starting point in the design of a wind tunnel. The purpose of a wind tunnel is to provide a uniform and controllable air flow in the test section that passes over the model. In an ideal case, the model should be tested at the same Reynolds and Mach number as the full-scale vehicle. This, of course, leads to a full-scale model, a very large tunnel, and very expensive models, or smaller pressurized or cryogenic tunnels, also with expensive models. Matching Reynolds and Mach number are mutually contradictory when using scale models. The Reynolds number is proportional to the ratio of inertia forces to viscous forces and the behavior of boundary layers and wakes is influenced by Reynolds number. The Mach number is proportional to the ratio of inertial to elastic forces, taken as the square root of $V^2/a^2 = M^2$. The change in gas properties such as density in passing through shock or compression waves depends on the Mach number. If the flow cannot be considered incompressible, the Mach number must be matched. Usually it is not necessary to produce the full-scale Reynolds number, but it must be of a reasonable value. Much low-speed testing involves take off and landing configurations where the Mach number is much less than 1. Both the lift curve slope and maximum lift coefficient are affected by Mach numbers as low as 0.2. This tends to require a tunnel speed approximately equal to the full-scale flight speed. In an unpressurized tunnel using air this means that the Reynolds number ratio of model to full scale is approximately equal to the scale ratio between the scale-model and the aircraft.

The question now is what is the minimum acceptable value of Reynolds number. Because much of low-speed testing is at high lift conditions, the effect of Reynolds number on airfoils at high lift must be considered. It is generally agreed that maximum lift and lift curve shape near stall will vary considerably with Reynolds numbers up to a million. There is, in fact, a whole series of airfoils for soaring gliders that are especially designed to operate at Reynolds numbers below 1,000,000 (Chapter 8).

Thus, the lower boundary for Reynolds numbers is in the range of 1,000,000–1,500,000 based on chord. At these values of Reynolds number, the model will have an extensive region of laminar flow, and the possibility exists of poor simulation owing to separation of the model's laminar boundary layer. It is assumed that laminar separations do not occur at full scale. Therefore, full scale can be simulated by artificial transition on the model (see Chapter 7). If the Mach number is taken as 0.2, then the tunnel velocity is about 150 mph. For this speed the Reynolds number is a little less than 1,500,000. Although the minimum Reynolds number cannot be rigidly defined, the above rational has been used to define a minimum Reynolds number of between 1,500,000 and 2,500,000 for low-speed tunnels.

For a mean geometric chord of 1 ft and aspect ratio of 8–9, then the span is between 8 and 9 ft. As the maximum span should be less than about 0.8 of the tunnel width, the width becomes 10–11.25 ft. This is why so many tunnels have been built in the 7 × 10–8 × 12 ft range. These tunnels have a maximum speed in the range of 200–300 knots. For rectangular solid-wall tunnels, Fig. 6.28 shows that for a small wing the wall correction factor will be minimum for a width to height ratio of about 1.5, again in 7 × 10 or 8 × 12 ft tunnels.

A model for a tunnel of this size is large enough so that the smaller parts are relatively easy to build.

For V/STOL models in a STOL descent case the speed will be near 70 knots due to model power limits or tip Mach number on propellers and rotors. The reduction in test speed will require a larger model to maintain reasonable Reynolds numbers. To minimize the wall corrections due to large downwash from these models, the span to width ratio must be smaller, between 0.3 and 0.5. Thus the V/STOL tunnels built in the 1960s were 20–30 feet wide.

For small research tunnels and student tunnels at universities the problem of building and model accurately will be most critical, assuming that students can hold an airfoil dimension to 0.01 in. It is desired to hold the model to 1–2% tolerance. For a 12% thick airfoil with 2% tolerance the thickness equals 0.50 in., and the chord is 4.16 in. Using a mean chord of 4.0 in. and aspect ratio of 8 the span is 32.0 in. As the maximum span is 0.8 of the tunnel width, the width is 40 in. Using a width to height ratio of 1.5 for a minimum wall correction factor, the height is 26.7 in. The cross-sectional area is then 7.4 ft^2, or 7.5 ft^2. A minimum test velocity would be about 100 fps or a dynamic pressure of 12 psf.

For a rectangular tunnel the tunnel width determines the model size and the Reynolds number at a fixed velocity. The cost of the tunnel shell and its required power tend to vary with the square of the test-section width. Since funds for a tunnel are usually fixed, the largest tunnel that the funds will buy is generally built.

The size of smaller tunnels is usually determined in the final analysis by the size of the room that will house the tunnel.

The Reynolds number per foot for a given size tunnel can be increased by building either a pressure tunnel or cryogenic tunnel using a cold gas such as nitrogen. These are special-purpose tunnels, and they are expensive to build and operate. The time required for model changes are long, because the test section must be isolated before entry to work on the model. The design of these tunnels will not be considered in this book.

Open–closed return. Another basic design consideration is whether the tunnel will be of the return or nonreturn (open circuit) type. Almost all of the small research tunnels are of the nonreturn type, usually because of the lower construction costs and low power consumption. Most of the larger tunnels are of the return type, the majority being single return with a few double return.

In Section 1.2 there is a listing of the advantages and disadvantages of each type of tunnel return.

Open–closed test section. Open jet tunnels will have a lower energy ratio than a closed jet wind tunnel owing to the jet entraining stagnant air as it passes from the contraction cone exit to collector inlet (Section 2.15). If the tunnel has an external balance, the balance usually has to be shielded from the air jet and one of the boundaries tends to be closed. Open throats do not work for an open circuit tunnel with a propeller in the diffuser unless the test-section region is enclosed in an airtight plenum or room. Open throat tunnels sometimes suffer from pulsations similar to vibrations in organ pipes (see Section 2.9). An open throat gives easy access to the model in small tunnels. In large tunnels scaffold of some type would be required to gain access to the model. The setting up and removal of the scaffold will require additional model change time. Since the jet length is usually kept short to reduce losses, there is the possibility that high lift models may deflect the wake enough to miss the collector. In general, the advantages appear to be with the closed throat tunnel, and most tunnels are of this type.

Closed throat tunnels that are not vented to the atmosphere at the test section, but at another location, will have the test section below atmospheric pressure. Thus they can suffer from leaks either through holes cut in walls for probes, wires, pipes, etc., or through the struts required to mount the model. These tunnels usually have a sealed room or plenum around the test section. When running, the plenum will be at the same static pressure as the test section. Most small open circuit tunnels are not built this way and they do suffer from leaks. This makes wood an ideal tunnel material because it is easy to patch.

The general layout for a tunnel, supposing that no extraneous factors seriously enter into the design, has reached a form generally agreed upon for reasons of construction economy and tunnel efficiency. This usually embraces a diffuser of three or four test-section lengths or more, and two sets of similar corners to save a little engineering and construction cost. The plane of the return passage is almost always horizontal, to save cost and make the return passage easier to get to. The vertical return is justified only when space is at a premium. No rule or general procedure has been agreed upon for the optimum shape of the return passage, round or rectangular. Factors governing the choice are given in the appropriate paragraphs.

Factors influencing the design of the test section, corners, return passage, and contraction cone are discussed on the following pages. The actual power losses are covered in Section 2.16. Further data on low-speed wind tunnel design may be found in Ref. 2.1.

2.1. THE TEST SECTION

As noted, the test requirements and the cost or space determines the size of the test section and the tunnel speed, and, hence, the required power. It will be recalled that the basic criterion was the desired Reynolds number. The cost is basically determined by the cross-sectional area of the test section, since building costs tend to vary with area.

Over the years many shapes have been used for test sections, such as round, elliptical, square, rectangular, hexagonal, octagonal, rectangular with filleted corners, flat ceiling, and floor with half round ends. The cost and power are directly determined by the cross-sectional area. The difference in losses in the test section due to the tunnel shape are negligible. Therefore, the shape of the test section should be based on utility and aerodynamic considerations.

For ease in installing models, changing models, installing ground planes, calibrating external balances, installing splitter plates for half models, or other modifications for nonstandard tests, nothing can match a test section with flat walls. The flat ceiling and floors simplify the installation of yaw turn tables for a three-strut mounting system and its image. The walls, ceiling, and floor allow easy installation of windows to view and photograph the model when flow visualization is used. As discussed earlier, a width to height ratio of about 1.5 will yield minimum wall corrections (Fig. 6.28).

Arguments have been made in the past for 7×10 ft tunnels because NASA has published detailed wall correction charts using Glauert's classical wall corrections. Using modern computational methods, wall corrections for any shape and size tunnel can be obtained (see Chapter 6). Thus other criteria should be used to determine the test section size.

As the air proceeds along the test section the boundary layer thickens. This action reduces the effective area of the jet and causes an increase of velocity. The velocity increase in turn produces a drop in local static pressure, tending to draw the model downstream. This added drag, called "horizontal buoyancy," is discussed in Chapter 6, where corrections may be found.

If the cross-sectional area of the jet is increased enough to allow for the thickening boundary layer, a constant value of the static pressure may be maintained throughout the test section. Unfortunately no exact design method is available that ensures the development of a constant static pressure. For a first approximation the walls of a closed jet should diverge about $\frac{1}{2}°$ each; finer adjustments may be necessary after the tunnel is built and the longitudinal static pressure is measured. Some tunnels whose test sections have corner fillets have these fillets altered until a constant static pressure is obtained. The advantages of such a flow are enough to justify a moderate amount of work in obtaining it.

To minimize secondary flow problems in the corners of rectangular contractions, a 45° fillet is often installed at the start of the contraction to form an octagonal shape. These fillets are carried through the test section to prevent boundary layer growth in the corners of the test section and down the diffuser tapering out at the end of the diffuser.

The length of the test section in common practice varies from one to two times the major dimension of the test section. The power losses in the test section are sizable (see Section 2.16) owing to the high speed, thus power can be saved by keeping it short. However, contractions do not have a uniform velocity distribution at the end of the contraction. Therefore, a constant area duct before the test section is required (see Section 2.7).

A practical detail in test-section design is the installation of sufficient windows for viewing the model, as shown in Fig. 2.1. In the course of testing it will become necessary to see all parts of the model: top, sides, bottom, and as much of the front as is reasonably practical. For safety reasons, windows in the test section should be made of plate safety glass. Plate glass will give minimum distortion when taking photographs and for lasers. If propellers and rotors are to be tested, glass manufacturers should be consulted on the best glass to use. What used to be called bulletproof glass (which never was bulletproof) is no longer sold under that name owing to product liability laws. It is a peculiar and interesting fact that despite the hazards of testing,

FIGURE 2.1. A large window in the test section is useful for photographs and observing the model. (Courtesy of University of Washington.)

more windows have been broken by overheating with photographer's lights than by model failure.

Adequate lighting is needed both to work on the model and for photographic purposes. For tunnels with fillets in the test section, lights can be built into the fillets.

2.2. THE DIFFUSER

The diffuser of a return wind tunnel usually extends from the downstream end of the test section to the third corner of the tunnel. It is divided into two parts by the tunnel fan. The second diffuser and fan section is often called the return duct or passage. The first diffuser usually extends to the first corner. Since the power losses in the tunnel vary as the velocity cubed, the purpose of the diffusers is to reduce the velocity by expanding the flow and recovering the static pressure. Since it is desired to reduce the velocity in the shortest possible distance to reduce losses, the diffuser is critical to the success of the tunnel. Diffusers are sensitive to design errors that may cause either intermittant separation or steady separation. These separations can be hard to find and can cause vibrations, oscillating fan loading, oscillations in test-section velocities (often called surging), and increased losses in the tunnel downstream of their origin. Diffusers are described by both their area ratios and an equivalent cone angle. The angle denotes an imaginary conical section with identical length and inlet and exit areas as the actual diffuser.

Diffusers are common to many fluid flows, and in many applications the equivalent cone angle can be quite large. In theory the only constraint on the angle is that the turbulent boundary layer does not separate. But this is for the case of uniform flow upstream of the diffuser. The flow leaving the test section is anything but uniform. There can be wakes from mounting struts and ground plane mounting struts; deflected wakes from the model, both laterally and vertically; and large separated wakes when the model is stalled in pitch and yaw. Also, poor corner vanes in the first corner may have adverse effects. The current practice uses an equivalent cone angle of 7° or less; however, the cone angle also depends on the area ratio and the area ratio determines the pressure recovery and pressure gradients, and, hence, the risk of separation (Ref. 2.2). Also, thick boundary layers at the diffuser entrance will increase the risk of separation. If a very long 5° diffuser is used to obtain a large contraction ratio, there is danger of a separation. Therefore, the total (both halves of the diffuser) tends to be limited to area ratios of five or six to one, half of the area ratio in each half of the diffuser. This area ratio limits the tunnel contraction ratio. To achieve larger contraction ratios a wide angle diffuser is used before the settling chamber. These are diffusers with an area ratio on the order of four to one and an equivalent 45° cone angle. Screens may be used to smooth out the velocity variations and maintain satisfactory flow (Refs. 2.3–2.13).

Methods of Fixing a Poor Diffuser

Although screens are often used to correct nonuniform velocities, their use in a diffuser is not recommended owing to the large pressure losses at high velocities and the resulting large power losses.

The use of splitter plates to reduce the diffuser angle is often suggested. However, any streamline in potential flow may be replaced by a solid boundary with no affect on the flow. The boundary growth along the splitter plate will reduce the diffuser angle slightly. The drag will help to make the velocity profile more uniform. The splitters will not make major changes in the overall adverse pressure gradient in the diffuser.

If the diffuser is separated or it is suspected that it is separated, the first problem is to locate the point of separation. In most cases tuft studies do not work owing to lack of windows and oblique camera angles for a camera inside the tunnel, although tufts may work at the start of the diffuser. A tuft on a fish pole with a person in the tunnel can be used, but the velocities in the first diffuser are so high that this method may be limited to the second diffuser. Another approach may be the use of fluorescent oil. A portable ultraviolet fluorescent tube could be used for illumination when the tunnel is shut down (Section 3.10).

The best way to fix a diffuser would be to redesign it and modify the tunnel. This may be possible in small tunnels, but in large tunnels such modifications to the structure will be impractical. Assuming that the location of the separation is known, the following fixes may cure the problem.

1. *Splitters.* These have been discussed previously and they are useful in preventing an incipient separation, reducing flow asymmetry, and smoothing out the velocity profile. If the separation is close to the start of the diffuser there may be problems with leading edge separation on the splitters when testing a powered model in yaw.

2. *Windmills.* Windmills are free-spinning propellers arranged to pump energy from areas of high dynamic pressure to those with low dynamic pressure. In some installations constant chord with zero twist and $\beta = 45°$ has been used with good flow improvement. One can make a theoretical case for highly tapered blades to make up for the variation of area with radius, but tests are lacking. Counterrotating windmills are probably unnecessary, since they will normally be used ahead of the fan and any rotation they add will be removed by the corner vanes and fan-straightener system. The windmill may reduce the larger turbulent eddies, but it increases the overall turbulence. The flow approaching the windmill will be nonuniform. This nonuniform flow will result in nonuniform blade loading at various azimuth positions, which will cause air flow pulsations at the blade rpm. Windmills will work best in circular or regular octagonally shaped diffusers.

3. *Vortex Generators.* Vortex generators are low-aspect-ratio wings set at an angle of attack near 15°. The tip vortex is used to pump the high

energy air from the free stream into the boundary layer. To be effective the span of the vortex generator must be of the same order as the boundary layer thickness. The vortex generator delays, or at best prevents, separation. Taylor (Ref. 2.14) suggests that the vortex generators be installed between 10 and 30 boundary layer thicknesses ahead of the separation point. Vortex generators are usually used either as corotating pairs (sense of the vortex are the same) or contrarotating pairs. The contrarotating pairs seem to be the most effective in delaying or preventing separation. See Refs. 2.15 and 2.16 for vortex generators and cylinders offset from the wall and perpendicular to flow to prevent flow separation in the diffusers. Vortex generators are also useful in improving fan efficiency when the fan spinner is nonrotating by improving flow on the inboard portion of the blades.

4. *Boundary Layer Control.* Boundary layer control slots are easily used only with open circuit tunnels. Here the test section is below ambient pressure and one need only build smooth entry slots at the beginning of the diffuser to have a device which may materially improve the flow in a wide-angle diffuser. Savage (Ref. 2.11) used this method to reduce the losses in a 17.6° diffuser to a value close to that for a 5° diffuser.

Applying suction to a closed return tunnel should be avoided because the removed air would have to be replaced. This could disturb the flow near the breather and may introduce dirt into the circuit. The boundary layer can also be energized by tangential blowing. This will not work on an open circuit tunnel because atmospheric pressure is the total pressure. In a closed return, tangential blowing either through tangential slots or possibly a series of round tubes may delay or prevent separation.

5. *Safety Screens.* Usually a strong wire safety screen of relatively large mesh is installed between the test section and the fan. Its purpose is to catch model parts, tools, etc., and thus protect the fan. The screen is often mounted at the leading edge of the first corner turning vanes in a closed return tunnel. The vanes are used to support the screen. The screen will increase the tunnel's losses but the protection to the fan blades is well worth the additional power. The screen can also be mounted to the trailing edge of the turning vanes. This may reduce the losses slightly but makes the attachment more complex. In an open circuit tunnel the screen will have to be mounted across the diffuser ahead of the fan.

The return passage will need access doors and windows, usually near the fan. The doors and windows should be sealed and of adequate size to allow for installation and removal of the fan blades, and drive motor, if it is placed in the fan nacelle. Built in lights near the fan are most useful. Provisions for pressure rakes about a half tunnel diameter ahead of the fan are nice. The rake will be used to adjust the first and second corner vanes to obtain a uniform velocity front at the fan (see Section 2.3). Drains at the lowest point in the tunnel are also useful since several times during the life of the tunnel it will have to be either washed or steam cleaned.

2.3. CORNERS

Although some tunnels have been built with 180° curved corners the added expense is usually not justified. Most corners consist of two 90° bends with a short duct between them. Corners are usually of constant area. To avoid large losses the corners are equipped with turning vanes. The shape of the vanes varies from bent plates to highly cambered airfoils. Provisions should be made to adjust the vanes either by pivoting the vanes or by trailing-edge tabs. If tabs are used, there should be some method of locking them to prevent their movement after the flow is acceptable. In the initial start up of a tunnel the vanes should be adjusted to ensure that the air is neither over nor under turned. The first two corners are the most critical in terms of losses (owing to high velocity) and the desire to have a uniform velocity at the fan.

The NASA Langley 4 × 7 m tunnel exhibited low velocities on the inner wall (when compared to the outer wall) starting aft of the test section and extending to the fan. Aft of the fan on the outer wall there was a reversed flow region that extended over about half of the duct length up to the third corner. Installation of large chord vanes with trailing-edge flaps (deflected toward the inner wall) just aft of the first corner improved the velocity profiles up to the third corner.

Over or under turning the flow at the corners can be checked by tufts on the trailing edge of vanes and a remotely controlled small camera with a flash. Oil flow on the floor may also give an indication. But the best method is a series of rakes to measure the velocity, especially if the tunnel is rectangular. Since most tunnels have constant area in the corners the rakes can be made of pipe or tubing. The top has a pad covered with medium hard rubber and the bottom pad is threaded to act as a jack. The total heads are connected to scanivalves, thus only the wiring and reference pressure tube need be brought out of the tunnel. Such a method was used at UWAL during a tunnel flow improvement program. The rakes are moved across the duct to determine the velocity profile. These data may also give some insight into flow separation and dead air in corners.

The losses in the corner vanes can be minimized by selecting an efficient cross-sectional shape and by using the best chord to gap ratio. The vanes using cambered airfoils and relatively blunt leading edges will be less sensitive to approaching airflow angularities than sharp leading-edge vanes. The vane camber lines can be designed by cascade theory to approximate any camber line in an infinite stream.

Assuming no change in area at the corners, then by continuity the velocity is constant around the corner. The drag in the corner is due to both skin friction and separation losses. It will appear as a drop in static pressure, ΔP. The loss is referred to the duct velocity by

$$\eta = \Delta p/q$$

where η = corner loss coefficient and q = dynamic head in the corner.

An abrupt corner without vanes may show a loss of 100% of the velocity head ($\eta = 1.00$). With carefully designed vanes a η of 0.15 is reasonable. The basic idea is to divide the corner into many turns of high aspect ratio. See Fig. 2.2.

In this application one may take the rate of change of momentum through a corner as $\rho h V \cdot V$ and equate it to the vane lift coefficient $\frac{1}{2}\rho c V^2 C_L$, where h is the vane gap and c is the vane chord, to determine that the vane lift coefficient is $2h/c$. Accordingly, to employ a reasonable C_L the gap–chord ratio should be 1 : 3 or smaller. Although it could be argued that a reduction in drag would be obtained by using the maximum vane chord possible (and hence the largest Reynolds number), experience has shown that the wake effects from each vane die out more quickly if many short-chord vanes are used in preference to fewer of large chord. This is especially true for the set of corner vanes just before the entrance cone.

Several vane profiles are shown in Fig. 2.3, and each is labeled with the loss experienced under test conditions by the various experimenters (Refs. 2.17, 2.18, 2.19, and 2.20) at Reynolds numbers of about 40,000. Since these data are limited and are for low Reynolds number, the data should be verified for large chord vanes. Equation (2.47) yields values slightly higher than Fig. 2.3 would indicate, but it is felt that the increase is justified for the usual installations.

Reference 2.21 reports that tests at $RN = 500{,}000$ 90° circular arc vanes with a gap–chord ratio of 1 : 4, a leading edge angle of 4–5°, and a trailing-

FIGURE 2.2. Corner turning vanes. The long tubes support static and total head tubes to measure the velocity distribution. Masking tape on floor has sticky side toward airflow to trap dust and dirt. (Courtesy of University of Washington.)

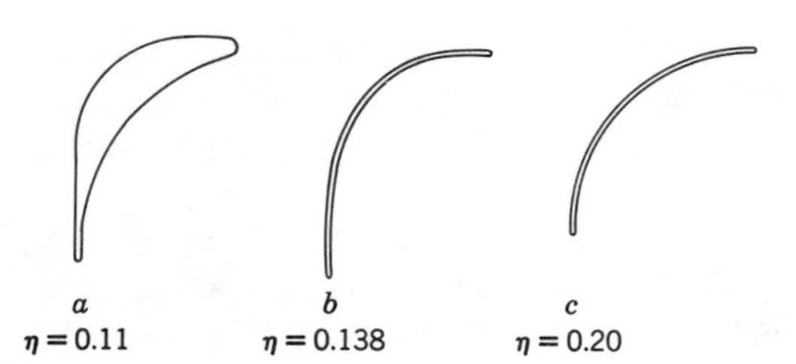

FIGURE 2.3. Corner vanes.

edge angle of 0 resulted in a pressure loss coefficient of 0.06 (excluding root losses). This is certainly a good value.

Any of these guide vanes can be used in conjunction with horizontal vanes to form a honeycomb.

Because the velocity is highest at the guide vanes just downstream of the tunnel test section, they are the most critical and should receive the most careful workmanship.

Data in Ref. 2.22 indicate that the corner vane illustrated in Fig. 2.3*b* probably has less loss than the value given and very probably is the most advantageous type to use.

The conversion of the fourth corner vanes to a large grid honeycomb will provide a useful tool for adjusting the upflow and cross-flow distribution in the test section. In 1976–1977 the University of Washington Aeronautical Laboratories, in cooperation with The Boeing Commercial Airplane Company, embarked on a program to improve the UWAL 8 × 12 ft low-speed wind tunnel, a double return tunnel (Ref. 2.23). An upflow–cross-flow mapping of the test section showed flow patterns similar to four solid-body rotations. This flow pattern resulted in a nonuniform upflow pattern and a cross-flow gradient in the region occupied by the vertical tail. This vortexlike flow pattern was found to exist also in single return tunnels, thus it is not unique to double return tunnels. In correcting this problem adjustable vane extensions were added to all the fourth corner vanes. Horizontal extensions with a larger chord than the vertical were also added. The extended chord allowed the upflow to be controlled. Thus the vanes had a large grid adjustable honeycomb at their trailing edge. There was a direct proportional relationship between a given vane location in the fourth corner and a corresponding point in the test section. This greatly aided the adjustment of the vane extensions. The test-section flow was very sensitive to the fourth corner vane extensions, and adjustments of 0.125 in. on a 24.0 in. chord vane extension were required. This adjustable honeycomb in the fourth corner provides a powerful tool for adjusting the flow quality in a test section. The lateral dynamic pressure distribution was made more symmetric, upflow variation was reduced from over 1° near the centerline to ±0.25°, and the cross-flow gradient was removed. The magnitude and variation of the u, v, and w turbulence components were reduced in magnitude and made more uniform.

The U.S. Air Force 16T tunnel at Arnold Air Force Station also had a nonuniform distribution of upflow and cross flow. The pattern was much more complicated than that at UWAL. It was discovered that the fourth corner vanes were under turning the flow. This facility bent the trailing edges in order to align them parallel to the tunnel centerline using a laser. Their average adjustment was about 4°, although some vanes required adjustment up to 13°. Of the 356 vanes, 300 were adjusted, and there was a significant affect on the flow.

Thus it appears that the tunnel design should make some provision for adjusting the flow leaving the fourth corner. An adjustable coarse honeycomb allows both cross flow and upflow to be adjusted. However, adjustments to the trailing edge of the vanes can also improve the flow. The fourth corner should be adjusted as a part of the initial tunnel check out, if required, to obtain better quality flow in the test section. The adjustments at the University of Washington took well over a hundred configurations of the fourth corner.

2.4. THE FAN SECTION

This section will deal with the fan nacelle, the fan, prerotation vanes, and flow-straightener vanes after the fan. Detail fan design will be covered in Section 2.5.

There is now general agreement that the wind tunnel fan should be located downstream of the second corner; the fan position between the first and second corners is now rarely used. First, however, let us rule out the positions that we may say are definitely undesirable. The fan develops its highest efficiency if it is located in a stream of a fairly high velocity, and its cost is at least partially proportional to its diameter squared. These two items rule out a fan in a very large part of the return passage or in the settling chamber. On the other hand, damage from a failing model and poor flow distribution make a position in the diffuser moderately risky. The argument for a position just downstream of the second turn is that the flow has by then been in a section of constant area for a considerable time and therefore should be relatively smooth when it meets the fan; also, of course, at this location the velocity is desirably high.

Three basic fan-straightener systems are in current use: (1) a fan with straightener vanes behind it (see Fig. 2.4); (2) a fan with prerotating vanes ahead of it, probably also having straightener vanes behind it; (3) counterrotating fans in which the second fan removes the rotation imparted by the first.

The counterrotating fan can remove all the twist for all tunnel speeds and power inputs. Since two fans can obviously be designed to develop more thrust than one, the counterrotating fans may become essential in high-

FIGURE 2.4. Nacelle-fan installation. The antitwist vanes are between us and the fan. (Courtesy University of Maryland.)

power installations. The drive is more complicated, however, as equal torque should be applied to both fans.

For tunnels of moderate size and power a single fan is usually quite satisfactory. If it is properly designed, a straightener system can be devised that will remove the twist for all power inputs and speeds. Such straighteners are discussed in the following paragraphs.

A variable-pitch fan is of great value even when a variable-rpm drive is available, since it gives much quicker speed control than varying the drive rpm. In tunnels with large contraction ratios the change in velocity distribution in the test section with change in fan-blade angle does not seem to be measurable. Also, when the drive motor is of the synchronous type the fan can be put in flat pitch for low pull-in torque and then opened out to develop the tunnel speed. This action may lead to greater power outputs from this type of motor, since the pull-in torque is often the limiting factor. For those tunnels that may be operated with two-dimensional inserts, the variable-pitch fan is definitely advantageous in achieving optimum operation.

The area ratio between the fan and the test section is usually between 2 or 3 to 1. If the ratio is made larger, there is the risk of a poor velocity profile before the fan and an increase in cost due to size. If the area ratio is smaller, the incoming velocity will be higher and the fan rpm will be larger to maintain reasonable blade angles. However, the tip speed is limited by the practice of keeping the tip Mach number low enough to avoid formation of shock waves. The fan-nacelle diameter in large tunnels is about 30–50% of the fan section diameter.

The fan motor is either mounted in the nacelle or outside the tunnel.

When the motor is in the nacelle, this usually requires cooling for the motor. The cooling air can often be ducted through the nacelle supports. If the motor is outside the tunnel, it can drive the fan either through a gear box in the nacelle or the more usual method of a long shaft passing through the second corner. This shaft will have to be held in careful alignment by bearings, and may or may not be enclosed in a fairing to reduce losses and the effect on the flow between the outside wall and the second corner turning vanes.

The fan may or may not have prerotation vanes upstream of the fan. The prerotation vanes are designed to produce a swirl opposite to the fan's swirl and hence zero swirl after the fan. This may not occur at all rpms. Thus, in most cases flow straighteners or antiswirl vanes are installed after the fan as a safety factor. The prerotation vanes are a stator and the fan spinner is attached to their inner end. These stators increase the velocity of the fan blade relative to the air stream. To avoid vibration between the prerotation vanes and the fan for N fan blades, do not use N, $3N/2$, or $2N$ prerotation blades. Similar constraints should be used between the fan blades and the nacelle supports. If prerotation vanes are not used, antiswirl or straightener vanes must be used downstream of the fan.

The major design problems in the tunnel, fan, and nacelle regions are structural, and the location of the motor will have a strong influence. As an example, if the motor is in the nacelle, then the nacelle struts must carry the torque and thrust of the fan to the tunnel shell, rather than only the thrust.

The nacelle should have a length to diameter ratio of about 3 with 30–40% of its length of constant diameter. The equivalent closing cone angle should be 5° or less. An excessive adverse pressure gradient over the rear portion will lead to separation and a persistant wake that may show in the test section. Expansion of the walls over the rear of the nacelle should be avoided if possible. The effective duct areas may yield an expansion angle of 10–12° for short nacelles. If the nacelle does separate, a vortex generator installation as described in section 2.2 may help. Their angle of attack should be relative to the local flow direction, which may change with fan speed. Extending the aft nacelle fairing may also help.

The wind tunnel fan is not like an airplane propeller. It is a ducted fan, and because it operates in a constant area duct, from continuity considerations there is no increase in velocity across the fan. The fan in a wind tunnel merely replaces the ΔP losses of the tunnel and model.

2.5. FAN DESIGN

There has been considerable discussion of the design of wind tunnel fans (Ref. 2.24), but a method proposed by Patterson (Ref. 2.25) is presented here because it considers the fan-straightener system as a unit and does not concern itself merely with the fan. Because rotation must be kept low if a fixed straightener system is to remove the twist for all conditions, consideration of the vanes becomes important, and their design will frequently necessitate a fan vastly different from what the ordinary criteria indicate. The fact that each section of this type of fan operates at constant efficiency, though of small merit as far as the fan is concerned, makes the usual graphical integration of the thrust and torque loading curves unnecessary, and hence the design is facilitated.

This theory, however, neglects the loss associated with the necessary tip clearance at the tunnel wall and the radial flow at the fan encountered as a result of the centrifugal action. The large boss recommended for use with a wind tunnel fan tends to lessen the latter effect. The tip-clearance loss will result in efficiencies slightly lower than indicated by this theory. The loss due to tip clearance adds to both the friction and the expansion losses that occur at the walls of a wind tunnel and indicates that instead of constant thrust the wind tunnel fan should perhaps have a graded thrust loading curve, greatest at the walls in order to best develop a uniform velocity front. This refinement is beyond the scope of this presentation.

Wallis (Ref. 2.26) states that "from a practical point of view the constant axial velocity design assumption appears to be as good as any other." He does add a warning regarding design lift coefficients increasing near the tip and the possibility of tip stall. It is interesting to note that using Wallis' method in the fan design for the German–Dutch low-speed wind tunnel (DNW) the original inflow velocity distribution was taken as a linear variation with the tip velocity 70% of the root velocity. After tests in a 0.10 scale tunnel the fan was retwisted to improve the velocity distribution in the fan outflow. After the design was completed a comparison was made between the assumed inflow velocity distribution and a uniform one. The blade geometry was held the same and the twist distribution was changed in this calculation. The twist distribution for uniform inflow is almost exactly the same as the twist for a nonuniform inflow (Ref. 2.27).

First let us consider the flow in a duct so that the terms and factors encountered later in the fan theory may be understood when applied.

Flow in a Duct

When Bernoulli's equation

$$p + \tfrac{1}{2}\rho V^2 = \text{constant}$$

is written between two locations in a duct, it applies only if the losses between the sections are zero. Naturally, in practice, they never are, and one or the other of the two terms at the second section must show a diminution corresponding to the loss in head. The law of continuity for an incompressible fluid, $A_1V_1 = A_2V_2$, where A and V are areas and velocities at two stations, makes it impossible for the velocity to fail to follow Bernoulli's rule, and hence the velocity head at the second location will be as predicted. But there will be a drop in static head Δp corresponding to the friction loss. This loss in pounds per square foot appears over the area A_2, so that the product of the two yields the drag of the section between 1 and 2. Multiplying the drag by the velocity yields the power lost. According to familiar experience, the drag of a surface varies with the dynamic pressure q, and it is customary to express the loss of the section in coefficient form, defining

$$k = \frac{\Delta p A}{\frac{1}{2}\rho A V^2} = \frac{\Delta p}{q} \tag{2.1}$$

It will be seen that the coefficient k compares with C_D in wing-drag calculations.

Throughout the wind tunnel the losses that occur appear as successive static pressure drops to be balanced by the static pressure rise through the fan. The total pressure drop Δh must be known for the design of the fan. If a model of the tunnel is available, the necessary pressure rise may be measured across the fan and extrapolated to full-scale Reynolds number. An alternative method is to calculate the energy ratio (see Sections 2.15 and 2.16) and find the fan pressure rise coefficient $k = \Delta h/(\frac{1}{2}\rho u^2)$, where u is the velocity through the propeller. It is now in order to consider several design features of a fan-straightener system such as the one shown in Fig. 2.5.

Factors Influencing the General Layout

It will be seen from Fig. 2.6 that high fan efficiencies are largely determined by proper selection of the advance ratio and utilization of L/D ratios of the order of 50. It remains to demonstrate the best methods for satisfying these criteria.

Large advance ratios imply lowered speeds of fan rotation, necessitating a drive motor of low rpm or a geared driving system. The desire for higher rpm for the driving motor indicates that the higher-speed regions of the wind tunnel are best suited for the location of the fan. Balancing that against the increase of nacelle drag as the local speed is increased, the best compromise

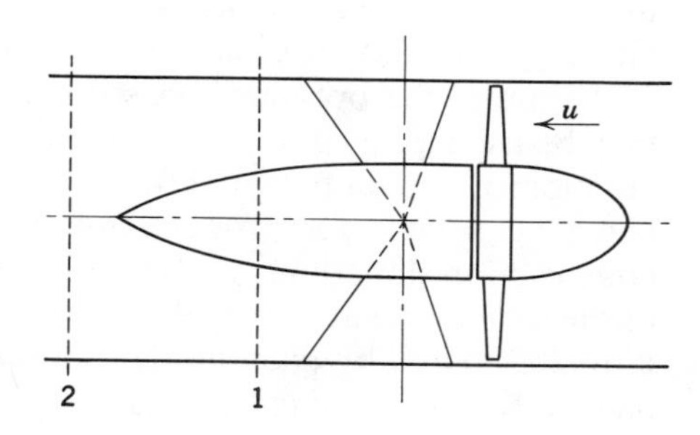

FIGURE 2.5. The fan-straightener combination.

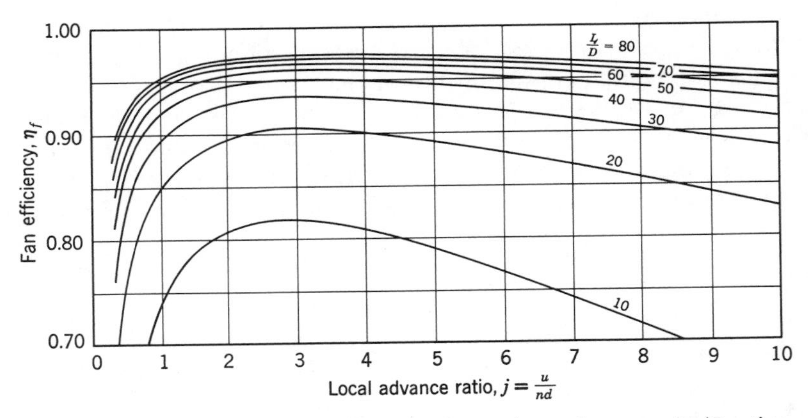

FIGURE 2.6. Approximate fan efficiencies for various advance and *L*/*D* ratios.

usually locates the fan downstream of the second corner after the test section. If the fan is to be driven by a motor outside the tunnel, the corner location offers a short shaft length.

Ratios of *L*/*D* as high as 50 and higher are obtainable only with "infinite" aspect ratio and moderately thin airfoils (Fig. 2.7). Infinite aspect ratio can be simulated by effectively endplating both the fan-blade root and the tip, endplating being accomplished by providing a large nacelle or "boss" for the root and maintaining a small tip clearance so that the tunnel wall becomes the tip endplate. The large nacelle is advantageous from other considerations, too. (See Section 2.4.) By decreasing the tunnel cross-sectional area at the propeller a higher velocity is achieved, and higher motor speeds are possible at the same advance ratio. The large boss also enclosed the fan root sections that must be thicker for structural reasons, leaving only the thin, highly efficient sections exposed to the airstream. Frequently it is possible to use an airfoil of constant thickness in the exposed portion, thus facilitating the design. Small gains are to be found from utilizing an *L*/*D* greater than 50, so that the actual airfoil selected is of secondary importance from an aerodynamic standpoint and structural considerations can be entertained. Type *E* of the RAF propeller sections is satisfactory, as is the slightly thicker type *D* (see Fig. 2.8). The ordinate of these airfoils are shown in Table 2.1.

Although the optimum boss diameter increases with advance ratio and may be as large as 0.6–0.7 times the fan diameter, smaller values of 0.3–0.5 are more practical for wind tunnel use. The very large boss requires a large and long nacelle for proper streamlining, which, in turn, involves costly construction difficulties and greater power losses from the diffusing action as the area of the air passage is increased. It would be possible to prevent the diffusing losses by shaping the tunnel so that the area throughout the fan-nacelle region remained constant. This is sometimes done despite the added

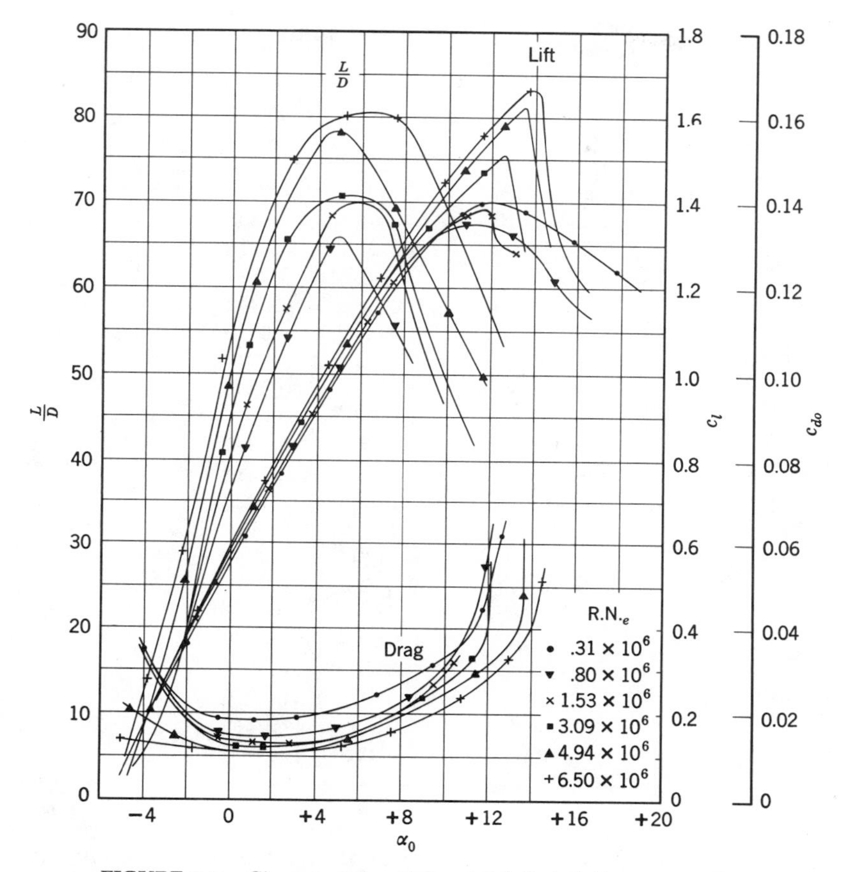

FIGURE 2.7. Characteristics of fan airfoil *D*, infinite aspect ratio.

expense. Certainly the equivalent conical diffusion angle should be kept to 7° or less.

The number of blades on the fan is somewhat arbitrary, for the product of the number of blades and their chord represents the total area and must be aligned with the thrust requirements. Several factors influence the selection of the number of blades. The minimum number probably is four; at least that

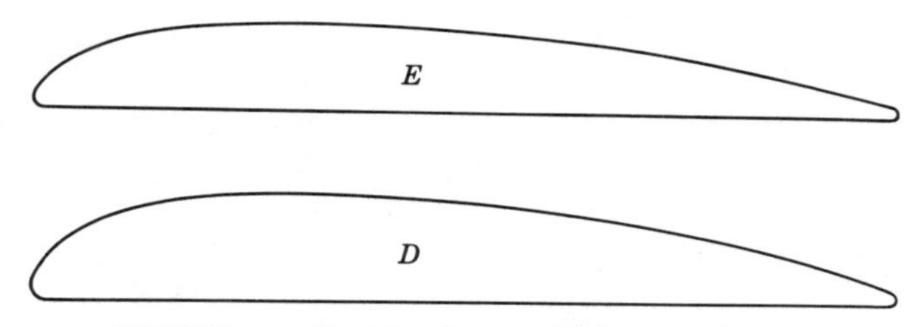

FIGURE 2.8. Satisfactory airfoils for fan sections.

TABLE 2.1. Ordinates of Fan Profiles *D* and *E*

Distance from Leading Edge	Height above Flat Undersurface	
	D	*E*
0	0.0135	0.0115
0.0125	0.0370	0.0319
0.025	0.0538	0.0442
0.05	0.0780	0.0610
0.075	0.0925	0.0724
0.10	0.1030	0.0809
0.15	0.1174	0.0928
0.20	0.1250	0.0990
0.30	0.1290	0.1030
0.40	0.1269	0.1022
0.50	0.1220	0.0980
0.60	0.1120	0.0898
0.70	0.0960	0.0770
0.80	0.0740	0.0591
0.90	0.0470	0.0379
0.95	0.0326	0.0258
1.00	0.0100	0.0076
L.E. rad	0.0135	0.0115
T.E. rad	0.0100	0.0076

number is needed to ensure little pulsation in the airstream. The maximum number of blades will doubtless be limited by strength considerations. The maximum value of the sum of the blade chords Nc must not exceed the local circumference at the root if excessive interference is to be avoided. The Reynolds number of the blade chord should be above 700,000 in order to keep the section drag low, and the tip speed should be low enough to avoid compressibility problems. Since the number of blades is not critical, a reasonable procedure is to estimate the number needed and examine the final design to see whether alterations are in order.

The Fan Advance Ratio, *j*

In the simple blade-element theory (Ref. 2.28), the angle of attack of a local section of the blade is simply the local blade angle minus the advance angle $\phi = V/2\pi nr$, where V = forward speed, n = rps, and r = section radius. This definition, which neglects both the induced indraft and rotation, is permissible only because a second assumption (that the airfoil coefficients should be based on an aspect ratio of 6.0) is made. With a wind tunnel fan, no indraft is possible, but rotation exists and the simple blade-element advance

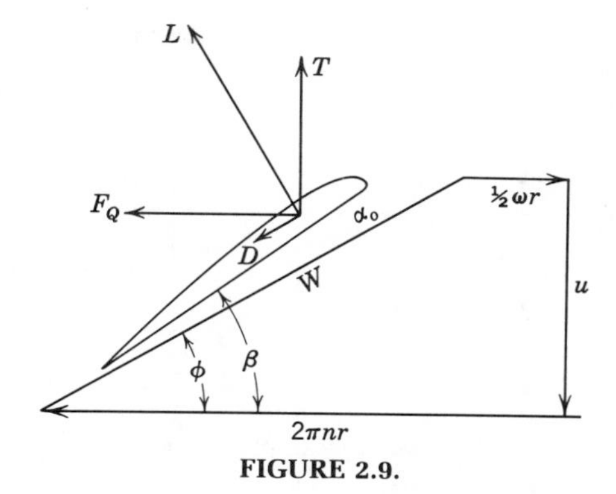

FIGURE 2.9.

angle is seriously changed. Figure 2.9 and Eqs. (2.24) and (2.25) demonstrate the proper interpretation of the advance angle for a fan in a duct.

The Rotation, *e*

It has already been demonstrated that in order to meet the requirements of the law of continuity there can be no increase of axial velocity in a duct of constant area. However, the fan imparts twist or rotation to the airstream and hence increases its absolute velocity. This added speed is removed, not turned, by the straightener vanes, and its energy appears as a rise in static pressure.

Increasing the diameter of the fan boss will decrease the amount of rotation for a given installation, as will increasing the fan rpm.

The rotation $e = \omega r/u$ will be largest at the fan boss.

The Straightener Vanes

Experiments at the NPL have shown that satisfactory antitwist or straightener vanes can be made by using the NASA symmetrical airfoils set with their chords parallel to the tunnel centerline provided that the amount of twist to be removed is small compared with the axial velocity. The limiting twist is that required to stall the vanes; that is, $e = \omega r/u = \tan \tau$ (where τ = angle of twist in the slipstream and ω = angular velocity in the slipstream at radius r) must correspond to an angle less than α_{stall} of a symmetrical section at infinite aspect ratio including multiplane interference. The interference is an advantage here because, with the type of straighteners to be employed, it decreases the lift curve slope by a factor of 0.75. That is, α_{stall} with interference is 33% above the free-air stall angle. (See Fig. 2.10.)

The chord for the proposed straightener may be found from

$$c_s = 2\pi r/N_s \tag{2.2}$$

where N_s = number of straightener vanes and c_s = chord of vane at radius r.

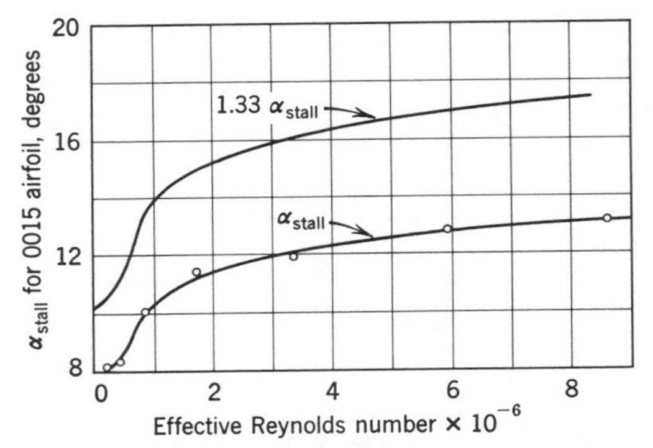

FIGURE 2.10. Effect of interference.

If a constant thickness ratio is assumed for the straightener vanes, the actual thickness at the wall would be large owing to the large chord. Hence it is advantageous to select a constant thickness (not thickness ratio). A reasonable value is that $t_s/c_s = 0.15$ at $x = r/R = 0.8$ (R = tunnel radius at propeller section). Hence from Eq. (2.2)

$$t_s/c_s = N_s t_s / 2\pi R x \tag{2.3}$$

If the number of blades is even, the number of flow straighteners should be odd and vice versa. This avoids an even multiplier between blades and vanes.

There will be a loss through the straightener, of course, and this loss will be greater than the skin friction of the vanes in free air, since the straightener is a diffuser, changing the rotational velocity ωr to static head. The pressure loss coefficient of a straightener composed of symmetrical NASA airfoil sections has been empirically determined as

$$k_s = 0.045(t_s/c_s) + 0.003 \tag{2.4}$$

Substituting from Eq. (2.2) we have

$$k_s = (0.045/2\pi r) t_s N_s + 0.003 \tag{2.5}$$

Fan-Straightener Theory

The theory for the design of a wind tunnel fan-straightener system is as follows:

Letting the total cross-sectional area at the plane of the fan be A_f and the area of the fan boss be A_b, the power output becomes

$$\text{Power out} = \Delta h \cdot (A_f - A_b) \cdot u = \eta_t \text{ bhp} \cdot 550$$

where η_t = total efficiency of fan and straightener system.

Hence

$$k = \frac{\eta_t \text{ bhp} \cdot 550}{\frac{1}{2}\rho u^3(A_f - A_b)} = \frac{\eta_t \text{ bhp} \cdot 550}{\frac{1}{2}\rho V_t^3 A_t} \frac{V_t^2}{u^2}$$

where A_t and V_t are the test-section area and velocity, respectively.
Applying the definition of the energy ratio and the law of continuity,

$$k = \eta_t \frac{1}{ER_1} \frac{(A_f - A_b)^2}{A_t^2} \tag{2.6}$$

The efficiency of the fan-straightener unit is derived from the basic relation

$$\eta_t = \frac{\text{Power out}}{\text{Power in}}$$

$$\text{Power in} = 2\pi nQ$$

where Q = torque and n = revolutions per second.
It will be convenient to consider the efficiency of a blade element in the development later, so rewriting η_t for an annulus of width dr at radius r we have

$$\eta_t = \frac{\Delta h \cdot 2\pi r \cdot dr \cdot u}{2\pi n \, dQ}$$

This procedure is possible since this method employs a constant efficiency over the entire cross section.
The elemental torque is

$$dQ = 2\pi r \cdot dr \cdot \rho u \cdot \omega r^2 \tag{2.7}$$

and, since $\Delta h = k \cdot \frac{1}{2}\rho u^2$ and $\Omega = 2\pi n$,

$$\eta_t = ku^2/2\Omega r^2\omega$$

If the local advance ratio is defined as

$$j = u/nd = u\pi/2\pi nr = u\pi/\Omega r \tag{2.8}$$

and the rotation of the flow e is expressed as a fraction of the axial velocity

$$e = \omega r/u \tag{2.9}$$

and

$$\eta_t = kj/2\pi e \tag{2.10}$$

Writing the loss in head due to the straightener as Δp_s and proceeding as in the derivation of Eq. (2.10), we find

$$\eta_s = k_s j/2\pi e \tag{2.10a}$$

We may determine e from

$$e = kj/2\pi\eta_t$$

And hence η_f becomes determined through

Fan efficiency = Total efficiency + Straightener efficiency loss

or, in symbols,

$$\eta_f = \eta_t + \eta_s \tag{2.10b}$$

Writing the elemental thrust as the pressure rise times the elemental area, we have

$$dT = \Delta p \cdot 2\pi r \cdot dr \tag{2.11}$$

Expressing the local radius as a fraction of the tip radius R by the relation $x = r/R$, and dividing the expression for the elemental thrust by $\frac{1}{2}\rho u^2 \cdot \pi R^2$ to reduce it to coefficient form, we have

$$\frac{dT_c}{dx} = \frac{\Delta p \cdot 2\pi r \cdot R}{\frac{1}{2}\rho u^2 \pi R^2} = \frac{2\,\Delta p \cdot x}{\frac{1}{2}\rho u^2} \tag{2.12}$$

The total pressure rise required of the fan and straightener is

$$\begin{aligned}\Delta h &= k_s \cdot \tfrac{1}{2}\rho u^2 \\ &= \text{Fan rise} + \text{Rotation} - \text{Straightener loss} \\ &= \Delta p + \tfrac{1}{2}\rho\omega^2 r^2 - k_s \cdot \tfrac{1}{2}\rho u^2\end{aligned}$$

Solving, the necessary rise through the fan is

$$\Delta p/\tfrac{1}{2}\rho u^2 = k + k_s - e^2 \tag{2.13}$$

And Eq. (2.12) becomes

$$dT_c/dx = (k + k_s - e^2) \cdot 2x \tag{2.14}$$

The elemental torque in coefficient form becomes

$$dQ_c = \frac{dQ}{\frac{1}{2}\rho u^2 \cdot \pi R^2 \cdot R} \tag{2.15}$$

$$dQ_c/dx = 4x^2 e \tag{2.16}$$

so that

$$Q_c = \int_{x_0}^{1.0} \frac{dQ_c}{dx}\, dx$$

and, finally

$$Q_c = (kJ/\pi\eta_t)(1 - x_0^2) \tag{2.17}$$

where x_0 = radius ratio at the root section and J = advance ratio of fan tip.

Equation (2.17) determines the input torque necessary and hence the power required to realize the total pressure rise Δh.

By approaching the problem in a slightly different manner it is possible to get a relation between j, η_f, and e such that the local L/D is determined.

We proceed as follows. The total pressure rise due to the fan is

$$\Delta h_f = \text{Static rise} + \text{Rotational dynamic head}$$

$$\Delta h_f = \Delta p + \tfrac{1}{2}\rho\omega^2 r^2 \tag{2.18}$$

and

$$\begin{aligned} \text{Power output} &= \Delta h_f \cdot 2\pi r\, dr \cdot u \\ &= 2\pi r\, dr \cdot u\, \Delta p + \tfrac{1}{2}\rho\omega^2 r^2 \cdot 2\pi r\, dr \cdot u \\ &= u \cdot dT + \tfrac{1}{2}\omega\, dQ \end{aligned}$$

The fan efficiency

$$\eta_f = \frac{u\, dT + \frac{1}{2}\omega\, dQ}{2\pi n\, dQ}$$

or, in coefficient form,

$$\begin{aligned} \eta_f &= \frac{u\, dT_c}{2\pi n \cdot dQ_c \cdot R} + \frac{1}{2}\frac{\omega}{2\pi n} \\ &= \frac{J}{\pi}\frac{dT_c}{dQ_c} + \frac{1}{2}\frac{ej}{\pi} \end{aligned} \tag{2.19}$$

Substituting from Eq. (2.14) and Eq. (2.16),

$$\eta_f = \frac{j}{2\pi e}(k + k_s)$$

Expressing the elemental thrust in a form similar to conventional wing coefficients, we have

$$dT = \tfrac{1}{2}\rho W^2 \cdot c\, dr \cdot NC_t \tag{2.20}$$

where $W = u/\sin\phi$ (twist neglected), N = number of blades, and

$$C_t = \text{thrust coefficient} = c_l \cos\phi - c_{d0}\sin\phi$$

and $$C_x = \text{torque force coefficient} = c_l \sin\phi + c_{d0}\cos\phi \tag{2.21}$$

Reducing Eq. (2.20) to the T_c form, we have

$$\frac{dT_c}{dx} = \frac{NC_t \cdot \frac{1}{2}\rho u^2 \cdot cR}{\frac{1}{2}\rho u^2 \sin^2\phi \pi R^2} = \frac{NC_t \cdot c}{\pi R \sin^2\phi}$$

$$\frac{dT_c}{dx} = \frac{yC_t}{\sin^2\phi} \tag{2.21a}$$

where $y = NC/\pi R$ by definition.

The corresponding elemental torque is

$$dQ = NC_x \cdot \tfrac{1}{2}\rho W^2 \cdot c\, dr \cdot r$$

$$\frac{dQ_c}{dx} = \frac{yxC_x}{\sin^2\phi} \tag{2.22}$$

Substituting in Eq. (2.19), the fan efficiency,

$$\eta_f = \frac{j}{\pi}\frac{C_t}{C_x} + \frac{1}{2}\frac{ej}{\pi} \tag{2.23}$$

From Fig. 2.9,

$$\tan\phi = \frac{u}{2\pi nr - \frac{1}{2}\omega r}$$

$$= \frac{u}{\pi nd - \dfrac{1}{2}\dfrac{\omega r}{u}\dfrac{u}{nd}\dfrac{\pi}{\pi}nd} \tag{2.24}$$

Hence from Eqs. (2.8) and (2.9)

$$\tan\phi = \frac{j}{\pi}\frac{1}{(1 - \frac{1}{2}ej/\pi)} \tag{2.25}$$

Therefore, from Eqs. (2.21), (2.23), and (2.25), we have

$$\eta_f = \frac{\frac{j}{\pi}\left(\frac{L}{D} - \frac{j}{\pi}\right) + \frac{1}{2}\frac{ej}{\pi}\left(1 - \frac{1}{2}\frac{ej}{\pi}\right)}{\frac{L}{D}\frac{j}{\pi} + 1 - \frac{1}{2}\frac{ej}{\pi}} \tag{2.26}$$

With η_f and j known, Eq. (2.26) can be employed to yield the L/D desired at each corresponding radius, but the values of c_l, c_{d0}, and α_0 cannot be determined accurately until the local Reynolds number is known. Hence it is necessary to determine an approximate Reynolds number as follows:

1. Using calculated L/D, read approximate lift coefficient $c_{l\,\text{approx}}$ in Fig. 2.7.
2. From $dT_c/dx = 2x(k + k_s - e^2)$ find dT_c/dx.
3. Calculate yC_t from

$$\frac{dT_c}{dx} = \frac{yC_t}{\sin^2 \phi} \tag{2.27}$$

4. Calculate y_{approx} from

$$y \cong \frac{yC_t}{c_{l\,\text{approx}} \cos \phi} \tag{2.28}$$

5. Get approximate c from

$$c_{\text{approx}} = \frac{\pi R y_{\text{approx}}}{N} \tag{2.29}$$

6.

$$RN = (\rho/\mu)cW \tag{2.30}$$

where μ = viscosity of the air. Having the Reynolds number, we now use the characteristic curves of the selected airfoil section to determine c_l, c_{d0}, and α_0.

The values of the advance angle ϕ may be determined from Eq. (2.25). The blade angle is determined from

$$\beta = \phi + \alpha_0 \tag{2.31}$$

Since dT_c/dx is known for each value of x from Eq. (2.14), C_t may be found from Eq. (2.21), $\sin \phi$ from Eq. (2.25), and y from Eq. (2.21a). From

$$c = \pi R y/N \tag{2.32}$$

the local chord may be computed.

Design Procedure

1. Select a desired overall efficiency η_t and add to it the estimated straightener loss (2–4%) to get the required fan efficiency η_f.

2. From the plot of approximate fan efficiencies versus advance ratios (Fig. 2.6) determine the required L/D and j range to attain η_f. If the available range is excessive, select the advance ratio for the tip speed as low as possible, as this will yield maximum rpm and minimum rotation. Determine n from

$$n = u/jd$$

Check to see that a tip speed is low enough so that compressibility losses will not be encountered.

3. Calculate k from Eq. (2.6), and e from Eq. (2.10). Check that e at the root is less than $1.33\alpha_{\text{stall}}$ from Fig. 2.10.
4. Calculate t_s/c_s and k_s from Eqs. (2.3) and (2.4).
5. Calculate η_s from Eq. (2.10a) and η_f from Eq. (2.10b).
6. Determine L/D from Eq. (2.16).
7. Calculate ϕ from Eq. (2.25).
8. Read approximate c_l from Fig. 2.7.
9. Find dT_c/dx from Eq. (2.14).
10. Calculate yC_t, y, c, and RN from Eqs. (2.27), (2.28), (2.29), and (2.30).
11. Using approximate RN, read accurate c_l, c_{d0}, and α_0 from Fig. 2.7, and get C_t from Eq. (2.21).
12. Calculate y from Eq. (2.21a).
13. Calculate c from Eq. (2.32).
14. Determine β from Eq. (2.31).
15. Determine Q from Eq. (2.17).

Example 2.1. A fan is required for a wind tunnel whose energy ratio is 5.0. The area of the test section is 56.4 sq. ft, and the testing velocity is 193 mph = 283 ft/sec. The wind tunnel diameter at the fan is 13 ft.

A boss diameter of 0.6 D and 12 blades are values selected for preliminary calculations. Hence $A_f - A_b = 133 - 47.8 = 85.2$ sq. ft, and $u = 284 \times 56.4/85.2 = 188$. Let $\eta_t = 0.93$, and $\rho/\mu = 58.00$.

STEP 1. Estimating the straightener loss at 3%, it is seen that the fan efficiency must therefore be 96%.

STEP 2. From Fig. 2.6 it is seen that $\eta_f = 96\%$ may be reached from $j = 2.2$ to $j = 4.8$, using $L/D = 50.0$, which is a reasonable value:

$$n = \frac{u}{JD} = \frac{188}{2.2 \times 13} = 6.58$$

$$\Omega = 2\pi n = 41.4 \text{ rad/sec}$$

$$V_{\text{tip}} = 2\pi nR = 2\pi \times 6.58 \times 6.5 = 269 \text{ ft/sec}$$

which is well below the approximate limit to avoid compressibility.

STEP 3.

$$k = \frac{\eta_t}{ER}\left(\frac{A_f - A_b}{A_t}\right)^2 = \frac{0.93}{5.00}\left(\frac{85.2}{56.4}\right)^2 = 0.425$$

$$e = \frac{kj}{2\pi\eta_t} = \frac{0.425}{2\pi(0.93)}j = 0.0729j$$

$$j = \frac{u}{nd} = \frac{u}{n\,Dx} = \frac{188}{(6.58)(13.0)x} = \frac{2.20}{x}$$

$$e_{\text{root}} = \frac{2.20}{0.6} \times 0.0729 = 0.267$$

$$\tau = \tan^{-1} 0.267 = 14.9°$$

This is below $1.33\alpha_{\text{stall}}$ from Fig. 2.10, using an estimated $RN = 3{,}000{,}000$.

STEP 4. The thickness of the straightener vanes (to be held constant) is $t_s/c_s = 0.15$ at $x = 0.8$. From Eq. (2.2) we have

$$c_{s(x=0.8)} = \frac{2\pi Rx}{N_s} = \frac{2\pi(6.5)(0.8)}{7.0} = 4.67 \text{ ft}$$

which makes $t_s = 0.15 \times 4.67 = 0.70$ ft. Hence from Eq. (2.3)

$$\begin{aligned} \frac{t_s}{c_s} &= \frac{N_s t_s}{2\pi Rx} = \frac{7 \times 0.70}{2\pi(6.5)x} = \frac{0.12}{x} \\ k_s &= 0.045(t_s/c_s) + 0.003 \\ &= (0.0054/x) + 0.003 \end{aligned} \tag{2.4}$$

The remaining steps are indicated and tabulated below.

x	0.6	0.7	0.8	0.9	1.0
j	3.67	3.14	2.75	2.44	2.20
e	0.267	0.229	0.200	0.178	0.160
t_s/c_s	0.20	0.171	0.15	0.133	0.12
k_s	0.0120	0.0107	0.0098	0.0090	0.0084

$\eta_s = \dfrac{k_s j}{2\pi e}$	0.026	0.023	0.021	0.020	0.018
η_t	0.930	0.930	0.930	0.930	0.930
η_f [Eq. (2.10b)]	0.956	0.955	0.951	0.950	0.948
j/π	1.17	1.00	0.875	0.777	0.700
$\frac{1}{2}ej/\pi$	0.156	0.1145	0.0875	0.0692	0.056
L/D [Eq. (2.26)]	40.6	38.8	36.1	36.5	37.0
$\tan \phi$	1.39	1.13	0.960	0.835	0.742
ϕ, deg	54.3	48.5	43.8	39.8	36.6
$c_{l\,\text{approx}}$	0.53	0.51	0.50	0.52	0.51
$k + k_s - e^2$	0.366	0.384	0.395	0.402	0.406
dT_c/dx	0.440	0.538	0.633	0.724	0.812
$\sin^2 \phi$	0.660	0.560	0.475	0.410	0.360
yC_t	0.290	0.301	0.301	0.296	0.292
$\cos \phi$	0.584	0.663	0.725	0.769	0.800
y_{approx}	0.937	0.890	0.755	0.742	0.715
c_{approx}	1.59	1.52	1.285	1.264	1.21
$W = u/\sin \phi$	231	251	273	293	313
RN_{approx}	2.13×10^6	2.22×10^6	2.04×10^6	2.08×10^6	2.21×10^6
c_l	0.55	0.53	0.51	0.51	0.51
c_{d0}	0.0140	0.0140	0.0150	0.0140	0.0130
α_0	−0.2	−0.2	−0.6	−0.5	−0.5
β, deg	54.1	48.3	43.2	39.6	35.8
C_t [Eq. (2.21)]	0.32	0.352	0.369	0.432	0.408
y	0.873	0.834	0.795	0.738	0.700
c	1.483	1.418	1.350	1.250	1.190

The usual requirement that the propeller blade section be thin (especially at the tips) does not rigidly hold in wind tunnel fans. The reasons are two: the airspeed at the fan is rarely very high and compressibility effects are not serious, and high enough L/D ratios are obtained so easily that straining for small increments through the use of thin sections is unnecessary. The thicker sections are stronger, too, but peculiar high-frequency vibrations that occur in many wind tunnel fans and the possibility of the propeller's being struck by airborne objects make it advisable to incorporate margins of safety of the order of 5.0 into their design. An advantage accrues from having removable blades, since a damaged blade may then be replaced without rebuilding the entire fan. However, the replacement blade must be mass balanced to match the first moment of the other blades, as a minimum.

Tests of fans designed by the above method indicate that actual efficiencies will be from 3 to 5% less than theoretical, owing to tip clearance and boundary layer effects at boss and tip.

If changes are made to the tunnel after it has been built, it may be necessary to make a fan revision. Though an entirely new fan would be best, flaps have been installed in several tunnels with satisfactory results and, of course, at much less cost than a whole new fan. The procedure is to rivet or screw flat sheet at the desired flap angle until the chord is satisfactory to meet the new condition.

It is interesting to note that modified aircraft propellers have been used successfully in many tunnels. The remote-control variable-pitch feature that most of these have is most desirable.

2.6. RETURN PASSAGE OR SECOND DIFFUSER

The second diffuser extends from the fan section to the third corner. As indicated in Section 2.2, this diffuser continues the expansion to the desired total area ratio. Again, the equivalent cone angle should be 5° or less. The fan at the entrance gives an almost constant total pressure profile.

There are two possible sources of trouble in the second diffuser. The first was discussed in Section 2.4 and that is flow separation on the aft portion of the nacelle.

The second problem in many rectangular tunnels is that the flow downstream of the fan may have nonuniform velocity distributions. In the flow improvement program at the University of Washington (Ref. 2.23) model tunnel studies showed very low velocities in the corners. The flow at the third corner was higher on the inside of the turn. However, the upflow–cross-flow distribution prior to the turn did not exhibit a discernable flow pattern similar to the test section. After the turn there were two distinct rotations in the upper and lower half of the duct. A model tunnel test section exhibited a similar cross-flow gradient as the full-scale tunnel.

Although this problem sounds like a secondary flow being produced by the velocity gradients prior to the turn, in all cases the sense of the rotation was reversed.

It was discovered that when the antiswirl vanes were uniformly deflected either positively to reduce fan swirl or negatively to increase fan swirl the corners were filled and the cross-flow gradient was reduced, but not eliminated. In the final tunnel modification the antiswirl vanes, which originally were simple flat plates hinged to the nacelle supports, were redesigned. The chord of the antiswirl vanes was increased. They were made into cambered airfoil sections using the nacelle struts for a leading edge. This modification filled the corners of the second diffuser and eliminated flow separation regions on the original flat plate vanes. The new cambered vanes were deflected to 12° and resulted in a much more uniform velocity distribution with all four corners filled with higher-energy flow. This flow improvement in the second diffuser and fourth corner also reduced the power consumption at all speeds by 2.0%.

2.7. THE CONTRACTION CONE

There are two problems in the design of the contraction cone. First is the presence of an adverse pressure gradient at the entrance and exit of the contraction. If either of these gradients becomes severe enough for the boundary layer to separate, there can be degradation of the quality of the test section flow, and an increase in the power required. Second, the surface streamlines of a rectangular contraction intersect the side walls. This leads to secondary flow in the corners with the attendant lower velocities and possibility of separation. The latter problem is alleviated by making the contraction octagonal. This is done by starting a 45° fillet at the start of the contraction cone and carrying the fillet through the test section and first diffuser.

Until the advent of the digital computer there was no wholly satisfactory method of designing contractions. The contraction was designed either by eye or by adaptions of approximate methods. Experience has shown the radius of curvature should be less at the exit than at the entrance. Most of the early work on contractions was based on potential theory. Once the wall shape was determined, then the regions of adverse pressure gradient were checked to make sure that they were not too sharp. The following paragraph briefly discusses four methods of contraction design using digital computers.

Chmielewski in Ref. 2.29 specified a distribution for a streamwise acceleration for a quasi-one-dimensional flow. Long constant area ducts were used, one on each side of the contraction, to ensure parallel flow. The velocities obtained from potential theory were checked by Stratford's separation criteria (Ref. 2.30). A characteristic feature of the velocity distribution was a velocity peak on the centerline of the constant area inlet that persists well into the contraction where the velocity then decreases, ending in a centerline velocity deficit. The velocity deficit continues for about one radius beyond the contraction exit. The wall velocity decreases at the inlet and increases at the exit. This leads to a cup-shaped velocity profile in the constant area exit duct. This type of velocity profile has been measured in wind tunnels. The rapid growth in the boundary layer thickness in the constant area inlet makes an inlet separation a possibility. The length of the contraction decreased as the inlet radius was decreased and the exit radius increased, for the contours used. If, however, the area reduction is too gradual, the boundary layer will be subjected to a mild pressure gradient over a long distance, increasing the risk of separation.

Morel in Ref. 2.31 considered uniform flow in the exit as being the basic requirement for a contraction. He also pointed out that as the contraction ratio increases beyond 4 the length will decrease for fixed exit requirements. Borger (Ref. 2.32) recommended a slight expansion near the contraction exit to improve the exit flow uniformity.

Mikhail and Rainbird (Ref. 2.33), by controlling the distribution of curvature, were able to control the wall pressures and gradients and flow uniformity at the exit. The length of the exit section was defined on the basis of 0.25% flow uniformity at the centerline at one radius from the exit. The contraction exit length (from the wall inflection point to constant area duct) is sensitive to the required length in the test section for a uniform velocity profile. For a contraction ratio of 8 the inlet section length varied from 0.15 to 1.00 times the inlet radius, while the test section settling length varied from 1.5 to 0.5 of the exit radius.

It is desirable to keep the length of the contraction as short as possible, the length of the contraction being defined as the sum of the settling chamber length plus the contraction plus the settling length of the exit section. The settling chamber length is required for honeycombs and/or screens to reduce turbulence, if they are to be used. A settling chamber length of 0.5 times the inlet diameter is often used.

Quite often for a new tunnel either a complete model or models of parts of the tunnel are built to check the design. Although the model tunnel can duplicate the velocities of the full-scale tunnel, the Reynolds number will be reduced by the scale factor. Thus, the boundary layer will be thicker in the model tunnel than in the full-scale tunnel. Despite this problem those facilities that have models of their tunnels have found them to be invaluable, both as a check of modifications to the full-scale tunnel and as a facility for experiments that would be too expensive to run in the large tunnel.

If a complete model tunnel is not built, often parts of the full-scale tunnel are simulated. Because the contraction section is critical to the flow quality in the test section, quite often models of this part are built. The model may include the fourth corner. This model will need a bellmouth at its entrance, and should include the test section to the first diffuser. The flow should be sucked through the model. This type model can be used to check the test-section settling length. The contraction's sensitivity to nonuniform flow can be checked also. The effect of honeycombs and/or screens in the settling chamber, or turbulence both at the entrance to the contraction and in the test section, can be determined.

2.8. COOLING

All the energy supplied to the propeller driving a wind tunnel finally emerges as an increase of heat energy in the windstream. This increases the temperature of the tunnel air until the heat losses finally balance the input. For low-power tunnels (and particularly those with open jets) this balance is realized at reasonable temperatures, the heat transfer through surface cooling and air exchange being sufficient. For tunnels with high-power inputs and high jet velocities this low-temperature balance no longer occurs. For example, the heat rise incurred by bringing air to rest at 450 mph is about 36°F. With an energy ratio of 8.0, the heat rise in the airstream would be 4.5°F per circuit, leading very shortly to prohibitive temperatures. Obviously, tunnels in this class require cooling arrangements to augment the inherent heat losses.

Additional cooling may be accomplished by four means: (1) an increase of surface cooling by running water over the tunnel exterior, (2) interior cooling by the addition of water-cooled turning vanes, (3) a water-cooled radiator in the largest tunnel section, or (4) a continual replacement of the heated tunnel air with cool outside air by means of an air exchanger.

Some high-speed tunnels use an air exchanger (Figs. 2.11 and 2.12) to replace the lower-energy boundary layer with cool outside air, having exchange towers to ensure adequate dispersion of the heated air and fresh air that is free from surface contamination. Assuming the previously mentioned rise of 4.5°F per circuit, a 10% exchange would limit the rise to 45°F, excluding heat losses elsewhere. (Ten percent is a lower than average amount of exchange.)

One difficulty associated with an air exchanger is that it puts the highest-pressure section of the tunnel at atmospheric pressure, and hence the jet pressure is below atmospheric. This leads to troublesome but by no means insoluble problems of sealing off the balance room. (The same low jet static pressure is present in open-circuit tunnels.) Another difficulty that must be considered is the possible effects of weather conditions on a tunnel with a large amount of air exchanged.

It should be mentioned that a breather slot at the downstream end of the test section can be used in conjunction with the air exchanger to get the jet up to atmospheric pressure and hence avoid balance sealing troubles. However, this arrangement with the air going in the breather and out the exchanger requires as much as 20% of the total power input.

The internal heat exchangers needed for high-powered wind tunnels require an immense amount of surface, a great deal more than is offered by all

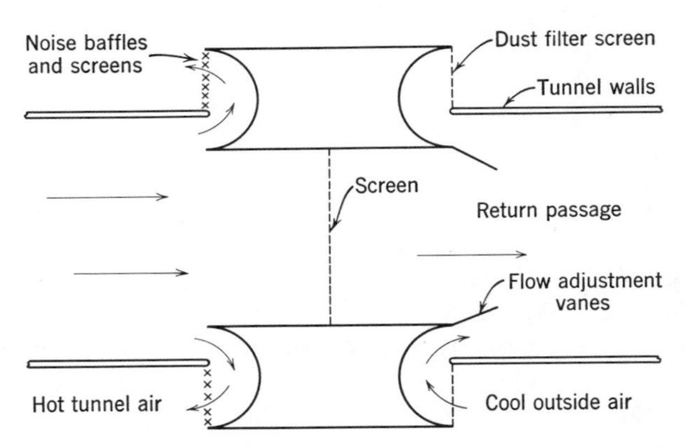

FIGURE 2.11. An air exchanger. The screen provides an extra pressure drop for additional flow through the air exchanger.

FIGURE 2.12. View of air exchangers from inside the return passage. (Courtesy United Aircraft Corp.)

four sets of guide vanes. Accordingly, a special installation is needed, and there is almost no way around a very large amount of drag. In a preliminary study reported by Steinle (Ref. 2.34) pressure drops of 8 to $18q$ were measured across exchangers; another design, unreported, had $4q$. Thus the heat exchanger must be placed in the largest section of the tunnel where the q is lowest; incidentally this helps because at this point the temperature of the stream is highest and heat exchange consequently simplest.

In view of its power cost a great deal of thought should go into the design of a heat exchanger, and it should be remembered that normal streamlining should be used here as well as elsewhere. The unknowns of internal and external boundary layer thicknesses make the problem of cooling through the walls quite difficult. In discussing internal cooling, Tifford (Ref. 2.35) agrees that a radiator has possible advantages over cooled turning vanes.

The obvious disadvantages of high temperatures in the wind tunnel include added trouble cooling the drive motor (if it is in the tunnel and does not have separate cooling), the rapid softening of the model temporary fillets, and increased personnel difficulties. Another deleterious effect is the drop in Reynolds number that occurs with increasing temperatures whether the tunnel is run at constant speed or at constant dynamic pressure. Figure 2.13 illustrates this effect.

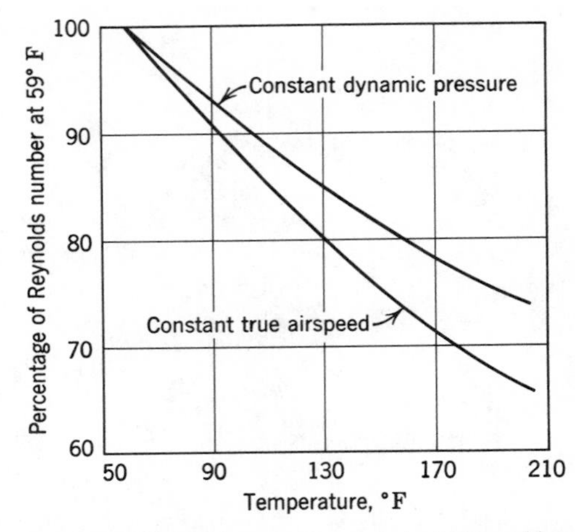

FIGURE 2.13. Effect of temperature on test Reynolds number.

Since most electric motors have high efficiencies, placing the motor outside the tunnel is probably not justified by the small amount of tunnel heating saved thereby, but the ease of motor repairs and tunnel repowering without a nacelle change are strong arguments for the external drive motor.

2.9. THE BREATHER—VIBRATION

If the tunnel is to be operated with an open jet, due consideration must be given to the possibility of pulsations similar to the vibrations in an organ pipe. This phenomenon, believed to be a function of jet length, can be quite serious.

The simplest solution, usually successful, consists of putting a slot (about 0.05 diam wide) into the diffuser which connects it to the atmosphere. Such an arrangement is called a "breather." If the slot is properly made and adjusted so that it is just large enough to prevent organpiping, the losses can be kept low. In some open jet tunnels alterations to the exit cone proved sufficient to prevent the vibration, but in others no satisfactory exit cones or breathers have been found that would permit operation above 200 mph.

The NASA Langley 4 × 7 m tunnel, when run in a closed-on-bottom-only test-section configuration, has exhibited periodic flow pulsations in the axial direction at three dynamic pressures. The magnitude of the velocity peaks increased with the addition of flapped vanes between the first and second corners (Section 2.3). The addition of triangular vanes with alternate deflections on the side walls and ceiling of the exit cone removed these pulsations at the higher dynamic pressures and reduced the width of the peak at the lowest dynamic pressure. The vanes are mounted inboard of the walls and ceiling. Model tunnel studies show that the remaining pulsation can be eliminated by modifying the reentry cone. The modification consists of a tapered cone forward of the original reentry cone.

Closed jet tunnels usually require breathers too, because the entire return passage is above atmospheric pressure, and some air may leak out. In turn the loss of air would drop the jet pressure below atmospheric unless it were replenished. The proper place for a closed jet tunnel breather is at the downstream end of the test section and like that for an open jet tunnel a slot about 0.05 diam wide usually suffices. It should be covered with a fine screen to prevent papers and such from entering the tunnel.

2.10. TEST SECTION FLOW QUALITY

For a test section that is clean, that is, no balance struts, fairings, etc., the flow outside the boundary layer would have the following ideal characteristics: the velocity profile at each station would be uniform; there would be no upflow or cross flow; and there would be no turbulence. Since it is doubtful that such flow can be obtained in a wind tunnel, the question then becomes what flow quality is acceptable at a reasonable tunnel performance level and operating cost.

Values for velocity variation across the test section are often quoted in the range of 0.20–0.30% variation from average. This gives a dynamic pressure variation of 0.4–0.6%. Values for angular variation are often quoted in the range of 0.1° from the average flow angle. The upflow variation is the critical flow angle, because the drag data must be corrected for upflow (see Sections 4.16 and 6.30).

The first requirement is steady flow, or how much unsteady flow can be accepted. Any time-dependent velocity fluctuations should be of small magnitude and at a low enough frequency so that they are not noticeable in balance or pressure measurements. In general, unsteady flow is a result of separated flow, either continuous or intermittant. This definition is used to separate unsteady flow from turbulence, which can arise from wakes of vanes, noise, etc., and occurs at higher frequencies. Unsteady flow caused by a flow separation that deflects the flow may be regarded as irrotational and inviscid outside the separated region. The only cure for these is to locate the source and eliminate it. This is not an easy task as the effects are usually detected some distance from the source. Obvious locations to look for separation are the first diffuser, first corner, fan nacelle, and contraction. If the tunnel has air leaks, the breather may cause separation. Usually leaks in a return tunnel will be near doors, drains, access holes for instrumentation, etc., and they can be sealed.

Other sources of unsteady flow can arise from the fan through nonuniform inflow. This will be at the blade frequency.

It is assumed that the number of fan blades, nacelle supports, and prerotation vanes are not even multiples of each other. Mechanical vibration of the fan will usually be at the fan rpm. Vibration of turning vanes due to turbulence or unsteady flow could occur over a wide frequency range. The vanes can be stiffened by horizontal plates, if required.

Nonuniform distribution of velocity can be caused by separation. It can also occur by either poor vane design or improperly adjusted vanes that cause the flow to over or under turn. See Section 2.3 for suggestions on checking this item.

The use of the antiswirl vanes to improve the flow in the second diffuser was covered in Section 2.6. Upflow and cross-flow distribution in the test section appear to be the result of poor flow into the third corner and out of the fourth corner. In this region the turning vanes are quite long, and in a great many tunnels the trailing edges have a spanwise bow. As discussed in Section 2.3, making sure that the flow is not over or under turned and converting the fourth corner vanes to a large grid adjustable honeycomb will do wonders to the upflow, cross flow, velocity, and turbulence in the test section.

There also must be a settling length after the contraction to allow the velocity to become uniform (see Section 2.7).

In sections dealing with the design of various parts of the tunnel and the preceding paragraphs, sources and possible remedies of poor flow in the test section have been discussed. A surprising number of tunnels suffer from one or more of these problems to various degrees. The problems can be cured or reduced by the methods that have been discussed, but this takes persistence and time. In almost all cases the basic problem is time. A new wind tunnel represents a large investment to management. It usually is behind schedule when built and thus the pressure to begin testing becomes irresistible. The down time for correcting these problems can be large. Also, a period of time must be set aside to allow for the complete recalibration of the test section after the modification. The authors cannot offer any solution to this part of the problem. The modification to the University of Washington tunnel, including the installation of the new parts and adjustment and calibration, consumed about three months of double-shift work. The work at the AEDC 16T tunnel took two weeks, the majority (two-thirds) of this time was with setting up and moving the scaffolding.

2.11. TURBULENCE REDUCTION: HONEYCOMBS AND SCREENS

Turbulence in the test section is reduced by the installation of honeycombs and screens upstream of the contraction. Screens reduce the axial turbulence more than the lateral turbulence. Screens have a relatively large pressure drop in the flow direction, which reduces the higher velocities more than the lower, and thus promote a more uniform axial velocity. Honeycombs have small pressure drops and thus have less effect on axial velocities, but owing to their length, they reduce the lateral velocities. The minimum length of a honeycomb should be 6–8 times the cell size. Both screens and honeycombs reduce lateral and axial turbulence, probably due to an exchange in energy between the axis as the turbulence tends toward isentropic turbulence downstream. Despite being located in the lowest speed portion of the tunnel, screens increase the required power of the tunnel. The General Motors full-scale automotive tunnel has quoted that a single screen with 58% porosity accounts for 25% of the circuit losses at high speeds.

A problem with screens is their amazing ability to accumulate dust. The dust is usually in a nonuniform distribution. Thus the screen's porosity and pressure drop will change, which in turn will change the velocity and angularity distribution in the test section in an arbitrary way with time. This problem will be aggravated when the tunnel is used for flow visualization studies using china clay, oil flow, or vaporized oil for smoke, or when oil is used for seeding lasers (see Chapter 3). When screens are used, they must be installed so that they can be cleaned, and the quality of the test section flow must be monitored.

For larger tunnels it is difficult to obtain either screens or honeycombs in widths adequate to span the tunnel ahead of the contraction cone. This means that screens must be spliced together. The splice is often accomplished by brazing widths of screens together, which must be done with extreme care. The individual screen mesh must be on top of each other, and should not be filled with the brazing material. If the splice is semisolid, it will introduce turbulence. A poor splice can be detected in the test section as a band of higher turbulence and possibly lower velocity. The same problem can occur with cables across the screen's face, with tension cables to the fourth corner to carry part of the screen load. Honeycombs are often mounted in a shadow-box-type structure and this also can give the same effect.

Ideally a tunnel should be designed with screens in mind: with slots ahead of the contraction cone for mounting screens. Then the screens are mounted on a support frame and slid into the tunnel. This also facilitates cleaning the screens. The slot must be sealed, of course.

If screens are added to an existing tunnel, the screens can be brazed to a support ring that has radial screws to pull the screen tight. For multiple screens this type of installation can be difficult. Screens also have been attached to cables that are then attached to the tunnel walls. It is not unknown for a screen to tear loose from a tunnel.

The radial force per foot of perimeter due to the pressure drop will be $C_p q d^2/8\delta$, where δ is the screen sag in feet, and the wire tensile strength may then be computed using the total wire cross-sectional area per foot of perimeter. Screens have also been used to correct extremely poor velocity distributions (Section 3.12). This is done by using a coarse mesh screen to support a fine mesh screen to damp out high velocity regions.

Screens used for turbulence reduction should have the projected open area to total area ratio, β, greater than 0.57. Screens with smaller ratios suffer from a flow instability that appears in the test section (Refs. 2.36, 2.37). At a Reynolds number based on a wire diameter of about 80, the wake from the wires is turbulent. Since this is a small-scale turbulence, it quickly damps out.

Most of the theoretical treatments of turbulence assume that the initial flow has isentropic turbulence. Most of the data for turbulence reduction are taken in flows that are either isentropic or very close to isentropic. This is done by inserting turbulence generators well upstream into a uniform flow (screens, an array of bars, or porous plates). After passing through this turbulence generator the turbulence approaches an isentropic state. In most cases the actual turbulent flow in a wind tunnel is not isentropic. In general the mechanism of turbulence and its manipulation is very complex and not completely understood. These facts often lead to the unpleasant result that the predicted reduction of turbulence by screens or honeycombs does not agree with the measured results. Nonetheless, the theory combined with data can be used to compare various turbulence reduction schemes in preliminary design trade-off.

For screens, many of the turbulent reduction theories are based on a pressure loss coefficient K, defined as pressure loss across the screen ΔP divided by the mean flow dynamic pressure q. Based on Ref. 2.38:

$$K = K_0 + \frac{55.2}{R_d} \tag{2.33}$$

where

$$K_0 = \left(\frac{1 - 0.95\beta}{0.95\beta}\right)^2 \tag{2.34}$$

$$\beta = \frac{\text{Projected open area}}{\text{Total area}} = \left(1 - \frac{d}{M}\right)^2 \tag{2.35}$$

d = wire diameter

M = mesh length

R_d = Reynolds number based on wire diameter, d

The turbulence reduction factor f is defined as the turbulence with manipulators installed divided by the turbulence without manipulators.

In Ref. 2.39 a study was made of screens and honeycombs. The porosity of the screens varied between 59% and 67% for mesh sizes from 4 to 42. The honeycombs had cell sizes from $\frac{1}{16}$ to $\frac{3}{8}$ in. and the length/cell size ratio varied between 6 and 8.

For the screens the calculated K from Eq. (2.33) when compared to measured values showed the variation of K with V within about 20% or less for β

between 29% and 62%. For $\beta = 67\%$, the value of K was in error by about 50%.

The turbulence reduction factors (f) for both axial and lateral turbulence also showed data scatter but no consistent pattern with velocity. The range of screen Reynolds number based on wire diameter varied from 70 to 300. As there was no trend of f with velocity, the average value over the speed range was used.

The upstream turbulence for the screens was close to isentropic. The average measured values of f, and average K from Eq. (2.33), showed good agreement in axial turbulence using Prandtl's equation (Ref. 2.40) and in the lateral direction using Dryden and Schubauer (Ref. 2.41):

$$f = \frac{1}{1 + K} \quad \text{for axial reduction (Ref. 2.40)} \tag{2.36}$$

$$f = \frac{1}{\sqrt{1 + K}} \quad \text{for lateral reduction (Ref. 2.41)} \tag{2.37}$$

When multiple screens are used, the turbulence reduction factors are the product of the individual screens. The pressure drop K is the sum of the individual screens. For two screens with spacing beyond the minimum either doubling the spacing or at the initial spacing rotating one screen 45° had little effect on the turbulence reduction factor. Multiple screens must have a finite distance between them so that the turbulence induced by the first damps out before the second screen. Spacing values based on mesh size of greater than 30 have been suggested as well as spacing based on a wire diameter of about 500. In any case, for multiple screens the requirement to clean the screens and the mechanical problems in installing the screens usually exceed these limits.

When a honeycomb was used (data for the $\frac{1}{4}$-in. cell), the axial reduction was the same as the 20 mesh screen. The lateral reduction was equal to three 20 mesh screens. The pressure drop for the honeycomb at 50 fps was slightly less than the calculated value at 50 fps for the 20 mesh screen. The honeycomb plus one 20 mesh is equal to three 20 mesh for axial reduction and about twice as effective in the lateral direction. The honeycomb was installed upstream of the screens as suggested in Refs. 2.42 and 2.43.

The flow behind the honeycomb was not isentropic (see Ref. 2.42 for a discussion of upstream turbulence effects). When the screen turbulence reduction factors were based on the turbulence behind the honeycomb, the turbulence reduction of Taylor and Batchelor (Ref. 2.44) with the value of α from Ref. 2.36 gave the best match:

$$f = \frac{1 + \alpha - \alpha K}{1 + \alpha + K} \quad \text{(Ref. 2.44)} \tag{2.38}$$

where

$$\alpha = \frac{1.1}{\sqrt{1 + K}} \tag{2.38a}$$

The theory in Ref. 2.44 accounts for a flow that is not normal to the screen upstream and is turned toward the normal downstream. Potential flow accounting for the boundary conditions was used to get the turbulence reduction factor. The α is the ratio of the flow angle normal to the screen downstream, ϕ, to the flow angle normal to the screen upstream, θ. The value of α varies between 0 and 1.

Data for pairs of screens with different honeycombs also illustrate a problem that occurs when using hot wires. The data in Fig. 12 (Ref. 2.39) had a 2-Hz bypass filter and the data in Fig. 13 had a 100-Hz filter. The data in Fig. 13 have much lower turbulence reduction factors in the axial direction. The lateral do not exhibit as large a change. The 2-Hz filter, it was stated, was passing a high-amplitude low-frequency noise source. What caused this noise was not stated. Many wind tunnels do have slow velocity fluctuations at frequencies much lower than those containing the turbulent energy. The bypass filter can mask these flows. The tendency to use filters to remove low-frequency data can mask the actual smoothness of the tunnel flow. The results of hot-wire anemometers depend strongly on the lower frequencies, thus either the effect of the frequency cutoff should be determined or the cutoff frequency should be stated when the turbulence level is given. The use of low-frequency cutoffs may explain some of the inconsistencies in published turbulence reduction data.

This low-frequency cutoff problem with hot wires may make the turbulence sphere and critical Reynolds number worth looking at again as a method of comparing the turbulence of various flows (see Section 3.15). If spheres with different diameters are used, one has a measure of the relative scales of turbulence and turbulence intensity with tunnel speed. Since the critical Reynolds number is 385,000, based on diameter, the tunnel speed is inversely proportional to the diameter.

2.12. AN APPROACH TO TUNNEL FLOW IMPROVEMENT

The first task is to eliminate separation in the first diffuser (Section 2.2). Make sure that the first two corners do not over or under turn, that the flow is uniform approaching the fan, and that the nacelle flow is attached (Sections 2.3 and 2.4).

The flow in the second diffuser should fill the duct (Section 2.6). The fourth corner turning vanes by the use of vane extensions and the horizontal splitter plates should be adjusted to yield uniform upflow and cross flow or no large solid-body-type rotation in the test section (Section 2.3). All of these should be done before honeycombs or screens are installed. The exception to the screens would be the wide-angle diffuser. This then should yield a tunnel with the best test-section flow for the minimum power.

Using an axisymmetric contraction cone with a contraction ratio of 11, Ramjee and Hussain in Ref. 2.45 have shown that the turbulence at the contraction cone exit is essentially independent of the screen Reynolds number based on either wire diameter or mesh size. For the longitudinal turbulence the values behind the screens ahead of the inlet varies from 0.029 to 0.052, while at the exit the variation was 0.0050–0.0054 for five different screens. The lateral variation at the inlet was 0.032–0.063, and 0.0060–0.0070 at the exit. The pressure loss coefficient for four of the screens varied from $K = 0.65$ to $K = 2.34$, based on Eqs. (2.33)–(2.35), and the porosity varied from 0.62 to 0.75 (the screen with porosity of 0.25 has been excluded). The screen with $K = 2.34$, inlet turbulence of 0.029, and exit turbulence of 0.0051 gave the same results at the exit or test section as the one with $K = 0.65$, inlet turbulence of 0.052, and exit turbulence of 0.0050. These are the longitudinal values. This implies that one should use screens with the smallest pressure loss coefficient, and if multiple screens are used, again, screens with the smallest loss should be adopted.

Ramjee and Hussain also determined that the ratio of rms exit to inlet turbulence as a function of contraction ratio (from Prandtl) is overpredicted in the longitudinal direction and underpredicted in the lateral direction. These predictions are

$$u \text{ component turbulence reduction} = 1/c^2$$

$$v \text{ component turbulence reduction} = 1/\sqrt{c}$$

The linear theory of turbulence reduction due to a contraction predicted an increase in the lateral component (Ref. 2.46), but this is not borne out by experiment, which shows a decrease.

Thus there does not at present appear to be a good method of predicting the effects of contraction ratios in turbulence reduction.

The axial and lateral turbulence should be determined at several speeds, at the start of both the contraction and the test section. This will determine the turbulence reduction due to the contraction. The effect of bypass filters on the turbulence measurements should be known (Section 2.11). If the turbulence level is too high, then the installation of a honeycomb followed by a screen or screens must be considered. The reduction of axial and lateral turbulence and the pressure drops at the contraction velocity range can be estimated by published values. A better approach, using hot wires, would be to measure these values with the same bypass filters that were used in the tunnel measurements.

If screens and/or honeycombs are installed, the reduction in turbulence should be checked as each device is installed. This will ensure that the minimum number of devices is used, thus holding losses to a minimum.

The following values for turbulence are suggested. The ideal value would, of course, be zero. The final accepted values will be a trade-off between installation costs and reduction in tunnel performance and power costs.

For tunnels intended for research use in boundary layers and boundary layer transition, the lateral values of the turbulence, which is usually the largest, must be kept as small as possible. Values of about 0.05% have been suggested.

Tunnels used for developmental testing can have larger turbulence values, perhaps as high as 0.5% in the axial direction, although some large government tunnels try for 0.1%. However, there is no general agreement as to the required absolute value. Another approach is as follows: The lower limit for the tunnel's Reynolds number range was on the order of 1.5 million. At these Reynolds numbers for a model in a smooth tunnel there will be an extensive region of laminar flow. Thus there is the possibility of poor flow simulation due to laminar separation on the model that does not occur at full scale. The usual practice to ensure full-scale simulation is to use trip strips on the model to fix the transition. Based on these practices the following suggestion is made: Test a smooth wing without flaps at lift coefficients around 0.2–0.5 without trip strips (Section 7.2) for natural transition using either oil flow or sublimination flow visualization (Section 3.10). If the natural transition occurs, depending on the airfoil section, around 30–60% of the chord, the turbulence level in the tunnel is acceptable. If this can be accomplished without screens or honeycombs, the power consumption will be less and the dirt problem with screens changing flow characteristics will be avoided (Section 2.11).

2.13. THE DRIVE SYSTEM

Since the thrust of the fan and the drag of the various tunnel components vary with the square of the fan rpm, it would appear that to maintain an even velocity front in the test section speed adjustments should be made by varying fan rpm rather than fan pitch. Although this conclusion is justified in short tunnels of low contraction ratio, in the larger tunnels, particularly those with dust screens and internal coolers to act as flow dampers, it is certainly not true. Indeed, many of the larger tunnels, which are equipped with both rpm and pitch change, use the latter as quicker and simpler. It does seem as though provision of both types of control is a good design procedure.

The various drive systems described subsequently can also be achieved by electrical methods that do not require rotating machinery. As an example, the use of a synchronous motor to drive a dc generator in item 1 can also be accomplished by high-current rectifiers to provide the dc. Thus, the described drive systems are best thought of as basic concepts. The actual details of how the drive is accomplished is best left to electrical engineers conversant in rotating power. This field, like the electronics field, changes rapidly.

Considering the drives capable of variable-speed control, we have the following:

1. *Generator and dc Motor.* A direct-current generator run by a synchronous motor and used to drive a direct-current motor in the tunnel electrically is a satisfactory system below 200 hp, the cost becoming excessive above that figure. It offers excellent speed control.

2. *Tandem Drive.* The combination of a dc motor for low powers and a single-speed induction motor for high powers is satisfactory for the range of 300–20,000 hp. With this arrangement the dc motor is used for low-power operation and for bringing the induction motor up to running speed.

3. *Variable Frequency.* A setup similar to the model motors described in Chapter 5 also may be used as a tunnel drive, normally below 3000 hp. This setup embraces a synchronous motor driving a dc generator whose output is used to run a dc motor which drives an alternator. The output of the alternator is used to drive the fan motor, which can be either a synchronous or an induction motor. This is a good system but quite expensive.

4. *Magnetic Coupling.* A synchronous motor can be used to drive a fan through a variable-speed magnetic coupling. This is one of the least expensive setups as far as first cost is concerned, and gives excellent speed control from zero to maximum velocity, since it is virtually "stepless."

5. *Multispeed Squirrel Cage.* An induction motor arranged to have several operating speeds may be used in conjunction with a variable-pitch fan to get a satisfactory drive. However, the upper power limit—around 2500 hp—and the high starting loads reduce its appeal.

6. *Wound-Rotor Induction Motor.* In general one cannot expect wide rpm changes, good control, or high efficiency from a wound-rotor induction motor, although such a motor has been used with reasonable success in combination with a variable-pitch fan. It does offer a low first cost and moderately small motor for tunnel installations.

7. *Doubly Fed Induction Motor.* This arrangement requires a variable-frequency power source, which is fed into the rotor of an induction motor. Its first cost is high, but it is probably the most widely used drive for very-high-power installations where efficiency is important.

8. *Internal-Combustion Drive.* The use of an internal-combustion engine is undesirable because of both high operating cost and lack of long-time reliability. In the few tunnels where they have been used the engineers invariably look forward to the day when the gasoline engine can be replaced by an electric drive. For reasons almost unknown (but surmised to be connected with the lack of a cooling airstream over the stationary engine and unskilled maintenance) reciprocating engines rarely deliver the life in tunnel use that they do on aircraft, 300 hours being a fairly typical figure per engine.

When an aircraft engine must be used, special care should be taken that the exhaust manifold be water jacketed or otherwise cooled. An annoying trouble with these engines (as if the above is not sufficient) is that their spark plugs foul up under the low-load operation frequently needed in a tunnel. See Section 2.4 for possible motor locations.

2.14. WIND TUNNEL CONSTRUCTION

The structural loadings on the various sections of a low-speed wind tunnel are usually less critical than the strength needed to avoid vibration, a significant exception being the assurance that the drive motor will stay in place should it lose one-half of its blades. The rest of the tunnel may be examined to withstand the maximum stagnation pressure with a safety factor of perhaps 4.0.

Since vibration of parts of the wind tunnel contributes to noise, discomfort of the tunnel crew, and possible fatigue failures, and usually adds to the turbulence in the wind stream, it is good practice to have the natural frequencies of all tunnel parts well above any exciting frequencies. Many of these parts of the tunnel are directly amenable to basic vibration theory; others must wait for treatment after the tunnel is built. At that time, for the small tunnels at least, flat panels can be checked with a simple shaker motor, and by means of a vibrometer or a similar device the natural frequencies can be determined. Any below the maximum fan rpm should be increased by stiffening the part. A special effort should be made to keep vibration out of the test section and balance supports.

All types of materials are used for tunnel construction: wood, plywood, thin metal, heavy metal (for pressure tunnels), cast concrete, gunnite, and plastics. Even though there is "general agreement" that low-speed wind tunnels require no cooling, heavy running in the summertime will usually make the tunnel operators wish they had some; in fact even 25-hp tunnels should be cooled. Many tunnels have an opening where cooler outside air can be blown into the tunnel during model changes. Some large metal tunnels spray water over the outside to cool the tunnel.

Fan blades for low-speed tunnels are frequently made of wood, although modified aircraft propellers are sometimes used with trailing edge flaps to provide a uniform pressure rise across the fan disk. Wood blades have excellent fatigue life owing to the cellular structure of wood. The wood should be straight grained and knot free. Damage to wood blades can be repaired by scarfing in a piece of wood. To avoid abrasion the leading edge at the tip can have a fiberglass insert. Often the last 2 or 3 in. of the tip are of balsa as this will shatter when an unwanted object wedges between the tip and tunnel, and can be easily replaced. There is no clear consensus on the matter of spare blades. Tunnels that use short-span, large-chord blades have used fiberglass and, more recently, composite materials such as carbon-fiber-reinforced epoxies. These blades may have spars and are filled with foam.

For small research and instruction tunnels plywood is possibly the best material. Particle board is another material used; however, it is not as stiff as plywood and can develop a sag when unsupported. A smooth surface finish is difficult to obtain on particle board; also, it chips easily and is difficult to patch. With wood, holes that get cut in the tunnel can be easily patched and epoxy fillers can be used to maintain a smooth interior surface. The plywood is placed inside a wood frame. The tunnel is often made in sections that bolt together. To prevent leaks, surgical tubing can be used. The end of one section has a groove routed to receive the tubing. This is just a large O ring. The fan area can be metal, as it is easy to roll up round sections. The contraction can be laminated wood or could be built up with a fiberglass and foam sandwich. When student labor is used, the design must consider the skills available. A door in the floor in the plane of the fan will be necessary so that blades may be taken out of their hubs and replaced.

The immense cranes needed during the construction of a metal tunnel are shown in Fig. 2.14. The construction of a poured concrete tunnel is shown in Fig. 2.15. Most large modern tunnels require many auxiliary buildings, as shown in Fig. 2.16.

The detailed design, including selection of material, erection, etc., of large tunnels is usually done by engineering firms that are familiar with all

FIGURE 2.14. Erection of the all-metal 12-ft pressure tunnel at the Ames Aeronautical Laboratory. (Official photograph, National Aeronautics and Space Administration.)

FIGURE 2.15. Forms for contraction cone and return duct for a poured reinforced-concrete tunnel. (Photograph courtesy of University of Washington.)

FIGURE 2.16. The three-interchangeable test-section DNW tunnel. The large building behind the tunnel is the storage area for the test-section parts (it is large enough to hold a 747). On the right is the office and three model check out rooms.

the building codes and have civil, mechanical, and electrical engineering expertise.

2.15. ENERGY RATIO

The ratio of the energy of the air at the jet to the input energy is a measure of the efficiency of a wind tunnel, though by no means a measure of the value of the tunnel for research. It is nearly always greater than unity, indicating that the amount of stored energy in the windstream is capable of doing work at a high rate before being brought to rest. The energy ratio, *ER*, is from 3 to 7 for most closed throat tunnels.

It is unfortunate that an exact agreement on the definition of energy ratio has not been reached. Some engineers use the motor and propeller efficiency, η, in their calculations; some do not. This disagreement results in three definitions, as follows:

1. The tunnel energy ratio, based on the tunnel losses

$$ER_t = \frac{(qAV)_t}{550\eta \text{ bhp}} \tag{2.39}$$

where q = dynamic pressure in the jet, lb/sq. ft; A = jet area, sq. ft; V = jet velocity, ft/sec; η = fan efficiency; and the subscript t refers to the test section.

2. The input energy ratio (for dc motors)

$$ER_I = \frac{(qAV)_t}{550EI/746} \tag{2.40}$$

where E = input voltage, I = input amperage.

3. The fan energy ratio

$$ER_f = \frac{(qAV)_t}{550 \text{ bhp}} \tag{2.41}$$

The value of ER_t is always greater than ER_f, which is always greater than ER_I. From a practical standpoint ER_I is by far the easiest to measure, since it entails only reading input meters. It sheds little light on the efficiency of the tunnel design, however. Although theoretically one could measure the pressure rise through the fan to obtain ER_t, in practice this operation is not easy. Complete motor-performance curves make possible the determination of ER_f.

Example 2.2. A wind tunnel with a test section 7 ft by 10 ft has an indicated airspeed of 100 mph at a pressure of 740 mm Hg and a temperature of 85°F. If input power is 25 amperes at 2300 volts, three-phase alternating current at an electrical power factor of 1.0, find (1) the true airspeed; (2) the Reynolds number of a 1-ft chord wing; (3) the input energy ratio (ER_I) of the tunnel, assuming the drag of the wing to be negligible.

Answer. 1. We first determine the air density ρ.

$$\rho = 0.002378 \frac{518}{(459 + 85)} \frac{740}{760}$$

$$= 0.002204 \text{ slug/ft}^3$$

(See Appendix for standard conditions.) Hence

$$\sigma = \rho/\rho_0 = 0.927$$

and

$$V_{\text{true}} = Vi/\sqrt{\sigma} = 100/0.962 = 103.9 \text{ mph}$$

2. From the Appendix the viscosity of the air is

$$\mu = [340.8 + 0.548(°F)]10^{-9}$$

$$= [340.8 + 0.548(85)]10^{-9}$$

$$= 387.4 \times 10^{-9}$$

and hence

$$RN = \frac{\rho}{\mu} Vl = \frac{0.002204 \times 103.9 \times 1.467 \times 1.0}{387.4 \times 10^{-9}}$$

$$= 868{,}000$$

3.

$$ER_f = \frac{qAV}{550 \text{ hp}} = \frac{\frac{1}{2}\rho AV^3}{550 \dfrac{\sqrt{3}\, EI}{746}}$$

$$= \frac{\frac{1}{2}(0.002204)(7 \times 10)(1.467)^3(103.9)^3}{550 \times \dfrac{(1.732)(25)(2300)}{746}}$$

$$= 3.721$$

2.16. POWER LOSSES

Eckert, Mort, and Jope at NASA Ames (Ref. 2.47) have put together a computer program with user manual to calculate the performance of wind tunnels. The user selects the desired flow conditions at the test section, and for a given circuit geometry the tunnel performance is calculated. The program will handle open circuit, single return, double return tunnels, and open or closed throat test sections. The program assumes that all component parts of the tunnel are properly designed. If there are design errors, it does not predict them, nor their effects on other parts of the tunnel, including test-section flow quality. The program was used to compare actual and published performance for seven tunnels and gave good agreement.

The loss analysis from the 1966 edition of this book is included in this section, because the authors feel that to really understand what is occurring physically an engineer must go through the equations and calculations by hand. In too many cases canned computer programs give no insight into the problem.

Wattendorf (Ref. 2.48) has pointed the way to a logical approach to the losses in a return-type wind tunnel. The procedure is to break the tunnel down into (1) cylindrical sections, (2) corners, (3) expanding sections, and (4) contracting sections, and to calculate the loss for each.

In each of the sections a loss of energy occurs, usually written as a drop in static pressure, Δp, or as a coefficient of loss, $K = \Delta p/q$. Wattendorf refers these local losses to the jet dynamic pressure, defining the coefficient of loss as

$$K_0 = \frac{\Delta p}{q}\frac{q}{q_0} = K\frac{q}{q_0} \tag{2.42}$$

or, since the dynamic head varies inversely as the fourth power of the tunnel diameter,

$$K_0 = KD_0^4/D^4 \tag{2.42a}$$

where D_0 = jet diameter and D = local tunnel diameter.

With the above definitions, the section energy loss $\Delta E = K\frac{1}{2}\rho A V^3$ may be referred to the jet energy by

$$\begin{aligned}\Delta E &= K\tfrac{1}{2}\rho A V^3 \frac{A_0 V_0 \cdot V_0^2}{A_0 V_0 \cdot V_0^2} \\ &= K\tfrac{1}{2}\rho A_0 V_0^3 \frac{D_0^4}{D^4}\end{aligned}$$

and, finally,

$$\Delta E = K_0 \cdot \tfrac{1}{2}\rho A_0 V_0^3$$

where A_0 = test section area and A = local area.

$$\text{Energy ratio} = \frac{\text{Jet energy}}{\sum \text{Circuit losses}} = ER_t$$

(see Section 2.15)

$$ER_t = \frac{\frac{1}{2}\rho A_0 V_0^3}{\sum K_0 \frac{1}{2}\rho A_0 V_0^3} = \frac{1}{\sum K_0} \tag{2.43}$$

The above definition of the energy ratio excludes the fan and motor efficiency.

The magnitude of the losses in a circular wind tunnel may be computed as follows:

In the *cylindrical sections* the pressure drop in length L is $\Delta p/L = (\lambda/D)(\rho/2)V^2$, and $K = \Delta p/q = \lambda(L/D)$. Therefore

$$K_0 = \lambda(L/D)(D_0^4/D^4) \tag{2.44}$$

For smooth pipes at high Reynolds numbers von Kármán gives (Ref. 2.49)

$$1/\sqrt{\lambda} = 2 \log_{10} R\sqrt{\lambda} - 0.8 \tag{2.45}$$

where D = local tunnel diameter, ft, V = local velocity, ft/sec, and $RN = (\rho/\mu)VD$.

Since Eq. (2.45) is tedious of solution, a plot is shown in Fig. 2.17.

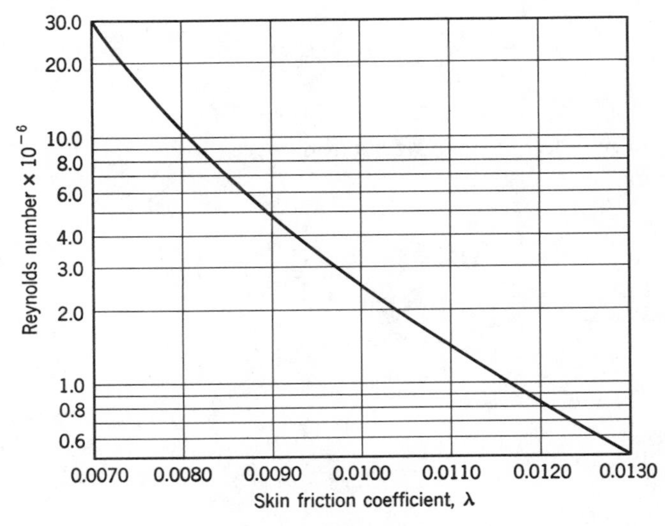

FIGURE 2.17.

For open cylindrical sections such as an open jet, a reasonable value for skin-friction coefficient is

$$\lambda = 0.08$$

For an open jet of length/diameter = 1.5, the loss becomes 0.08 × 1.5 = 12% as compared to about one-tenth that value for a closed jet.

In the *divergent sections,* both wall friction and expansion losses occur. The combined loss of the two is summed up by

$$K_0 = \left(\frac{\lambda}{8 \tan (\alpha/2)} + 0.6 \tan \frac{\alpha}{2}\right)\left(1 - \frac{D_1^4}{D_2^4}\right)\frac{D_0^4}{D_1^4} \tag{2.46}$$

where α = divergence between opposite walls, D_1 = smaller diameter, and D_2 = larger diameter.

It will be seen that the smaller expansions yield smaller losses up to the point where the skin friction of the added area becomes excessive. This, it will be seen by differentiating Eq. (2.46), occurs when

$$\tan (\alpha/2) = \sqrt{\lambda/4.8}$$

For reasonable values of λ the most efficient divergence is therefore about 5°. However, space limitations for the tunnel as well as the cost of construction may dictate that a slightly larger divergence be employed at an increase in cost of operation.

It will be noted that the losses in a divergent section are two to three times greater than the corresponding losses in a cylindrical tube, although the progressively decreasing velocity would seem to indicate losses between that of a cylindrical section with the diameter of the smaller section and that of one with the diameter of the larger section. The reason for the added loss is that the energy exchange near the walls is of such a nature that the thrust expected from the walls is not fully realized. Effectively, a pressure force is thereby added to the skin-friction forces.

As noted in Sections 2.2 and 2.6, the effective cone angle also depends on the diffuser area ratio, which determines the pressure recovery and pressure gradient. The diffuser's adverse pressure gradient will lead to separation if too large. Since the return passage should have a more uniform flow from the fan than the first diffuser from the test section, the expansion angle is sometimes larger. An additional argument for this type of design is that the disturbance caused by a model in the test section may limit satisfactory diffuser angles below smooth flow values.

In the *corners,* friction in the guide vanes accounts for about one-third of the loss, and rotation losses for the other two-thirds. For corners of the type shown in Fig. 2.3 the following, partly empirical relation is reasonable, being based on a corner pressure drop of $\Delta p/q = 0.15$ at $RN = 500{,}000$:

$$K_0 = \left(0.10 + \frac{4.55}{(\log_{10} RN)^{2.58}}\right) \frac{D_0^4}{D^4} \tag{2.47}$$

In the *contraction cone* the losses are friction only, and the pressure drop is

$$\Delta p_f = \int_0^{L_c} \lambda \frac{\rho}{2} V^2 \frac{dL}{D}$$

where L_c = length of contraction cone.

$$K_0 = K \frac{D_0^4}{D^4} = \frac{\Delta p_f}{q} \frac{D_0^4}{D^4} = \int_0^{L_c} \lambda \frac{dL}{D} \frac{D_0^4}{D^4} \frac{D_0}{D_0}$$

$$= \lambda_{\text{ave}} \frac{L_c}{D_0} \int_0^{L_c} \frac{D_0^5}{D^5} \frac{dL}{L} \tag{2.48}$$

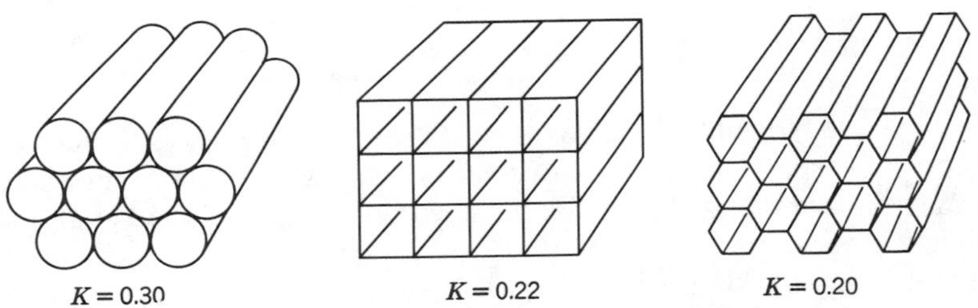

FIGURE 2.18. Some honeycombs and their losses.

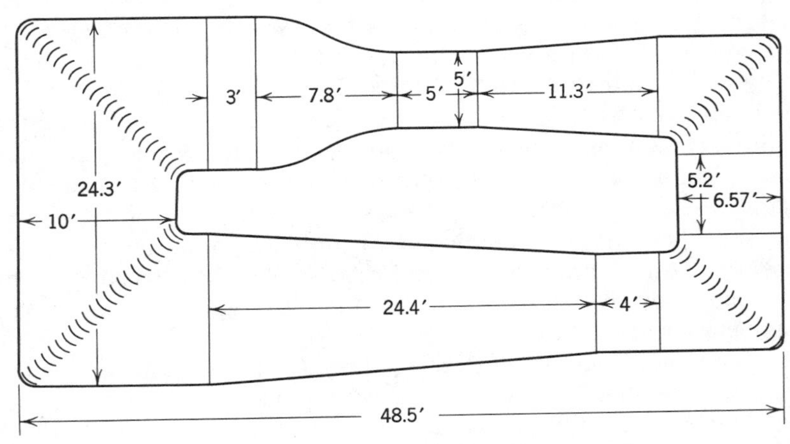

FIGURE 2.19. Layout of tunnel used as example.

Assuming a mean value for λ,

$$K_0 = 0.32\lambda L_c/D_0 \tag{2.49}$$

Since the total loss in the contraction cone usually runs below 3% of the total tunnel loss, any errors due to approximations in the cone losses become of small importance.

Losses in *honeycombs* are given in References 2.39 and 2.50. Values of K suitable for use in Eq. (2.42a) are given in Fig. 2.18 for honeycombs with a length/diameter = 6.0, and equal tube areas. Roughly speaking, the loss in a honeycomb is usually less than 5% of the total tunnel loss.

The losses incurred in the single return tunnel of Fig. 2.19 based on a tunnel temperature of 100°F (ρ/μ = 5560) and a test section velocity of 100 mph are as tabulated.

Section	K_0	Total Loss (%)
1. The jet	0.0093	5.1
2. Divergence	0.0391	21.3
3. Corner	0.046	25.0
4. Cylinder	0.0026	1.4
5. Corner	0.046	25.0
6. Cylinder	0.002	1.1
7. Divergence	0.016	8.9
8. Corner	0.0087	4.7
9. Corner	0.0087	4.7
10. Cylinder	0.0002	0.1
11. Cone	0.0048	2.7
	0.1834	100.0

$$\text{Energy ratio, } ER_t = 1/\Sigma K_0 = 1/0.1834 = 5.45$$

Probably this figure should be reduced about 10% for leaks and joints.

The effect of varying the angle of divergence or the contraction ratio for a tunnel similar to the one of Fig. 2.19 may be seen in Figs. 2.20 and 2.21.

The possibility of attaining higher energy ratios has several promising leads. One fundamental is the increase of efficiency that accompanies larger Reynolds numbers. That is, a large tunnel similar to a small tunnel will have the greater efficiency of the two.

Reduction of the losses in the divergent passage is limited, as previously stated, to a certain minimum angle between opposite walls. The use of this minimum angle would, however, yield smaller losses than are customarily encountered. Corner losses may be reduced through the use of two relatively

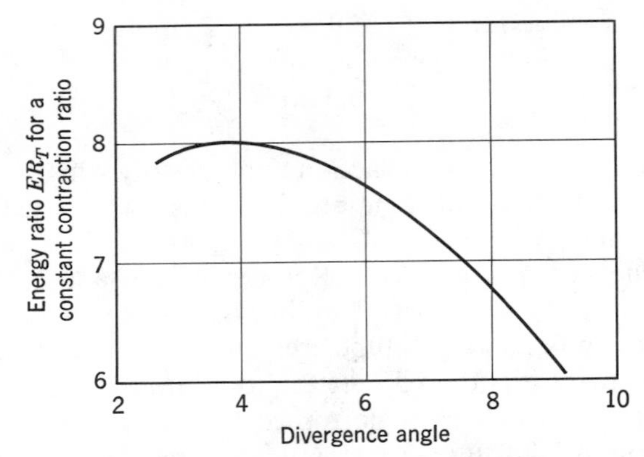

FIGURE 2.20. Effect of divergence angle on energy ratio.

untried innovations. The first is to break the four 90° turns into several vaned turns of less than 90° (Ref. 2.12). The second is to employ potential elbows (Ref. 2.51) for the turns. Increasing the contraction ratio through a longer return passage will also increase the energy ratio but at an added cost in tunnel construction. Increased length before the first turn is particularly effective.

An entirely different approach, particularly useful for high-speed tunnels, is to reduce the power required for a given speed by reducing the air density through partly evacuating the entire tunnel. This procedure greatly complicates model changes, since the tunnel pressure must be relieved before the

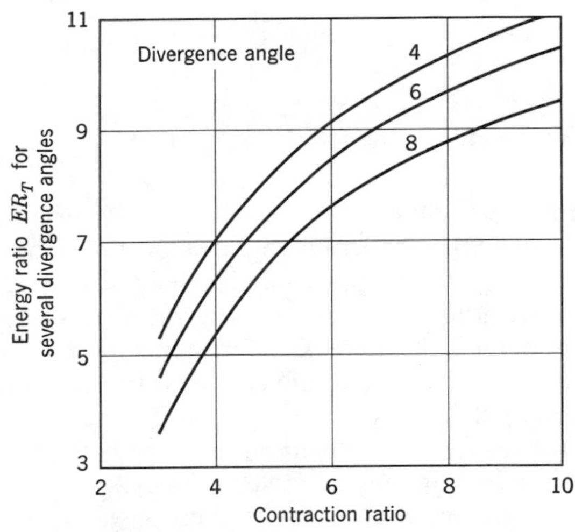

FIGURE 2.21. Effect of contraction ratio on energy ratio for several divergence angles.

tunnel crew can enter. (Pressure doors that seal off the test section from the rest of the tunnel can be used; this reduces the pumping between runs.) Since the power required is a function of ρV^3, by reducing the pressure to one-fourth its former value, the speed may be increased by the ratio $\sqrt[3]{4}$ to 1 for the same power input. Stated differently, a 59% higher Mach number will be attained with the lower pressure.

When figuring power requirements for a proposed design, consideration must also be given to (a) the power required to overcome model drag under the most extreme cases and (b) the power required to overcome the increased tunnel losses due to stalling of parts of the diffuser from the model effects.

For (a) the power required to fly a model whose span is 0.8 tunnel diam, $AR = 5$, and $C_D = 1.00$ is probably sufficient.

Item (b) for conventional tests is covered in (a) above, but for wingtip mounting or section tests as much as 150% *more* power may be needed if the diffuser is seriously stalled and large rotational and diffusion losses are created.

Losses for a well-designed open circuit tunnel are given below. These theoretical values were confirmed by model and full-scale tests (Ref. 2.52):

Section	K_0	Total Loss (%)
1. Inlet including six screens	0.021	14.0
2. Entrance cone and test section	0.013	8.6
3. Diffusor	0.080	53.4
4. Discharge at outlet	0.036	24.0
	0.150	100.0

$$\text{Energy ratio, } ER_t = 1/0.15 = 6.67$$

2.17. TEST SECTION INSERTS

Many tunnels have auxiliary two-dimensional test sections which fit inside the normal test section in order to provide testing facilities for shorter-span models at considerable savings in model cost. These jets are usually either endplate or contraction types.

The endplate jet insert consists of two flat plates sealed at tunnel floor or ceiling (Fig. 2.22) with a space between them for mounting and testing a constant-chord two-dimensional airfoil model. In some installations the model supports come up inside the endplates, and the rear pitch strut is used to hold and move the wake survey rake. Since the drag of the model changes with many factors, and since drag changes can make more air pass *around* instead of through the test section, special provision must be made to control

FIGURE 2.22. Endplate type of insert to form two-dimensional test section. (Courtesy Convair San Diego.)

the test dynamic pressure. One procedure is to use the customary double piezometer to hold constant the quantity of air that enters the original test section, and ascertain that a constant fraction of *that* air passes through the new test section by reading pitot-static tubes placed outside it. Variations can be made by remotely movable trailing edge flaps on the endplates.

A second approach is to contract the whole tunnel test section by an additional contraction section. This has the advantages of greatly increasing the contraction ratio and making the installation of turbulence screens easy, and of providing sure and positive two-dimensional conditions. On the other hand, a much smaller quantity of air now passes through the tunnel for a particular test speed, since the test-section area is reduced, and in many cases the performance of the tunnel fan is greatly impaired.* Still another source of trouble is the problem of diffusing the overcontracted passage. Somewhere some extremely rapid diffusion is going to have to take place, and the task is exceedingly difficult to do with even passable efficiency. In one installation a two-dimensional insert converted a 9-ft-diameter tunnel to

* A convenient way to think of the operation of the fan is to discuss the pressure rise it can produce in the units of the dynamic pressure approaching it. The reduction of quantity mentioned above will hence reduce the amount of power the fan can absorb.

a $2\frac{1}{2}$-ft by 9-ft test section; the problem of the separated diffuser arose, but vortex generators on all four walls very nearly cured it (see Fig. 2.23).

Several other items are of interest when inserts are under consideration. One of these is the question of whether to mount the insert horizontal so that the model is vertical or vice versa. The horizontal insert offers the opportunity of using liquid seals around the model mounting endplates, but, on the other hand, the model is somewhat more difficult to get to for adjustments. In some instances a severe pressure differential will occur with this type of insert, and great difficulty is experienced in keeping any but the heaviest liquids in the seals. Additional breather slots will usually alleviate this condition. Force measuring is also made more difficult by the large and indeterminable tares of the model mounting endplates, so much so that it is sometimes preferable to obtain the lift by pressure distributions on the model or on the floor and ceiling, and drag by a momentum rake, or to leave a small gap between model and walls.

In conclusion, it does not appear possible to recommend one type of two-dimensional jet insert as superior, with the final note that, aerodynamics aside for a moment, the endplate type of insert is far easier to install and remove.

FIGURE 2.23. Rotor blade section mounted in a two-dimensional insert. Note vortex generators on downstream wall. (Courtesy Georgia Institute of Technology.)

2.18. SAFETY

Though it may seem strange to bring the question of safety into a discussion of wind tunnels, the long roster of injured indicates it is not to be overlooked. Accidents in tunnel use include fires, falls, injuries from sharp-edged models, and personnel being locked in a tunnel when it is started.

Starting with the last, anybody who enters a part of a wind tunnel not readily visible to the tunnel operators without firm understanding regarding the restarting time is an idiot. But since it is not the practice to injure idiots, many tunnels preface a start with a blast on a horn, and a 5-s wait for incumbents to punch stop switches installed in test sections, return passages, and near the fan blades. Another approach is to install an automobile ignition switch at the test section entrance. The switch is in the tunnel start circuit. Still another approach is to use light beams (similar to burglar alarms) in various parts of the tunnel circuit which prevent the tunnel from starting when cut. When it is necessary for someone to be in the tunnel while it is running, very clear signals or understanding must again be obtained.

Falls are unfortunately frequent in wind tunnels. Their rounded surface, often coated with oil or ice, and the precipitous slope of entrance cones have resulted in bruises and even broken arms and legs. The authors speak feelingly at this point. In view of the danger associated with the entrance cone and settling chamber, pitot-static tubes, thermocouples, and the like should be wall mounted.

Fires in wind tunnels seem to be almost the rule rather than the exception: a broken propeller can spark a dust screen into fire; a trouble light can make plenty of trouble; or building forms can in some way become ignited. Since the tunnel is closed, special care should be taken in selecting fire extinguishers. And special care should be taken also to see that fire extinguishers emitting poisonous vapors are not easily tripped. It may well be that the world's record for the hundred-yard dash rightfully belongs to a tunnel engineer who inadvertently activated the carbon dioxide system in one of the largest east-coast tunnels.

The advent of the sharp edges on metal models is a new and potent hazard; the authors do not know a tunnel engineer who has not suffered from this source. It is only good sense to protect the tunnel crew from these sharp edges, using either tape, wood, or plastic slats.

Safety may seem like a puerile subject, but it looses that appearance *afterward*.

REFERENCES

2.1. P. Bradshaw and R. C. Pankhurst, The Design of Low-Speed Wind Tunnels, NPL ARC 24041, 1962.

2.2. J. M. Robertson and H. R. Fraser, Separation Prediction for Conical Diffusers, *Transactions ASME,* Series D82, 201, 1960.

2.3. A. T. McDonald and R. W. Fox, Incompressible Flow in Conical Diffusers, *Tech. Rep. No. 1,* Army Research Office (Durham) Project 4332, 1964.

2.4. L. R. Reneau, J. P. Jhonson, and S. J. Kline, Performance and Design of Straight, Two-Dimensional Diffusers, *Trans. ASME, J. Basic Engineering,* **89,** 141–156, 1976.

2.5. J. R. Henry, C. C. Wood, and S. W. Stafford, Summary of Subsonic-Diffuser Data, *NACA RM* L56F05, 1956.

2.6. C. A. Moore, Jr. and S. S. Kline, Some Effects of Vanes and Turbulence in Two-Dimensional Wide Angle Subsonic Diffusers, *NACA TN* 4080, 1958.

2.7. D. L. Cochran and S. J. Kline, Use of Short Flat Vanes for Producing Efficient Wide-Angle Two-Dimensional Subsonic Diffusers, *NACA TN* 4309, 1958.

2.8. G. B. Schubauer and W. G. Spangenberg, Effect of Screens in Wide-Angle Diffusers, *NACA TR* 949, 1949.

2.9. A. A. Townsend, Equilibrium Layers and Wall Turbulence, *J. Fluid Mechanics* **11,** 97, 1961.

2.10. M. M. Gibson, The Design and Performance of a Streamline Diffuser with Rapid Expansion, *ARC* **21,** 126, 1959.

2.11. S. B. Savage, A Short Low-Speed Wind Tunnel with Wide-Angle Diffuser, McGill University AE-2, 1960.

2.12. G. Darrius, Some Factors Influencing the Design of Wind Tunnels, *Brown-Boveri Review,* July–August 1943.

2.13. G. Sovran and E. D. Klomp, Experimentally Determined Optimum Geometrics for Rectilinear Diffusers with Rectangular, Conical or Annular Cross-Section, *Fluid Mechanics of Internal Flow,* G. Sovran (ed.), Elsevier Publishing Co., Amsterdam, 1967, pp. 270–319.

2.14. H. D. Taylor, Design Criteria for and Applications of the Vortex Mixing Principal, *United Aircraft Corp. Report* M-15038-1, 1948.

2.15. Y. Sendoo and M. Nishi, Improvement of the Performance of Conical Diffusers by Vortex Generators, ASME Paper 73-WA/FE-1, Nov. 1973.

2.16. M. Sajben and C. P. Chen, A New Passive Boundary Layer Control Device, AIAA and SAE 12th Propulsion Conference, 1976.

2.17. G. Krober, Guide Vanes for Deflecting Fluid Currents with Small Loss of Energy, *TM* 722, 1932.

2.18. A. R. Collar, Some Experiments with Cascades of Airfoils, *R & M* 1768, 1937.

2.19. G. N. Patterson, Note on the Design of Corners in Duct Systems, *R & M* 1773, 1937.

2.20. D. C. McPhail, Experiments on Turning Vanes at an Expansion, *R & M* 1876, 1939.

2.21. C. Salter, Experiments on Thin Turning Vanes, ARC *R & M* 2469, October 1946.

2.22. K. G. Winters, Comparative Tests of Thin Turning Vanes in the RAE 4 × 3 Foot Wind Tunnel, *R & M* 2589, 1947.

2.23. S. Shindo, W. H. Rae, Jr., Y. Aoki and E. G. Hill, Improvement of Flow Quality at the University of Washington Low Speed Wind Tunnel, Paper 78-815, AIAA 10th Aerodynamic Testing Conference, 1978, pp. 336–343.

2.24. A. R. Collar, The Design of Wind Tunnel Fans, *R & M* 1889, August 1940.

2.25. G. N. Patterson, Ducted Fans: Design for High Efficiency, *ACA* **7,** July 1944.

2.26. R. A. Wallis, *Axial Flow Fans, Design and Practice,* George Newnes, London, 1961.

2.27. M. Seidel (ed.), Construction 1976–1980, Design, Manufacturing, Calibration of the German–Dutch Windtunnel (DNW), *DNW,* 45–50, 1982.

2.28. W. C. Nelson, *Airplane Propeller Principles,* Wiley, New York, p. 9.

2.29. G. E. Chmielewski, Bounday Layer Considerations in the Design of Aerodynamic Contractions, *J. Aircraft,* **11,** 8, August 1974.

2.30. B. S. Stratford, The Prediction of Separation of the Turbulent Boundary Layer, *J. Fluid Mechanics,* **5** 1, 1–16, 1959.

2.31. T. Morel, Comprehensive Design of Axisymmetric Wind Tunnel Contractions, *J. Fluids Engineering, ASME Transactions,* 225–233, June 1975.

2.32. G. G. Borger, The Optimization of Wind Tunnel Contractions for the Subsonic Range, *NASA TTF* 16899, March 1976.

2.33. M. N. Mikhail and W. J. Rainbird, Optimum Design of Wind Tunnel Contractions, Paper 78-819, AIAA 10th Aerodynamic Testing Conference, 1978, pp. 376–384.

2.34. W. C. Steinle, The Experimental Determination of Aerodynamic Total-Pressure Losses for Heat Exchanger Surfaces Considered for the 7 × 10 Foot Transonic Wind Tunnel, *DTMB Aero Report,* 1951.

2.35. A. N. Tifford, Wind Tunnel Cooling, *JAS,* March 1943.

2.36. G. B. Schubauer, W. G. Spangenburg, and P. S. Klebanoff, Aerodynamic Characteristics of Damping Screens, *NACA TN* 2001, January 1950.

2.37. P. G. Morgand, The Stability of Flow Thru Porous Screens, *JAS* **64,** 359, 1960.

2.38. D. G. DeVahl, *The Flow of Air Through Wire Screens, Hydraulics and Fluid Mechanics,* R. Sylvester (ed.), Pergamon, New York, 1964, pp. 191–212.

2.39. J. Scheiman and J. D. Brooks, Comparison of Experimental and Theoretical Turbulence Reduction from Screens, Honeycomb, and Honeycomb-Screen Combinations, *JAS,* **18,** 638–643, August 1981.

2.40. L. Prandtl, Attaining a Steady Air Stream in Wind Tunnels, *NACA TM* 726, October 1933.

2.41. H. L. Dryden and G. B. Schubauer, The Use of Damping Screens for the Reduction of Wind Tunnel Turbulence, *JAS,* **14** 221–228, April 1947.

2.42. R. I. Loehrke and H. M. Nagib, Experiments on Management of Free-Stream Turbulence, *AGARD Report* 598, September 1972.

2.43. R. I. Loehrke and H. M. Nagib, Control of Free-Stream Turbulence by Means of Honeycombs: A Balance Between Suppression and Generation, *J. Fluids Engineering,* 342–353, September, 1976.

2.44. G. I. Taylor and G. K. Batchelor, The Effect of Wire Gauze on Small Disturbances in a Uniform Stream, *Quarterly Journal of Mechanics and Applied Mathematics,* **II** 1–29, March 1949.

2.45. V. Ramjee and A. K. M. F. Hussain, Influence of the Axisymmetric Contraction Ratio on Free-Stream Turbulence, *J. Fluids Engineering,* 505–515, September 1976.

2.46. G. K. Batchelor, *The Theory of Homogeneous Turbulence,* Cambridge University Press, 1967, pp. 58–75.

2.47. W. T. Eckert, K. W. Mort, and J. Jope, Aerodynamic Design Guidelines and Computer Program for Estimation of Subsonic Wind Tunnel Performance, *NASA TN* D-8243, October 1976.

2.48. F. L. Wattendorf, Factors Influencing the Energy Ratio of Return Flow Wind Tunnels, p. 526, 5th International Congress for Applied Mechanics, Cambridge, 1938.

2.49. T. von Kármán, Turbulence and Skin Friction, *JAS,* January 1934.

2.50. H. E. Roberts, Considerations in the Design of a Low-Cost Wind Tunnel, Paper presented at 14th Annual Meeting of the Institute of Aeronautical Sciences, January 1946.

2.51. John J. Harper, Tests on Elbows of a Special Design, *JAS,* **13,** 1946.

2.52. T. N. Krishnaswamy, Selection of the Electric Drive of the 14 × 9 Wind Tunnel, *Journal of the Aeronautical Society of India,* **7,** No. 2, 1955.

CHAPTER 3

Instrumentation and Calibration of the Test Section

After a tunnel is constructed, the next step is to determine its flow characteristics and, of course, to change any that are not satisfactory. First, however, it is necessary to discuss the quantities that we shall be measuring and the instruments that experience has shown are the best to do the job. Besides those instruments needed for calibration we shall also discuss others needed for testing.

The low-speed airstream is defined when we know its distribution of dynamic pressure, static pressure, and total pressure, and its temperature and turbulence. We may then compute its velocity and the Reynolds number for a particular model. Much of our interest, then, is centered on determining pressures which can be measured by either simple fluid manometers or more complex signal conditioners, transducers, amplifiers, analog-to-digital converters, and stored on tapes or disks of many sizes, shapes, and forms.

3.1. MANOMETERS

One of the oldest methods of measuring pressures, and one of the easiest to build, is a manometer (Fig. 3.1). A manometer does not measure absolute pressure but a change in pressure either from a reference pressure such as atmospheric, or the difference between two pressures. A simple U-tube

FIGURE 3.1. An adjustable-angle multiple manometer. Sometimes a dial-type thermometer is mounted directly on the manometer for fluid temperature data. (Courtesy Convair, San Diego.)

manometer can be made from two pieces of straight glass tubing connected by tubing at the bottom, or by bending glass tubing into a U shape. The tubing is filled with a fluid, and the difference in fluid height in the two tubes is measured, usually by an attached scale. The density or specific gravity of the fluid used must be known. Then the change in pressure can be calculated as follows:

$$\Delta P = K(\Delta h \times \text{SpGr} \times \text{weight of water}) \times \sin\beta \tag{3.1}$$

where

K = constant for dimensional homogenity.

Δh = the difference in height of fluid between reference tube and tube in question.

SpGr = the specific gravity of the fluid corrected for temperature, usually measured by a hydrometer.

Weight of water = the weight of a cubic unit of water.

$\sin\beta$, if the manometer is not vertical, $\Delta h \times \sin\beta$ = vertical height; β = angle between the horizontal and manometer tubes.

Often when pressure coefficients are required, multitube manometers are used. It is also possible to put the test-section total and static pressure on the manometer. Then the pressure coefficient, C_p, is merely the Δh of desired pressure divided by the Δh of the dynamic pressure.

In the past, wind tunnels have used a wide array of manometers. These have ranged from the simple U tube using a ruler or strip of graph paper to measure fluid heights to large banks of 50–150 tube manometers with constant-level reservoirs to maintain the reference fluid height, to manometers that can be precisely read. The U-tube and precision manometers could read one or two pressures, and the large multitube manometers were used to determine pressure distributions over the model surfaces. The simple, precision, and small multitube manometers usually are read and recorded by an individual. The large multitube manometers were photographed, and the tubes were read later by various techniques.

With the advent of the digital-data-acquisition system the use of manometers has declined. The use of pressure transducers, multistep, multiport valves (scanivalves) have many advantages over banks of multitube manometers. When manometers were used to record a large number of pressures, a large amount of tunnel time was required to connect, phase out, and leak check the manometer. The large bundle of tubing from the model to the manometers outside the tunnel could modify the support tares and interference and thus affect the balance data. Using scanivalves and pressure transducers allows the pressure ports to be connected to the scanivalves, phased out, and leak checked prior to the model entry into the tunnel, saving large amounts of tunnel time. The small-diameter wires for both stepping the valves and the transducers are much easier to handle, thus reducing the tare and interference problem. The manometers using photographs were faster in acquiring the data by orders of magnitude. But the time required to develop film and read the tubes so that the data could be reduced is also orders of magnitude longer than the digital system. Thus the use of pressure transducers to measure pressures has tended to replace the manometer.

The major disadvantage, and in some cases risk, with the use of scanivalves is the usual lack of data visibility, and the possibility of either the transducer slipping calibration or the valve leaking and being undetected. Strangely enough, these mishaps only occur on those units that are measuring critical pressures (Section 3.8).

There are still a few cases where a manometer is useful. One is in instruction where the properly set up manometer will show the student the shape of pressure distribution, such as the chordwise distribution of pressure on a wing, or the momentum loss in the wake (Fig. 4.50). The second is during a test where it becomes necessary to get a quick look at a few pressures at a few model conditions.

Both the manometer and the scanivalve must be phased, that is, the engineer must know which pressure port on the model is connected to which manometer tube or scanivalve. This task requires meticulous bookkeeping

TABLE 3.1

Liquid	Specific Gravity (approx.)	Remarks
Alcohol	0.800	Will absorb water
Kerosene	0.800	
Water	1.00	Poor meniscus owing to surface tension[a]
Methylene chloride	1.30	Attacks rubber
Bromobenzene	1.50	Quite volatile
Carbon tetrachloride	1.59	Attacks rubber; hard to color; toxic
Acetylene tetrachloride	1.59	Attacks rubber
Tetrabrome-ethane	2.97	Attacks rubber and vinyl[b] plastic; toxic
Mercury	13.56	Oil meniscus for best result; attacks brass and solder; toxic

[a] Can make acceptable with photographic wetting agents.
[b] Polyethelene may be used.

and is very time consuming. After all the model ports are connected and phased, they must be checked for both plugs or leaks. A plug will show as a very slow response to a change in pressure. A leak will show as a change in pressure after a pressure is applied to the model port and held. If either of these occur and cannot be repaired, that model pressure port must be considered no good.

Some common manometer fluids and their approximate specific gravities are shown in Table 3.1. Care must be taken with those fluids that are toxic. If such fluids must be used, one should consult with environmental health personnel on the necessary precautions. Alcohol and water may be dyed with water-soluble dyes. Kerosene may be dyed with oil-soluble dyes. The use of dye makes a clear fluid easier to see; however, dyes may, with time, discolor the tubes.

3.2. THE PITOT-STATIC, PITOT, AND LONG STATIC TUBES

Pitot-Static Tubes

The most common device for determining the total head and the static pressure of a stream is the pitot-static tube, an instrument that yields both the total head and the static pressure. A "standard" pitot-static tube is shown in Fig. 3.2*a*. The orifice at *A* reads total head ($p + \frac{1}{2}\rho V^2$), and the orifices at *B*

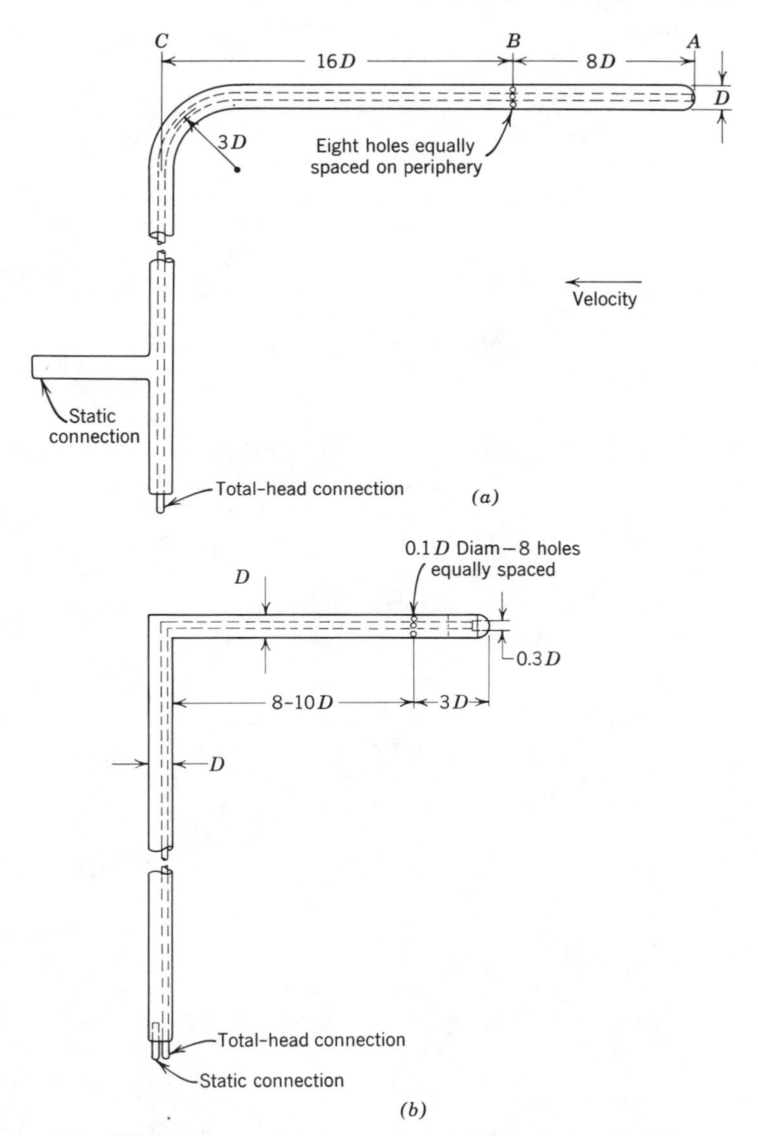

FIGURE 3.2. (*a*) "Standard" pitot-static tube. (*b*) Prandtl design.

read the static pressure, p. If the pressures from the two orifices are connected across a manometer or pressure transducer, the pressure differential will, of course, be $\frac{1}{2}\rho V^2$, from which the velocity may be calculated. (For determining ρ, see Example 3.1.)

The pitot-static tube is easy to construct, but it has some inherent errors. If due allowance is made for these errors, a true reading of the dynamic pressure within about 0.1% may be obtained.

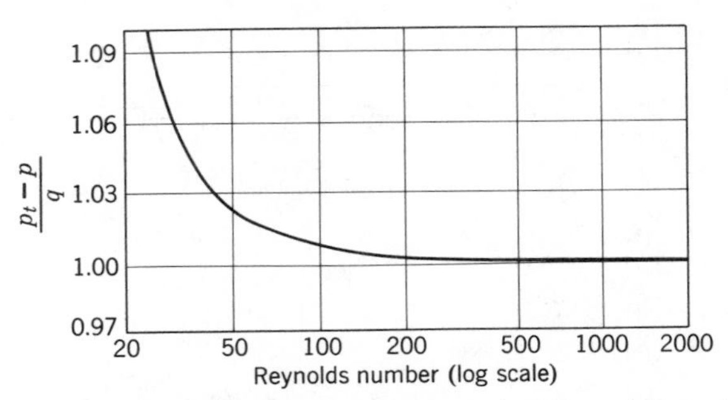

FIGURE 3.3. Performance of pitot tube at low Reynolds number.

It has been amply demonstrated that a total-head tube with a hemispherical tip will read the total head accurately independent of the size of the orifice opening as long as the yaw is less than 3°. A squared-off pitot tube will go to higher angles without error, but both square- and round-tip pitot tubes suffer errors if they are used at too low Reynolds numbers or too close to a wall. Corrections for squared-off pitot tubes under these conditions are in Figs. 3.3 and 3.4 and Ref. 3.1.

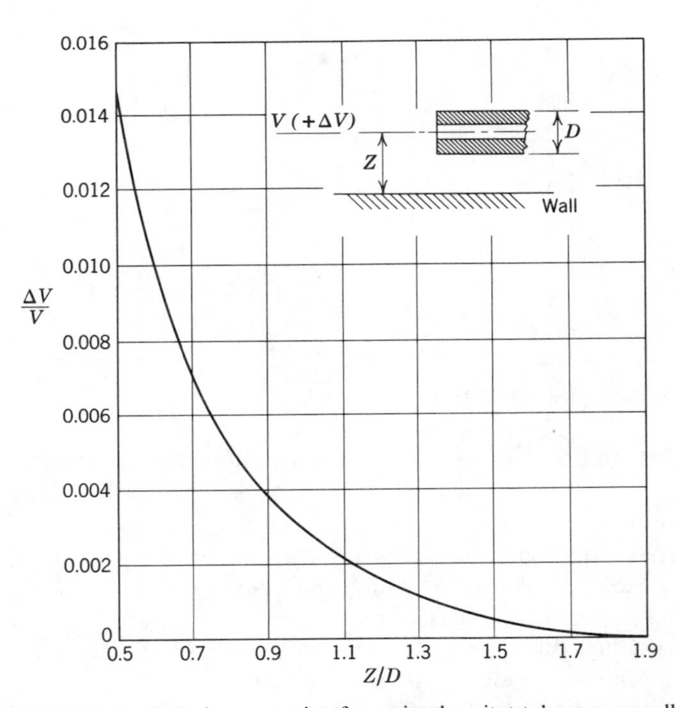

FIGURE 3.4. Velocity correction for a circular pitot tube near a wall.

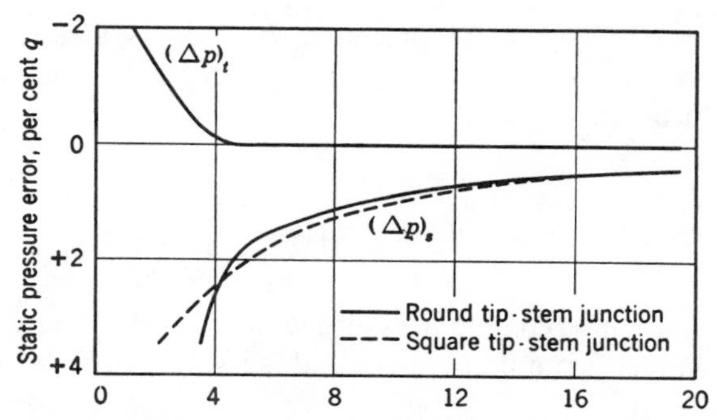

FIGURE 3.5. Static orifice distance from tip base or from stem centerline, diameters (*AB* and *BC* in Fig. 3.2*a*); see Example 3.1.

The static holes suffer from two effects: (1) The crowding of the streamlines near the tip reduces the pressure along the shank of the pitot-static tube so that the static pressure at the static orifices will be low. The amount of error is seen in Fig. 3.3. (2) A high-pressure region exists ahead of the stem that tends to make the indicated static pressure too high.

The two effects may block each other out if the static holes are properly located. The "standard" pitot-static tube does not employ this principle because it would require the static holes to be so close to the tip that small deviations in tip construction or damage to the tip could make a relatively large error in the static reading.

Hence: (1) If a new pitot-static tube is to be built, either it may be designed as per Fig. 3.2*a* and its static pressure readings corrected as per Fig. 3.5, or the Prandtl design may be used. The Prandtl design (Fig. 3.2*b*) should require no correction but should be checked for accuracy. (2) Existing pitot-static tubes should be examined for tip and stem errors so that their constants may be found.

If a long static tube is available, the static pressure can be determined along a longitudinal line in the test section. Then the pitot-static tube can be placed on this line and the static pressure can be calibrated.

Example 3.1. A pitot-static tube whose static orifices are 3.2*D* from the base of the tip and 8.0*D* from the centerline of the stem reads 12.05 in. of water on a manometer for a particular setting of the tunnel. If the test section is at standard pressure and 113°F, find (1) the dynamic pressure and (2) the true airspeed.

First the pitot-static tube error must be found.

(a) *Tip error.* From Fig. 3.3 it is seen that static orifices located 3.2*D* from the base of the tip will read 0.5% q too low.

(b) *Stem error.* From Fig. 3.3 it is seen that static orifices located 8.0*D* from the stem will read 1.13% q too high.

(c) *Total error.* The static pressure therefore will be 1.13 − 0.5 = 0.63% q too high, and hence the indicated dynamic pressure will be too low. The data should be corrected as follows:

$$q_{\text{true}} = 1.0063 q_{\text{indicated}}$$

$$V_{\text{true}} = 1.0031 V_{\text{indicated}}$$

1. Accordingly the dynamic pressure will be 1.0063 × 12.05 in. of water, or 12.13 in. of water. From the Appendix this is 12.13 × 5.204 or 63.12 lb/sq. ft.
2. The density is

$$\rho = 0.002378(518/572)$$
$$= 0.002154 \text{ slug/ft}^3$$

The true airspeed is hence

$$V_t = \sqrt{\frac{2(63.12)}{0.002154}}$$
$$= 242.09 \text{ fps} = 165.06 \text{ mph} = 143.34 \text{ knots}$$

The accuracy of a standard pitot-static tube when inclined to an airstream is shown in Fig. 3.6.

The pitot static tube can only be used in free air. If it is placed near a model, the model's static pressure distribution will be sensed by the static ports, and the reading will not be the free-stream velocity. This is why tunnel dynamic pressure calibrations are made without a model in the test section. The same problem exists on an aircraft where great care has to be taken in the location of a static pressure source so as to avoid a location where the static pressure varies with lift.

Pitot Tubes

A pitot tube measures total pressure. The shape of the tube affects its sensitivity to flows inclined to the tube axis. Pitots with hemispherical noses begin to show errors in total pressure near a 3° flow inclination. Pitot tubes with a sharp square nose begin to show errors near 8° flow inclination. This can be improved by chamfering the nose.

A Kiel tube (Ref. 3.2) can exceed flow angles of 30° (Figs. 3.7 and 3.8). Simplified versions of a Kiel tube can be made (Fig. 3.9) that are good up to about 15° or so. Often when large flow angles are present, a flow probe that can be rotated is used. This probe usually has a yaw head or claw that is used to determine when the probe is parallel to the flow.

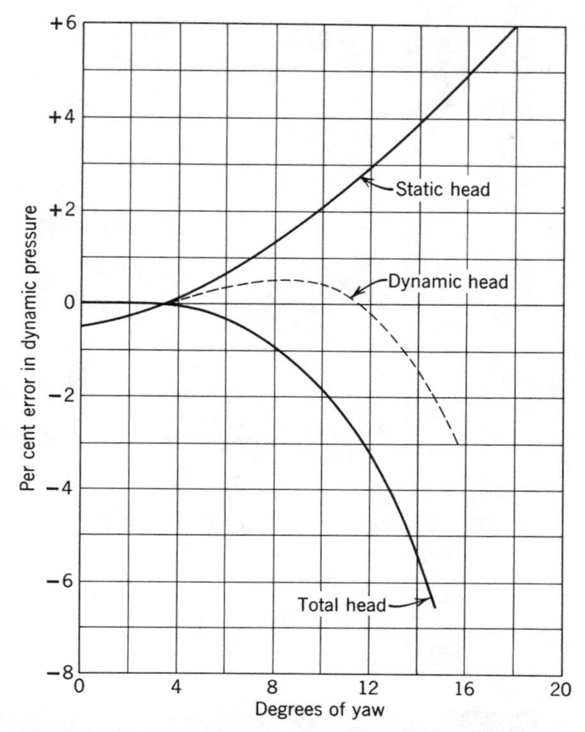

FIGURE 3.6. Performance of standard pitot tube in yaw.

It is a relatively easy task to measure a pitot's sensitivity to flow angle by use of a flow of known angularity. The pitot is then pitched to determine its sensitivity to flow angularity.

Long Static Tube

As the name implies, this is a long tube that extends through the test section. The tube is often suspended on a series of wires that are used to align and tension the tube. The tube is equipped with a number of static rings. These rings have four or more static ports around the circumference that are mani-

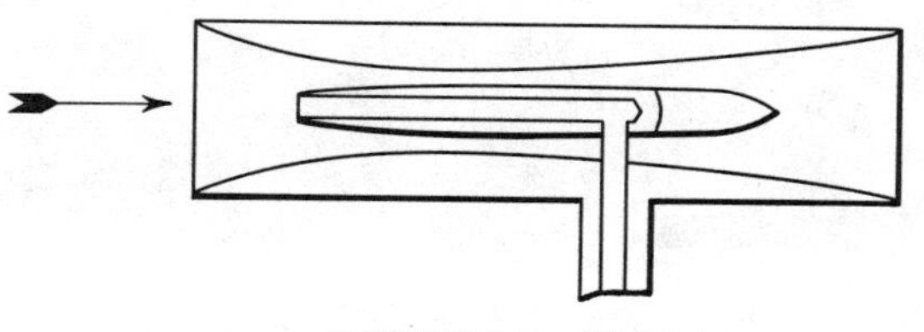

FIGURE 3.7. Kiel tube.

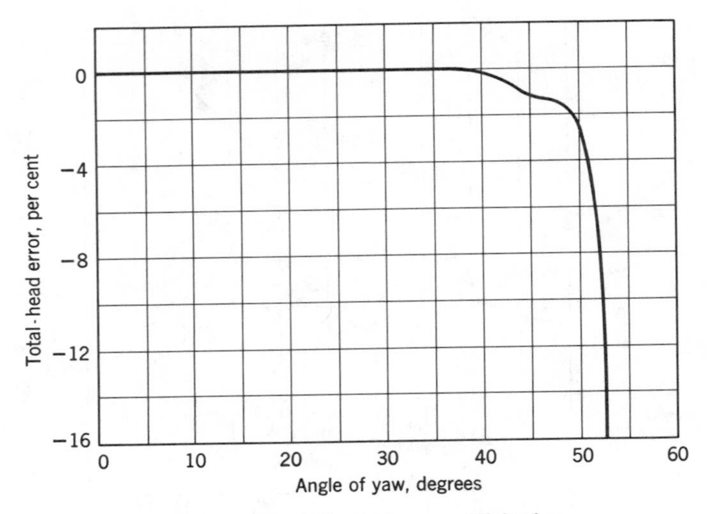

FIGURE 3.8. Effect of yaw on Kiel tube.

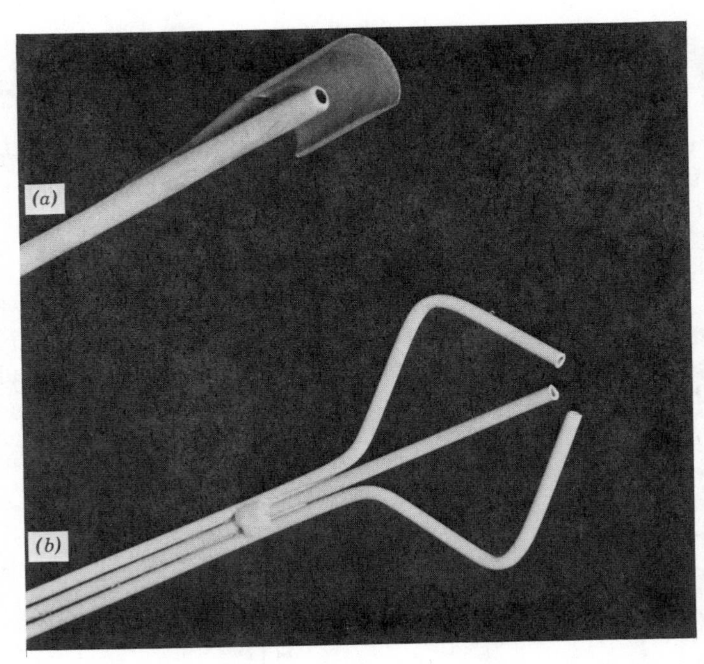

FIGURE 3.9. (*a*) Simple shielded total head tube. (*b*) Claw. (Photograph courtesy of University of Washington.)

folded together. Assuming that the flow angularity is small, then aft of the nose the flow will be parallel to the tube. This then yields a static pressure distribution throughout the length of the tube.

3.3. STATIC TEMPERATURE

The test-section static temperature is required to determine the true test-section velocity through the use of Bernoulli's incompressible equation. It should be noted that the test-section static pressure is also required to obtain the mass density through the equation of state. The true velocity is needed for the calculation of Reynolds number and of the velocity ratio in flutter model testing, to match the advance ratio in propeller power testing and helicopter rotor testing, etc.

The static temperature is essentially constant through a boundary layer in incompressible flow. Thus, the easiest method of measuring static temperature is through the use of a flush, wall-mounted temperature probe. It is desirable that the probe be of the remote-indicating type so that the temperature may be read at a convenient location. The probe should be located in the test section in a region where its chance of damage is minimal. The temperature probe should have an output in millivolts so that it can be recorded by a digital-data-acquisition system if one is used.

3.4. FLOW DIRECTIONS

Yawheads

A yawhead is a sphere that has two or four static ports, often 90° apart, on the forward face of the sphere (Fig. 3.10). If the ports are at exactly 45° to the centerline of the support, and the flow is parallel to the support, then $(P_a - P_b)/q = \Delta P/q = 0$. If there is flow angularity, then $\Delta P/q \neq 0$, and the value of $\Delta P/q$ will be a function of the flow angle. In practice, a yawhead must be

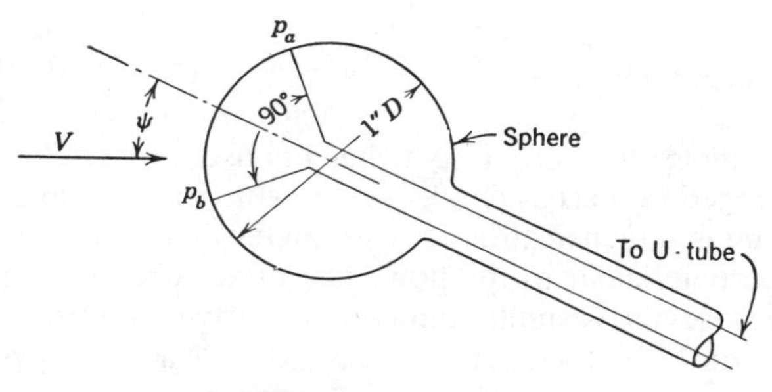

FIGURE 3.10. A yawhead.

calibrated. Assuming the flow has no angularity, the probe is pitched or yawed about its center through an angle range both positive and negative. This is done for the yawhead in the "upright," or normally used position, and then the yawhead is rotated 180° or to an "inverted" position. This will result in two curves of $\Delta P/q$ (upright and inverted) versus angle for the yawhead. If the static ports are symmetrical to the support axis, the two curves will lie on top of each other. If there is an error in the static port location, the curve will be displaced by twice the error. The true curve lies half-way between the two measured curves. If the flow used to calibrate the yawhead is not parallel to the yawhead support axis at the zero angle, the true curve will not pass through the zero angle. If it is desired to have the yawhead calibration independent of free-stream dynamic pressure, $(P_a + P_b)/(P_a - P_b)$ can be used.

Quite often a yawhead has five static ports rather than the two just described. One port is at the leading edge of the sphere and measures total pressure, and the other four ports are in pairs from two yawheads so that both the yaw or cross-flow and pitch or the upflow can be measured with the same device. Pitot-static tubes that use a hemispherical nose often have two yaw heads built into the nose. In this case, one instrument at any place in the test section will measure total pressure, static pressure, dynamic pressure, upflow, and cross flow. This is quite useful for determining flow conditions in a test section.

Claw

A claw will also measure flow angularity and is simpler to build. In its simplest form a claw consists of two parallel pieces of tubing that are bent 45° away from their common axis, and then bent back 90° toward their common axis (Fig. 3.9). The two heads of the probe are cut off square about two diameters from the center line. Often a third tube is added to measure total pressure, and the two claws can be made to simultaneously measure both cross flow and upflow. The calibration technique for a claw is the same as that for a yawhead. Claws are more delicate than yawheads because the two tubes used to measure ΔP can be easily bent, thus changing the calibration.

Other Pressure Devices

Sometimes a simpler version of a yawhead is used. One device consists of five tubes arranged in a cross configuration with one tube in the center and two pairs of tubes attached to it at right angles. The center tube is cut off with its end perpendicular to the flow. The other tubes are cut off at a 45° angle. A second device is similar but consists of two parallel tubes cut off at an angle. Both of these devices are calibrated similar to a yawhead or claw.

Nonpressure Devices

Other devices that can be used to measure flow angles are hot wires, dual split film sensors, small airfoil or wedges, and vanes.

Hot Wires

If two hot wires are mounted in an "x" or right-angle configuration, one wire can measure the velocity in the x or u direction and the second wire can measure the velocity in the z or w direction. From the u, w velocities the upflow can be determined. Another pair of wires can be used to measure the u and v velocities and yield the cross flow. Hot wires are delicate and can be damaged easily. A typical "x" hot wire will have two wires 0.05 in. long of 0.00015 in. in diameter separated by 0.005 in. Hot wires and the associated electronics are commercially available.

Single or Dual Split Film

A single split film will be parallel to the axis of a small glass rod. A dual split film will have a second pair of split films at 90° to the first, aligned along the rod circumference. These devices, similar to hot wires, can be used to measure both velocity and flow angles. The split film is suitable for measuring mean velocities but does not have the frequency response of hot wires for turbulence measurements. Thin-film sensors and the associated electronics are commercially available.

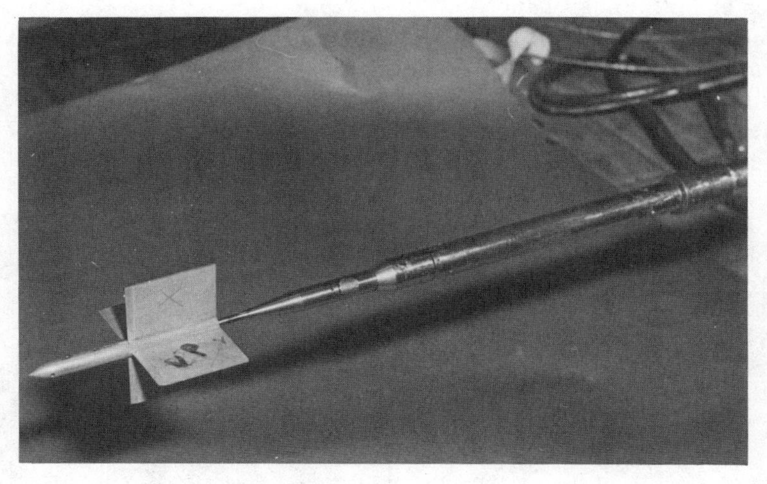

FIGURE 3.11. Vane-type flow angularity probe. Uses a five-component balance, and can resolve about 0.006°. Owing to its high natural frequency, probe can make continuous motion surveys. (Photograph courtesy of Propulsion Wind Tunnel, USAF/AEDC.)

Airfoils and Vanes

Flow angularities can also be measured using small wings, attached to sting balances. The flow angle calibration is obtained by the rotation of the drag polar (see Section 4.16). Small vanes of various configurations can also be used (Fig. 3.11; Ref. 3.3). Often these are attached to low-friction potentiometers to read the angle, or to balances. These devices measure the flow angle of a finite region rather than a point.

To measure velocities in highly deflected wakes, use has been made of an instrumentation rake that contains pitot tubes and static tubes and a device to measure flow angularity. This rake is then pitched to null the flow angle probes, thus aligning the rake with the flow. A potentiometer or digital shaft encoder can be used to record the angle and the velocity is obtained from the total and static pressures.

3.5. RAKES

Total Head Rake

The profile drag is often measured by the use of a drag wake rake (See Section 4.18). If the momentum loss in the wake is to be determined by pressure measurements, a bank of total head tubes is often used. The rake also should have two or more static tubes offset from the total head tubes to obtain the local static pressure. The length of the rake must be adequate to encompass more than the width of the wake. Often, the tube spacing is greater at both ends of the rake than in the center by a factor of 2. The spacing of the tubes must be exact so that the momentum profile can be accurately determined. Generally, the total tubes are made of 0.0625 in. thin-wall tubing (Fig. 3.12).

The static tubes must be offset from the plane of the total tubes to avoid interference effects on the static pressure. Their purpose is to determine the

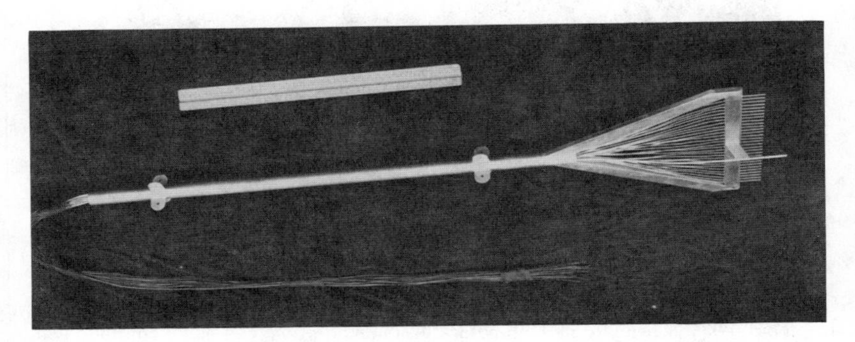

FIGURE 3.12. A total head rake. Note single static-pressure tube. (Courtesy Aerolab Development Corp.)

static pressure in the wake. Because the static pressure can be affected by both the total head tubes and the base of the rake, they must be carefully calibrated so that the error in static pressure is known. Since the static tubes must have a hemispherical nose shape, it is often possible to adjust this shape to reduce the error to zero (Ref. 3.4).

As an alternative to a wake rake, a mechanical traverse mechanism can be used. By the use of various shaft encoders the location of a probe can be determined with a high degree of accuracy. The sensor can be a pitot and, preferably, a static tube, hot wire, or thin film.

When measuring the momentum loss in the wake by any method, care must be taken to ensure that the whole width of the wake is measured.

Boundary Layer Mouse

Measurements in the boundary layer are often made to detect the transition between laminar and turbulent flow or to find the local skin-friction coefficient. Many of these measurements require a knowledge of the velocity profile in the boundary layer. Within the boundary layer the static pressure is essentially constant and the total pressure varies. There are several ways in which the velocity profile can be obtained.

The oldest method is by use of a boundary layer mouse. This device is a series of total head tubes, often with oval or flat inlets. To obtain the velocity profile at the surface requires the total head tubes to be spaced closer together than their diameters. Thus, the total head tubes are placed on an inclined plane to obtain the required close vertical spacing. The boundary layer mouse often has a static orifice to measure the static pressure, or the static pressure can be measured by a surface port. In use the mouse is attached to the model. The boundary layer mouse measures the velocity profile over a finite span of the model, rather than a single spanwise station (Fig. 3.13; Ref. 3.5).

The velocity profile can also be measured by use of a vertically traversing mechanism whose position off the surface can be quite accurately determined by a digital optical encoder. The traverse mechanism can carry a single total head probe, a hot wire, or a split film. Very good agreement has been shown between pitot probes, hot wires, and thin film when supported by a traverse mechanism. The traverse mechanism is supported by the model (Fig. 3.14, Ref. 3.6).

In general, it is better to support the boundary layer mouse or the traverse mechanism from the model rather than the tunnel walls. This avoids two problems. First, when the walls are used for support, the probes must be moved when the model is pitched and then reset to obtain a very close proximity to the surface for data recording. The second problem with a wall support is that most models tend to move slightly and often oscillate when under loads owing to balance deflections.

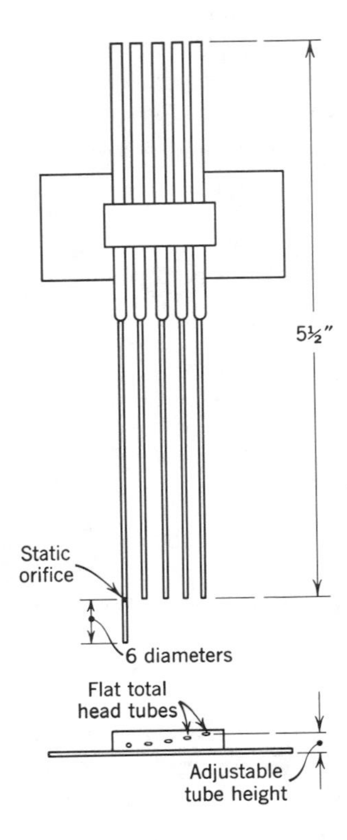

FIGURE 3.13. A boundary layer mouse.

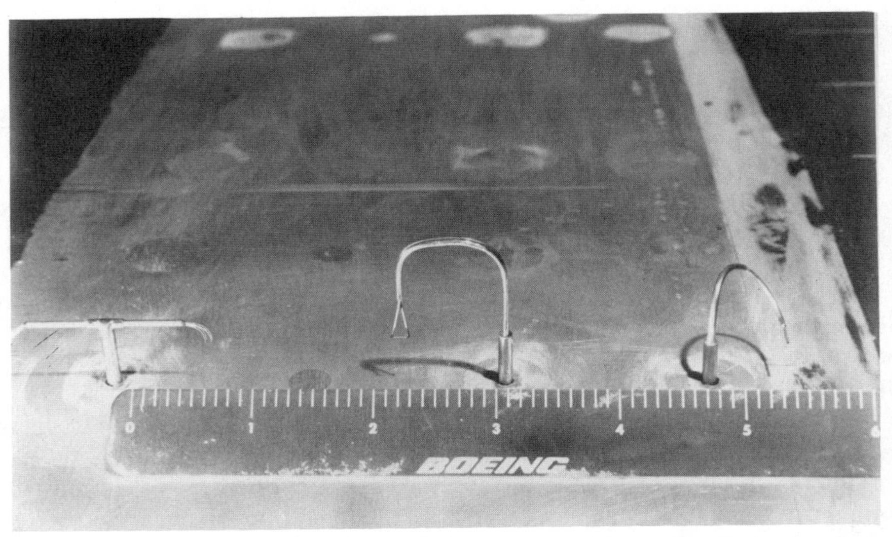

FIGURE 3.14. Flow probes. Left to right: hot wires, dual split film, simple pitot. Probes moved by mechanical transverse below model. (Photograph courtesy of Boeing Aerodynamic Laboratories.)

If the probes are being used to detect the transition between laminar and turbulent boundary layers, extreme care must be taken to ensure that the probe itself is not disturbing the flow.

Many methods are in use to determine the location of the transition region. They include:

1. Plotting the velocity gradient in the boundary layer and determining whether the flow is laminar or turbulent by the slope of the gradient (Figs. 3.15 and 3.16).

2. Crossplotting 1, determining the beginning of transition as the point where the velocity is lowest (Fig. 3.17).

3. Reading the static pressure at a small height above the surface, determining the transition by a slight dip in the plot of static pressure versus percent chord.

4. Reading the dynamic pressure at a small height above the surface and noting the minimum value of q from a plot of q versus percent chord.

5. Reading the velocity at a small height above the surface with a hot-wire anemometer, and noting the transition as a region of unsteadiness in the output.

6. Reading the velocity at a small height above the surface with a hot-wire anemometer or thin-film gage, and noting start of transition as the point of minimum velocity (Fig. 3.17).

7. Carefully emitting smoke from flush orifices, and noting transition by the dispersal of the smoke stream (may be difficult at high velocities).

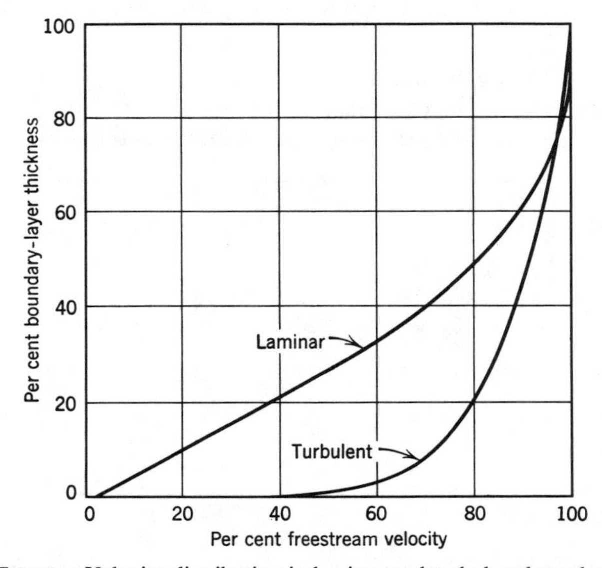

FIGURE 3.15. Velocity distribution in laminar and turbulent boundary layers.

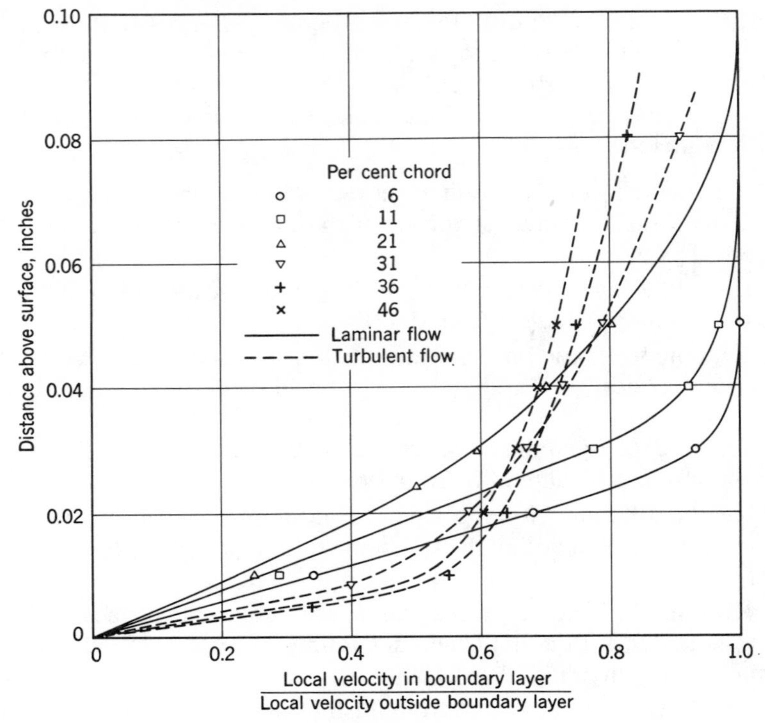

FIGURE 3.16

8. Painting the model with special chemicals that evaporate slowly. The evaporation will proceed most rapidly where the flow is turbulent (Section 3.10).

9. Listening to the boundary layer with an ordinary doctor's stethoscope connected to a flat total head tube, which is moved progressively along the surface toward the trailing edge. As long as the flow is laminar one

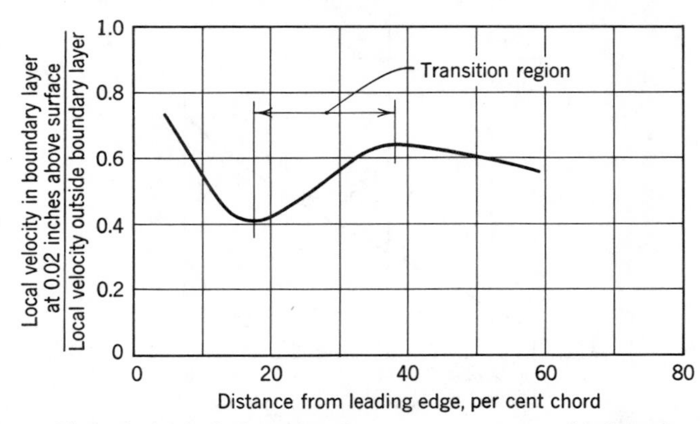

FIGURE 3.17. The velocity in the boundary layer at a constant small height above the surface.

hears a soft *sh-sh-sh-sh*. When it is turbulent, a distinct roar is heard. This same input fed into a transducer becomes quite graphic on an oscilloscope.

Preston Tube

This is a method used to experimentally measure the wall coefficient of skin friction. This is done by measuring both a static pressure and total pressure at the same chordwise location. The total pressure is measured by a pitot tube that touches the surface. This can be done because for unseparated turbulent flow there is a region near the wall on the order of 10% of the boundary layer thickness in which the flow depends on the local wall skin friction τ_w, the density ρ, the kinematic viscosity ν, and a length. Preston took one-half of the pitot tube diameter as the length (Ref. 3.7).

$$\frac{\tau_w d^2}{4\rho\nu^2} = f\left(\frac{(P_0 - P_w)d^2}{4\rho\nu^2}\right) \tag{3.2}$$

where P_0 = pitot pressure and P_w = wall static pressure at same point in flow.

In a pipe using four pitots with inside to outside diameter ratios of 0.600 (the outside diameters varied), the value of f was found. The wall skin friction τ_0 is found from $\tau_w = (P_1 - P_2)D/4L$. Here P_1 and P_2 are wall static pressures separated by length L in a pipe of inside diameter D. Preston found that $y^* = f(x^*)$ or

$$y^* = \log_{10}\left(\frac{\tau_w d^2}{4\rho\nu^2}\right) = -2.628 + \tfrac{7}{8}\log_{10}\left[\frac{(P_0 - P_w)d^2}{4\rho\nu^2}\right] \tag{3.3}$$

where y^* is a dimensionless shear stress for incompressible, isothermal flow and x^* is a dimensionless pressure difference for incompressible isothermal flow.

In 1964 Patel published the results of tests with 14 different circular pitot probes using three different pipe diameters (Ref. 3.8). The limits on pressure gradients for the calibration are also given. Patel obtained empirical equations for

$$y^* = \log_{10}\left(\frac{\tau_w d^2}{4\rho\nu^2}\right) = f\log_{10}\left[\frac{(P_0 - P_w)d^2}{4\rho\nu^2}\right] = f(x^*) \tag{3.4}$$

over three ranges of y^*:

1. $3.5 < y^* < 5.3$.
2. $1.5 < y^* < 3.5$.
3. $y^* < 1.5$.

These ranges correspond to fully turbulent flow, transition flow, and viscous sublayer, respectively.

For transition flow (2) or $1.5 < y^* < 5.3$,

$$y^* = 0.8287 - 0.1381x^* + 0.1437(x^*)^2 - 0.0060(x^*)^3 \tag{3.5}$$

In the viscous sublayer (3) or $y^* < 1.5$,

$$y^* = 0.37 + 0.50x^* \tag{3.6}$$

using the classical law of the wall in the viscous sublayer and by defining an effective center of the pitot as

$$y_{\text{eff}} = K_{\text{eff}}(d/2) \tag{3.7}$$

Patel obtained a value of $K_{\text{eff}} = 1.3$ for a round pitot tube. The velocity calculated from Preston tube data U_p is the true velocity in the undisturbed boundary layer at y_{eff}.

The work of Preston and Patel is for round pitot tubes. There are other calibrations for oval or flat Preston tubes in the literature (Ref. 3.9–3.13).

3.6. HOT-WIRE ANEMOMETRY

Hot wires are used to obtain instantaneous velocity measurements in turbulent flows as well as mean velocities and, with multiprobes, flow angularity.

The probe or sensor is a fine wire (diameter of a few microns) stretched between two supporting needles on the probe (Fig. 3.14). Current passed through the wire raises its temperature above the adiabatic recovery temperature of the gas. The hot wire responds to changes in total temperature and mass flux (T_0 and ρU).

In subsonic applications where the density is high and the fluid temperature is low and constant, the problem of heat in the support needle (end losses) and radiation effects can be ignored, and the wire's response basically is a function of velocity alone. Under these conditions by the use of appropriate calibration and measuring the voltage across the bridge, both mean and turbulent velocities are obtained. If desired, the output from the hot wire can be recorded on FM tape for data reduction and analysis.

Electronic hot-wire circuits include a feedback system to maintain the wire at a constant wire resistance. This is a constant-temperature anemometer. Feedback can also be used to maintain a constant-current anemometer. Constant-temperature anemometers are easiest to use in subsonic or incompressible flow, while the constant-current anemometers are preferred in compressible flow. Frequency response to 50 kHz is easily obtained.

In incompressible flows, the hot wire costs less and is easier to use than a laser anemometer. The hot wire does intrude into the flow and the laser does not. The hot wires are extremely fragile, and they often break, usually just after calibration and before use. If the flow field is dirty, the risk of the wire breaking is high.

If turbulence measurements are not required, a single or dual split film can be used. These units are more rugged and can be used to measure both mean velocities and flow angles (Fig. 3.14).

3.7. LASER VELOCIMETRY

The most common laser Doppler velocimeter (LDV) uses optics to split the laser beam into two parallel beams that are focused to cross at the point where measurements are to be made (a dual-beam system). Owing to wave interference, a fringe pattern in an ellipsoid-shaped volume at the beam intersection is formed. A second lens assembly (the receiver) with a small aperture is focused on this fringe region to collect light from seed particles crossing the fringes. This light is fed to a photodetector that is used as the input to the sophisticated electronic signal processor that measures modulated frequency.

In a dual-beam system the frequency from the scattered light from the two beams are superposed on the photodetector's surface. The mixing process in the photodetector then gives the difference in the frequency from the two beams. All other frequencies are too high to detect (this is called an optical heterodyne). If β is the angle between one of the beams and the bisector of the two beams and λ is the wavelength of the laser light, then it can be shown that when the particle velocity is much less than the speed of light the modulator frequency is

$$f_D = \frac{2u_x \sin \beta}{\lambda} \tag{3.8}$$

where u_x is the velocity parallel to the plane of the two beams and perpendicular to the bisector of the beams. Thus the relationship between the flow velocity and f_D is linear and a function of half the beam angle and the wavelengths of the laser light. Perhaps it is easier to think of the system working in the following manner.

When a particle moving with the flow passes through each fringe, it is illuminated, which causes a series of pulses from the photomultiplier. As the distance between the fringes is known, the time to cross the fringes is measured, and this yields the particle velocity. It should be noted that the fringe spacing in micrometers is equal to the calibration constant in meters/second/megahertz. When a large number of samples are taken, the signal processor and computer calculates both the average velocity and instantaneous velocity, which can be used to obtain the turbulence or velocity variation.

To calibrate the laser Doppler velocimeter the wavelength of the laser is required. This is known to an accuracy better than 0.01%, thus only the beam crossing angle is required. As an alternate calibration method the velocity of the edge of a rotating disk at constant speed can be measured by the laser Doppler velocimeter for a direct calibration. This also can check the extent to which the fringes are parallel by varying the point along the optical axis where the wheel intercepts the measuring volume.

One way to measure flow angles is the following: (1) set the plane of the crossed beam parallel to the tunnel centerline; (2) by rotating the optics, measure the velocity at $\pm 45°$ to the reference setting of 1. Because the measured velocity is parallel to the plane of the beams, complete the rhombus (measured velocity forms two legs) to obtain the magnitude and flow direction of the flow at the measuring point. Thus, by measuring the flow angle along the aerodynamic center of the tail, the downwash and velocity at the tail can be obtained. The average values can be obtained by multiplying the local angle or velocity by the local tail area, and dividing the sum by the tail area. As some lasers will lase at different color or wavelengths, two color beams can be split and one can measure the $+45°$ and the other the $-45°$ velocity. This method allows both components to be measured simultaneously, but there are complications in the optics and two photodetectors are required.

There are two basic modes that can be used with a LDV, the forward scatter mode and the back scatter mode. In the forward scatter the receiving optics and photodetector would be on the opposite side of the test section from the transmitting optics. In the back scatter mode both the transmitter and receiver are on the same side of the tunnel. The advantages/disadvantages of both modes are listed below.

Forward Scatter Mode

Advantages

1. Can use low-power lasers, such as helium neon, output 0.5–30 mv. These units have a high level of reliability.
2. The laser does not need cooling.
3. The signal to noise ratio is large.
4. Can measure higher air speeds but may require more laser output at higher speeds.

Disadvantages

1. Will need windows on both sides of the test section that are relatively flat optically to avoid beam misalignment.
2. When making traverses through the test section, the receiver must track the measuring volume. This may require both the transmitter and receiver to be on a steady base. In many tunnels the test section

is elevated above the ground and the floors at the test section level often vibrate with the tunnel.

3. Often the model will block the transmitted beam. This occurs, for example, when measuring downwash in the tail region.

In general, forward scatter systems are better for small tunnels, and measuring flow in the jet without a model.

Back Scatter Mode

Advantages

1. The transmitter and receiver can be placed on the mount that is used to traverse the flow and there is no problem tracking the measurement volume.
2. Only one window is needed that is relatively flat optically to avoid beam scatter.
3. There are less problems with the model blocking signals, although in certain cases problems will occur.

Disadvantages

1. A much higher power laser is needed (continuous-wave argon), up to 20 W output. The higher power requires water cooling with associated plumbing problems.
2. The higher power increases the danger from the beams and their reflections. Usually it is necessary to have the laser operation site inspected and approved by laser or environmental safety personnel. An output of over 0.5 W is defined as high power.
3. The signal to noise ratio is much lower, which may increase the time required to make measurements. The signal to noise ratio is a function of laser power.

The output of the photodetector is a frequency, and this must be converted to a voltage or some number proportional to velocity. There are many ways that this can be done, but no apparent universally accepted optimum method. Because of problems with low data density (few data points), the use of a computer to analyze the data is most advantageous. Using a mini computer with direct memory access up to 500,000 16-bit words per second can be transferred to memory. This rate is fast enough for a LDV but, if necessary, a separate buffer could be used. If more data are to be taken than the memory can store, mass storage devices, such as disks, tapes, etc., can be used. This implies the use of a dedicated computer for use with the LDV. A mass storage device can be used as an alternate, but the data transfer rates are slower and the advantages of on-line data reduction are lost. Thus, there is a strong tendency to use a dedicated computer of adequate size and speed for LDV data reduction.

A particle generator may be needed for adding particles of a specific size (usually 0.1 μm to 10 μm) and density into the flow. Particle parameters must be controlled to get adequate signal strength and to ensure that particles travel at instantaneous local velocity. At low speeds natural dust may provide needed particles. Particle seeders are commercially available.

Problems in turbulent flow may occur when (1) intensity is large with respect to mean flow and (2) mean flow velocity changes rapidly with either time or location. Particles may not be able to maintain instantaneous velocity of flow (worse at low air density). If particle seeding is required, the size and type and the keeping of particles at instantaneous flow velocity must be resolved.

A typical seeding method is atomization, the process of generating liquid droplets with compressed air. Usually light oils are used. The resulting aerosols will contain some large droplets (10–15 μm). These probably do not follow the flow field but they do tend to deposit themselves on the walls and windows, thus adding a cleaning chore for the tunnel crew.

The use of small-size solid particles in the air can also be used. There are problems with keeping the original particle size and feeding the powder into the air at a constant rate. There is also the abrasion or sand blast problem on the tunnel fan blades and the possibility of the powder working its way into seals on shafts and bearings. As with the oil, if lots of running is done, there is a cleanup problem.

Laser velocimeters are used to measure momentum losses in wakes to obtain drag, measure circulation to obtain lift, map flow fields, trace vortex paths, measure turbulence, etc. Figure 3.18 shows a flow survey in the flap cove and flap region of a GAW-1W airfoil and flap at 16.0° angle of attack. Reducing the original plot to a reproduction size for the book looses some of the detail in the cove and on the leading edge of the flap. The plot has been produced by computer graphics.

Laser velocimeters, optics, and electronic signal processors are all commercially available.

If the LDV is to be used as a flow-visualization-type diagnostic tool, the precision usually obtained in both locating the measured flow volume and the velocity can be relaxed. This is possible because the relative magnitude of changes are required, rather than absolute values. The LDV itself, if not permanently mounted at the tunnel, must be capable of being installed and operational in 4–8 h. The LDV traversing system must move the measurement volume quickly at the expense of accurate location. The accuracy of measurement velocities can be relaxed in a trade-off for rapid acquisition of the required samples.

The data system and computer should give the results, usually in plotted form for traverses, in 5 min or less after the run. In essence, when used in this manner, the LDV is competing with other flow visualization techniques

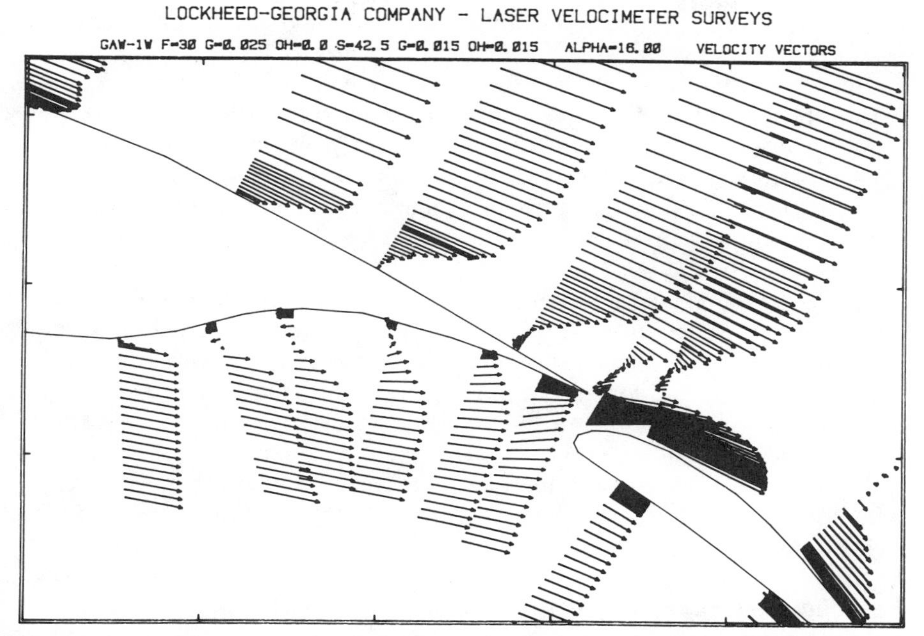

FIGURE 3.18. Computer-generated plot of laser velocimeter survey in region of single slotted flap. (Photograph courtesy of Lockheed Georgia Company.)

such as china clay, tufts, etc., and to be useful it must operate within their required time (Refs. 3.14–3.17).

3.8. PRESSURE TRANSDUCERS

The most commonly used pressure transducers are of the diaphram type. Generally they are of the differential pressure type and a preselected reference pressure is applied to the reference side. (Absolute pressure types are also available.) They come in a wide range of pressure values; however, 2.5 psid and 5.0 psid are most commonly used. They require a bridge power supply similar to strain gages and the output voltage varies with pressure.

A series of these transducers are made to fit into scanivalves. Up to five of these valves can be ganged to a stepping motor. The valves can be obtained with 24, 36, or 48 ports. The stepping motor rotates one of two ground plates that connect each input tube in sequence to a plenum. Through its control, the scanivalve can be made to step to each inlet port and hold for a predetermined time and then step to the next port. The hold time is adjustable but constant for each port and is required to allow the plenum pressure to stabilize. The plenum has a small volume, but small-diameter tubing is also used in the model. The time to stabilize is a function of the tubing length and pressure.

The scanivalve does not have to step through all the ports but can be set to step through 12 ports if that is all that are used and then rapidly step to home, or the first port. Often the first and last ports are connected to the same pressure source. This is done to make sure the valve has stepped properly by comparing these pressures, and the second port is connected to a fixed known pressure to monitor the transducer's calibration during a run, as the reading of the port will be a constant value.

The reference pressure is manifolded to all the transducers and usually is led out of the tunnel with the electrical leads. This tube is used to periodically check the transducer calibration and recalibrate if necessary, usually once a shift. This is done by removing the tube from the reference and applying a known pressure to the transducer. In pressure units the transducer output is usually compared to the applied pressure.

The transducers are calibrated using a primary standard in a dead weight tester or a calibrated secondary standard. These units most often apply pressure to the reference side of the transducer, but this is not mandatory. As the reference pressure is usually the test-section static, which is above the model static pressure other than in the stagnation region, the calibration is, in the same sense, as the recorded pressure.

When the transducer is calibrated outside the model, the bridge voltage used and the calibrated amplifier filter setting and gain are recorded. The same values are used during the test. As an alternative, the bridge voltage can be recorded and the bridge output measured. The amplifier gain is then entered into the data-reduction process. The bridge output is adjusted to zero for zero pressure or the voltage for zero pressure is recorded. Then a series of pressures are applied and the pressure and output voltage are recorded. These data are fed into a curve-fitting routine to determine the calibration curve. The transducers tend to be linear, but often a third degree curve fit is used.

Often when a transducer is independently calibrated and then mounted in the model or scanivalve, there are slight but measurable shifts of the calibration curve, apparently caused by clamping the transducer. The transducer will repeat its calibration when mounted over a period of time. Thus, it appears that the pressure transducer should be calibrated mounted in the scanivalve or the model. Calibration must be checked if the transducer clamp is loosened and then retightened.

When only a few pressures are required, a pressure transducer is used for each pressure.

Rapid Response Pressure Transducers

These are basically piezoresistive-type semiconductor pressure gages that physically are quite small, and they are commercially available. These pressure gages are able to measure high-frequency pressure fluctuations. The transducers require associated electronics located near the gage to optimize their response. The transducers can be obtained in shapes that can be used to measure total and static pressures. The transducers are intended to be used for the measurement of pressure with time, but time-averaged pressures can also be obtained.

These pressure gages tend to be differential transducers, and the reference side can be connected to either a known pressure or to a source of approximately time-averaged pressure, whose fluctuations are required. Since the units are sensitive to temperature, constant current excitation will minimize the effect of temperature.

These transducers can be used to measure turbulence and noise. The gages can be damaged by particles in the flow, but they are stronger than hot wires. They also have another advantage over hot wires in that they respond only to pressure, rather than pressure and mass flux.

Similar to hot wires, it may be advantageous to record the output on FM tape for later analysis. If specific frequency information is required, spectrum analyzers and correlators may be used, or digital spectral analysis techniques can be used. For total noise or turbulence measurements, rms-to-dc converters can be used. In general, these transducers can be used for measurements similar to those made with hot wires or laser velocimeters and thus have the same advantages and disadvantages. The piezoresistive gages designed to measure surface pressures can be built into a model to measure surface pressure variations.

Typically, these gages will require periodic calibration during the test. A novel way to calibrate a large number of this type gage was developed by the U.S. Army at Ames Research Center. The usual way is to disconnect the reference source, and use it to calibrate the gages. The Army's method was to use a scanivalve in reverse. The transducer in the scanivalve was replaced with a dummy that was connected on the reference side to the model's reference pressure manifold. A laboratory standard pressure measuring unit was also connected between the dummy transducer and the model manifold to measure the absolute pressure in the manifold. The pressures used for calibration came from the reservoir through pressure regulators that were connected to the scanivalve ports as was the desired reference pressure used when running. This allowed a rapid stepping of calibration and running reference pressure without disconnecting any tubing.

The same Army group built an instrument that would accept the output of up to 10 transducers. These outputs in pressure units were applied to the y axis of a CRT and the transducer location was applied to the x axis. The CRT then displayed in real time a continuous line representing the pressure profile of the selected set of transducers giving data visibility similar to the old manometers.

For a rotor blade using an absolute pressure transducer, a clear plastic cylinder was built to surround the model. By careful attention to model construction and transducer location this cylinder could be sealed to the model. Then all of the transducers could be simultaneously calibrated end to end. With this method the excitation voltage could be set to produce a common slope calibration for on-line quantitative monitoring of blade response. This system did not work for differential pressure transducers owing to difficulty in sealing one side of the transducers.

3.9. DATA ACQUISITION

The path from some physical parameter to a stored digital number will be followed. Initially we will look at these functions as they were carried out by hand, and then follow the same process by an analog–digital system.

Using Example 3.1, three measurements were needed to determine the velocity from a pitot-static tube. The hand process for total pressure would be to (a) identify the manometer tube that is connected to the total pressure port, (b) read the height of the tube with the proper sign, and (c) write the sign and height in the correct place on a data sheet.

This is exactly the same process that a data system performs. First the desired parameter must be located. Assuming that five scanivalves of 36 ports or scans are used. Then the data system must know the valve and scan position that is connected to the total pressure port, say valve three, as each scanivalve has its own transducer and each transducer has its own data channel. Then the data channel identifies the valve or transducer, that is, channel 13. If the total pressure is on the fifth port or scan, the address for total pressure is then channel 13, scan 5. This address always exists in the system when using scanivalves. This is the same as finding the manometer tube.

The next item is to read the value and sign, which is not as simple as the manometer tube. In general, transducers produce electrical signals; the amplitude depends on the value of the physical parameter. The signal can have different forms, such as a current variation, voltage variation, resistance variation, etc. The level of the signal is generally quite low and requires amplification. Amplifiers act only on input voltages, thus some transducers must have their output changed to a voltage. This is done with a signal conditioner.

The signal from the transducer consists of two parts: the usual signal and signals that represent noise. The noise can have an electrical cause or an aerodynamic cause. If, to improve the accuracy of the measurement, the noise is not wanted, it often can be filtered out by electric analog filters. Extreme care must be taken to ensure that part of the required signal is not lost in this process.

The sign and amplitude or voltage of the analog signal that comes out of the amplifier and/or filter represents the value of total pressure. To use this value in calculations it must be expressed as a number, that is, it must be in digital form. This is accomplished by an analog-to-digital converter (ADC). The ADC sends the digital number and sign to the computer for storage. The total pressure is then stored in its file at the correct address, that is, channel 13, scan 5. The file contains additional information, usually report number, run number, angle of attack, and yaw angle. The data system has followed the same process as was done by hand: identify the manometer tube (address), read the sign and value in digital form (sign plus voltage to ADC for digital form), and write value and sign on data sheet (store with correct address). Analog-to-digital converters are usually very expensive pieces of equipment, so unlike the amplifier and filter, which also are expensive, one is not supplied for each channel; usually there is only one or two for the system.

The data system also does not consist of one data channel. The number of channels in the system is determined by the estimated number of parameters that will have to be measured over the life of the system. A system may have 20 or 100 channels. It should be noted that the use of a 36-port scanivalve allows 36 parameters to be measured on one channel. Thus, the number of parameters that can be acquired does not necessarily equal the number of available channels.

Assuming that the system has 40 channels and one analog-to-digital converter, how can the data be recorded quickly? This quandry is solved by the use of a multiplexer. A multiplexer is an electronic device that, in a given sequence, switches each channel for a short but finite time to the ADC. The sequence of switching is important to keep track of the address. Thus, the multiplexer starts with the first channel and steps through the 40 channels.

Multiplexers were also used in the days when things were done by hand and slide rules. They were either rotary switches or a series of switches that were used to connect different signals to an expensive device that allowed the signals to be read in digital units. If the evolution of multiplexers is followed, it would go like this: very slow manual switches to faster relays to reed relay switches in a neutral gas to the current multiplexers. The reed switches could handle up to 10^3 switching operations per second. The current use of integrated circuits and semiconductor analog switches has increased the rate to 10^5 per second.

Closely tied to the faster multiplexers has been the decrease in time required for the ADC's to read and convert the voltages.

With the higher speed of the ADC's and multiplexer switching it is possible to take an almost simultaneous sampling of all of the channels as long as the sample time is short. This led to systems that take a large number of samples of very short time. Each sample is then summed and averaged and a standard deviation is often determined.

The evolution of the digital computer should not be neglected. The early computers were large expensive main frame computers that required large staffs to operate. These machines were feasible to use only if there were a large group of users who shared the computers on a batch mode operation.

The input was in the form of cards that were delivered to the computer and the output was picked up at a later time. The cards consisted of both the programs and the data and the output was tab data, output cards to make additional tab copies and, later, magnetic tape.

The advent of the mini computer really made the on-site wind tunnel data system practicable. The cost was low enough to allow the mini computer to be dedicated to supporting wind tunnel tests only. Programs resided in the computer and the ADC fed its results directly into the computer. The computer was equipped with adequate disk storage so that the run data in millivolts could be stored, converted to engineering units, stored, and finally converted to the final coefficients and stored. The output now was tab data and plots. These systems can even plot the data in very close to real time, as it is acquired. The mini computers did not require large staffs similar to the large main frame, and, in fact, do not even require an operator.

There is a problem in low-speed tunnels with the wind tunnel wall corrections. They are a function of the wing lift only, thus for tail-on runs the equivalent tail-off run must be available. Forcing the tunnel to make these runs in the correct order often increases the test time due to model changes (see Chapter 6). The problem can be circumvented by applying quasicorrect wall corrections to the real-time plots. These plots are then used to direct the test, but not for analysis.

The mini computers and their offspring, the micro, also allowed the setting of model attitude and acquisition of the data to be automated. These capabilities now exist on micro units that can acquire, convert to digital form, store, and reduce data at slower rates, and for a limited number of channels. There have been improvements also in electronic components required for each channel. Bridge voltage power supplies and bridge balancing impedences are more stable and are designed to keep the electrical noise as low as possible. Amplifiers can cover frequencies down to almost zero or direct current. They are more stable (less drift) and have better noise levels and improved common node rejection. Programmable power supplies of secondary standard quality can be used to calibrate the amplifier, usually about every 4 h to check drift. The calibration is stored and used for all subsequent readings until the next calibration. The use of electric analog filters can reduce electric noise to reasonable levels.

The aerodynamic noise from the tunnel and model are more difficult to remove. The frequency of this noise is often below the lower filter limit of the amplifier. These can be removed by an analog very-low-frequency by-pass filter. The units are expensive, and one may be required for each channel. They also require long response time to filter the low frequencies and thus negate the advantages of the high-speed digital acquisition system. It is possible to use special processors that use numerical processing in real time to remove these signals. There are also cases where it may not be desirable to filter out these signals. In Chapter 2 the problem of filtering out low-frequency signals is discussed regarding misleading results on turbulence measurements (see Section 2.11).

3.10. FLOW VISUALIZATION

It is difficult to exaggerate the value of flow visualization. The ability to see flow patterns on a model often gives insight into a solution to an aerodynamic problem. Flow visualization can be divided into two broad categories. The first is surface flow visualization when the visualization medium is applied to the surface such as tufts and oil flows. The second type is off the surface, such as smoke and streamers.

Methods of Recording

There are basically four methods of recording flow visualization tests. The first and best, but least permanent, method is for the engineer to observe with his or her eyes. Because of depth perception one can see a three-dimensional picture. However, there is no permanent record to which we can refer in the future. It is possible, however, to sketch the flow as it is observed. To do this efficiently one should have prepared a basic drawing of the model on which streamlines or separated regions will be sketched when the tunnel is running. The other three common methods of recording the results of flow visualization are by film, either still or movie, or television camera and magnetic tape. It must be realized that all three of these methods are using a two-dimensional medium to often record a three-dimensional phenomena. This is especially true when using smoke or helium bubbles to trace flow streamlines past the model. All three of these mediums can be used in either black and white or color. The photographic methods, while requiring more time for developing and printing for stills, when compared to video, yield higher resolution. Video has the advantage of instant replay.

Ultraviolet Fluorescence Photography

Ultraviolet fluorescence photography is used when the medium used for visualizing the flow has been treated with a dye that radiates in the visual spectrum when excited by long-wave ultraviolet light (Ref. 3.18). The wavelength of the long-wave ultraviolet light is 320–400 millimicrons, and it is transmitted by optical glass. There are three sources that can be used to produce the ultraviolet light. These are special fluorescent tubes (black lights), mercury vapor lamps, and electronic photo flash units.

In wind tunnel use, the first two light sources are used to enable the test engineer to observe the flow, and the flash units are used to take still photographs of the flow when desired. Fluorescent tubes and mercury vapor lamps in general do not have a high enough light intensity to allow photographs without a very long exposure. Because fluorescent material emits light in the visual range, the tunnel test section must be shielded from visible light. Both the mercury vapor lamps and the flash units also emit visible light, thus they must be equipped with an exiter filter that will transmit ultraviolet light and absorb visible light. Two filters that accomplish this are Kodak Wrattan filter No. 18A or Corning Glass No. 5840.

For larger tunnels the flash lamps are studio units marketed for commercial photographers. The flash lamps should be able to handle 2000 W/s per flash as rated by photographers; the units come with power supplies that can store energy in capacitors and have the necessary trigger circuits. The reflectors for the flash units should be 10–14 in. in diameter to be efficient. The Corning glass exciter filter comes in 6.5 in. squares, four of which can be glued together and built into a frame to cover the reflector. As an alternative, one glass filter can be used with the flashbulb without a reflector. This would require approximately one additional *f* stop. For research tunnels where the camera to subject distances are small, standard flash units, and a Wrattan 18A exciter, which can be obtained in a 3.5 in. square, can be used.

Photographic film is sensitive to blue and ultraviolet light. The light reaching the camera will contain both the visible fluorescent radiation and reflected ultraviolet radiation. To prevent the ultraviolet from exposing the film, a barrier filter is attached to the camera lens. The barrier filter can be a Kodak Wrattan filter No. 2A, 2B, 2C, or 2E. These can also be obtained in 3.0 in. square gelatin sheets.

It is difficult to specify the filter combinations in advance as they depend on many factors that are specific to a given application. For large tunnels, which are expensive to operate, it is usually worth the time to make a mock-up of the tunnel and use this to determine the filters, exposure, and *f* stop for the film and fluorescent material. The mock-up should duplicate the distances from the light source or sources, and from the camera to the model. Also, the mock-up should duplicate the glass or plastic windows that will be used in the wind tunnel. Some acrylic plastics and safety plate glass are designed to absorb ultraviolet radiation. The brightness of fluorescent dyes depends on the amount of ultraviolet radiation, and any loss through windows from the source should be minimized to avoid use of large apertures with their smaller depth of field.

In both the tunnel test section, as well as the mock-up, a series of exposures at various *f* stops should be made to determine the desired exposure. The negative that appears acceptable on proof prints, plus the negative of at least one smaller aperture should be enlarged and used to determine final exposures. The negative is exposed to a small bright light source by the fluorescent dye, thus the size of the final image for a tuft will change with exposure. This effect cannot be seen on normal contact size proof prints.

Focusing the camera with a fluorescent light source is usually no problem because this can be done using either the black light fluorescent lights or the mercury vapor lamp with an exciter filter as a light source.

It should also be noted that black lights that can be fitted to standard fixtures are several orders of magnitude cheaper than the mercury vapor light with its power supply and exciter filter.

It is also possible to photograph the fluorescent material with some video cameras during the flash from the light source.

Surface Visualization

The simplest and most often used surface method is tufts. This is basically light flexible material that is attached to the surface, and aligns itself with the local surface flow. Tufts do affect the aerodynamic forces.

The two basic methods of attaching tufts to the surface are by Scotch tape or by glue. When tape is used, the tufts are usually made on a tuft board. The tuft material is strung back and forth around pins, then the tape is applied to the tufts and the tuft material is cut at the edge of the tape. This gives a length of tape with tufts attached that is applied to the model. The model surface should be cleaned with naphtha or other solvents to remove oil (Fig. 3.19).

When tufts are glued to the model, a nitrocellulose cement such as Duco is used, thinned 50% with acetone or methyl ethyl ketone. Often 10% pigmented lacquer is added both to obscure the portion of the tuft under the glue and to make the glue dots visible by using a contrasting color.

The glue technique is used for minitufts. These tufts have the least effect on the aerodynamic data and thus can be left on the model. The tuft material is monofilament nylon that has been treated with a fluorescent dye. Two sizes are used: 3 denier (diameter = 0.02 mm, 0.0007 in.) and 15 denier

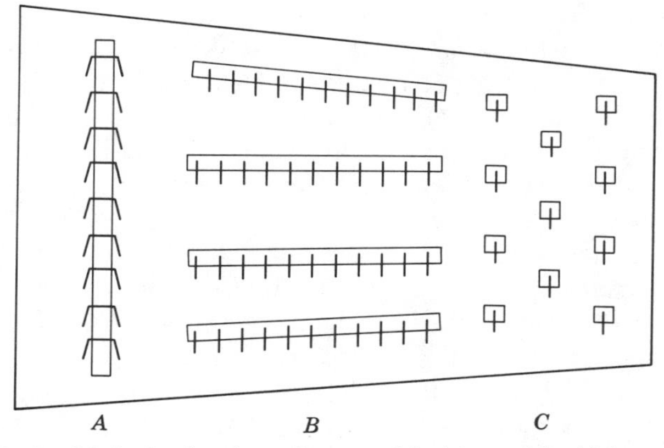

FIGURE 3.19. Methods of taping tufts to model. *A* is used for high-speed tests.

(diameter = 0.04 mm, 0.0017 in.). The dye used was Leucophor EFR Liquid in a concentration of 1% in water with 2% acetic acid added. The tuft material is wound on an open wire reel and immersed in the dye for 15 min at 82.2°C (180°F) with frequent agitation. After drying for at least 1 h the tuft material is wound onto small spools. During this step the material should be wiped with tissue pads to remove loose fluorescent powder that can transfer to the model surface in irregular patterns.

There are also polyester and cotton sewing threads, such as "Clark's O.N.T." mercerized cotton No. 60, which has been treated with a fluorescent material. These are larger and more visible than the minitufts. The thread is a multiple-strand material and tends to unravel with time.

The thread can be attached with either tape or glue. The minitufts are viewed and photographed in ultraviolet light, and the threads can be photographed in either ultraviolet or white light.

When gluing tufts to a model a square grid is used (of about 0.75 × 0.75 in.). The tuft material is taped to the wing under surface and then wrapped around the wing in a chordwise direction. The material is in the chordwise direction on the top and moves diagonally across the wing on the bottom surface. As an alternate, the tuft material can be taped at both the leading and trailing edges. After the tuft material is applied it is glued using a hypodermic syringe with a fine needle (a coarse needle can be partially closed with pliers). As the desired size drop of glue forms on the needle, touch it to the surface and pull away quickly. After the glue dries the tufts are cut just ahead of the glue spot of the next tuft. The model surface should be cleaned before the tufts and glue are applied, using Freon or chlorinated hydrocarbons (Ref. 3.19).

The monofilament nylon minitufts acquire static charges. These can be neutralized by the use of antistatic solutions or the antistatic material used in home dryers.

Tufts affect the aerodynamic loads on a model. In Fig. 3.20 the lift curve near stall shows the effect of various tufts on the data. The glued mini and No. 60 thread consisted of about 900 tufts. The two taped tufts consisted of about 300 tufts. The data are an average of five runs for each set of tufts. The minitufts and the glued No. 60 thread have the minimum effect on lift. The effect of the tape can be seen by comparing the two sets of No. 60 thread tufts. The six-strand floss tufts are similar to the tufts made out of yarn. The three different tufts can be seen at $\alpha = 27.3°$ in Figs. 3.21–3.23.

China clay is a suspension of kaolin in kerosene. The fluid is applied with a paint brush, usually with the model set at the desired attitude. The tunnel is started as quickly as possible after the model is painted. When the mixture has dried, photographs are taken, usually Polaroid-type positives, after the tunnel is shut down because the pattern does not change rapidly with time. This is similar to the Fales method with the kaolin substituted for lamp black (Fig. 3.24).

The oil flow run used 40W motor oil treated with a very small amount of

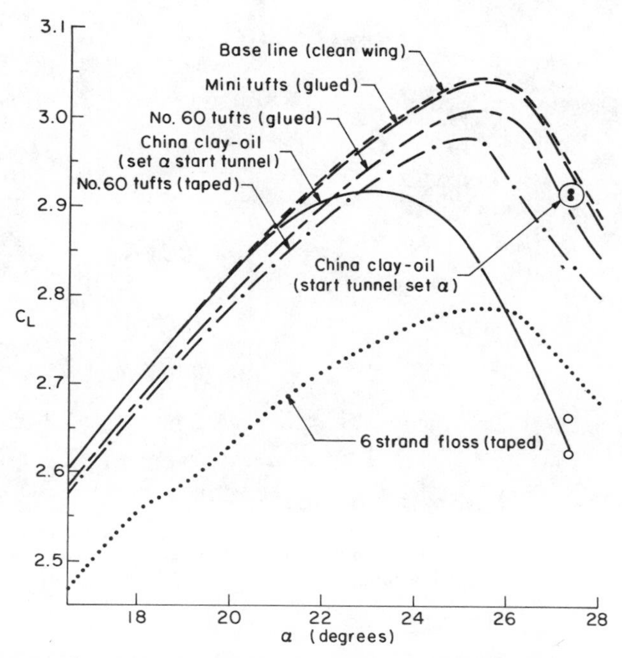

FIGURE 3.20. Effect of various tufts, china clay, and oil flow on lift curve near stall.

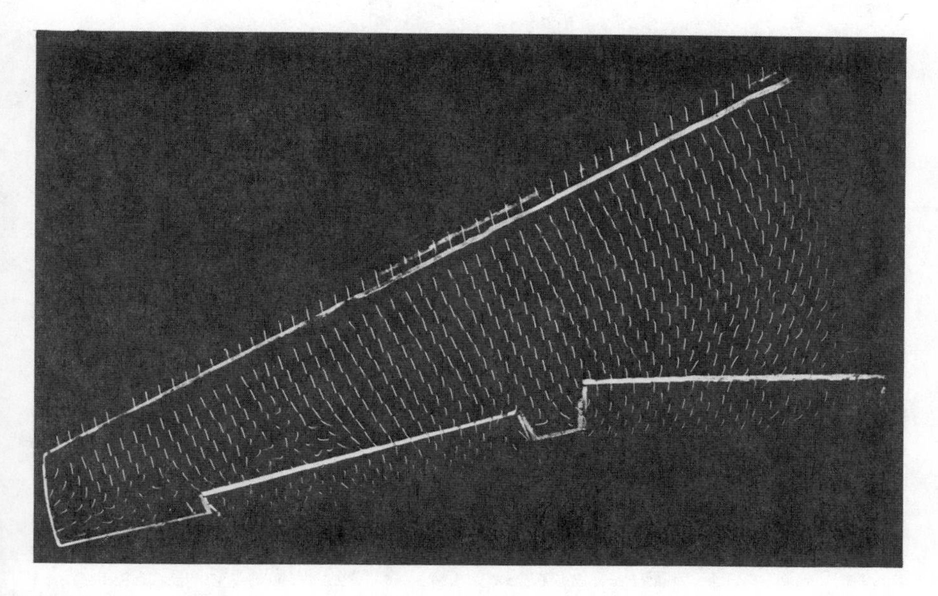

FIGURE 3.21. Fluorescent minitufts, ultraviolet light source. $\alpha = 27.3°$. Wing is outlined with a fluorescent felt marker pen. Compare stalled region near tip with Figures 3.22–3.25. (Photograph courtesy of University of Washington.)

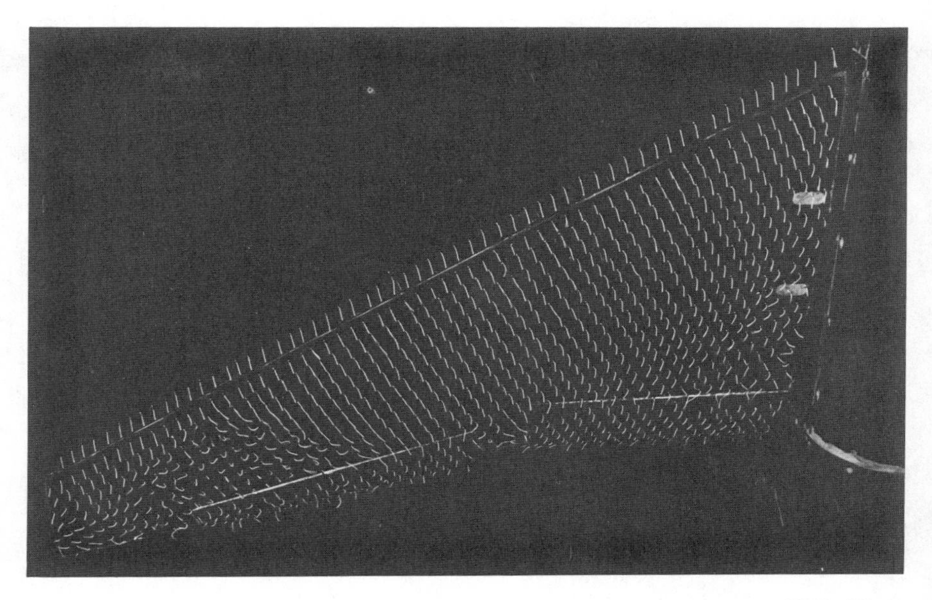

FIGURE 3.22. No. 60 thread tufts glued to wing, ultraviolet light source. $\alpha = 27.3°$. (Photograph courtesy of University of Washington.)

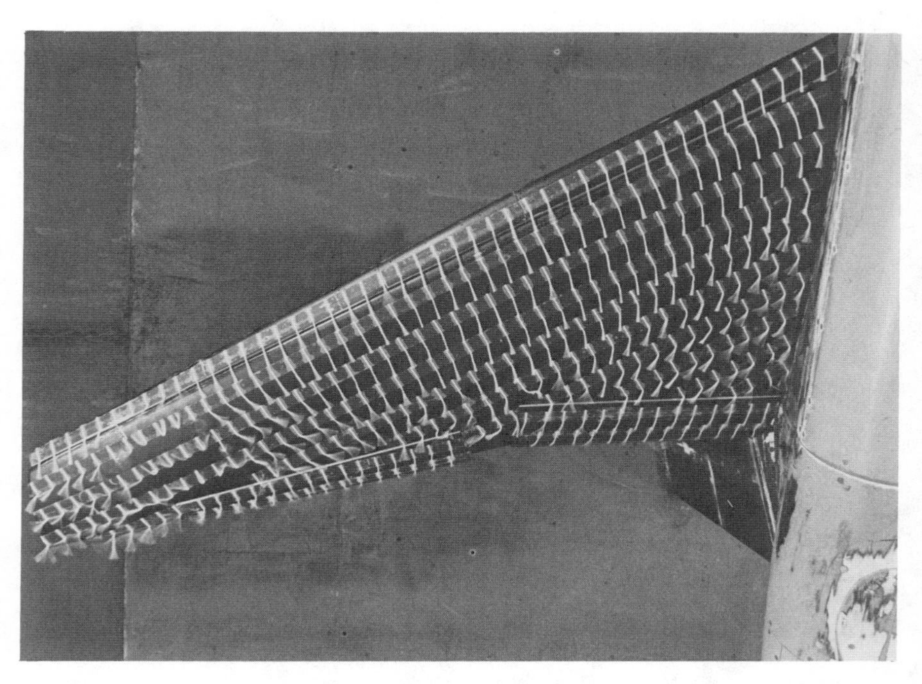

FIGURE 3.23. No. 6 floss (crochet yarn), white light source. $\alpha = 27.3°$. Tufts taped to wing as in method *B* of Fig. 3.19. (Photograph courtesy of University of Washington.)

FIGURE 3.24. China clay applied at $\alpha = 0°$, tunnel started, and brought close to speed, model pitched to $\alpha = 27.3°$. (Photograph courtesy of University of Washington.)

fluorescent dye. If the oil flows too slowly, it is thinned with naphtha, and if it is too thin, 60-70W oil is added. The viscosity of the mixture is adjusted by trial and error for each application. The oil flow requires fluorescent lighting, and photographs can be taken with the tunnel off, but the available time is short as the oil will flow under gravity (Fig. 3.25).

The generally accepted practice when using the older, large, six-strand floss was not to take data when tufts are applied to the model. Figure 3.20 explains why this has been the practice. Both the lift curve slope and maximum lift are greatly reduced. The argument for minitufts is that their effect on the data is minimal, hence they can be left on the model. Oil and china clay also have minimum adverse aerodynamic effects. During this comparison test of tufts at the model's minimum drag (about a lift coefficient of 1.0) the model drag decreased as tufts from the mini to No. 6 floss were added to the left wing with the horizontal tail on and the reverse happened with the tail off. The tufts apparently change the wing's span load distribution. However, old practices die slowly, so force data are often not taken during surface flow visualization. This, then, can lead to an improperly established flow field and the possibility of misleading flow visualization, especially near stall, and oddly enough, near minimum drag. If, however, force data are taken before and during the flow visualization run, the error will be avoided.

The usual method for china clay is to set the model at the desired angle of attack and bring the tunnel up to speed. On a wing with a slotted leading edge and/or trailing edge flaps, this can result in erroneous aerodynamic data and flow visualization due to flow separation in the slots at low Reynolds number

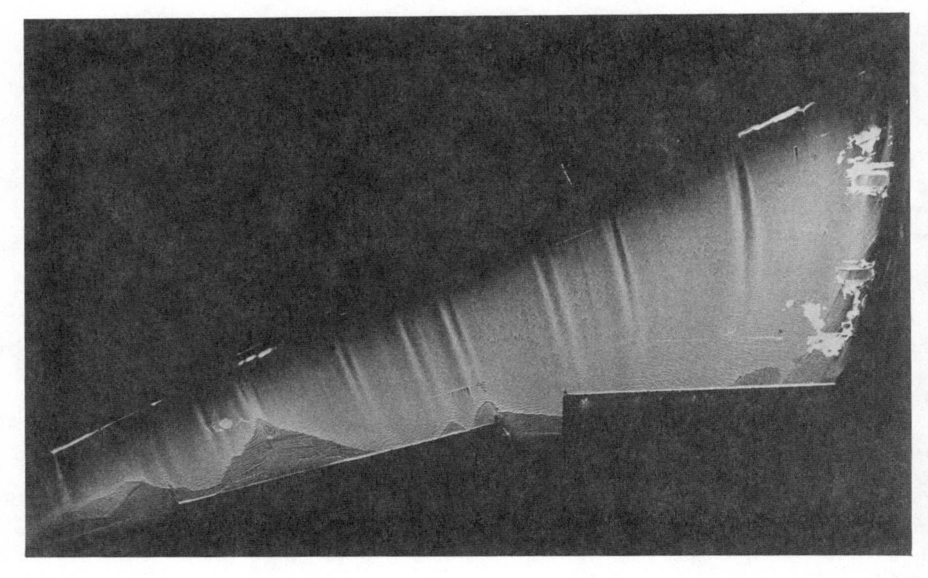

FIGURE 3.25. Oil applied at $\alpha = 0°$, tunnel started, and brought close to speed, model pitched to $\alpha = 27.3°$. (Photograph courtesy of University of Washington.)

during the tunnel acceleration. This is shown by the oil flow and china clay data points beyond $\alpha = 27.3°$ in Fig. 3.20. Similar data were obtained on the clean wing. Figure 3.26 shows a china clay flow visualization for this test method. Similar results were obtained with oil flow.

Figures 3.24 and 3.25 were obtained by setting an α well below stall, starting the tunnel, and at 50–60 mph, pitching the model to the desired α as the tunnel was accelerating to $36q$. For these figures the force data agreed with the clean wing data and the flow is similar to Figs. 3.21 and 3.22 for the mini and No. 60 thread tufts.

The series of photographs in Figures 3.21–3.25 give the reader a comparison of flow visualization using tufts, oil and china clay for the same model at the same conditions, with Fig. 3.26 showing the flow with a stalled leading edge slat.

The transition point between laminar and turbulent flow cannot be determined by tufts and is difficult with china clay. However, both oil flow and sublimation methods will determine the transition point. In one sublimation technique a mixture of naphthalin and a carrier such as fluorine, acetone, or methyl ethyl ketone is sprayed on the model using a standard air spray gun. Note the last two can remove many paints. The operator must wear a respirator mask when doing this. The mixture will leave the wing white, therefore it works best on a black or dark wing. The turbulent boundary layer will scrub the mixture off. In Fig. 7.4 the natural transition is shown on a wing using naphthalin, Fig. 7.5 shows the transition using oil flow, and Fig. 7.6

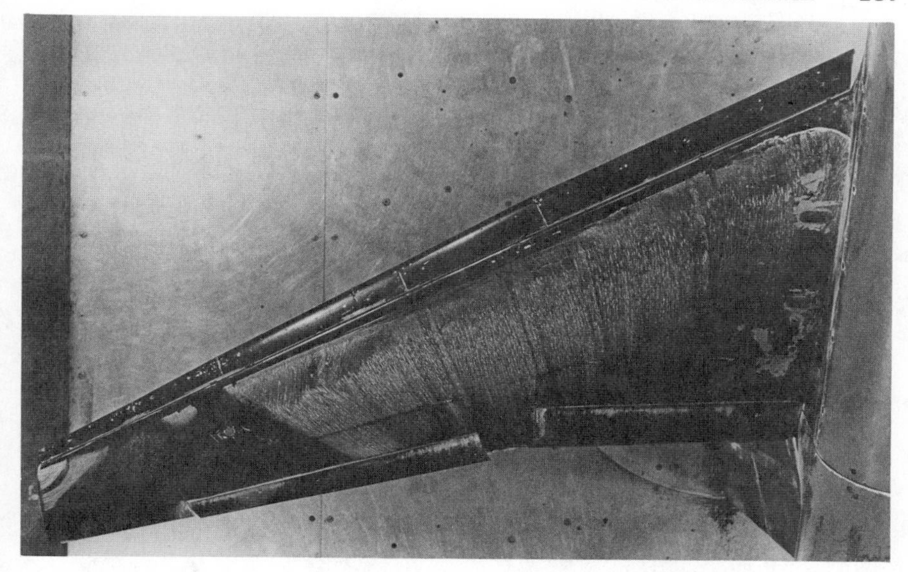

FIGURE 3.26. China clay applied at $\alpha = 27.3°$, tunnel started, and brought to speed. Leading edge slotted flap is stalled. Compare to Figs. 3.21 and 3.24. (Photograph courtesy of University of Washington.)

shows a fixed transition. When oil flow is used, a subcritical bump can cause a wake, which can confuse the transition location. This does not occur with sublimation. The most common material for oil flow is petroleum lubricating oils. These materials are messy to clean up afterward, both on the model and even more so in the tunnel. Another material that works as well as oil when treated with a fluorescent dye is poly glycol. At high C_L's or high surface velocities this material may have too low a viscosity, making it difficult to use. This material can be cleaned up with soap and hot water. When cleaning the tunnel after extensive oil flow runs, a portable set of ultraviolet fluorescent tubes is most useful (Refs. 3.20–3.22).

Off Surface Visualization

The least expensive method is a tuft wand. This is a long tuft on a pole, and it is useful for tracing flow near the model. If it is necessary to put a person in the tunnel, he or she must wear goggles to protect the eyes from dust. It goes without saying that the person should not disturb the flow about the model.

To study the flow, especially the tip vortices, a tufted wire grid is used (Fig. 3.27).

To trace streamlines, helium-filled soap bubbles that have neutral buoyancy can be used. The bubbles are inserted ahead of the model and are photographed with a high intensity light that passes through the tunnel as a

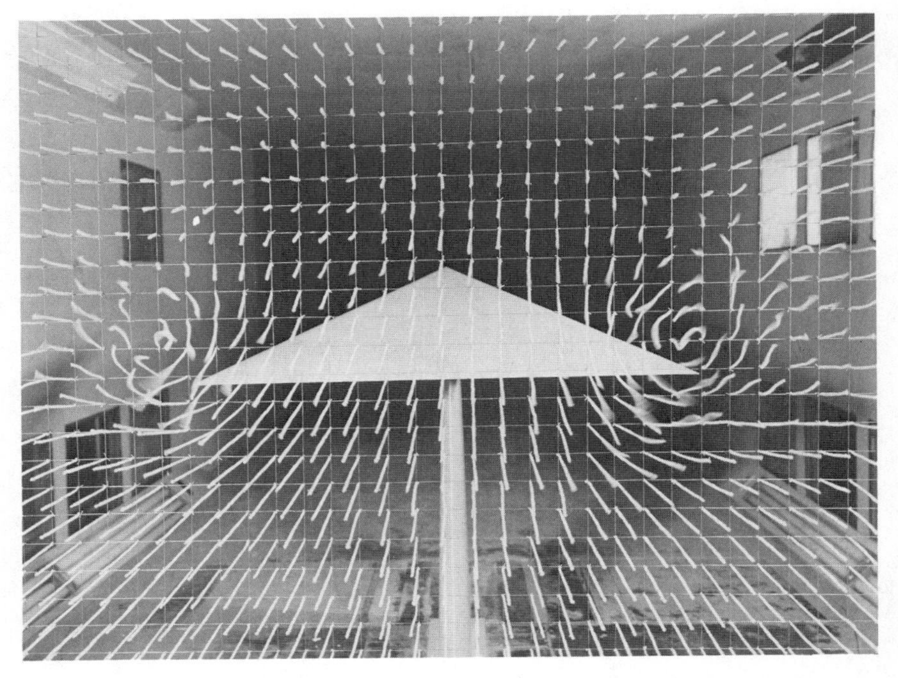

FIGURE 3.27. Flow visualization by the grid-and-tuft method. (Official photograph National Aeronautics and Space Administration.)

plane of light. With proper photographic exposure time the bubbles appear as streaks (of course, the test section is darkened). Maximum tunnel speed for the helium bubbles is about 60 mph. The bubble generators are available commercially.

Another method is smoke, which can be produced in a number of ways. Burning of damp straw, rotten wood, and tobacco to produce smoke are not too satisfactory as those who have tried know. The same is true of pyrotechnic smoke devices (smoke bombs). Chemical methods of producing smoke include both titanium tetrachloride and tin tetrachloride, which produce smoke when brought into contact with damp air. Both compounds are corrosive. A mixture of anhydrous ammonia and hydrogen sulfide also produces smoke, odors, and, if the air is damp, sulfuric acid. Steam and liquid nitrogen produces a good dense smoke with no aftereffects.

A method used at the University of Washington that works at velocities of over 100 mph is a vaporized petroleum product called Type 1962 Fog Juice, which is used in theatrical productions (Fig. 3.28; Ref. 3.23). Other tunnels have had success using very light oils. The smoke generator consists of about 75 in. of 0.060 in. outside diameter (wall = 0.010 in.) stainless-steel tubing. For a probe the tubing is placed inside a 0.375 in. diameter steel tube and held by a collet about 8.0 in. from the end of the tubing, which is bent 90°

FIGURE 3.28. Smoke flow visualization. Helicopter rotor operating in flow breakdown region. (Photograph courtesy of University of Washington.)

about 3 in. from the end. Ceramic beads are used to insulate the stainless tubing from the outer tube. To vaporize the fluid, 10–15 amps are applied to the stainless tubing from the collet to a point about 60.0 in. away inside a nonheat-conducting handle. This allows the stainless tube to expand.

The power unit consists of a variac whose output is connected to the 230 V windings of a 1.5 kVA 115 : 230 transformer. The variac is used to control the temperature on the stainless tubing by applying 0–50 V. The tubing has about 3 ohms resistance, so the current is limited to a maximum of 15 amps.

The fuel reservoir is airtight and has a pressure regulator used to set plant air pressure at about 30 psi to feed the fuel to the probe and a needle valve to control the fuel flow. Plastic tubing connects the reservoir to the probe.

It takes some experience to obtain the desired volume of smoke, which is affected by the air pressure, fuel flow, and voltage. The following values are approximate. With the power switch off, set the variac at 70%, apply 30 psi to the reservoir, and crack open the needle valve. When a small stream of fluid comes out the end, turn on the power. When turning off, cut the power, and when a stream of fluid leaves the tube, shut the needle valve off. This is done to reduce carbon formation in the stainless tubing. If hot fluid is emitted, the temperature is too low; either increase the voltage or reduce the fuel flow. If the smoke pulsates, it is generally a sign that the air pressure is too low.

Another method of producing small discrete filaments of smoke at low velocities is described in Ref. 3.24. This is the smoke wire technique and it appears to be limited to flows where the Reynolds number, based on wire diameter, does not exceed 20, or velocities from 6 to 18 fps. The limit is based on preventing the wake from the wire disturbing the flow behind the wire, and the limit has been determined by experiment. This method uses a small diameter wire that is coated with an oil. The best results were obtained using Life-Like model train smoke, which consists of a commercial-grade mineral oil with small amounts of oil of anise and blue dye added. The liquid-coated wire has 40–80 V ac or 40–60 V dc impressed across it. As the wire is heated fine smoke streak lines form at droplets on the wire (approximately 8 lines/cm for a 0.003-in.-diameter wire). As the wire is heated it expands, and thus sags. This can cause problems with the accurate placement of the streak lines. To avoid this the wire was prestressed to about 1.5×10^5 psi, which is near the yield point of type 302 stainless-steel wire; thus the wire must be handled carefully.

As the oil-treated wire produced smoke for periods of up to 2 s, a timing circuit was used for the lights and cameras used to photograph the smoke. The circuit for the timer is given in Ref. 3.24. This timing device operates from the application of current to the wire. When alternating current was applied to the wire the heating–cooling provided variable density streak lines that have, essentially, timing marks.

Since the smoke wire is limited to low Reynolds number tests it is probably best suited to use in small tunnels as it is difficult to run large tunnels at the required low velocities (see Chapter 8).

Color Displays of Wake Flows

In this method the total pressure distribution is mapped by a total head probe and a pressure transducer. The changes in total pressure are then plotted in color. There are two approaches in use to make these color plots.

The first system by Crowder (Refs. 3.25, 3.26) uses a total pressure probe that extends through the ceiling or wall of the tunnel. The probe is driven externally to the tunnel along a segment of an arc. At the end of each swing along the arc the length of the probe is changed (polar coordinate motion). Thus the probe surveys an area of the tunnel. The pitot is connected to a pressure transducer. The output of the transducer, through suitable electronics, is fed back to a light-emitting diode at the rear of the pitot. For three adjustable ranges of pressure the LED will produce three colors. A camera is mounted downstream of the model. With the tunnel blacked out the camera by remote control takes a time exposure of the probe sweeping out the area in the flow. With an adjustment of the test-section ambient light the model can be superimposed on the flow field photograph.

Other than the electronics to supply the proper current to the LED this system requires no additional data processing and the results are available when the film is developed, or after the run if instant film is used. The multicolor picture then shows the variation in total pressure from which wakes, vortices, etc., can easily be determined. This method trades off resolution in pressure for an increase in spatial resolution of the moving LED.

The second method is similar, with a probe being used to survey an area of the tunnel flow. The probe location and the output of the transducer (hot wire, thin film, total pressure, etc.) is recorded on magnetic tape. The tape is then processed through a computer to yield a digital output. These data are then processed through a second computer program that color codes the data into a set of equal voltage increments. This program creates three standard files for the white light primary colors: red, green, and blue. These files are used in another computer program to drive a color video tube to produce the colored flow field survey (Ref. 3.27).

Holography, Interferometry, Flow Visualization

Lasers with coherent light can be used to produce holographs that can be used for density measurements and flow visualization. These systems have been used to Mach numbers as low as 0.20.

Basically, the laser beam is split into collimated object and reference beams. The object beam passes through the flow and is combined with the reference beam on a photographic film. When the film is developed and illuminated by the reference beam, the information contained in the object beam is reconstructed.

If holographic information is desired, the interference fringe pattern is obtained on the film by exposing the film to the object beam twice. The interference pattern is reconstructed by illuminating the film with the reference beam. To obtain density measurements requires both calibrations and careful optical alignment.

Another variation of holographic technique is to use a double pulsed laser hologram of particles in the flow field. This yields the velocity at many points in the flow field simultaneously (Ref. 3.28).

3.11. SPEED SETTING

The great majority of wind tunnel testing is done at a constant dynamic pressure. It is not practical, however, to insert a pitot-static tube in the test section to measure dynamic pressure because the model pressure field will affect the readings.

The tunnel speed is usually determined by measuring either static or total pressure in the settling chamber ahead of the contraction cone and a static

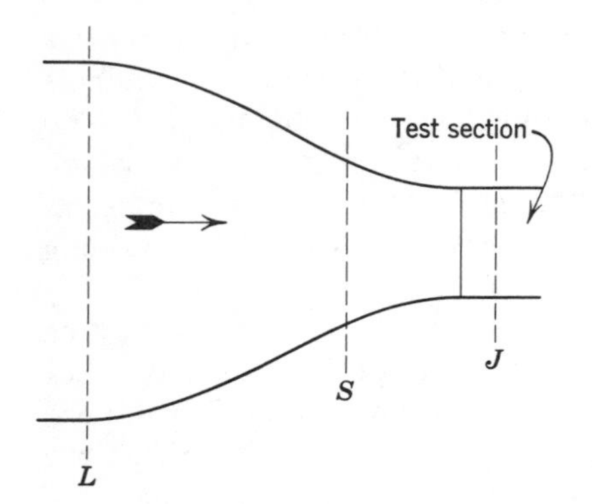

FIGURE 3.29.

pressure ahead of the test section (Fig. 3.29). If honeycombs or screens are used in the settling chamber, the pressure should be read downstream of these devices. Using the subscripts L for the bellmouth or settling chamber and S for the region before the test section, Bernoulli's equation between the two stations will be

$$P_L + q_L = P_S + q_S - K_1 q_S \tag{3.9}$$

where $K_1 q_S$ allows for the small loss in total pressure between the bellmouth and the test section, and P = the static pressure and q = the dynamic pressure. Assuming incompressible flow the continuity equation between the two stations is (A is the area):

$$A_L V_L = A_S V_S \tag{3.10}$$

Squaring, multiplying by $\rho/2$ and solving for q_L:

$$\frac{\rho}{2} V_L^2 = q_L = \left(\frac{A_S}{A_L}\right)^2 \frac{\rho}{2} V_S^2 = \left(\frac{A_S}{A_L}\right)^2 q_S = K_2 q_S \tag{3.11}$$

Rewriting Bernoulli's equation and substituting for q_S above:

$$P_L - P_S = q_S(1 - K_1 - K_2) \tag{3.12}$$

If station S does not have the same cross-sectional area as the test section (station J), the continuity equation can be used to yield:

$$q_S = \frac{\rho}{2} V_S^2 = \left(\frac{A_J}{A_S}\right)^2 \frac{\rho}{2} V_J^2 = K_3 q_J \tag{3.13}$$

Thus

$$P_L - P_S = (1 - K_1 - K_2) K_3 q_J \tag{3.14}$$

In this form the constant K_1 can include the small additional loss in total pressure from station S to the jet.

The actual calibration of the tunnel is accomplished by running the tunnel at various values of $P_L - P_S$ while measuring the dynamic pressure in the test section with a calibrated pitot-static tube. This evaluates $(1 - K_1 - K_2)K_3$. Because the velocity is not uniform throughout the test section, it is necessary to survey the test section. As a minimum, the area in the vertical plane of the model trunnion should be surveyed. A better approach would be to survey a volume of the test section that is occupied by the model. If the pitot-static tube used has a hemispherical nose, it can be equipped with two yawheads at 90° to each other, and the distribution of the upflow and cross flow can be obtained simultaneously.

The survey of the test section can be done with a simple pitot-static tube if there is a method to position it throughout the test section, or a rake of pitot statics can be used. If there is no method of remotely positioning the pitot-static tube, the survey of the test section becomes a very tedious operation.

If a volume of the test section is surveyed and there are large differences in the distribution of the dynamic pressure, it may be desirable to use a weighted average of the measured dynamic pressure.

When the dynamic pressure calibration is completed, there is a relation established between the indicated dynamic pressure $(P_L - P_S)$ and/or the actual dynamic pressure q_J. There usually are a series of these calibrations for different test-section configurations. If the tunnel has an external balance, calibrations are often made with and without the balance struts and their fairings. If a ground plane is used often, calibrations are made for various heights of the ground plane.

It also is desirable for each test-section configuration to have the distribution of the total pressure, static pressure, upflow and cross flow throughout the region occupied by the model. When the tunnel has an air exchanger or heat exchanger, the temperature distribution should also be measured.

The static sources for the tunnel speed control should be, ideally, either a ring around a tunnel station or at least a portion of a ring on the two side walls or the ceiling and floor. If the ceiling and floor are used, then the effect of the wing's bound vortex tends to cancel out. The static source should consist of either a series of static sources or a sealed tube flush with the surface with many holes evenly spaced along its length. The tubes or multiple static sources of the ring should be manifolded together to yield an average static pressure at the station. If total pressure is used in the contraction cone, it is desirable to have multiple sources also.

The two static pressures in the simplest case can be connected to a U-tube manometer that is used to run the tunnel. It is more desirable to use either a bellows-activated mechanical balance with preset weights and a null voltage output when on dynamic pressure, or a high-quality differential pressure transducer. Then the electrical output can be transmitted to any location convenient for the tunnel operator, and if a digital data system is used, the output is fed to the computer.

One note of caution: extreme care must be taken to make sure that there are no leaks in the tunnel dynamic pressure system. In large tunnels, the leak checking can be a time-consuming process. Thus, if possible, the system should not be disturbed once it has been leak checked. It is often desirable to record the static pressure at either of the two tunnel stations. Additional static sources for this use should be provided rather than tapping into the tunnel dynamic pressure system.

It is also desirable to have a simple method of periodically checking the system for leaks. One way to do this is to use a calibration wing that can be installed easily. The wing can be run through a pitch series at several dynamic pressures. If the slope of the lift curve does not change, there is no change in the dynamic pressure calibration. This has the further advantage of also checking the tunnel upflow, because if the drag polar does not rotate, the upflow has not changed (Section 4.16).

3.12. VELOCITY VARIATION IN THE JET

The variation of the dynamic pressure, $q = (\rho/2)V^2$, may be measured across the test section by means of a pitot-static tube. The local velocities may then be obtained from

$$V = \sqrt{2q/\rho} \tag{3.15}$$

For the velocity survey the pitot-static tube is moved around the jet, and the dynamic pressure is measured at numerous stations. The velocities as calculated from the dynamic pressures or the pressures themselves are then plotted, and the points are connected by "contour" lines of equal values. The variation of q in the working range of the jet should be less than 0.50% from the mean, which is a 0.25% variation in velocity.

A plot of the dynamic pressure distribution in a rectangular test section is shown in Fig. 3.30. Of interest is the asymmetry usually found, and the maximum variation well above satisfactory limits. The survey should have been carried to the walls.

The correction of an excessive velocity variation is not as serious a problem as the correction of excessive angular variation. There are more methods of attack, for one thing, and less probability that the variation will change with tunnel speed, for another.

It is not correct to think of the tunnel as having uniform flow. The same particles of air do not reappear in a plane of the testing section. Slowed by the wall friction, the particles closest to the walls are constantly being overtaken and passed by the particles of the central air. The greater loss near the walls would be expected to yield a lowered velocity near the perimeter of the test section, and doubtless this would occur if the contraction cone did not tend to remove such irregularities.

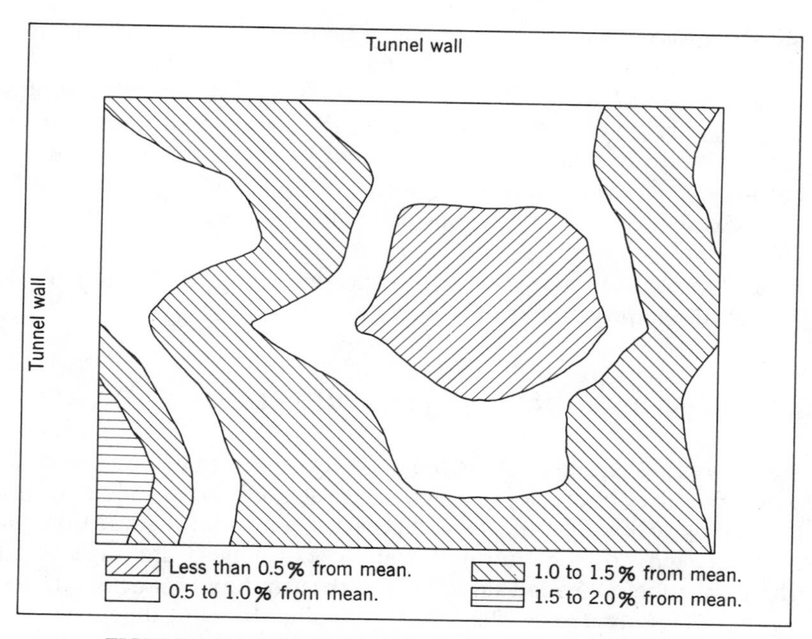

FIGURE 3.30. Distribution of test-section dynamic pressure.

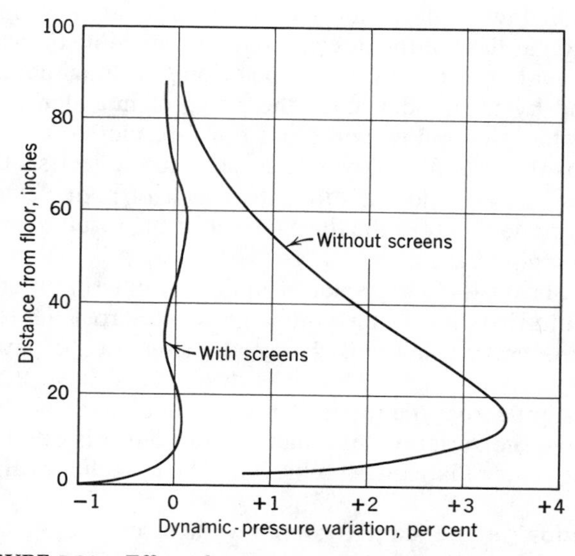

FIGURE 3.31. Effect of screens on velocity distribution in the jet.

If satisfactory velocity distribution is not obtained, there still remain several minor adjustments for improving the situation. As the unsatisfactory flow may be caused by local flow separations, or improperly set turning vanes, or if the velocity variation is annular, the source may be the propeller. Problems in these parts of the tunnel should be corrected (see Chapter 2). Screens may be added in the largest section of the tunnel, so located radially that they cover the sections that correspond to high-velocity regions in the jet. The improvement in velocity distribution by such screens is shown in Fig. 3.31. The loss in energy ratio they cause is quite small and is far outweighed by the improvement in testing conditions.

3.13. LONGITUDINAL STATIC-PRESSURE GRADIENT

The static-pressure gradient along the test section must be known in order to make the necessary buoyancy corrections. (See Sections 6.3 and 6.9.) It may be obtained by reading the local static pressure with a pitot-static tube that is progressively moved from entrance cone to exit cone. Care must be taken that the pitot tube is headed directly into the wind and that no extraneous static pressure is created by the bracket holding the pitot tube.

Perhaps a more convenient method is to use a long static tube (Section 3.2).

3.14. ANGULAR FLOW VARIATION IN JETS

The variation of flow angle in the jet can be measured by many devices (Section 3.4). Regardless of the device used it is desirable to map the upflow and cross flow in a series of traverse planes along the longitudinal axis in the region occupied by the model, over the range of intended dynamic pressures. Often, when the upflow and cross flow are plotted as flow direction vectors, regions of vortexlike flow can be seen in the test section. As indicated in Chapter 2, such flow is often the result of poor velocity distributions in the return duct before the third corner or the result of improperly set fourth corner turning vanes.

A variation across the wing span of upflow results in an effective twist being imparted to the wing. A cross-flow gradient across the test section in the region of the vertical tail will change the slope of the yawing moment versus side slip or yaw angle. Thus it is desirable to have the upflow and cross flow constant across the tunnel. This is difficult to achieve. It would be desirable to have the variation less than $\pm 0.10°$, but it is often necessary to accept the best values that can be achieved. The maximum variation should be held to $\pm 0.20°$.

The correction of the data for upflow is discussed in Sections 4.16 and 6.30.

3.15. TURBULENCE SPHERE

The disagreement between tests made in different wind tunnels at the same Reynolds number and between tests made in wind tunnels and in flight indicated that some correction was needed for the effect of the turbulence produced in the wind tunnel by the propeller, the guide vanes, and the vibration of the tunnel walls. It developed that this turbulence caused the flow pattern in the tunnel to be similar to the flow pattern in free air at a higher Reynolds number. Hence the tunnel test Reynolds number could be said to have a higher "effective Reynolds number."

For a sphere it has been experimentally verified that the Reynolds number at which the drag coefficient decreases rapidly depends strongly on the degree of turbulence in the wind tunnel. The Reynolds number at which the reduction occurs decreases with increasing tunnel turbulence. This can be seen on physical grounds because a high turbulence leads to transition at a lower Reynolds number. The point of separation shifts aft on the sphere yielding a smaller wake and less drag. Flight measurements on spheres (Ref. 3.29) show that in the free atmosphere the critical Reynolds number is independent of the turbulence structure, the Reynolds number being 3.85×10^5. This value is larger than many wind tunnels achieve, although many low-turbulence tunnels approach this value. In the atmosphere the turbulent eddies are so large relative to the sphere that they do not affect the thin boundary layer of the sphere.

Before the now common use of hot-wire anemometry a turbulence sphere was used to measure the relative turbulence of a wind tunnel (Fig. 3.32). The critical Reynolds number for the sphere can be measured two ways. One method is to plot the measured C_D based on cross-sectional area versus Reynolds numbers (Fig. 3.33). From the plot read the Reynolds number in the tunnel for $C_D = 0.3000$. The second method is to take the average of the four pressures on the aft surface of the sphere and subtract this value from the stagnation value at the leading edge of the sphere yielding a ΔP. Plot $\Delta P/q$ versus Reynolds number for the sphere and find the Reynolds number for $\Delta P/q = 1.220$ (Fig. 3.34). The pressure method has certain advantages. It needs no drag balance with the associated balance calibration and no evalua-

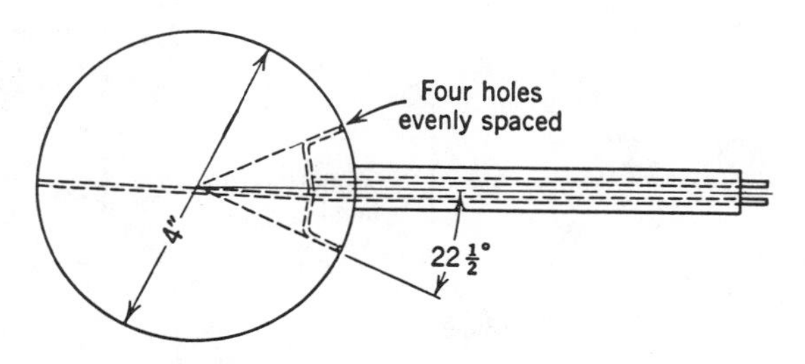

FIGURE 3.32. Turbulence sphere.

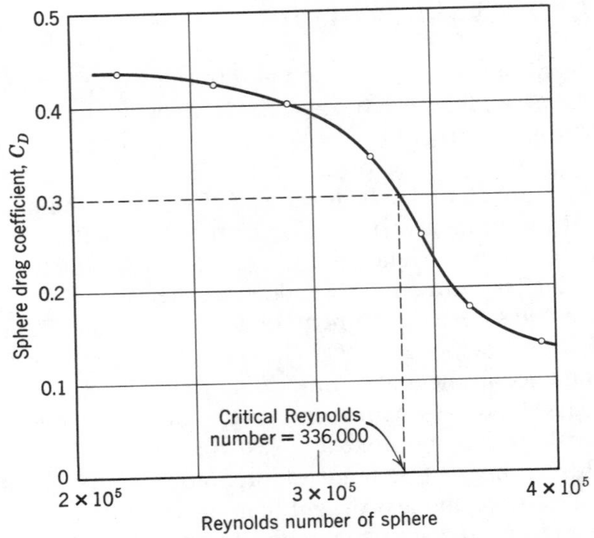

FIGURE 3.33. Variation of drag coefficient with Reynolds number for a sphere.

tion of support tares for the portion of the support sting exposed to the airstream (Section 4.17). Also, the sphere support sting can be stiffer as no deflection is needed by the drag balance.

The critical Reynolds number as defined by the sphere by either force or pressure measurements is then used to define a turbulence factor for the tunnel by comparing the tunnel's critical Reynolds number to the atmospheric free air Reynolds number:

$$TF = 385{,}000/RN_{\text{tunnel}} \tag{3.16}$$

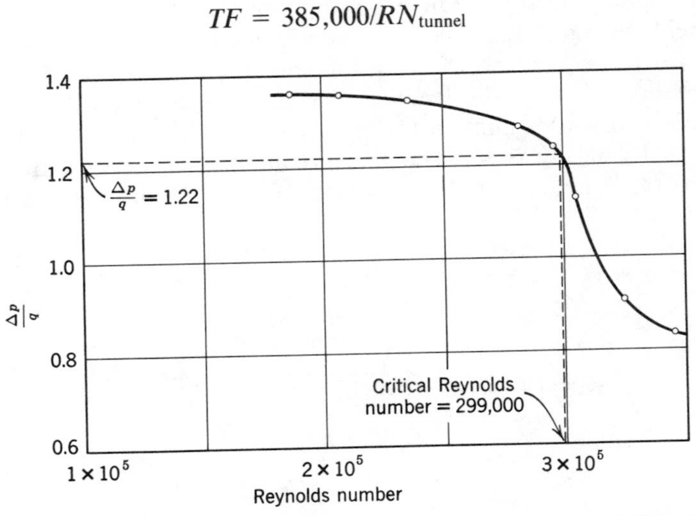

FIGURE 3.34. Variation of pressure coefficient with Reynolds number for a sphere.

Then the effective test Reynolds number is defined by

$$RN_e = TF \times RN_{\text{test}} \tag{3.17}$$

The use of a turbulence sphere yields what may be thought of as an average value of tunnel turbulence. It does not give any information on the magnitude of turbulence in either the axial or lateral direction. The use of a turbulence sphere may, however, prove to be a simple method of monitoring any change in tunnel turbulence. Its use requires no prior calibrations and the installation and running in a tunnel can be designed to be simple and quick.

The relation between the critical Reynolds number of a sphere and turbulence is shown in Fig. 3.35. This is from the work of Dryden, Kuethe, et al. (Refs. 3.30 and 3.31).

Turbulence spheres can be made from cue, duck, and bowling balls. Several sizes are needed to enable the turbulence factor to be measured at different tunnel air speeds.

A brief examination of Eq. (3.16) might lead to the conclusion that the higher the turbulence the better the tunnel, as the effective Reynolds number of the test would be higher. This correction is not exact and if the tunnel has excessive turbulence, the model may have a premature transition from laminar to turbulent flow, which can be critical for laminar flow airfoils.

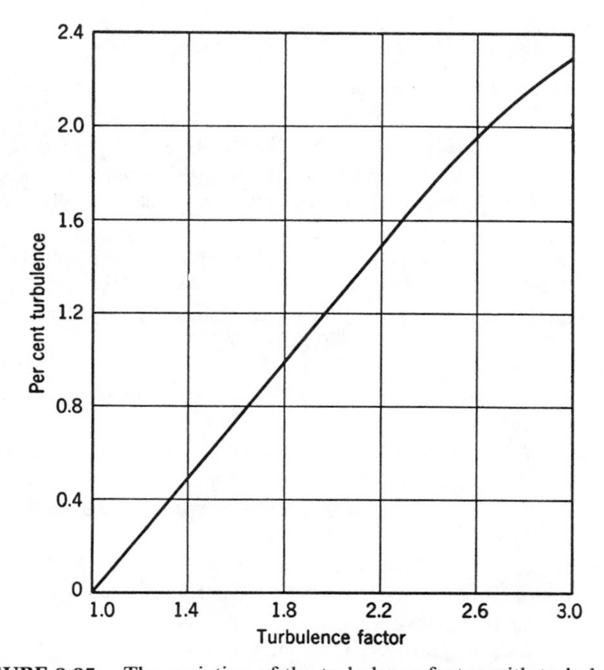

FIGURE 3.35. The variation of the turbulence factor with turbulence.

However, low-speed models are often equipped with trip strips that fix the transition point on the model (Section 7.2), which may reduce the requirement for extremely low turbulence. This requirement for low test-section turbulence is not as severe for small student tunnels as it would be for larger tunnels that are used for research or development tests. See Section 2.11 for turbulence reduction by the use of screens and honeycombs.

3.16. SLIP RINGS

To transfer data from a rotating rotor, propellers, or rotating balance requires the use of slip rings. Poor signals are obtained if extreme care is not taken to keep the slip rings clean. Assuming that the slip ring assembly is carefully manufactured, it can be kept clean by flushing the slip ring brush area on a periodic basis. The Army at Ames had trouble-free operation (over 20 h) by using a mixture of Exxon 2380 (a synthetic oil) mixed 1 part to 9 parts of Freon. This solution was forced into the contact area and allowed to drain.

Slip rings are made of "coin" silver (a copper/silver alloy) and silver graphite brushes with about 30 psi brush pressure. The ring surface must be concentric with the shaft and have a polished surface.

REFERENCES

3.1. E. R. Spaulding and Kenneth G. Merriam, Comparative Tests of Pitot-Static Tubes, *TN* 546, 1935.

3.2. G. Kiel, A Total-Head Meter with Small Sensitivity to Yaw, *TM* 775, 1935.

3.3. W. Luchuk, C. F. Anderson, and M. L. Homan, 1981 Calibration of the AEDC-PWT Aerodynamic Wind Tunnel (4T) at Mach Numbers from 0.1 to 1.3, *AEDC TR* 82-10, 1982.

3.4. L. N. Krause, Effects of Pressure Rake Design Parameters on Static Pressure Measurements for Rakes Used in Subsonic Free Jets, *TN* 2520, 1951.

3.5. A. Silverstein, Determination of Boundary-Layer Transition on Three Symmetrical Airfoils in the NACA Full-Scale Wind Tunnel, *TR* 637, 1939.

3.6. G. W. Brune, D. A. Sikavi, E. T. Tran, and R. P. Doerzbacher, Boundary Layer Instrumentation for High-Lift Airfoil Models, Paper 82-0592, AIAA 12th Aerodynamic Testing Conference, 1982, pp. 158–165.

3.7. J. H. Preston, The Determination of Turbulent Skin Friction by Means of Pitot Tubes, *JRAS,* **58,** 109–121, February 1954.

3.8. V. C. Patel, Calibration of the Preston-Tube and Limitations on Its Use in Pressure Gradients, *J. Fluid Mechanics,* **23,** 185–208, 1965.

3.9. A. Quarmby and H. K. Das, Measurement of Skin Friction Using a Rectangular Mouthed Preston Tube, *JRAS* 73, 288–330, March 1969.

3.10. A. Quarmby and H. K. Das, Displacement Effects on Pitot Tubes with Rectangular Mouths, *Aeronautical Quarterly,* 129–139, May 1969.

3.11. F. A. MacMillan, Experiments on Pitot-Tubes in Shear Flow, *ARC R & M* 3028, 1957.

3.12. J. M. MacMillan, Evaluation of Compressible-Flow Preston Tube Calibrations, *NASA TN* D-7190, 1973.

3.13. J. M. Allen, Reevaluation of Compressible-Flow Preston Tube Calibrations, *NASA TM* X-3488, 1977.

3.14. M. Lapp, C. M. Penney, and J. A. Asher, Application of Light Scattering Techniques for Measurements of Density, Temperature, and Velocity in Gasdynamics, Aerospace Research Laboratories Report 73-0045, April 1973.

3.15. F. Durst, A. Melling, and J. H. Whitelaw, *Principles and Practice of Laser-Doppler Anemometry,* Academic Press, London, 1976.

3.16. J. F. Meyers and W. V. Feller, Processing of the Laser Doppler Velocimeter Signals, Paper, 5th International Congress on Instrumentation in Aerospace Simulation Facilities, Pasadena, September 1973.

3.17. NATO-AGARD Conference on Applications of Non-Intrusive Instrumentation in Fluid Flow Research, AGARD CP-193, St. Louis, France, May 1976.

3.18. *Ultraviolet & Fluorescence Photography,* M-27, Eastman Kodak Co., Rochester, New York.

3.19. J. P. Crowder, Add Fluorescent Mini-Tufts to the Aerodynamicists' Bag of Tricks, *Aeronautics and Astronautics,* **18,** 54–56, November 1980.

3.20. J. D. Main-Smith, Chemical Solids as Diffusable Coating Film for Visual Indication of Boundary-Layer Transition in Air and Water, *ARC R & M* 2755, 1950.

3.21. G. W. Brune, Topology in Aerodynamics—An Aid for the Interpretation of Flow Patterns, Boeing Document D6-51878.

3.22. W. Merzkirch, *Flow Visualization,* Academic Press, New York, 1974.

3.23. S. Shindo and O. Brask, A Smoke Generator for Low Speed Wind Tunnels, *Technical Note* 69-1, University of Washington, Dept. of Aeronautics and Astronautics, February 1969.

3.24. S. M. Batill and T. J. Mueller, Visualization of Transition in the Flow over an Airfoil Using the Smoke-Wire Technique, *AIAA Journal,* **19,** 340–345, March 1981.

3.25. J. P. Crowder, E. G. Hill, and C. R. Pond, Selected Wind Tunnel Testing Developments at Boeing Aeronautical Laboratory, AIAA Paper No. 82-0458.

3.26. J. P. Crowder, Quick and Easy Flow Field Surveys, *Aeronautics and Astronautics,* **18** 38–39, 45, October 1980.

3.27. A. E. Winkelmann and C. P. Tsao, A Color Video Display Technique for Flow Field Surveys, Paper 83-0611, AIAA 12th Aerodynamic Testing Conference, 1982, pp. 305–311.

3.28. F. C. Jahoda, Pulse Laser Holographic Interferometry, *Modern Optical Methods in Gas Dynamic Research,* D. S. Dosanjh (ed.), Plenum Press, New York, 1971.

3.29. C. B. Millikan and A. L. Kline, The Effect of Turbulence, *Aircraft Engineering,* 169, August, 1933.

3.30. H. L. Dryden and A. M. Keuthe, Effect of Turbulence in Wind Tunnel Measurements, *NACA Report* 342, 1929.

3.31. H. L. Dryden, et al., Measurements of Intensity and Scale of Wind-Tunnel Turbulence and Their Relation to the Critical Reynolds Number of Spheres, *NACA Report* 581, 1937.

CHAPTER 4

Model Force, Moment, and Pressure Measurements

The purpose of the load measurements of the model is to make available the forces, moments, and pressures so that they may be corrected for tunnel boundary and scale effects (Chapters 6 and 7), and utilized in predicting the performance of the full-scale airplane.

The loads may be obtained by any of three methods: (1) measuring the actual forces and moments with a wind tunnel balance; (2) measuring the effect that the model has on the airstream by wake surveys and tunnel-wall pressures; or (3) measuring the pressure distribution over the model by means of orifices connected to pressure measuring devices.

These methods are considered in detail in the following sections.

4.1. BALANCES

Besides lift, drag, and pitching moment, the airplane is subjected to rolling moment, yawing moment, and side force. This makes a total of six measurements in all: three forces, mutually perpendicular, and three moments about mutually perpendicular axes. The wind tunnel balance must separate these forces and moments and accurately present the small differences in large forces, all without appreciable model deflection. Furthermore, the forces and moments vary widely in value. It is seen that the balance becomes a problem that should not be deprecated; in fact, it might truthfully be said that the balance design is among the most trying problems in the field.

In order to picture the situation most clearly, an impractical wire balance based on readings made with spring scales is shown in Fig. 4.1. The model, supposedly too heavy to be raised by the lift, is held by six wires, and six forces are read by the scales A, B, C, D, E, and F:

1. Since the horizontal wires A, B, and F cannot transmit bending, the vertical force (the lift)—$L = C + D + E$.
2. The drag $D = A + B$.
3. The side force $Y = F$.
4. If there is no rolling moment, scales C and D will have equal readings. A rolling moment will appear as $RM = (C - D) \times b/2$.
5. Similarly, the yawing moment $YM = (A - B) \times b/2$.
6. The pitching moment $M = E \times c$.

Exact perpendicularity between the components must be maintained. For instance, if the wire to scale F (Fig. 4.1) is not exactly perpendicular to wires A and B, a component of the drag will appear (improperly, of course) as side force. A similar situation exists in regard to lift and drag and lift and side force. Since the lift is the largest force by far in conventional wind tunnel work, extreme care should be taken to ensure its perpendicularity to the other components.

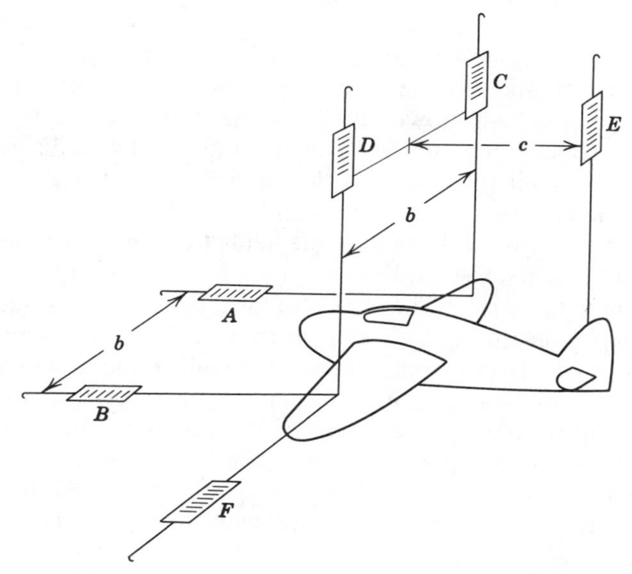

FIGURE 4.1. Diagrammatic wind tunnel balance.

Before we proceed to more complicated balances, Fig. 4.1 should be studied until a clear picture is obtained of the forces and moments to be measured.

Two fundamental types of balances are in general use: external balances, which carry the loads outside the tunnel before they are measured, and internal balances, which fit into the models and are arranged to send data out through electrical wires. Both types have their advantages, and few tunnels can get along with either one alone.

Four types of external balances are in general use, each possessing certain advantages over the others. These balances are named from their main load-carrying members—wire, platform, yoke, and pyramidal—and are discussed in the following paragraphs.

4.2. WIRE BALANCES

One of the earliest types of wind tunnel balances was a wire balance similar to Fig. 4.1. The spring scales were not used for the balance output since their deflections could change the angle of attack. Usually the model was mounted inverted so that lift adds to the weight to prevent unloading the wires. With this type of balance the accurate determination of the large tare drag of the wires is difficult. The wires tend to break, which can lead to the loss of the model, and, in general, this type is no longer in use.

4.3. EXTERNAL BALANCES

Currently, most external balances are of the strut type. These balances support the model and provide for changing the angle of attack and yaw, and transmit the model loads down into a system of linkages that separate them into their proper components. Such an apparatus is shown diagrammatically in Fig. 4.2, and a linkage system is shown in Fig. 4.3. The general massiveness of a balance structure may be seen in Fig. 4.4.

Tracing the pathway followed by the loads from model to measuring unit (Fig. 4.2), we see first the model is supported on two front load members or "struts," and a tail strut.* The struts, in turn, connect to the inner part of a floating ring frame that is free to turn (model yaw), and a mechanism is provided to raise or lower the tail strut to produce model pitch. The outer part of the floating frame is held in place by a system of struts that are specially designed to be strong in tension and compression but very weak in bending. These struts separate the components of the load by means of a linkage system and feed them into the measuring units. Above the floating frame is a fairing turntable on which the windshields for the load members

* This is the most frequently used arrangement. Others are discussed subsequently.

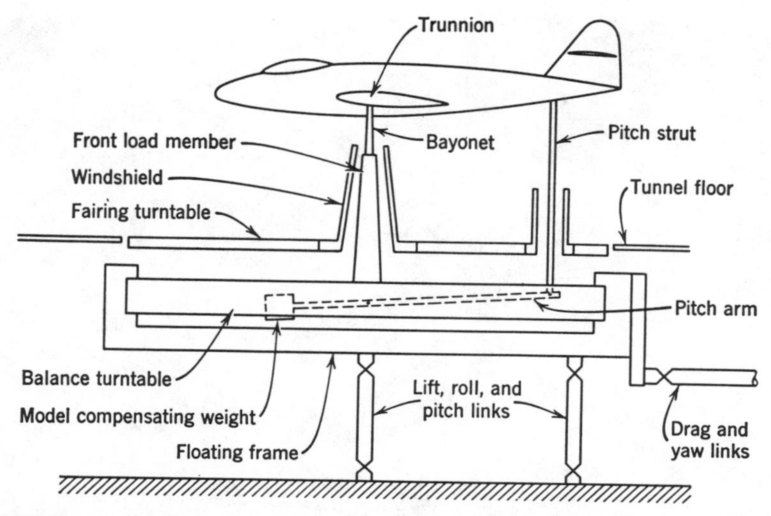

FIGURE 4.2. Diagrammatic sketch (greatly simplified) of some balance components.

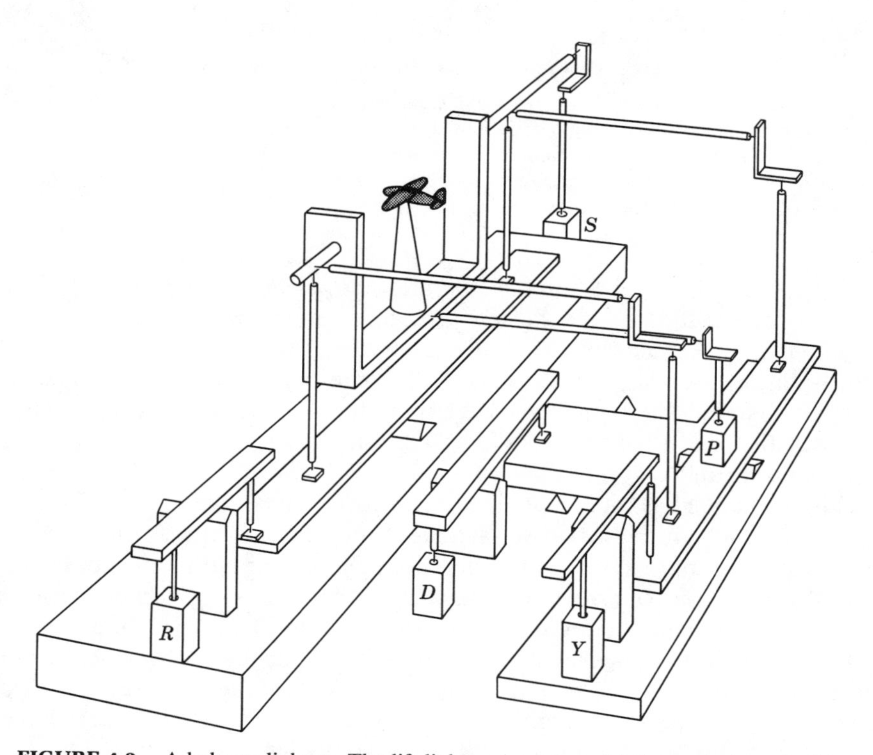

FIGURE 4.3. A balance linkage. The lift linkage (not shown) is beneath the roll table.

FIGURE 4.4. The massiveness of a wind tunnel balance is well illustrated by this photograph of a balance designed for a 150-mph wind tunnel with a 9-ft-diameter test section. During this early setup in the factory the load members have been dropped in place without going through their respective windshield support bases. As shown, the balance has approximately 45° negative yaw. (Courtesy Georgia Institute of Technology.)

are mounted. The load turntable operates the fairing turntable by closing microswitches, and, as the fairing turntable rotates, the windshields are gear-driven to remain parallel to the airstream. In some balances the tail-strut fairing is moved up and down to keep the exposed length of tail strut always constant. The windshields are in addition insulated, and upon contact with the load members they activate fouling lights so that the trouble may be noted and corrected.

Thus a balance has three main identification features: the manner (number of struts) by which the model is fastened to the floating frame, the type of linkage system that separates the components, and the type of measuring unit. Normally the accuracy of a balance is a function of the deflections permitted and the number of measuring units needed to read one component. Thus, if three lift readings must be added to get lift, the error from measuring units alone may be three times that which will occur if the three readings are added in the linkage system and fed into one measuring unit. Linkages have no relative motion and are hence virtually frictionless.

The linkage system by which the force and moments are separated nave gradually worked into three* different fundamental types, each possessing some advantages over the others. These are named *platform, yoke,* and *pyramidal,* according to the manner in which the main system is assembled. They will be discussed further in the following paragraphs. In addition to the above types of balances, there is also an internal type, which fits into the model and is widely used in high-speed applications. (See Section 4.12.)

4.4. PLATFORM BALANCE

The platform balance (Fig. 4.5) utilizes either three or four legs to support the main frame. For the three-legged type, the forces and moments are:

$$\begin{aligned}
\text{Lift} &= -(A + B + C) \\
\text{Drag} &= D + E \\
\text{Side force} &= -F \\
\text{Rolling moment} &= (A - B)(l/2) \\
\text{Yawing moment} &= (E - D)(l/2) \\
\text{Pitching moment} &= C \times m
\end{aligned}$$

Platform balances are widely used. Rugged and orthogonal, they may be constructed and aligned with a minimum of difficulty. But they also have disadvantages: (1) the moments appear as small differences in large forces, an inherently poor arrangement; (2) the balance resolving center is not at the model, and the pitching moments must be transferred; and (3) the drag and side force loads put pitching and rolling moments on the load ring. These interactions must be removed from the final data.

4.5. YOKE BALANCE

The yoke balance (Fig. 4.6) offers an advantage over the platform balance in that moments are read about the model. However, the inherent design of the yoke leads to bigger deflections than the platform balance, particularly in pitch and side force. Because the balance frame must span the test section in order to get the two upper drag arms in their proper positions, the yaw lever arm is exceptionally long. The high supporting pillars are subject to large deflections. Once again the final forces must be summed up: the drag is the addition of three forces, and the lift is the sum of two. The yoke balance

* There are many others.

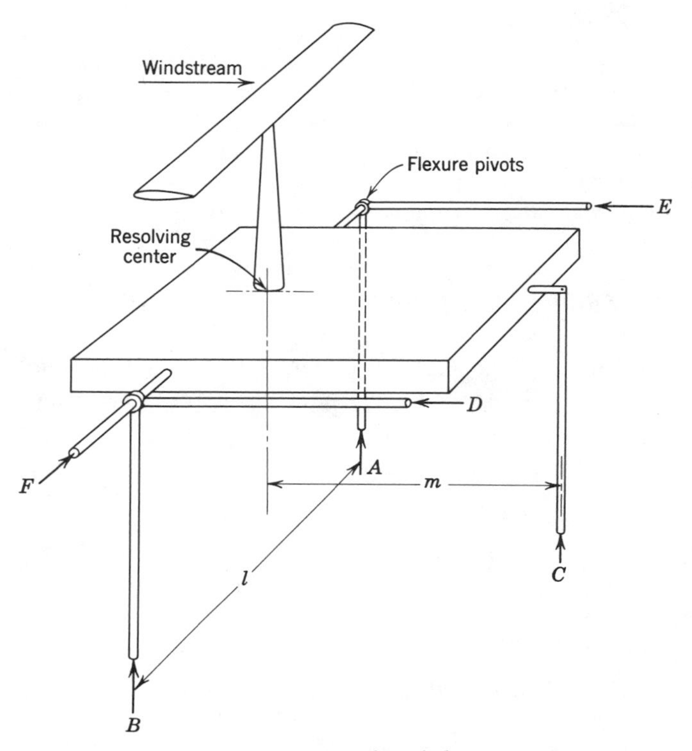

FIGURE 4.5. Platform balance.

brings out the pitching moment in the drag system instead of in the lift. For the yoke balance, the forces and moments are:

$$\text{Lift} = -(A + B)$$

$$\text{Drag} = C + D + E$$

$$\text{Side force} = -F$$

$$\text{Rolling moment} = (B - A)(l/2)$$

$$\text{Pitching moment} = -E \times m$$

$$\text{Yawing moment} = (D - C)(l/2)$$

4.6. PYRAMIDAL BALANCE

The complaints usually heard against the platform and yoke balances are largely overcome by the ingenious engineering of the pyramidal type. However, as usually happens, additional difficulties are added.

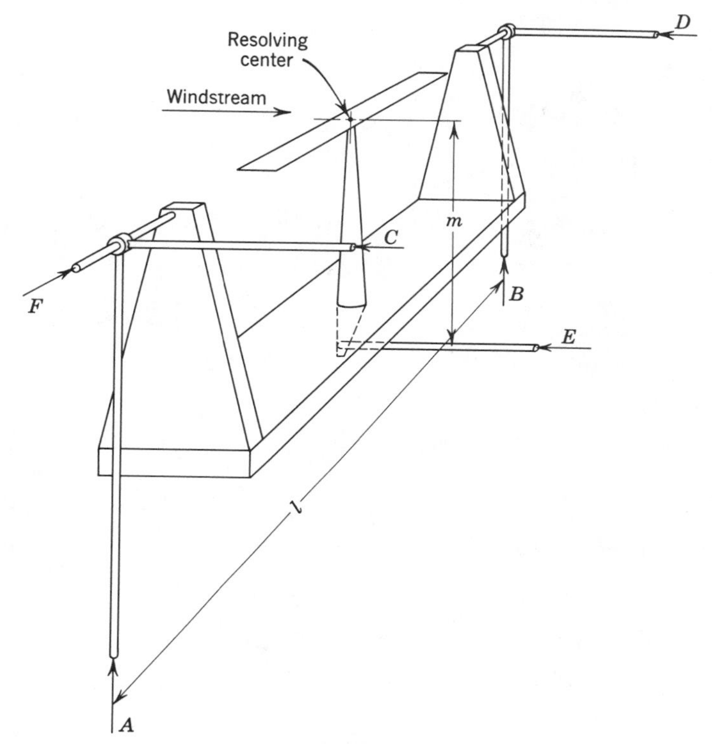

FIGURE 4.6. Yoke balance.

These are the advantages: the pyramidal balance reads the moments about the resolving center, and the six components are inherently separated and read directly by six measuring units. No components need be added, subtracted, or multiplied. The difficulties involved in reading the small differences in large forces are eliminated, and direct reading of the forces and moments simplifies the calculations.

Several criticisms of the pyramidal balance are warranted. The alignment of the inclined struts is so critical that both the construction and the calibration of the balance are greatly complicated. Furthermore (and this appears quite serious), deflections of the inclined struts may so change their alignment that the moments are not accurate. This effect must be thoroughly investigated during the calibration of the balance.

The manner in which the pyramidal balance separates the moments is not simple, and it behooves the student to approach the setup using an elementary truss system. Consider a truss in which two legs are jointed (Fig. 4.7). The force D, acting through the pin joint O, produces only tension in OE and compression in OF. No force is registered at A. However, the force G, not acting through O, produces bending in OE, and OE would collapse unless

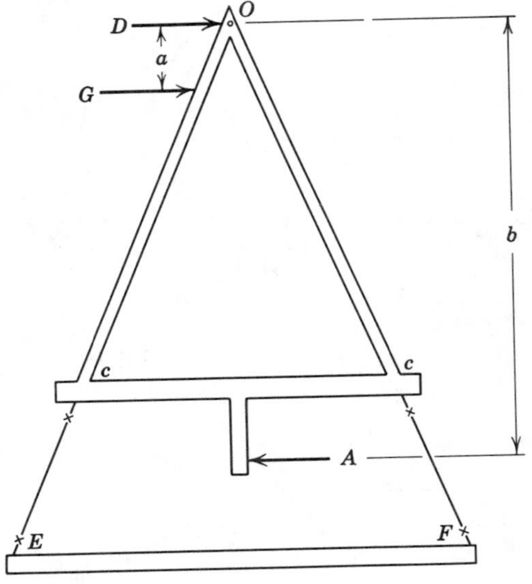

FIGURE 4.7

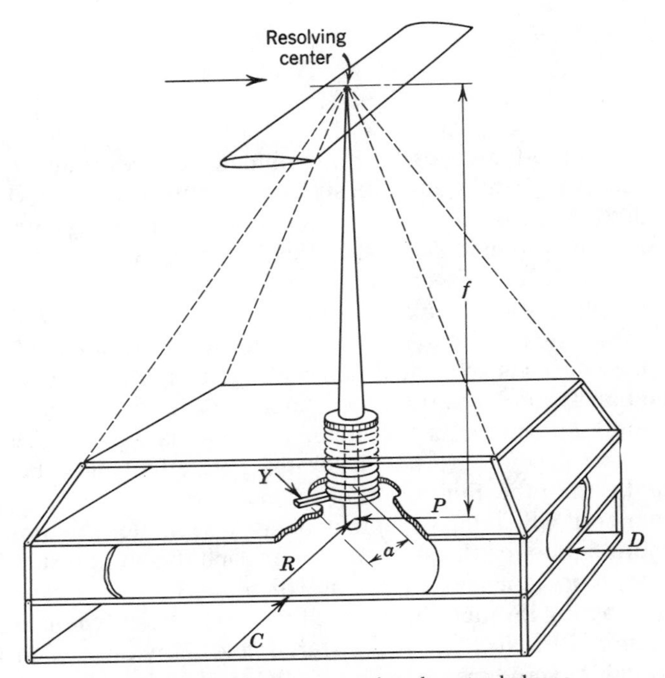

FIGURE 4.8. Pyramidal or virtual center balance.

the force $A = aG/b$ were present. If G and b are known, the size of the force A determines the point of action of G. In this manner, if G were a known drag force, its pitching moments about the resolving center O would be determined by the force A.

Though the previous example illustrates the principle of the pyramidal balance, in actual practice a considerable revision is required. In order to prevent the legs of the pyramid from being in the airstream, they are cut off at what would be c in Fig. 4.7. The truncated legs are then carefully aligned so that their extensions pass through a common point. The complete setup is illustrated in Fig. 4.8. The forces and moments are:

$$\begin{aligned} \text{Lift} &= \text{total weight on lowest table} \\ \text{Drag} &= D \\ \text{Side force} &= -C \\ \text{Pitching moment} &= -P \times f \\ \text{Rolling moment} &= R \times f \\ \text{Yawing moment} &= Y \times a \end{aligned}$$

4.7. BALANCE DESIGN

We cannot begin a discussion of wind tunnel balance design without a note of warning to those who might undertake the job with little experience. A wind tunnel balance is an immensely complicated apparatus, and its design and construction are much better left to balance engineers than to tunnel engineers. Scarcely a wind tunnel exists that has not been held back from use by long delays resulting from balance calibration, and sad indeed have been many tunnel engineers who found that they were saddled with research on balances rather than on airplanes. Buy the balance, then, if it is at all possible, and insist that it be set up and calibrated at the factory. If delays occur there, the tunnel is available for additional calibration, improvements to flow quality, and a host of other problems particularly besetting a new tunnel (see Chapter 2).

Any wind tunnel balance is of necessity a compromise between the required maximum load capability of all components and the accuracy required for minimum loads. The solution to this problem is much simpler for an internal balance, since several balances can be designed and built for different load ranges and one selected for the desired loads. A risk with an internal balance is that the model loads may inadvertently exceed the balance capacity. This may result in a failed balance. For an external balance a careful study must be made to try to account for every type of model and test that the balance will be used for over a period of years. In this type of study there is no substitution for wide and long experience in wind tunnel testing.

The expected loads that a balance must carry are primarily a function of the size of the tunnel test section and speed. If the size of a tunnel is doubled, the model size and the loads are increased fourfold for a fixed model to tunnel size ratio, model attitude, and tunnel speed. It should be noted that in this case the Reynolds number goes up by a factor of 2 and model volume by a factor of 8.

As a rule of thumb the maximum model span will be equal to or less than 0.8 of the tunnel width for a full model. For a half model, the model span will be equal to or less than 0.7 of the tunnel height.

A model of an aspect ratio of 9 and a span equal to 0.8 of the tunnel width in an 8 ft × 12 ft tunnel will lead to an area of 10.24 sq. ft. If the model is tested at a dynamic pressure of 60.0 psf, then $qS = 614.4$ lb. If one wishes to measure drag to 0.0001 in cruise condition, or one drag count, the drag balance must be able to resolve 0.06 lb. If one assumes the model is a powered model of a propeller-driven airplane, the maximum dynamic pressure when power testing will be close to 15.0 psf and $qS = 153.6$ lb, and for one drag count the drag balance must be able to read 0.015 lb. The same model, power off, past stall could have a C_D of 1.00 and at $qS = 614.4$ lb, which would yield a drag load of 614.4 lb. Thus, the drag balance should be able to resolve loads from 0.015 to 600 lb, a ratio of 40,000 to 1. Another way to look at the requirement is that the minimum reading should be 0.0025% of full scale. These values are typical of what is required for a general-purpose low-speed wind tunnel balance.

The ratio of maximum to minimum values of the drag balance also has implications for a computerized data system. If the desired ratio is 40,000 : 1, then, since computer data systems are binary, to reach this ratio requires 16 bits plus sign, which is 65,536. If a 14-bit (16,384) analog-to-digital converter is used, then the minimum drag value that can be detected for 600 lb full scale is 0.037 lb or two times larger than required. This problem is relieved, as discussed in Section 3.9, by taking multiple passes at the data and averaging the value. However, the minimum detectable value is still 0.037 lb. The sensitivity of the balance and data system can be checked, as described in Section 4.15.

Although in the previous example the requirement of one drag count $C_D = 0.0001$ was used, the authors do not want to leave the impression that measuring to this level is easy or routine. One drag count is the goal for cruise drag, but in most cases it is difficult to get model repeatability when configuration changes are made to much better than 2 or 3 counts unless very special care is taken during model changes.

An external balance can be designed and built to obtain these ranges of load-resolving capability. This type of balance is quite large physically, and when designed to measure each of the six components separately such a range of values is obtainable when the balance is carefully aligned, which is possible because of its size (see Section 4.16).

This range of resolution is much more difficult to obtain with an internal strain gage balance and in the previous example for a propeller-powered model might require the use of two internal balances: one for power-off and a second for power-on testing. Unfortunately, testing the same model on two balances is not efficient and would come close to requiring two separate tests to be run. The two advantages of an external balance are its great resolving power and its ability to hold its calibration over very long periods of time. The disadvantages can be its size, and often high initial cost and time required to align or reduce the interactions between the six components. It is possible to calibrate the interactions and not align the balance (Section 4.13).

The advantage of a sting or internal balance is a lower initial cost, although this may be negated by building several with load ranges to meet testing requirements over a period of time. A second advantage is that often one balance can be used in several tunnels, assuming that the balance can be adapted to different stings. With special stings, an internal balance can often test to higher angles of attack than an external balance, which is most useful for some tests of fighters that are often required to operate at extreme angles of attack well beyond stall.

Basically, no balance system will handle all possible tests for all possible flight vehicles. But an external balance proves to be the most useful because of its large load range and versatility in being adapted to uses that were never thought of when it was designed.

Table 4.1 lists some suggested values in terms of maximum coefficients that would be desirable to achieve with a balance.

For a half-span model it should be noted that the values given for the longitudinal components, that is, C_L, C_D, and C_m, apply, but the balance component is now $C_L = C_Y$, $C_D = C_D$, $C_m = C_n$. This should not affect the yaw balance since the three moment balances often have the same ranges with external balances. The requirement that the side force balance carry the full lift load may have an affect on the sensitivity of side force. A second problem with half models is that the center of lift of a half wing will not be at the balance moment center. In fact, it will be below the moment center, possibly one-half the distance from the centerline of the tunnel to the floor. This can lead to large rolling moments and can affect the load range of the roll balance. These comments apply to an external balance only. A sting balance for a half model would, in most cases, be designed for that purpose.

The minimum values in terms of forces and moments should be based on the smallest model expected at the lowest expected dynamic pressures and

TABLE 4.1. Probable Maximum Coefficients Developed by Full and Half-Span Models

C_L	+4.0	−2.0	C_n	+0.20	−0.20
C_D	+1.0	−1.0	C_l	+0.20	−0.20
C_m	−1.0	+1.0	C_Y	+1.0	−1.0

the maximum value from Table 4.1 from the largest model at the highest expected dynamic pressure. If a compromise must be made in the balance design, care must be taken to ensure that the balance is more than adequate for the ranges of model size and dynamic pressure at which most testing is expected to take place. This means that if one needs a higher resolution in side force for a full model than is available when the side force balance maximum load is equal to the lift load, and half-model tests are not a primary requirement, then the half-model requirements can be relaxed. It should be noted that for an 8 ft × 12 ft tunnel model span equal to 0.8 tunnel width and an aspect ratio of 9 yielded a wing area of 10.24 sq. ft. A half model with a span equal to 0.70 of the tunnel height would have an area of 13.94 sq. ft for a full wing, since the Reynolds number will increase by the square root of the area ratio, or by 1.16. The main advantage of a half model lies in the reduction of model costs for the wing and, especially, the flap system.

Care should also be taken not to overemphasize a small model at low dynamic pressure in determining the minimum loads. Quite often the small models are for research purposes and are used to determine trends, not values, for a flight vehicle. It would be almost the height of absurdity to size the minimum values of the balance for the NASA 80 × 120 V/STOL test section (a part of the 40 × 80 tunnel), for a 6 sq. ft model at a dynamic pressure of 10 psf.

The balance requirements previously discussed are for tunnels intended for research and development testing and would range in size from 7 ft × 10 ft (70 sq. ft) and upward. Small tunnels intended for student use can have the tolerances relaxed, and they can be further relaxed for simple demonstration tunnels.

Special-purpose tunnels for nonaerodynamic testing will have an entirely different set of values. Table 4.2 lists the balance range for the General Motors Automotive Tunnel. It should be noted that automotive engineers are not as interested in developing large lift coefficients as are aeronautical engineers.

In addition to the loading table, ranges for the pitch and yaw angle must be given. Pitch angle range will vary with the rearward distance of the pitch

TABLE 4.2. Mechanical Balance Load Ranges for General Motors Automotive Tunnel

Component	Low Range	High Range
Lift	517 lb	1506 lb
Drag	360 lb	1012 lb
Side force	517 lb	1506 lb
Pitching moment	2360 ft-lb	8114 ft-lb
Yawing moment	2360 ft-lb	8114 ft-lb
Rolling moment	2360 ft-lb	8114 ft-lb

TABLE 4.3. Permissible Measuring Errors in the Various Aerodynamic Coefficients[a]

	Low Angle of Attack	High Angle of Attack
Lift	$C_L = \pm 0.001$, or 0.1%	$C_L = \pm 0.002$, or 0.25%
Drag	$C_D = \pm 0.0001$, or 0.1%	$C_D = \pm 0.0020$, or 0.25%
Pitching moment	$C_m = \pm 0.001$, or 0.1%	$C_m = \pm 0.002$, or 0.25%
Yawing moment	$C_n = \pm 0.0001$, or 0.1%	$C_n = \pm 0.0010$, or 0.25%
Rolling moment	$C_l = \pm 0.001$, or 0.1%	$C_l = \pm 0.002$, or 0.25%
Side force	$C_Y = \pm 0.001$, or 0.1%	$C_Y = \pm 0.002$, or 0.25%

[a] For balance design requirements the actual loads should be figured using the smallest model expected to be tested and the lowest dynamic pressure. In some cases the tolerances can be relaxed somewhat.

strut from the front struts but should in any event provide for ±40°. Usually yaw from −40° to +190° is allowed for. Two degrees per second is a good rate of change for both pitch and yaw.

Associated with the balance design is the question of which axis system to use in designing the balance in order to measure forces and moments. There are three axis systems that are called wind, body, and stability axis. These systems are defined as follows. For the wind axis, forces and moments are measured parallel and perpendicular to the centerline of the tunnel. Thus lift is perpendicular to the longitudinal centerline and drag is parallel to it. Body axis moves with the airplane, thus the axes will pitch, yaw, or roll with the model. Stability axis rotates with the model in yaw but not in pitch.

Almost all external balances use wind axis and internal balances use body axis. Some external balances use the stability axis. See Section 6.30 for equations to transfer wind axis data to either stability or body axis.

The desired accuracies are similarly attacked by first preparing a permissible error list from the aerodynamicist's viewpoint (Table 4.3).

A critical maximum error condition may arise during power testing of complete models because the dynamic pressure of the tunnel may then be unusually low.

Depending on the amount of money available, a number of accessories may be incorporated that can materially reduce both the operating time and the work-up. A control panel for a wind tunnel using a digital computer to control the data acquisition is shown in Fig. 4.9.

4.8. MODEL MOUNTING

Any strut connecting the model to the balance will add three quantities to the forces read. The first is the obvious drag of the exposed strut; the second is the effect of the strut's presence on the free air flow about the model; and the

FIGURE 4.9. Tunnel operator's control panel with computer data-acquisition control. (Photograph courtesy of University of Washington.)

third is the effect of the model on the free air flow about the strut. The last two items are usually lumped together under the term "interference," and their existence should make clear the impossibility of evaluating the total tare by the simple expedient of measuring the drag of the struts with the model out.

The earliest attachments were by means of wires or streamline struts. The ruling criterion was to add the smallest possible drag and then either estimate it or neglect it. With the advent of the image system of evaluating the tare and interference (Section 4.17), these effects were evaluated by measurements rather than being estimated.

The mounting struts employed at first still tended toward the minimum drag criterion and were airfoil shape. Later, however, many mounting struts of polygonal cross section were made. The idea behind this trend was that the Reynolds number of the mounting struts would always be very low and they might therefore have not only a large drag but also a drag that varied widely under minute changes in shape or Reynolds number (see Fig. 4.10). To minimize the change in drag with Reynolds number, the transition point between laminar and turbulent flow should be fixed by some permanent method such as staking the transition point with a center punch. The main advantage of struts with a larger cross section is that they will reduce the deflection of the model.

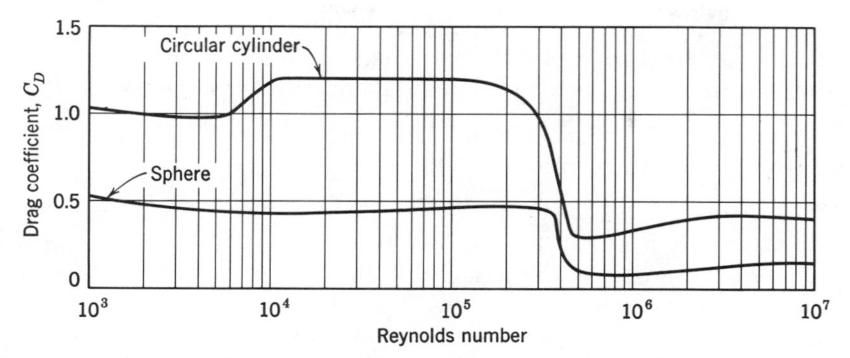

FIGURE 4.10. The variation of the drag coefficient of circular cylinders and spheres with Reynolds number.

Only a minimum of strut is exposed to the airstream, the remainder being shielded by fairings not attached to the balance. In this way the tare drag of the mounting is decreased, sometimes being only 50% of the minimum drag of an average wing. It is not advisable to try to decrease the tare drag of the "bayonets" by continuing the windshields up close to the model because a fairing close to the model can increase the interference effects more than it decreases the tare. The proper balance between amount of exposed strut and proximity of the windshield to the model may be found by having adjustable sleeves at the windshield top. The sleeve location at which $C_{d0\,\min}$ for model plus tare and interference is a minimum is the best, since this indicates that the tare plus interference is a minimum too.

Some balances yaw the model support struts oppositely to the model, so that the struts always remain parallel to the airstream and hence contribute the smallest possible effect when the model is yawed. Another useful arrangement is to have several sets of supports of varying size from which the smallest can be selected according to the load range.

One feature sometimes considered necessary for the ordinary support system is a diaphragm seal that prevents flow from around the balance up between the supports and shields into the tunnel. There are two types of pressures that may cause this flow. The first is due to the basic tunnel design, which not infrequently results in a test-section static pressure below the atmospheric pressure, and hence in a pressure differential, sometimes quite large, between the balance chamber and the test section.

The second pressure is that resulting from the attitude of the model. This is much smaller than the first and can be eliminated simply by a light diaphragm seal. Closing off the balance room in no way changes the necessity for the support column seal.

The attachment fittings usually come into the wing at about the 30–50% chord point. In complete airplanes, the most rearward center of gravity location may be used to give maximum room for the fittings. If a model of a multiengined airplane is to be tested, the mounting strut interference will be smallest if the struts do not attach at a nacelle point.

The various arrangements of mounting are discussed below.

Single-Strut Mounting

This arrangement is by far the simplest. Only a single windshield is needed, and it need not move as the model is rotated in yaw. The single strut is satisfactory for models and nacelles (see Fig. 4.11) and may be used in conjunction with wingtip supports to evaluate tare and interference. The single strut usually mounts in the body of the model. The lateral spacing of the trunnion pins must be as large as possible to carry the rolling moments and to prolong the life of the trunnion pin bushings when the model has asymmetric stall (this structure is buried in the model body). When the strut is made with a relatively large cross-sectional area, it is fairly rigid in torsion (yaw) and a fair-lead may be incorporated in its leading edge for wires for scanivalves for pressures, power to drive the horizontal tail, hinge moment data, etc. Because the wires are internal to the strut, additional tare and interference runs are not required for such wires, and because the upper end of such a strut is large, it is difficult to use for wing alone tests.

FIGURE 4.11. Single strut with pitch arm. Both strut and pitch arm rotate in yaw. (Photograph courtesy of University of Washington.)

FIGURE 4.12. Fork-and-pitch-arm-type mounting strut. Both fork and pitch arm rotate in yaw. (Photograph courtesy of University of Washington.)

Single-Strut with Fork

An increase in resistance to roll deflections may be gained by splitting the single strut into a fork at the top (Fig. 4.12). This method may have less torsional (yaw) rigidity, as the forward struts that carry the trunnion pins are usually of a small cross section. With small-cross-section struts, the tare and interference effects are less. This system usually requires additional tare and interference studies if wires, etc., are taken into the model. Both the single strut and single strut with fork have a single wind shield that is not a part of the balance. The structure inside this wind shield can be quite robust, thus limiting the support system deflections to the struts that are exposed to the air stream. The balance fairing and its enclosed balance structure can be designed to be removable below the tunnel floor, thus allowing a stiff, non-pitching strut to be attached to the balance. This strut with a flat plate that yaws can be installed flush with a splitter plate above the wind tunnel floor for half models and building/structural shape tests.

Single-Strut Only

In this design the model pitching and/or yaw mechanism is inside the model (Fig. 4.13).

FIGURE 4.13. Single strut with pitch and yaw mechanism inside model. (Photograph courtesy of Texas Engineering Experiment Station, The Texas A&M University System.)

Two-Strut Mounting

The two-strut mounting surpasses a single strut for rigidity in both torsion (yaw) and roll, but adds the complication that the windshields must be moved and rotated as the model is yawed. Figure 4.14 shows a setup with the mounting struts side by side; in Fig. 4.15 they are employed in tandem.

Three-Point Mounting

The conditions of rigidity, tare, and interference evaluation and ease of varying the angle of attack are all met satisfactorily by the three-point supporting system. This system is the most complex and requires that two and sometimes three windshields be arranged to yaw with the model (Fig. 4.16). The rear strut introduces side forces that complicate the yawing moment measurements of a yawed model.

Wingtip Mounting

When it becomes necessary to determine the pressure distribution of regions close to the mounting struts, the models are sometimes mounted from the

FIGURE 4.14. F-86D on two-strut support. (Courtesy North American Aircraft Corp.)

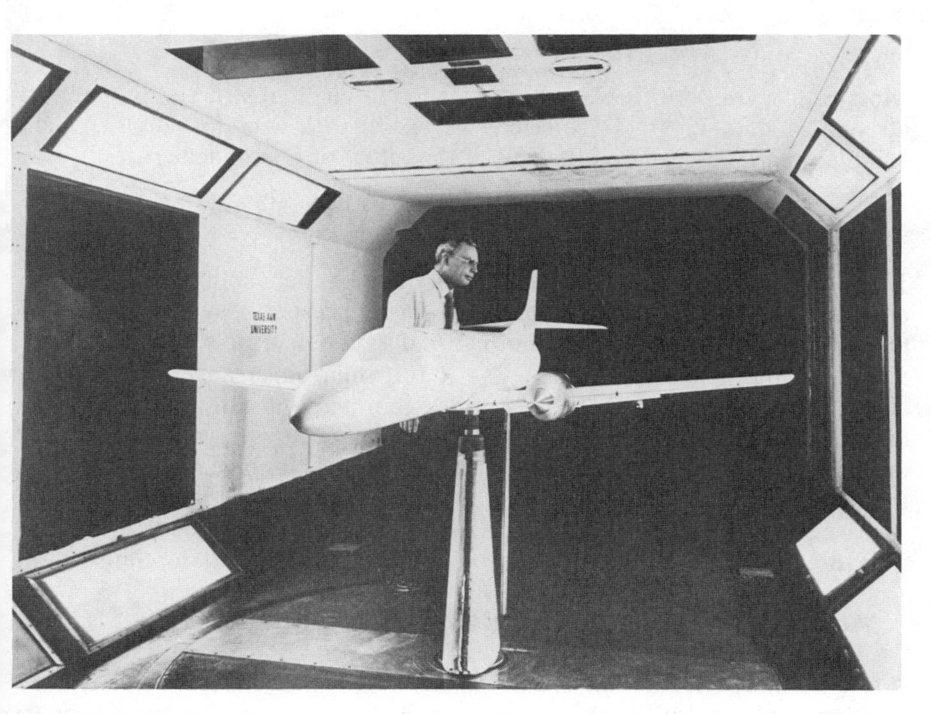

FIGURE 4.15. Two-strut tandem. (Photograph courtesy of Texas Engineering Experiment Station, Texas A&M University System.)

FIGURE 4.16. DC-9 on three-strut mount. (Courtesy of McDonnell Douglas Corporation.)

wingtips, leaving the fuselage and nacelles in air unobstructed by support fittings. Models of larger scale may be tested with wingtip mounting, and valid comparisons can be obtained of the effect of component parts.

Half Models

The largest scale models may be tested by having them split down the plane of symmetry, only one-half of the model being present. Asymmetric flow is prevented by a large plate at the plane of symmetry, or by mounting the model on the tunnel floor. See Fig. 4.17. Such an arrangement, though obviously unsuited for yaw tests, yields accurate pitch, lift, and downwash data at the maximum Reynolds number. The increase in Reynolds number is not large over a full model, about 20%. The main advantage appears to be in the model construction costs.

Care should be taken that the horizontal tail does not approach the tunnel wall too closely or stability at the stall will appear much too optimistic.

Mounting from the Tunnel Roof

A few balances mounted above the tunnel support the model in an inverted position for "normal" running. (See Fig. 4.18.) This arrangement seems to

FIGURE 4.17. A combination of a ground board and reflection model test. (Courtesy Convair San Diego.)

be a holdover from early wire balances that supported the model similarly so that the lift forces would put tension in the wires. No particular advantage seems to accrue from inverted testing. On the contrary, such a balance position hinders the use of a crane to install models, and the terminology of testing "normal" and "inverted" becomes obscure.

There also can be extra costs associated with this method in obtaining a rigid base on which to mount the balance.

Mounting from a Tail Sting

Engineers using small supersonic tunnels found that struts normal to the flow such as those used in low-speed tunnels caused excessive blocking in

FIGURE 4.18. Ceiling-supported model. Inverted testing is quite common in Europe. (Photograph courtesy of British Aerospace P.L.C.)

supersonic tunnels, and now almost invariably use a sting mount (Fig. 4.19). In order to use the same models in low-speed tunnels (which are cheaper) the sting mount is employed. Sting mounts are fine for those airplanes having jet engine exits at the fuselage tail, since this furnishes a place for a sting. Tare and interference can be difficult to evaluate when the sting diameter requires a change in the aft fuselage contour.

When an internal balance is used, especially for a heavy, high-Reynolds-number model, the axial balance load range may have to be excessively large, since it will have to carry $\sin\alpha$ times the model weight, or 400 lb for an 800-lb model at $\alpha = 30°$. A drive system for a sting support for a small tunnel is shown in Fig. 4.20.

The use of large-scale panel models for investigation of control surfaces is discussed in Sections 5.7–5.10. These panels require mounting arrangements different from those for wings and complete models. Several mountings are discussed below.

Mounting on a Turntable (see Fig. 5.11)

When the model is mounted on a turntable flush with the tunnel wall, the forces and moments on the turntable are included in the data and are difficult

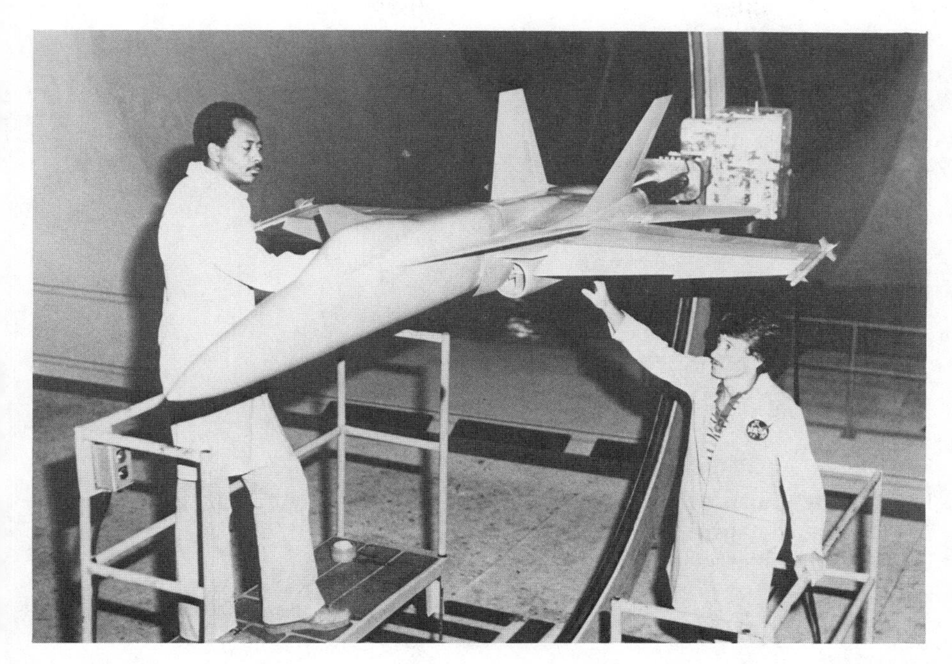

FIGURE 4.19. F-18 in NASA Langley 30 × 60 ft tunnel. Model sting mounted on quadrant for high angle of attack test. (Photograph courtesy of NASA, Langley, and Dynamic Engineering, Inc.)

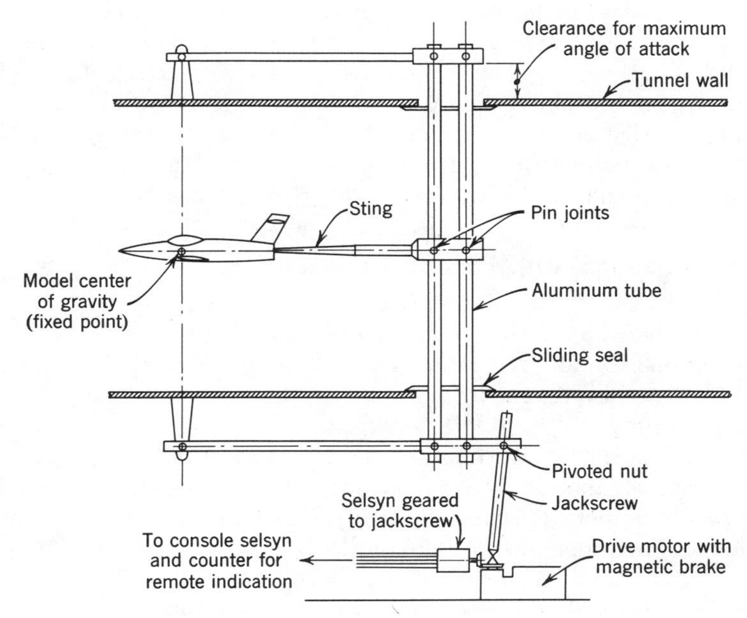

FIGURE 4.20. Parallelogram linkage for angle of attack control used in a 40 in. × 40 in. low-speed tunnel.

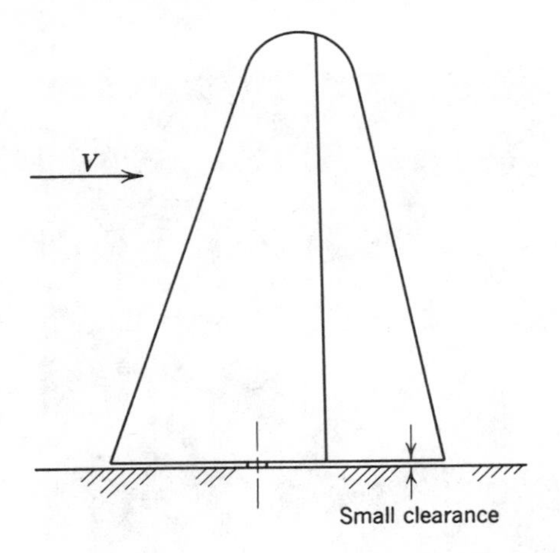

FIGURE 4.21. Panel model mounted on short strut.

to separate. Fortunately, for the type of tests usually sought with this arrangement, the absolute value of the drag is not needed, and the effect of the endplate on lift is negligible.

Mounting on a Short Strut

Mounting the panel model on a short strut (see Fig. 4.21) has the advantage of decreasing the tare drag of the setup, but it is hard to evaluate the effect of the slot. Theoretical considerations indicate that a slot of 0.001 span* will decrease the effective aspect ratio enough to increase the induced drag by 31%; a slot of 0.01 span will cause an increase of 47%. The effect of viscosity (not included in the above figures) will tend to decrease the error listed above, but the degree of viscous effect has not been clearly established. If the slot can be held to less than 0.005 span, its effect will probably be negligible; few engineers believe that 0.02 span is acceptable.

Mounting as a Wing With an Endplate

Mounting the panel as a wing with a small endplate to assist in keeping the spanwise lift distribution as it should be is shown in Fig. 4.22. No endplate of reasonable size will prevent tip flow; hence the spanwise load distribution with this mounting will be greatly in error.

The last paragraphs draw attention to the advantage of having a yoke-type balance frame, whether or not the balance is a yoke balance. The presence of lateral brace members to which bracing wires may be attached is a great convenience. Such members are obviously necessary for wingtip mounting.

* This span is, of course, the complete wing span, not the panel span.

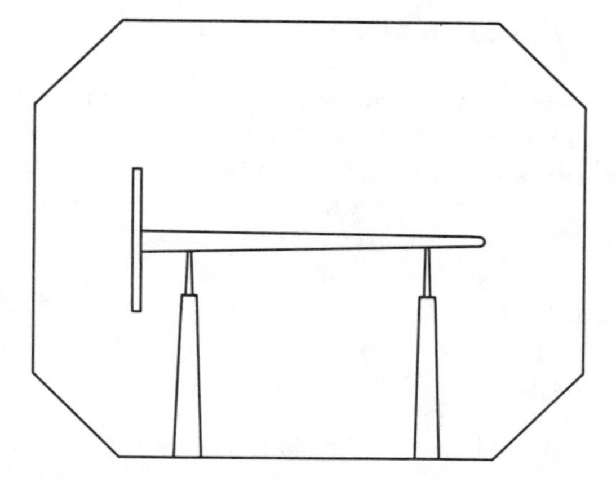

FIGURE 4.22. Panel mounted as a wing with endplate.

Some balances have a ring that completely encircles the tunnel jet. Though the ring offers a number of brace points, the part of the ring above the test section interferes with the installation of the image system.

4.9. DEFLECTIONS

One of the most troublesome problems of wind tunnel balances is rigidity. Deflections in the balance may move the model from the resolving center and invalidate the moment data or nullify the balance alignment so that part of the lift appears as drag or side force.

The answer to the problem is obvious: either the deflections must be kept down to where they are negligible, or they must be evaluated and accounted for in the work-up. Of course, keeping them down is preferable.

The largest source of deflection is the mounting system. This must be long to reach out of the test section and thin to avoid excessive interference. Both these requirements are in direct antithesis to the criterion of minimum deflection. The only way the wind tunnel engineer can meet this problem is to utilize materials of high modulus of elasticity for the strut. The desire for the shortest mounting strut possible is a strong argument for the selection of a rectangular or elliptic jet shape. Deflections in the balance frame may be diminished by having a deep and rigid framework. None of the common measuring units have deflections large enough to be serious, and so they rarely cause this type of trouble.

The effects of deflections are evaluated during the process of calibrating the balance, and corrections, if necessary, are given to the computing staff for inclusion in the data work-up. It is also not unknown for the model itself to have internal deflections.

4.10. BALANCE LINKAGES AND PIVOTS

Basically an external balance consists of a large number of levers that are designed to have minimum deflection under maximum load. This increases their weight and often their size. The joints between the levers are usually pivots with very small angular motion. It is required that the pivots have very low friction (ideally zero friction) to avoid hysteresis in the balance when the directions of the loads are reversed. Early pivots in wind tunnel balances were knife edges. Because knife edges can be damaged by shock loads and can only carry loads in a compressive direction, they have been replaced by flexure pivots. It should be noted that ball and roller bearings make very poor pivots where the angular motion is very small and loads must be carried perpendicular to the shaft axis of the bearings. Under these conditions both the balls and their races can easily develop flat spots and a resulting large amount of friction when rotated.

The advantages of flexures are as follows:

1. They can be designed to withstand loads in any direction with no lost motion between the coupled members.
2. They are essentially frictionless, thus eliminating hysteresis effects.
3. They will withstand relatively rough treatment.
4. They are virtually wearproof; thus their characteristics remain constant over an indefinite period.

Flexures are usually of two types. The first is a composite or rod flexure. These are used to transmit loads along their axis with small angular rotation and the critical design load is compressive. As the length of the rod often must be adjusted, the ends consist of fine right- and left-handed threads that can be clamped in a nut that is split in two and bolted together. These flexures are machined out of a solid bar stock (Fig. 4.23). The second is a restrained flexure or *X* flexure. These are almost frictionless pivots, and if the rotation is small, the center of rotation is fixed. They can be made by machining a bar into a *Z* shape on a mill. The bar is cut into strips and then assembled in an *X* shape with tapped holes in the top and bottom for assembly to other parts. With electric discharge milling, they can be made from one piece. *X* flexures are used in pairs.

The *X* flexure can be less than half as stiff as a rod flexure and allows twice the rotation angle. When the *X* flexure is loaded at right angles to its axis, two of the flexure strips are in compression and two are in tension when acting as a pivot. Thus the change in stiffness of the tension strips is compensated for by the opposite change in stiffness of the compression strips (Ref. 4.1).

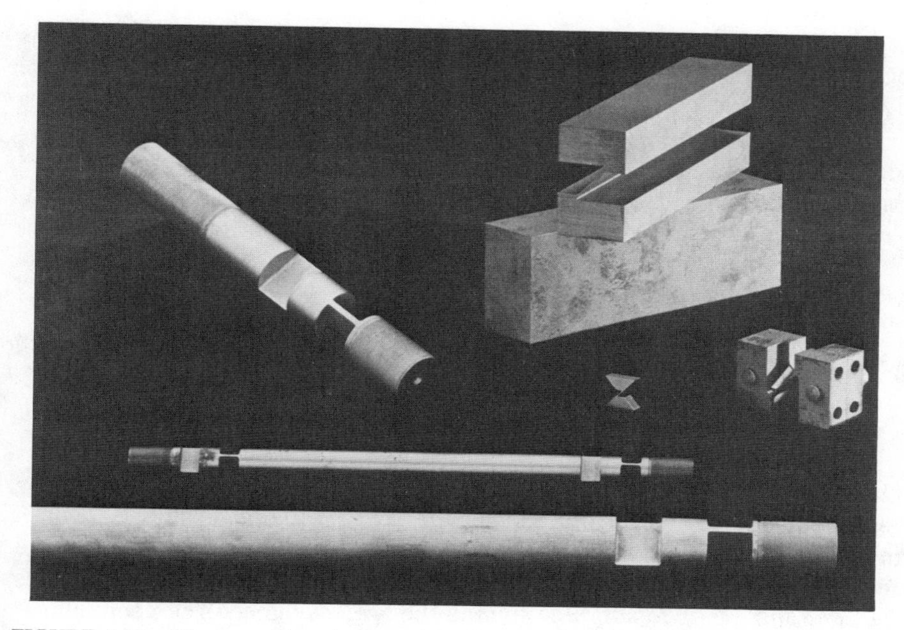

FIGURE 4.23. Flexures. Threads on rod flexure allow length adjustment. *Z* flexure stock made on milling machine. Two halves of *X* flexure are cut from stock and pinned together. (Photograph courtesy of University of Washington.)

4.11. BALANCE MEASURING DEVICES

The device used to measure the output of the balance should have the following characteristics:

1. The curve of the applied load versus the indicated load should be invariant. Assuming that the curve is linear and passes through zero this means that the slope should remain constant, especially with time. If the slope of the curve changes (slips calibration), it often is most difficult to detect. This problem has become more serious with digital data systems. These systems use amplifiers with dc response and these amplifiers are prone to drift. This requires checks on the amplifiers, usually every 4 h. It also is desirable to make end to end checks by the use of check weights at the same time the amplifier is checked.

2. When the applied load is removed gradually, the measuring device should return to zero. The zero-shift problem is not as serious as slipping calibration. In most tunnels it is a standard policy to check the ''balance zero'' for shift at the end of each run or run series. If the zero shift is too

large, the runs are repeated. Care must be taken with the end of run zero to make sure that the tunnel velocity is also zero.

3. The measuring device should display no hysteresis. In general, this is not a problem with the measuring units. Often when hysteresis appears it is due to some portion of the measuring system being loose or slipping.

4. If an automatic digital data system is used, the output of the measuring device must be compatible with the data system. This applies to external balances that use unit weights to balance out a large part of the applied load, thus the measuring unit only was subjected to a small part of the total load to improve accuracy. The values of the unit weights must be recorded in some manner by the data system.

The following types of measuring units have been used on wind tunnel balances.

The Automatic Beam Balance

The automatic beam balance is shown in Fig. 4.24. It consists of a weighing beam that has an electrically driven rider. When the beam drops down, a contact is made that causes the driving motor to move the rider in the direction that will balance the beam. A counter on the motor shaft locates the rider and may be calibrated to read the force weighed. The pendulum *H* (see Fig. 4.24) can be adjusted to balance out the destabilizing component due to the weight of the beam. To avoid the troubles usually associated with mechanical contacts it is preferable to use a linear variable differential transformer on the end of the weigh beams. The output signal may be fed into an amplifier, which activates the poise motor. This type of force measuring balance has the advantage that the measurement is always made in the balance position and thus there is no displacement required. The disadvantage is that they can be slow in reaching a balance condition.

If unit weights are used, they are applied to the balance beam and must be recorded in some manner by a data system. To convert balances to a com-

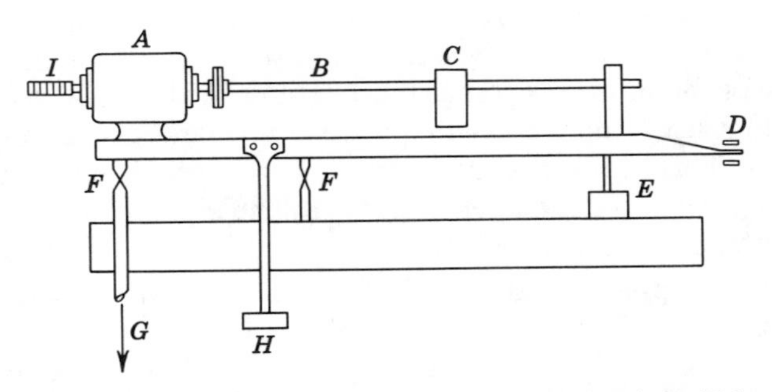

FIGURE 4.24. Beam balance. *A*. Driving motor. *B*. Threaded rod. *C*. Rider. *D*. Reversing contacts. *E*. Dashpot. *F*. Flexure pivot. *G*. Applied load. *H*. Pendulum weight. *I*. Counter.

puter-sensitive output, shaft encoders can be attached to the threaded rod. This may require some gearing so as not to exceed the rotational range of the shaft encoder.

Hydraulic Capsules

The hydraulic capsule is a device that measures forces through the pressures they exert on pistons of known area. They are not exactly null, but the amount of deflection of the piston is so small as to be negligible. The resulting pressure is a function of the size of the load and is measured through accurate pressure gages. A pressure transducer would be used to produce a voltage output that is compatible with digital systems.

Electric Measuring Devices

There are several methods for measuring forces or pressures electrically, most of them depending on amplifying the effect that tiny deflections have on the capacitance, inductance, or resistance of the unit. For example, the resistance of a carbon pack varies as the pressure on it, and the current it passes for a fixed voltage may be used as an index of the load. The amount of current needed to keep the core of a solenoid in a fixed location is an index of the load on it. The change in capacitance of a plate condenser with small deflections of the plates may again indicate a load. The resistance of a wire changes with the tension of the wire, and the current passed for a fixed voltage may indicate the tension. And so ad infinitum. A hundred different setups may be possible. It should be borne in mind that through amplification the most minute changes may be noted and that remarkable accuracy is possible.

Electric strain gages of the wire gage type have been tried as measuring devices in several external balances with a satisfactory degree of success. The electromagnetic arrangement that measures the forces by the amount of current needed to maintain zero deflection in the unit has also been successful. These systems often have a unit weight addition setup in order to extend their range with maximum accuracy.

Any measuring system must be checked under vibratory load to obtain the damping and balance that yield the greatest accuracy.

By far the most widely used of the electric measuring devices is the electric wire strain gage. Basically, the wire strain gage consists of a resistance made of very fine wire cemented to a flexure. The load, by deflecting the beam by a minute amount, stretches the wires glued to the beam and changes their resistance and the amount of current that will flow through them for a fixed applied voltage. The current, being proportional to the load, thus becomes the indicator of the size of the load.

The above, elementary as it is, serves to illustrate the principle of the gage itself, but many refinements are needed before a useful electric strain gage

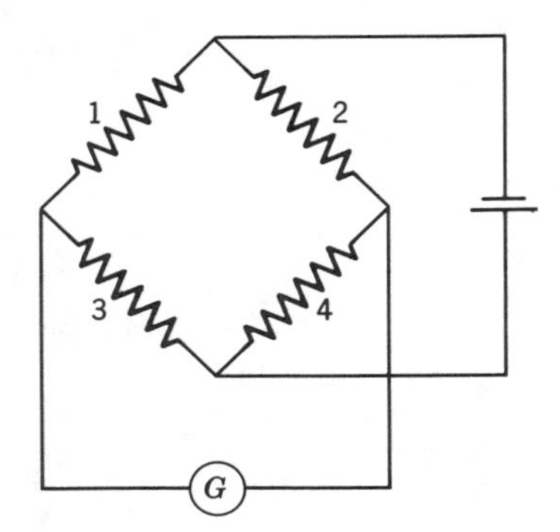

FIGURE 4.25. Simple electric strain gage circuit.

can be realized. Temperature compensation is accomplished by gluing a second gage on the opposite side and connecting it to a Wheatstone bridge circuit as shown in Fig. 4.25. It will be seen that, as long as both resistances (marked 1 and 2 in Fig. 4.25) are equally increased or decreased, there will be no change in the potential measured by the potentiometer. Thus the flexure-gage combination is insensitive to heat.

There is a wide range of commercial strain gages available, which vary widely in size, backing material, and sensitivity to temperature. The gages can be affixed to the flexure material with a variety of bonding agents, ranging from simple cellulose cement to the more exotic adhesives, some of which require special curing techniques. The current practice with the widespread use of solid-state electronics is to use bridge voltages of 2.5 or 5.0 V. There are signal conditioners commercially available that supply the bridge voltage and balancing impedence. There can be one, two, or four active gages in a bridge, but generally either two or four gages are used to minimize the effect of temperature. Four-gage bridges also increase the sensitivity and accuracy. The bridge output is amplified before converting to computer-sensible bits in an analog-to-digital converter.

It will be necessary to bring out strain gage leads from rotating machinery, on some occasions, using slip rings (Section 3.16).

4.12. INTERNAL STRAIN GAGE BALANCES

Strain gages are used in internal balances and some external balances. Some external balances also use load cells that are basically strain gage devices.

Some general facts concerning selection and mounting of gages are:

1. Adequate strain must be provided in the measuring elements under the design loads (approximately 600–700 microinches per inch).
2. Match gages for gage factor and resistance.
3. Use the longest gage possible.

4. Follow recommended mounting procedures (surface cleaning, baking time, etc.).
5. Use extreme care in locating gage and in applying curing pressure. A gage that has slipped will introduce errors.
6. Good solder joints are a must. A "cold" solder joint will not function properly.
7. All mounting surfaces must be carefully machined.

As far as internal balances go, there are two basic types: force balances and moment balances. In any six-component system there will be three force units and three moment units.

Force-measuring elements employ either a cantilever beam or a column arrangement. An eccentric column provides greater sensitivity but also allows more deflection as does a single cantilever. The choice might well depend on the particular balance size and arrangement needed for a specific model. The axial force "cage" shown in Fig. 4.26*a* is one of the most common types. This unit can be made very sensitive by sizing the flexures, but, since the model is attached to the cage, it is subjected to the relatively large normal forces. One might thus expect (see Figs. 4.26*b* and 4.27) an obvious interaction to occur because of the deflection of the cage flexures. A unit designed at DTMB, shown in Fig. 4.26*c*, has reduced this kind of interaction to a minimum. All forces except axial load are carried by the webs as shear or direct tension or compression. A rod transmits the axial force to a cantilever beam mounting the gages.

The arrangement for normal force readout is shown typically in Fig. 4.28. In this case, the wiring is arranged so that the difference between two moments is read electrically. Since the normal force is equal to the difference of the two moments divided by the distance between gages, the unit may be calibrated directly in terms of applied normal force. It is important that both gage stations have the same section properties, I/y, and matched gages. The greater the gage spacing the more accurate the normal force readout. If M_f and M_r are the front and rear moments, then the normal force N is given by $(M_r - M_f)/d$, where d is the spacing between gages. It is noted that the same arrangement may be used to measure side force.

Pitching or yawing moments may be measured by the same gage arrangement discussed above except, as shown in Fig. 4.28, the bridge is connected as a summing circuit. The differential circuit employed for normal force will also yield moment if the moment reference point is between the two gage stations for then,

$$M_{\text{ref}} = M_f + \frac{x_{\text{ref}}}{d}(M_r - M_f)$$

where x_{ref} is measured from the front gage station. Another way is indicated in Fig. 4.29 where the pitch gages are "stacked" and located between the

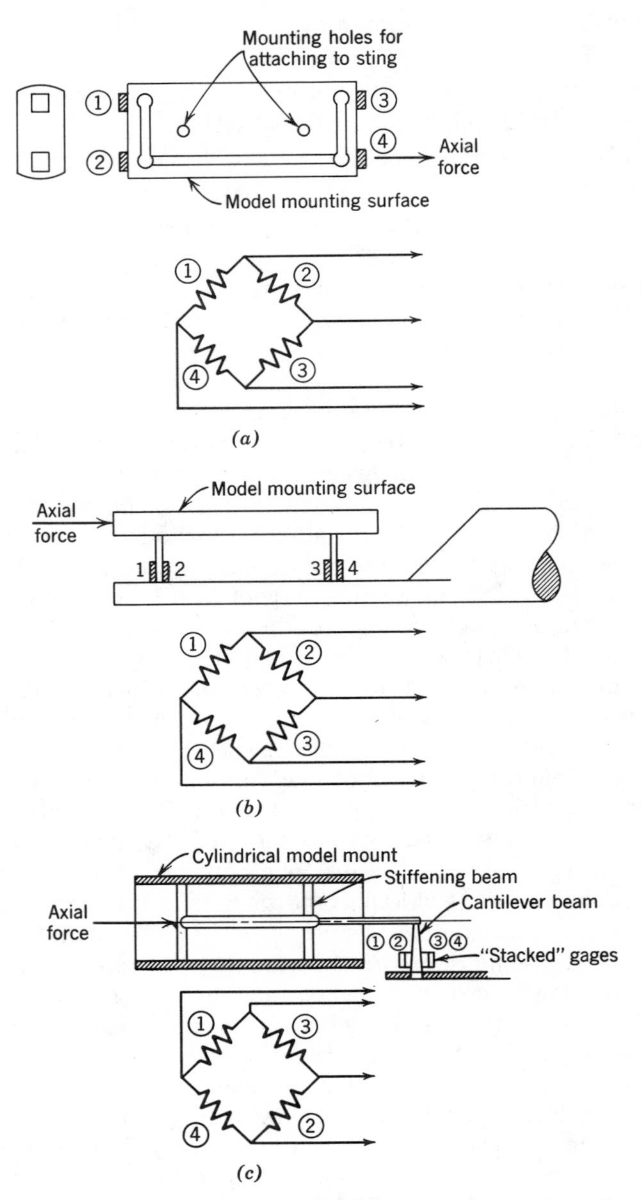

FIGURE 4.26. Axial force gages.

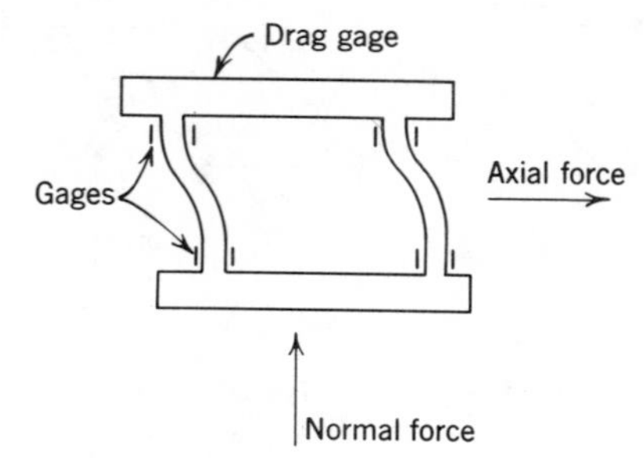

FIGURE 4.27. Normal force interaction on axial force.

normal force gages. For rolling moment a torque tube or a double beam type with gages mounted on the side faces of the beams can be utilized. There are many mechanical variations in internal balance design but the basic arrangement of strain gages attached to flexures is common to all.

With the advent of electric-discharge-milling capability many internal balances are made out of one piece of material. This eliminates problems with hysteresis in mechanically attached joints. If the balance is made of separate parts and fastened together, great care must be used to avoid slippage, which leads to hysteresis, and in obtaining the desired alignment. Internal balances have also been made successfully of built up parts welded together with an electron-beam-welding process (Ref. 4.2).

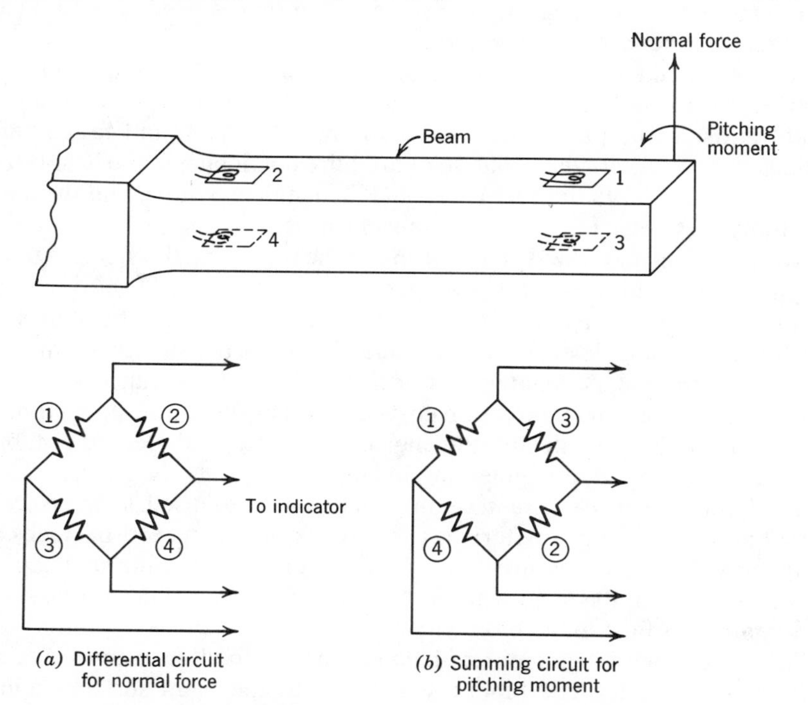

FIGURE 4.28. Normal force and pitching moment gage arrangement.

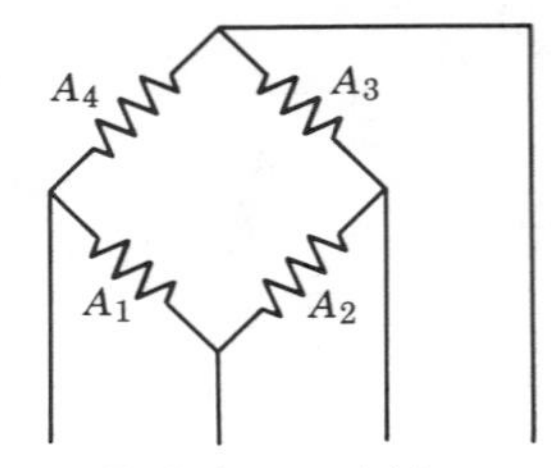

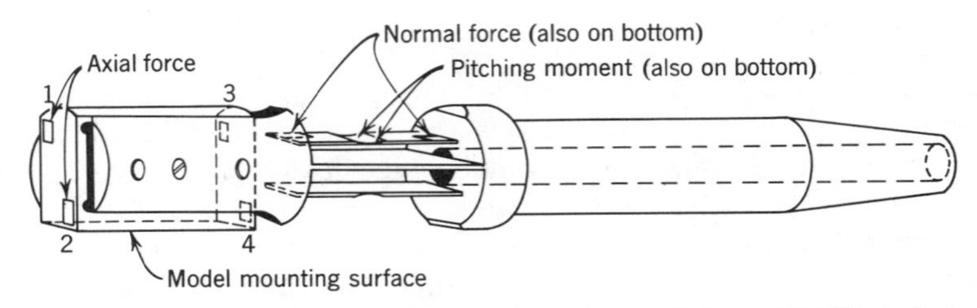

FIGURE 4.29. Three-component internal strain gage balance (NASA design).

The choice of material is a matter of engineering judgment and both aluminum and steel are used. A common steel that is used is AISI 4340M. The heat treatment of this steel requires quenching, thus there can be distortion of machined parts, including breaking of the parts due to thermal shock. For this steel it is better to machine after heat treatment. If vacuum melt 4340M is used, the physical properties remain the same but fatigue life is increased. Two other steels that do not require quenching in the heat treating (they are cooled in the furnace) are 17-4PH, a stainless steel, and the maraging family of steels. Both of these present no machining problems and will not warp when heat treated. The purpose of heat treating the steels is to raise the yield point and increase the fatigue life.

Since the flexures in a balance are designed on the basis of bending strain, advantages in some designs may be achieved by the use of aluminum alloys. The cost of the balance material is of little relative importance when compared to the design and fabrication costs, as well as the tunnel down time due to a balance failure. Basically, the choice of material and method of fabrication is a matter of good engineering judgment.

In addition to three- and six-component balances used to measure the overall aerodynamic characteristics of models, many special balances normally using strain gages are built. These can include wing and tail root bending moments; loads on wheels, doors, and nacelles; and the most common control surface hinge moments.

Control surface hinge moment balances can be built in many ways. One method is to build gaged brackets that bolt to the main surfaces and the control surface. Each bracket is bent to set the control at the desired angle.

Each of the brackets must be calibrated. This method is useful on models that have limited space. Another method is to use a precision-sealed ball bearing in the outboard hinge. The inboard end of the control surface has a strain gage beam rigidly attached to a pin. This pin slides into a close fitting hole in a metal control surface. To set control surface angles a hole is drilled through the control surface for a roll pin and the pin is attached to the hinge moment balance (one hole for each control surface angle). The use of a beam balance that is rigidly attached to the main surface and uses a friction clamp to hold a hinge pin on the control surface is not too successful, because the angle of the control surface will often change due to slippage of the friction clamp (see also Section 5.1).

4.13. SOURCES OF BALANCE INTERACTIONS

No balance is capable of perfectly measuring the loads that it was intended to measure. By this is meant that the lift load only makes a change in the lift-load-sensing element, with no change in the other five components due to lift load. There are two general sources of errors in balances. The first arises from misalignment of the balance parts and is caused by manufacturing tolerances, in both the parts and their assembly. These are linear or first degree. The second arises from the elastic deformation of the various parts. These are second degree and nonlinear. There could be higher-order terms due to plastic deformation of balance parts, but in this case the parts have improper dimensions and there is an error in the balance design.

Initially we shall not consider problems in the balance sensing elements such as temperature effects on strain gages, nor possible effects from the electronics used to obtain signals from the balance.

The purpose of calibrating a balance is to arrive at a set of equations that can be used to determine the loads applied by the model through the output signals. These are called the balance interaction equations. In the past external balances were adjusted to make the interactions as small as possible. When these small interactions were nonlinear, they were replaced by linear approximations. This was done to minimize the personnel and time required to obtain the model forces.

With the advent of dedicated mini digital computers it is possible to be more sophisticated in applying corrections to the raw balance sensor outputs. Balances are designed and built to meet various test requirements. But all balances show certain mechanical similarities. For any balance sensor to be able to produce an output requires a deflection in the balance. This deflection must be an elastic deflection so that the sensors' output is repeatable.

Curry illustrates (Ref. 4.3) the sources of errors in a balance by considering a load sensor (Fig. 4.30) that is intended to measure axial force. The effect of the normal force N, the pitching moment M, and the axial force A

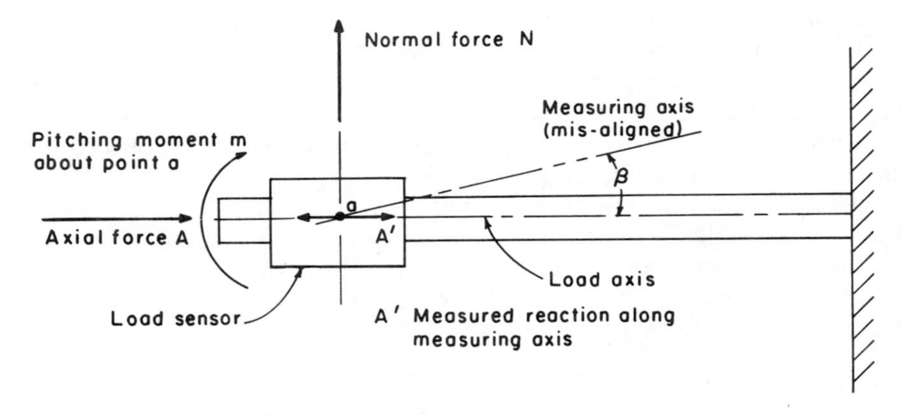

FIGURE 4.30. Schematic of axial balance load sensor.

on the load sensor is determined. The sensor will have an output A', and this is to be measured along the load axis.

Now if there is an angular displacement β between the measuring axis and the load axis, components of the applied normal force will appear along the measuring axis. Thus the sensor output A' becomes

$$A' = A \cos \beta + N \sin \beta \tag{4.1}$$

Balance design criteria dictate that the balance be rigid to support the model, and that the parts be built to close tolerances to minimize misalignments. Thus β can be considered to be a small angle. If β is in radians, then $\cos \beta = 1$, $\sin \beta = \beta$, and Eq. (4.1) becomes

$$A' \approx A + N\beta \tag{4.2}$$

Now assume that due to manufacturing tolerances there is a small initial misalignment of the measuring axis relative to the load axis. This misalignment is constant, and the angle will be β_0.

When the normal force acting through the center of the sensor is applied, the sensor and its support will deflect as a cantilever beam with an end load. This results in a misalignment due to normal force β_1. The application of a pitching moment about the point may cause the beam to deflect in a different mode, causing an angular displacement β_2. The deflection curve of the deflected beam and sensor are, of course, different with these two loads.

Thus β_1 and β_2 are independent of each other and are only functions of the normal force and the pitching moment. Since β_0 is constant and β_1 and β_2 will vary with the loads, they can be written as

$$\beta_0 = C_1; \quad \beta_1 = C_2 N; \quad \beta_2 = C_3 M$$

The total angular displacement will be the sum of the misalignment and the deflections due to loads:

$$\beta = \beta_0 + \beta_1 + \beta_2 = C_1 + C_2N + C_3M \tag{4.3}$$

Substituting Eq. (4.3) into Eq. (4.2) yields

$$A' = A + C_1N + C_2N^2 + C_3NM \tag{4.4}$$

The sensor output A' is thus a function of the axial load, the initial misalignment (the first-degree term), and displacements due to the other loads (second degree terms).

If the axial force sensitivity constant is defined as k_A (output units per unit of axial force), then the axial output $\theta_A = k_AA'$. Thus Eq. (4.4) becomes

$$\theta_A = k_AA' = k_A(A + C_1N + C_2N^2 + C_3NM) \tag{4.5}$$

Solving for the axial load

$$A = \frac{\theta_A}{k_A} - (C_1N + C_2N^2 + C_3NM) \tag{4.6}$$

The terms in the bracket are the balance interactions. The term θ_A/k_A is the sensitivity term and it is usually called raw data, the uncorrected output from the axial force sensor. The C_1 term is the initial angular displacement, a first-degree term, and it is generally due to tolerances in manufacture and assembly. The C_2 and C_3 terms are the second-degree terms and are due to elastic deformation. The interactions are not dependent on the component sensitivities but are only functions of balance geometry and material properties.

In a six-component balance each of the six measuring components may be considered to have six degrees of freedom in overall deflections. Each degree of freedom may be influenced by more than one applied load component (as A' was influenced by N and M). In addition to overall deflections, deflections within the supporting members and flexures that comprise the internal configuration of the sensing element also contribute to the interactions. These deflections should be considered in a more comprehensive analysis. In analyzing the deflections and their resulting interaction terms, one source of deflection may be considered at a time since superposition applies. A six-component balance has 27 interactions.

It must be noted that when using a six-component balance for pitch runs all six components must be recorded since the balance interaction equations for each component contain terms from all six components. If the lateral components are not recorded, there will be errors in the longitudinal components.

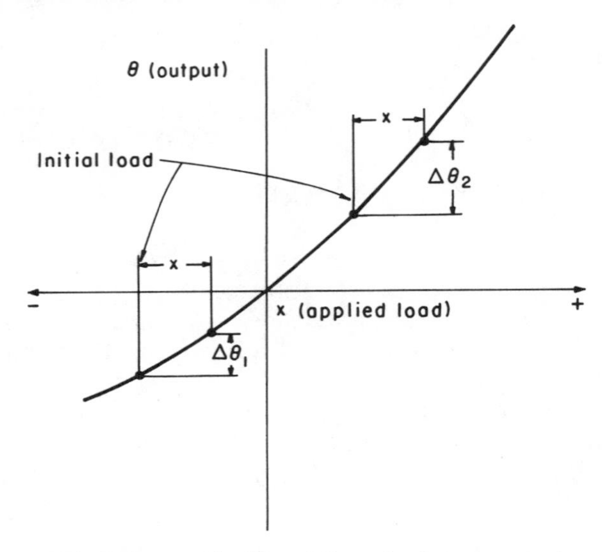

FIGURE 4.31. Nonlinear balance load sensor output.

The second-degree interactions (N^2 and NM) arise because of the elastic deformations, thus they are functions of the modulus of elasticity, which in turn is a function of temperature. If strain gages are used for sensors, they also are affected by temperature. This implies that the balance sensitivity is dependent on temperature.

The second-degree interactions are nonlinear. This means that the raw data values depend on both the applied load and the initial load (Fig. 4.31).

When the balance is calibrated, applied loads are put on the six components, both individually and in combinations adequate to find the interactions. This requires 21 load configurations (Ref. 4.3). During the calibration the values of the raw data should be plotted and faired to obtain smooth curves, or various computer curve-fitting routines can be used. The effects of temperature and initial loads must also be determined. The six interaction equations, similar to Eq. (4.6), contain the applied loads on both sides. This makes the algebraic solution difficult, if not impossible. These equations can be solved by iterative methods using the sensor-measured raw data as a first approximation. Matrix methods can also be adapted for the solutions. For a detailed discussion of balance calibrations see Refs. 4.3 and 4.4.

4.14. CALIBRATION OF INTERNAL STRAIN GAGE BALANCES

Internal or sting balances are generally calibrated outside the tunnel. These calibrations use a calibration rig upon which the balance is mounted (Fig.

4.32). The balance can be rotated through 360° and can be pitched in the vertical plane. The pitching motion is used to keep the balance horizontal at all times and allows the balance and sting deflection to be measured. The rolling provision allows positive and negative loads in lift and side force to be applied with hanging weights. The rig usually has positive stops at 0°, 90°, 180°, and 270° to facilitate these loads. Loads can also be applied at other roll angles, as models on sting balances are often rolled rather than yawed. Calibration bars are used to apply the loads. These bars are indexed so that various values of moments can be applied while holding the normal or side force constant (Refs. 4.3, 4.4).

At zero load the calibration bar is leveled. As loads are applied at each load station the bar is again leveled and the deflection of the sting and calibration bar is recorded. Leveling is done to ensure that each load is applied in the same direction. The deflection measurements are used to correct indicated angles of attack during the test (Fig. 4.33). When all the data have been acquired, the balance is rotated 180° and the process is repeated with loads of the opposite sign. An adequate number of both simple component and combination loads must be applied to define both the first- and second-degree interaction constants.

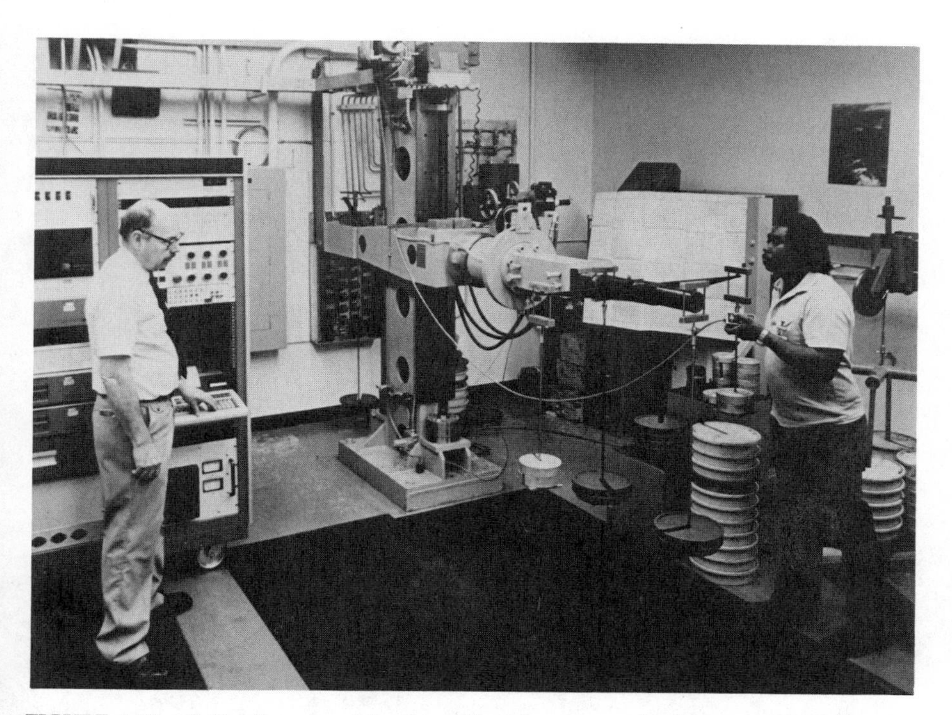

FIGURE 4.32. Calibration mount for internal strain gage or sting balances. This balance is being subjected to combined loads of lift, rolling moment, and pitching moment to determine second-order interactions. (Photograph courtesy of Boeing Aerodynamic Laboratories.)

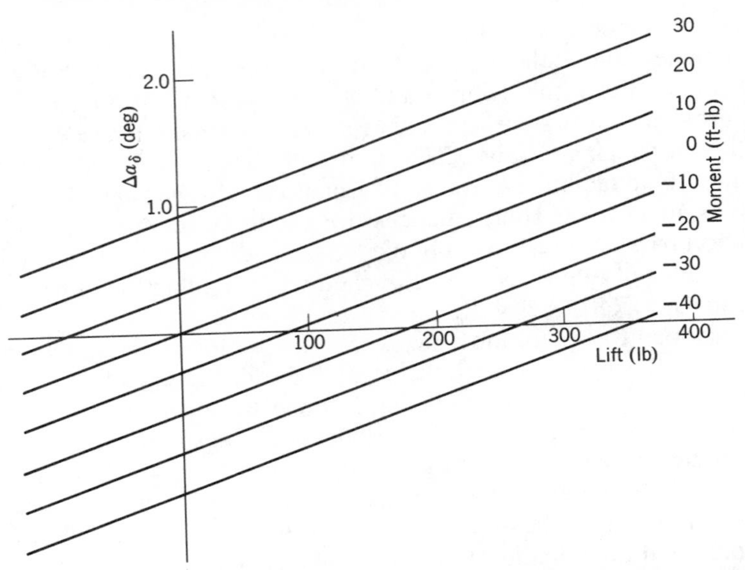

FIGURE 4.33. Angle of attack correction grid for use with an internal strain gage balance.

The loads applied to the balance must be positively correlated to the model. If a plane on the balance is defined as the normal plane, then, when the model is mounted to the balance, the wings are perpendicular to the normal plane. This alignment is usually accomplished by pins, keys, or splines.

The calibration procedure serves several purposes, such as:

1. To proof load the balance.
2. To determine calibration slopes for each component.
3. To determine component sensitivity.
4. To determine balance interactions.
5. To determine deflections under load.
6. To check repeatability of load data.

The balance stability (variation with time) is important, and if such a shift occurs, it may well be due to poor gage bond. The *sensitivity,* or minimum load response, should be checked by observing the minimum load that will produce a response. The *hysteresis* is the degree of repeatability; *linearity* is the variation of the reading-load curve from a straight line.

Apparent strain caused by changes in the balance temperature at zero load (zero shift) can be caused by temperature gradients across the balance structure, and by the average temperature level of the sensing elements, such as strain gages. For balances using strain gages the zero shift due to temperature can be reduced by the following: use of factory-compensated gages; use of bridges with four active arms and thermal-sensing-compensating wire in the bridge circuit; and finally insulating, heating, or cooling of the balance to minimize both temperature excursions and gradients, if possible.

When strain gages are used, the following should be kept in mind. To the extent that the data read out system may be considered as an infinite impedance, with no shunting paths in balancing circuits, the interactions in strain gage balances may be considered independent of the data system. If, however, the balance is calibrated to determine the interactions on a data system that does not have infinite impedance, then during a test the data system used must be compatible, impedancewise, with the calibration data system. Also, the effects of shunting paths across arms of the bridges or across "half bridges" will modify the individual arm or half-bridge sensitivity to the quantities they sense. Thus the interaction quantities that these elements sense may fail to cancel to the same extent that they would with either no or different shunting paths, thus contributing to differences in the interactions.

4.15. CALIBRATION OF EXTERNAL BALANCES

External balances are usually attached to a large mass of concrete and thus are calibrated in place, while internal balances are usually calibrated outside of the tunnel. Some of the tunnels built in the 1970s that are designed to use both external and sting balances have designed the external balance so that it can be removed from the tunnel for calibration in a laboratory. However, a large number of external balances are still calibrated in the test section.

Let us start by making clear the immensity of a wind tunnel balance calibration: with a competent crew the first calibration of a new balance will take 3 months *at least*. This figure supposes that adequate shop facilities for all sorts of changes are available.

Calibration embraces loading the elements of the balance to see whether they read what they should, ascertaining the deflections of the setup, loading the balance in combined cases that simulate the conglomeration of loads the model will put on it, loading with the balance yawed, measuring the natural frequency of the balance so that resonance will not be encountered, and applying fluctuating loads so that it can be determined that the balance reads the mean.

All the above require considerable added equipment, and it is a good idea to make permanent as much of it as possible, since calibration checks will be needed many times during the life of the balance. First in this list comes a loading tee (Fig. 4.34).

Another form of balance tee is shown in Fig. 4.35. This tee allows each of the six components to be applied independently.

The calibration tee must provide a method to accurately attach cables for applying the loads. A reasonable tolerance is 0.005 in. on their location. The

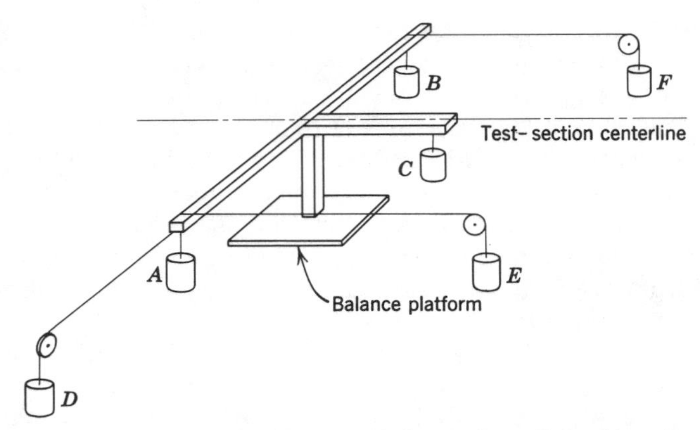

FIGURE 4.34. Balance loading tee. Weights added at A, B, or C should produce no drag or crosswind force, and a weight moved from E to F should produce no indicated change in drag.

FIGURE 4.35. External balance loading tee. Each component can be independently loaded or combined loads can be applied. Shown are loads to side force and rolling moment. (Photograph courtesy of University of Washington.)

cable attachment can be knife edge hooks or ball and socket joints with the balls swaged onto aircraft cable for large balances. Smaller balances can use piano wire.

With either of the tees (Figs. 4.34 and 4.35) provisions have to be made to ensure that the loads applied by cables are parallel or perpendicular to the tunnel centerlines for a wind-axis balance. First the balance must be set at zero yaw. Then alignment lines are scribed on the floor using toolmakers' blue (or lines could be drawn on masking tape). A plumb bob can be used to align the pulleys. Next the cables are attached and an engineering level (optical) and 0.01 rulers can be used to level the cables, which must be loaded to about one-half of the full load. After leveling, the horizontal alignment should be checked. Pulleys should have a large radius, about 8 in. minimum, and should be built for this purpose. The groove for the cable should be cut so that the cable lies on the top edge of a groove narrower than the cable diameter for positive alignment. The pads holding the pulleys should be adjustable in the vertical and horizontal planes. The most difficult load is positive lift. One approach is to lay a strong steel beam across the top fillets and use a pivoted lever to load lift. This reduces the weights and either a cable or rod with ball joints can be used. This can be plumbed optically or by use of a precision bubble level.

A set of calibrated weights will be needed; if it is decided not to buy a full set, they may usually be borrowed from the local state highway department. A half dozen dial gages for measuring deflections will also be needed.

The first step in calibration is a complete operational checkout of the measuring units and data read out system. For a new balance each component should be loaded to the maximum load and balance deflections checked. The amount of deflection is a function of the balance design (see Sections 4.4–4.6). These loads are also proof loads, although this usually is not a problem. Most balances have balance stops to protect the balance in case of a failure. These can be set at this time. A repaired balance should be put through the same loading cycle. Quite often new or repaired balances require two or three full load loading cycles before they settle in and give good repeatability.

The traditional method of calibrating external balances has been to adjust the balance to minimize the interactions. This was done to minimize the labor of reducing the data by hand. This practice resulted in the balance readings being very close to the applied loads. Thus, when the coefficient data on repeat runs or between tests did not check, the raw balance readings proved useful in finding the source of the error. If balance readings agree in trends with the coefficients, then a check of the model configuration usually elicits from the customer "the model is just the same, but" and the problem is solved.

With the ready availability of digital computers it is possible to merely calibrate the balance and then measure the interactions and correct the data through a complex but quick computer program. This latter method has been common for internal balances (see Section 4.14). Since smaller tunnels used for instruction usually do not have dedicated computers fed by data systems, but rather have the data entered by hand, the older method may still have some advantages.

The balance alignment procedure for minimum interactions is:

1. For each component load and adjust the component's slope to 1 : 1 or until load is equal to the output reading.

2. Load each component and reduce the interactions on the other five components (this assumes a six-component balance). The best way to proceed is to first make sure that lift is perpendicular to drag and side force, using lift as the load. At the same time make sure the lift load passes through the moment center by checking pitching moment and rolling moment. Next, make sure drag is perpendicular to side force. This may require a recheck of lift, depending on the balance. This task is tedious on a new balance, but a feel for the balance is soon acquired that makes the job easier. When the interactions are minimized, the remainder is due to deflections that must be present to make the measurements.

3. During the work in 2, balance output will have to be plotted. A plot similar to Fig. 4.36 is useful to determine the error on the component due to its own load. Balance loads should be applied from zero to full scale to zero. Plots of component loaded versus the other five components will show zero shifts and hysteresis. Depending on the balance design, it may be necessary

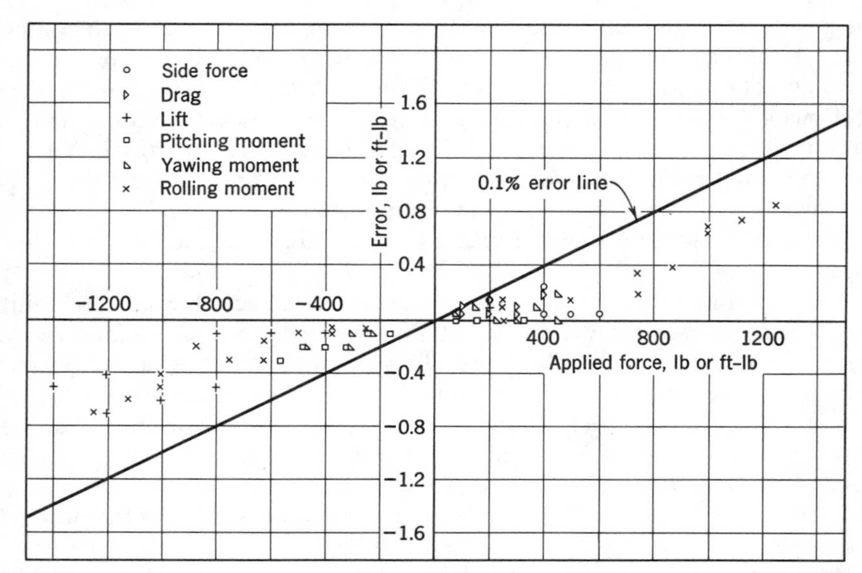

FIGURE 4.36. Plot of actual error versus applied load for the six force systems when each component is loaded separately. The straight line indicates the maximum permissible error of 0.1% of the applied load.

to check the interactions with the balance yawed. This is to ensure that the balance resolving center coincides with the model trunnion. The trunnion is the point about which the model is pitched and yawed. If these two points do not coincide, the data reduction will have to account for the discrepancy.

4. When minimum interactions have been achieved, repeated loadings should be made to check the balance repeatability. The setting of the safety stops should be rechecked. Combined loadings should also be applied to determine the magnitude of the second-degree interactions (see Section 4.13). It is also desirable to apply very small loads to check the balance sensitivity. At UWAL a small aluminum bell crank mounted on a torsional flexure pivot was built for this purpose. The vertical arm was connected to the balance by fine piano wire. The horizontal arm contained the weight pan and a bubble level. This device was supported on a scissor jack stand. The jack stand was used to level the wire to the balance and level the bell crank. Then loads of 0.01 lb were applied to drag and side force, and moments of 0.10 in.-lb were applied to the yawing moment. The pan and gravity were used for lift, pitching, and rolling moments.

5. The natural frequency of the balance about the three axes should be determined along with the effect of the model weight on the frequencies. The natural frequencies are a function of the balance design. The balance also should be checked with fluctuating loads to ensure that it measures the mean value.

Example 4.1. The precise balance calibrations with second-degree terms are often not needed for small tunnels used for instructional purposes. This is especially true if the data are not read by a small data logging system and/or if the computer used to reduce the data has limited capacity.

The balance is loaded as any balance and the interactions are plotted. It will be assumed that the balance is a three-component balance that can be adjusted. When a load is applied to lift, plots of lift load (on y axis) versus the readings on lift, drag, and pitching moment (on x axis) are made. If the curves are slightly nonlinear, they are replaced by a linear approximation. When this has been done for all three components, the slopes of the nine curves are taken as $K_B = \Delta L_R/\Delta M_B$. The subscript B is the load applied to the balance and the subscript R is the balance output reading (both are in engineering units). The subscripts on the K's define the slope where 1 = lift, 2 = drag, 3 = pitching moment. The first number is the balance reading (sub R) and the second number is the balance load (sub B).

Thus, for the lift reading one has

$$L_R = K_{11}L_B + K_{12}D_B + K_{13}M_B$$

or, in matrix form,

$$\{F_R\} = [K_{ij}]\{F_B\}$$

where F can be a force or a moment as required. This matrix equation can be inverted to give

$$\{F_B\} = [K_{ij}]^{-1}\{F_R\}$$

which is the required equation to determine the forces and moments applied to the balance by the model from the balance output readings.

As an example:

$$\begin{aligned} K_{11} &= \Delta L_R/\Delta L_B = 1.000 & K_{21} &= 0.0221 & K_{31} &= 0.162 \\ K_{12} &= \Delta L_R/\Delta D_B = 0.00622 & K_{22} &= 1.00 & K_{32} &= -0.00338 \\ K_{13} &= \Delta L_R/\Delta M_B = -0.330 & K_{23} &= 0 & K_{33} &= 1.0000 \end{aligned}$$

Then the $[K_{ij}]$ matrix is

$$\begin{bmatrix} 1.000 & 0.00622 & -0.3300 \\ 0.0221 & 1.000 & 0 \\ 0.162 & -0.00338 & 1.000 \end{bmatrix}$$

and $[K_{ij}]^{-1}$

$$\begin{bmatrix} 0.949 & -0.00482 & 0.313 \\ -0.0212 & 1.000 & -0.00692 \\ -0.1536 & 0.00416 & 0.948 \end{bmatrix}$$

Then the balance calibration equations are

$$\begin{aligned} L_B &= 0.949L_R - 0.00482D_R + 0.313M_R \\ D_B &= -0.0212L_R + 1.000D_R - 0.00692M_R \\ M_B &= -0.1536L_R + 0.00416D_R + 0.948M_R \end{aligned}$$

If one is trying to adjust the balance for minimum interactions, then from the K_{21} term or the $\Delta D_R/\Delta L_B$ term it can be seen that the lift and drag are not perpendicular. The K_{31} or the $\Delta M_R/\Delta L_B$ term shows that the balance moment center is aft of the geometric vertical centerline of the balance. Thus the curves can be used to provide information on the required adjustments.

4.16. BALANCE AERODYNAMIC ALIGNMENT

By definition lift is perpendicular to the remote velocity and drag is parallel to it. In an ideal tunnel with the flow parallel to the test-section boundaries, it would only be necessary to align the external balance so that lift is perpendicular to the ceiling and floor and drag parallel to the ceiling and floor.

Unfortunately, most tunnels do not have absolutely perfect flow in the test section. Usually there is some up or down flow (usually called upflow) and some cross flow. There is also a local upflow due to the air flowing over the fairing, which shields much of the balance mounting struts from the airstream. Because upflow affects the accuracy of drag, it is more critical than cross flow for a full model. With a floor–ceiling-mounted half model the cross flow would be most critical. The following discussion will be for a full model.

The tunnel flow angularity usually is not uniform across the tunnel in the region occupied by the wing, and it may vary with the dynamic pressure. For a given wing planform, an average value of the upflow can be obtained for a given dynamic pressure and the balance aligned so that drag is parallel and lift perpendicular to the flow. But for any other wing or dynamic pressure the balance would not be properly aligned. Furthermore, the problem of applying loads during the balance calibration when the loads are not level will just make difficult tasks more difficult. The usual procedure is to align the balance so that lift is perpendicular to the test-section ceiling and floor and drag parallel to them or to an internal balance reference surface (Sections 4.14 and 4.15). Then the upflow is measured and corrected for in the data work up.

One thing simplifies the alignment: the lift for most tests is from 5 to 25 times larger than the drag, and it is usually sufficient to align so that no lift appears in the drag-reading apparatus without checking to see whether any drag appears in the lift-reading mechanism beyond ascertaining that the drag system is perpendicular to the lift.

The balance alignment to the tunnel flow is generally accomplished by running a wing both normal and inverted from zero lift to stall. To ensure equal support strut interference for both normal and inverted runs, dummy supports, identical to the conventional ones, are installed downward from the tunnel roof. They are then as shown in Fig. 4.37. The data from both normal and downward lift are plotted as lift curves (C_L versus α), polars (C_L versus C_D), and moment curves (C_L versus C_m). The negative lifts and moments are plotted as though they were positive. (See Fig. 4.38.) The angular variation between the lift curves is twice the error in setting the angle of attack and, as shown, indicates that the α is set too low. That is, when the balance angle indicator reads -1°, the model is really at 0° to the average wind. The polar shows that the lift is not perpendicular to the relative wind, part of it appearing as drag. Here the balance is tipped aft in reference to the relative wind, for a component of the lift is decreasing the drag when the lift is positive and increasing it when the lift is negative (Figs. 4.39 and 4.40).

The same procedure outlined above for a wing must be followed for each complete model: runs with the image system in; model both normal and inverted. These runs yield the true angle of zero lift and alignment correction. The additional runs needed for tare and interference are discussed in the next section.

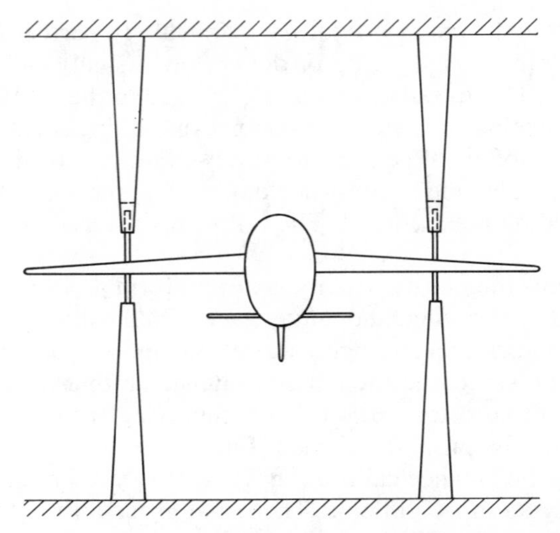

FIGURE 4.37. Arrangement for determining tare and interference simultaneously.

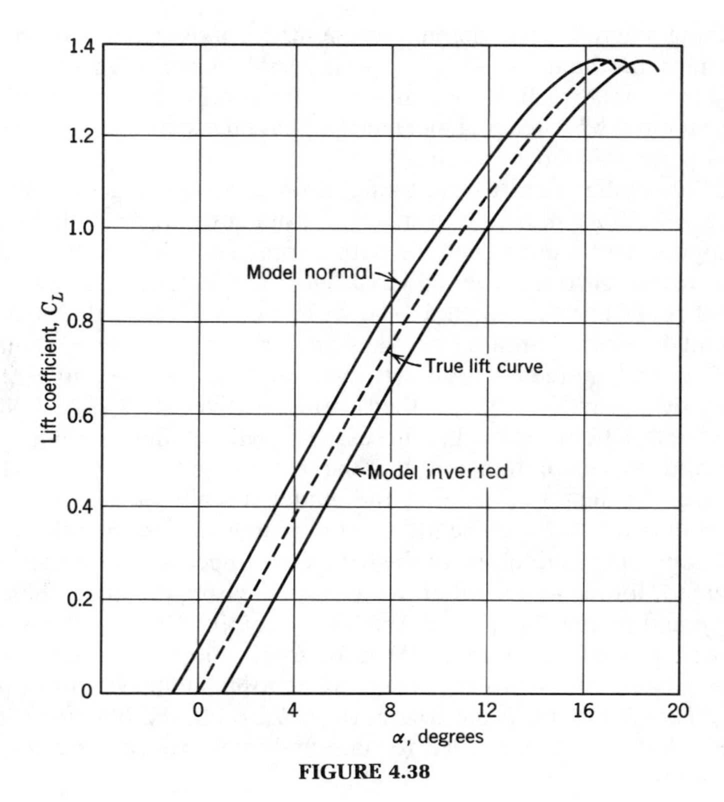

FIGURE 4.38

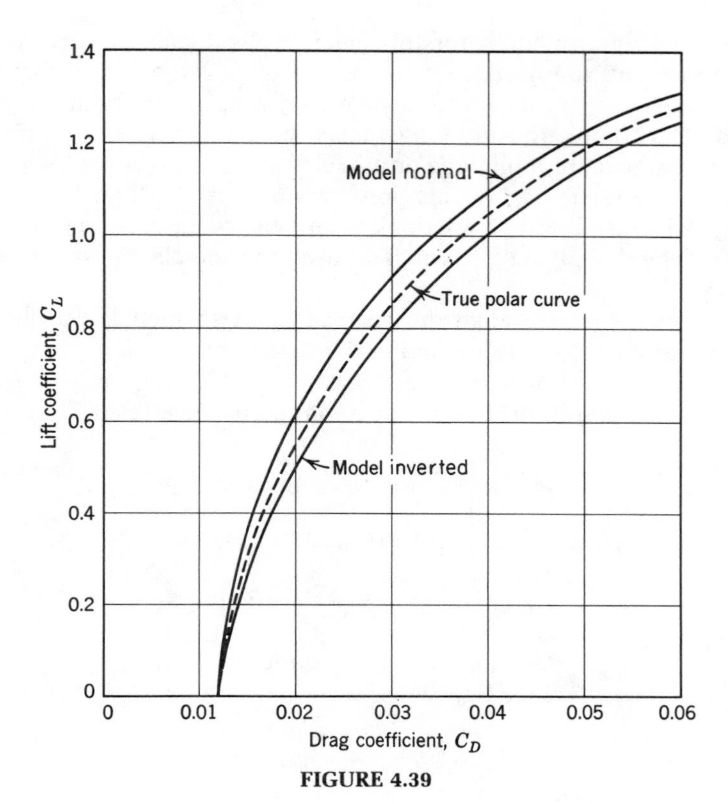

FIGURE 4.39

FIGURE 4.40

The tunnel time required for this balance alignment plus the additional time for tare and interference runs can be quite large. Also, many wind tunnel models are not designed so that they can be run both upright and inverted. Thus, in many wind tunnel tests these values are taken from data obtained from a calibration wing or model, and the upflow values are determined over the range of dynamic pressures used in the tunnel.

If angular and velocity variations in the airstream are large, the above alignment would apply only to wings whose span and chord approximate the test wing.

It is impracticable to align the balance for each model, and hence the misalignment correction is applied in the data workup as follows:

Suppose that the polars of the normal and inverted runs appear as in Fig. 4.39. With the wing in the normal position the balance reads

$$C_{D\,\text{indicated}} = C_{D\,\text{true}} - C_{L\,\text{indicated}}\,(\tan \alpha_{\text{up}}) \tag{4.7}$$

where α_{up} = angle of misalignment (Fig. 4.40). Hence

$$C_{D\,\text{ind}} - C_{D\,\text{true}} = -C_{L\,\text{ind}} \tan a_{\text{up}} \tag{4.8}$$

The correct C_D ($C_{D\,\text{true}}$) lies halfway between the $C_{D\,\text{normal}}$ and $C_{D\,\text{inverted}}$ curves. Let the difference between the curves at some C_L be ΔC_D. Then

$$\Delta C_D = C_{D\ \text{normal}} - C_{D\ \text{inverted}} \tag{4.9}$$

$$C_{D\,\text{ind}} - C_{D\,\text{true}} = \Delta C_D/2 \tag{4.10}$$

and, if the difference between the curves is read at $C_L = 1.0$, the angle of misalignment, α_{up} may be found from

$$\tan \alpha_{\text{up}} = (\Delta C_D/2)_{C_L=1.0} \tag{4.11}$$

The correction to the drag coefficient is then

$$C_{D\,\text{true}} = C_{D\,\text{ind}} + (C_{L\,\text{ind}})\tan \alpha_{\text{up}} \tag{4.12}$$

True C_L is close enough to $C_{L\,\text{ind}}$ so that usually no correction to C_L is needed.

When α_{up} is a small angle, tan α_{up} in Eq. (4.11) is often replaced with the angle in radians. Rather than calculating tan α_{up} in Eq. (4.8) at one value of C_L such as 1.0, it is better to measure ΔC_D at several values of C_L, plot ΔC_D or $\Delta C_D/2$ versus C_L. Then fair in a linear curve or use linear regression to obtain the slope or tan α_{up}.

Two more important points in regard to the evaluation of the alignment correction remain. First, in order to have the tare and interference effects identical for both the model normal and inverted runs, the image system must be installed and the dummy struts arranged as in Fig. 4.37.

A second problem with the tunnel upflow and dynamic pressure calibration is that in some tunnels these values can be an undetermined function of time. Some tunnels will expand and contract with the weather, and wooden tunnels can be affected by humidity. This may cause a change in tunnel flow quality. A more serious change in tunnel flow can occur in tunnels equipped with screens to reduce turbulence. Screens tend to get dirty as the tunnel is run, and since they are located in the tunnel bellmouth, they can have a large effect on both the tunnel upflow and dynamic pressure. A simple way to check on this problem, besides visually checking screens and cleaning them, is to run a calibration wing at several check q's in an upright position. If there is no change in the lift curve slope, the q calibration is the same. No rotation of the polar, either corrected or not corrected for upflow, means the upflow is the same. Another method to track upflow is to install a yawhead or similar flow-measuring instrument in a fixed location using a mount that maintains probe alignment. The probe is best run without a model in the tunnel. This method is not as accurate or as sensitive as the calibration wing.

The engineer who upon finding a change in the data for a second series test of a certain model proclaimed that $\alpha_{Z.L.}$ was "not where he left it" was not entirely without scientific backing.

The method of evaluating upflow with upright and inverted runs is the same with a sting balance. The sting balance can simplify inverting the model if the sting is capable of rolling 180°.

Although the error in angle of attack can be determined from the lift curve of Fig. 4.38, the upflow angle is often so small that it is difficult to obtain it from the lift curve. Usually the upflow angle is taken from Eq. (4.11) or a plot of Eq. (4.8) and added to the model's geometric angle of attack. For a model mounted in a normal upright condition with upflow in the tunnel

$$\alpha = \alpha_{\text{geom}} + \Delta\alpha_{\text{up}} \quad \text{(in degrees)} \tag{4.13a}$$

$$\Delta C_{D\,\text{up}} = \Delta\alpha_{\text{up}} C_{LW} \tag{4.13b}$$

with $\Delta\alpha_{\text{up}}$ in radians and C_{LW} = wing lift coefficient. The above corrections to the angle of attack and the drag coefficient are not the total corrections. There are additional corrections owing to the constraint to the tunnel flow due to the walls or wall corrections (see Chapter 6).

When the variations in upfiow are very large in the region of the wing as measured by a yawhead, laser, or similar instrument, the following method can be used. The upflow at several spanwise stations of the wing is multiplied by the chord at that location. The product is then plotted versus the wing span and the area under the resulting curve is divided by the wing area to yield an average value for the wing. A similar method can be used to determine an average for the wing's downwash at the tail. This method is not as accurate because it is difficult to get the same accuracy with a flow probe as is given by a good balance.

When evaluating the tunnel upflow–balance misalignment, the model support tares in both upright and inverted runs must be known. The upflow for the tare and interference is obtained with model and image as shown in Fig. 4.37. This yields a tunnel upflow balance alignment without the local upflow effect of the balance fairing or wind screen. These balance alignment data are then used to evaluate the support and interference tares as detailed in the next section. When the model is then run both upright and inverted without the image system, the local upflow from the wind screens is also accounted for.

When the tunnel upflow and balance alignment value is determined, that is, tan α_{up}, the value should be checked by applying it via Eq. (4.12) to both the upright and inverted runs. If tan α_{up} is correct, the upright and inverted curve should collapse and form a single or true curve as in Fig. 4.39.

As can be seen, the determination of balance alignment and upflow is a time-consuming effort if it is done for each model that is tested. This is why it often is done with a calibration model.

In order to correct for misalignment of the side-force balance, two runs must be made, with both the tare and the interference dummies in. The model in normal position should be yawed in one direction and then inverted and yawed in the same direction relative to the tunnel. The correct side-force curve will be halfway between the curves made by model normal and model inverted. The inversion is necessary to nullify effects of the model's irregularities.

Side-force corrections as outlined are rarely made, since they entail a set of dummy supports that can be yawed; moreover extreme accuracy in side force is not usually required. The principles of the correction, however, are important.

It should be recalled that changes in the shape of a polar curve may be due to scale effects and that comparisons of various tests of similar airfoils must be made from readings at the same effective Reynolds number. (It has been shrewdly noted that, if the section selected is one of the more "popular" types that have been frequently tested, it is nearly always possible to find some results that will "agree" with yours.)

4.17. TARE AND INTERFERENCE MEASUREMENTS

Any conventional wind tunnel setup requires that the model be supported in some manner, and, in turn, the supports will both affect the free air flow about the model and have some drag themselves. The effect on the free air flow is called *interference;* the drag of the supports, *tare*. Although tare drags could be eliminated entirely by shielding the supports all the way into the model (with adequate clearances, of course), the added size thus necessitated would probably increase the interference so that no net gain would be achieved.

The evaluation of the tare and interference is a complex job, requiring thought as well as time for proper completion. The student invariably suggests removing the model to measure the forces on the model supports. This procedure would expose parts of the model support not ordinarily in the airstream (although the extra length could be made removable) and would fail to record either the effect of the model on the supports or the effect of the supports on the model.

First let us consider a rarely used method that evaluates the interference and tare drag separately. Actually the value of the sum of the two will nearly always suffice without our determining the contribution of each, but, besides being fundamental, this long method may offer suggestions for determining interference for certain radical setups. The procedure is as follows:

The model is first tested in the normal manner, the data as taken including both the tare and the interference. In symbolic form we have:

$$D_{\text{meas}} = D_N + I_{\text{LB}/M} + I_{M/\text{LB}} + I_{\text{LSW}} + T_L \tag{4.14}$$

where D_N = drag of model in normal position, $I_{\text{LB}/M}$ = interference of lower surface bayonets on model, $I_{M/\text{LB}}$ = interference of model on lower surface bayonets, I_{LSW} = interference of lower support windshield, and T_L = free air tare drag of lower bayonet.

Next the model is supported from the tunnel roof by the "image" or "mirror" system. The normal supports extend into the model, but a small clearance is provided (Fig. 4.41*a*). The balance then reads the drag of the exposed portions of the supports in the presence of the model. That is,

$$D_{\text{meas}} = T_L + I_{M/\text{LB}} \tag{4.15}$$

For the interference run the model is inverted and run with the mirror supports just clearing their attachment points (Fig. 4.41*b*). We then get

$$D_{\text{meas}} = D_{\text{inverted}} + T_{\text{U}} + I_{\text{UB}/M} + I_{\text{USW}} + I_{M/\text{UB}} + I_{\text{LB}/M} + I_{\text{LSW}} \tag{4.16}$$

where D_{inverted} = drag of model inverted (should equal the drag of the model normal, except for misalignment) (see Section 4.16) and the symbol U refers to the upper surface.

Then the mirror system is removed and a second inverted run is made. This yields

$$D_{\text{meas}} = D_{\text{inverted}} + T_U + I_{\text{UB}/M} + I_{M/\text{UB}} + I_{\text{USW}} \tag{4.17}$$

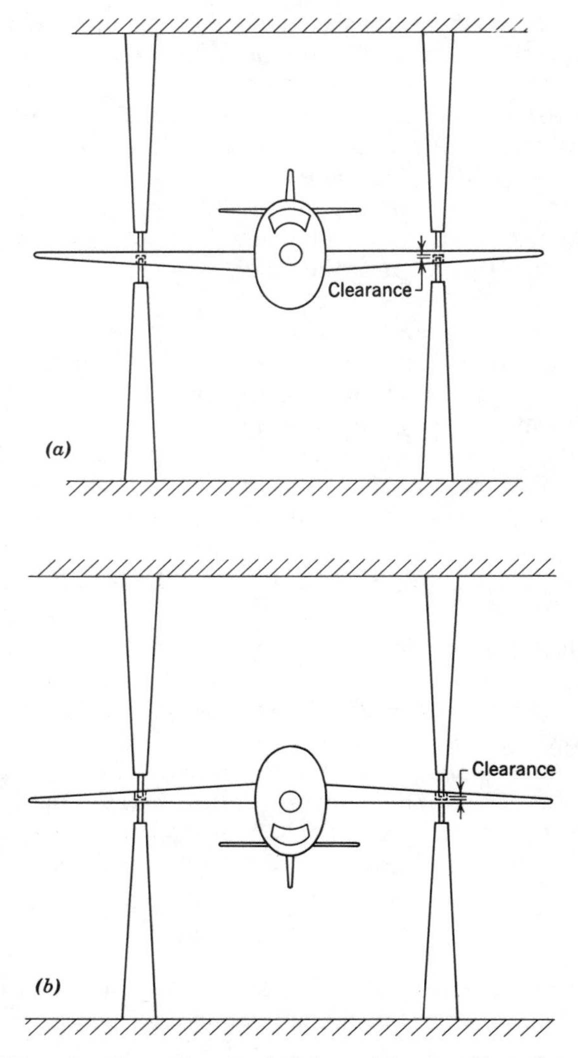

FIGURE 4.41. Mirror (or "image") method of determining the effect of the supports on the model.

The difference between the two inverted runs is the interference of supports on the lower surface. That is, Eq. (4.16) minus Eq. (4.17) yields

$$I_{\mathrm{LB}/M} + I_{\mathrm{LSW}} \tag{4.18}$$

By subtracting Eqs. (4.15) and (4.18) from the first run [Eq. (4.14)], the actual model drag is determined if the balance is aligned. As explained more fully in Section 4.16, the difference between runs made in the normal and inverted position with the mirror system in can be used to find the proper corrections for alignment.

The support tare and interference effects can be found in three runs instead of four by using a slightly different procedure. In this case the normal run is made, yielding

$$D_{\text{meas}} = D_N + T_L + I_L \tag{4.19}$$

where $I_L = I_{M/\text{LB}} + I_{\text{LB}/M} + I_{\text{LSW}}$. Next the model is inverted and we get

$$D_{\text{meas}} = D_{\text{inverted}} + T_U + I_U \tag{4.20}$$

Then the dummy supports are installed. Instead of the clearance being between the dummy supports and the model, the exposed length of the support strut is attached to the model, and the clearance is in the dummy supports (Fig. 4.37). This configuration yields

$$D_{\text{meas}} = D_{\text{inverted}} + T_L + I_L + T_U + I_U \tag{4.21}$$

The difference between Eq. (4.20) and Eq. (4.21) yields the sum of the tare and interference $T_L + I_L$. One actual setup is shown in Figs. 4.42 and 5.5.

The second procedure has the advantage that the dummy supports do not have to be heavy enough to hold the model, nor do they require any mechanism for changing the angle of attack.

A third method of evaluating the tare and interference, sometimes employed where an image system is impracticable, consists of adding extra dummy supports on the lower surface and assuming their effect to be identical with the actual supports. Sometimes there is danger of mutual interference between the dummies and the real supports.

Doubtless the increase of runs necessary to determine the small tare and interference effects and the concern expressed about the difference between those effects on the upper and lower surfaces seem picayune. Yet their combined effect represents from 10 to 50% of the minimum drag of the whole airplane—clearly not a negligible error.

It should be noted that the tare and interference forces vary with angle of attack and with model changes. They must be repeatedly checked and evaluated, particularly for major changes of wing flaps and nacelle alterations close to the support attachment. With many models every configuration must have its own support interference evaluated, a long and tiresome test procedure.

Because the evaluation of tare and interference requires large amounts of time, they often are approximated by evaluating them on a calibration model, and these values are then applied as approximate tares to other models.

For models of propeller-driven aircraft where the model is powered by

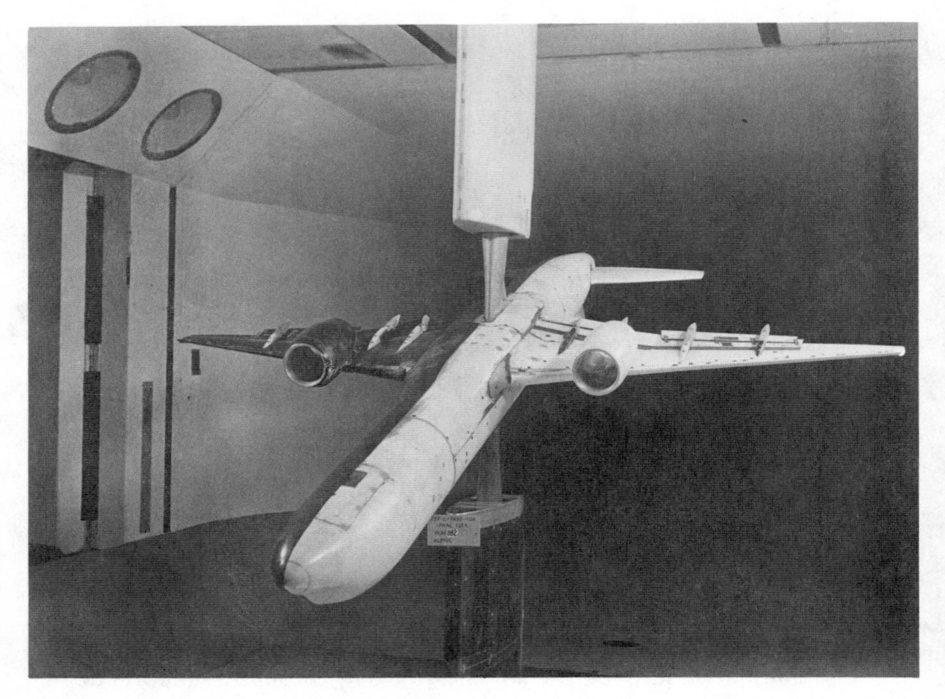

FIGURE 4.42. Tare and interference runs on Boeing 767 mounted on single strut and pitch arm. The model is painted light blue on right side and black on left side for flow visualization and identification purposes. (Photograph courtesy of Boeing Aerodynamic Laboratories.)

variable-frequency electric motors (see Chapter 5) it is necessary to bring both power and water-cooling lines into the model. These usually are too large to be contained inside the mounting strut. If a special fairing is required to get the wires, etc., into the model or if they are taped to the mounting strut, the tares for these items must be evaluated. The easiest method is to make a set of pitch and yaw runs with the same model configuration, with and without the wires. The difference between these runs is then added to the standard strut tares.

A note of caution in regard to Figs. 4.37 and 4.41: the images are supposed to be mirror images. However, if the model struts enter the model fuselage (Section 4.8 and Figs. 4.11 and 4.13), the trunnion, which usually is on the balance moment center and the tunnel centerline, is not at the centerline of the model. When the model is inverted, the exposed portion of the image strut and the distance from its fairing to the model must be exactly the same as the upright model. This often requires the length of the image fairing exposed to the tunnel flow to be either longer or shorter than the standard, or lower, fairing. Thus, it generally is a good policy to make the image fairing longer than the normal fairing and let it extend through the tunnel ceiling.

In the three-run method of taking tare and interference there is a possibility of air flowing in and out of the gap between the mounting strut and the fairing if the standard struts do not have a seal at their tops (Section 4.8). To get the correct tare and interference the image system must duplicate exactly the geometry in this region.

For a support system that uses two supports on the wing and a tail support, an alternate method is sometimes used. The reason for this alternate treatment is that the length of the tail support varies as angle of attack is changed. This factor so complicates the dummy arrangements that a system is usually employed which does not require a complete dummy tail support.

The procedure is as follows: Consider the second method of evaluating the tare and interference. When the image system is brought down to the inverted model, a short support is added to the then upper surface of the model where the tail support would attach in a normal run. The piece attached corresponds in length to the minimum exposed portion of the tail support and increases the drag of the model by the interference and tare drag of a tail support on the model's lower surface. For angles of attack other than that corresponding to minimum length of exposed tail support, the drag of the extra exposed tail support length must be evaluated and subtracted.

A rear support windshield that moves with the rear support to keep a constant amount of strut in the airstream could be employed as long as the added interference of the moving shield is evaluated by a moving shield dummy setup.

The evaluation of the tare, interference, and alignment of a wing-alone test follows the procedure outlined above, except that further complication is introduced by the presence of a sting that must be added to the wing to connect it to the rear strut of the support system. The tare and interference caused by the sting may be found by adding a second sting during the image tests. As may be noted in Figs. 4.43 and 5.5, the attachment of the sting to the rear support includes a portion of the strut above the connection, and the dummy sting has a section of support strut added both above and below its connection point. This complication is needed to account for the interference of the strut on the sting, as follows.

When the wing is held at a high angle of attack, there will be an obtuse angle below the sting. When the wing is inverted and held at a high angle relative to the wind, there will be an acute angle below the sting, for the rear strut will then be extended to its full length. To eliminate this difference between the normal and inverted tests the support strut is extended above the sting attachment point, so that the sum of the angles between the sting and the support is always 180°. The image sting has the same arrangement. Note that although the angles between the sting and the rear support vary with the angle of attack, the image sting is always at right angles to its short rear support strut. Furthermore, the image rear support strut does not remain vertical but changes its angle with the wing. The error incurred by failing to have the sting image system simulate the exact interference and rear strut angle is believed to be negligible.

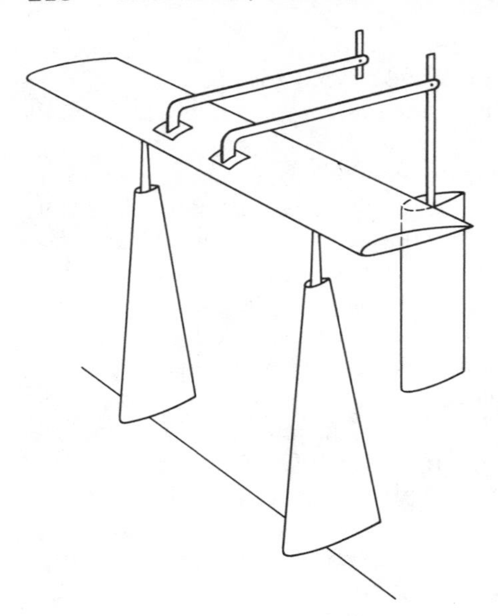

FIGURE 4.43. Setup for determining tare and interference of sting.

Tare and interference for the tail strut alone may be evaluated as shown in Fig. 4.44, or for a sting mount as shown in Fig. 4.45. The setup for determining the tare and interference for a fork support is shown in Fig. 4.46. Here the model is supported externally and a small clearance left where the struts come into the wing. This measures the drag of the fork plus the effect of the wing on the fork, but not the effect of the fork on the wing, which experience has shown to be small. Hence, the tares are used as approximate tares where the desire to save tunnel time precludes taking the usual tare and interference runs.

Figure 4.47 shows the results of a wind-alone test for an NACA 0015 wing of AR = 6.0. The wing in this particular test was small, and the corrections for tare, interference, and alignment are correspondingly large, but the variation of the corrections is typical. The following points are of interest:

1. The correction for tare and interference decreases as C_L (and α) increases.
2. The incidence strut drag decreases with increasing α. (The amount of strut exposed decreases with α.)
3. The alignment correction increases with C_L.

A large amount of interference may arise from air that bleeds through the windshields that surround the support struts to protect them from the windstream. These struts frequently attach at points of low pressure on the model, and, if the shield is brought close to the model, a considerable flow may be induced that will run along the model. This flow may stall the entire

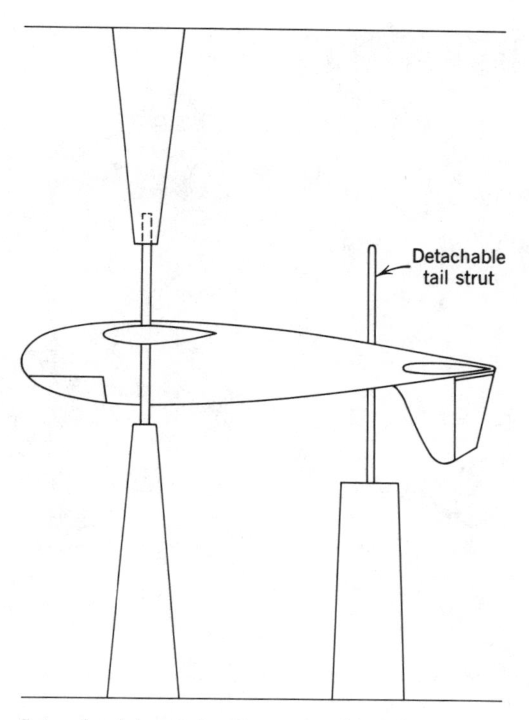

FIGURE 4.44. Setup for determining the tare and interference of the tail support.

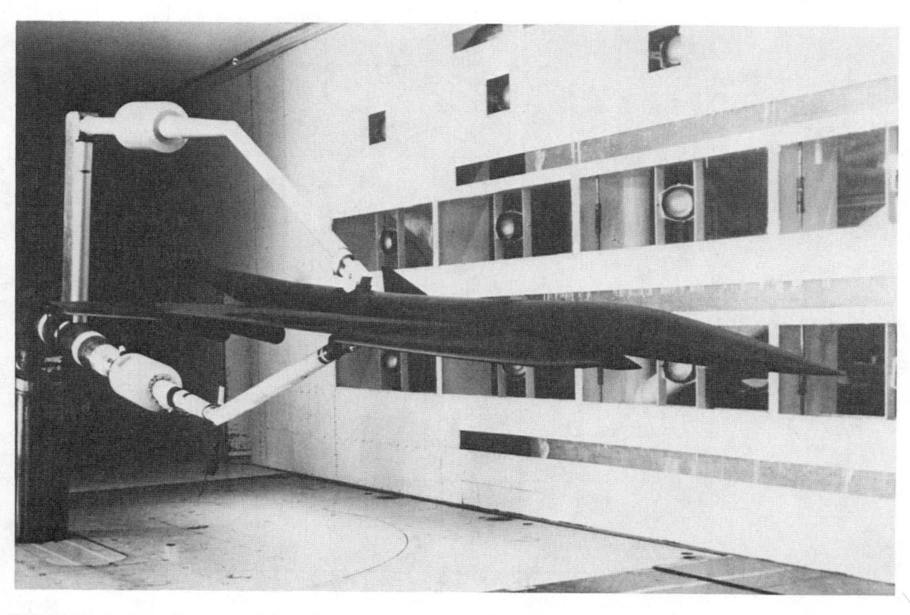

FIGURE 4.45. Tare and interference: supersonic transport with surface sting support in NASA Langley 4 × 7 m tunnel. (Photograph courtesy of NASA Langley and Dynamic Engineering, Inc.)

FIGURE 4.46. Tare and interference determination for a fork-type support. (Courtesy University of Washington.)

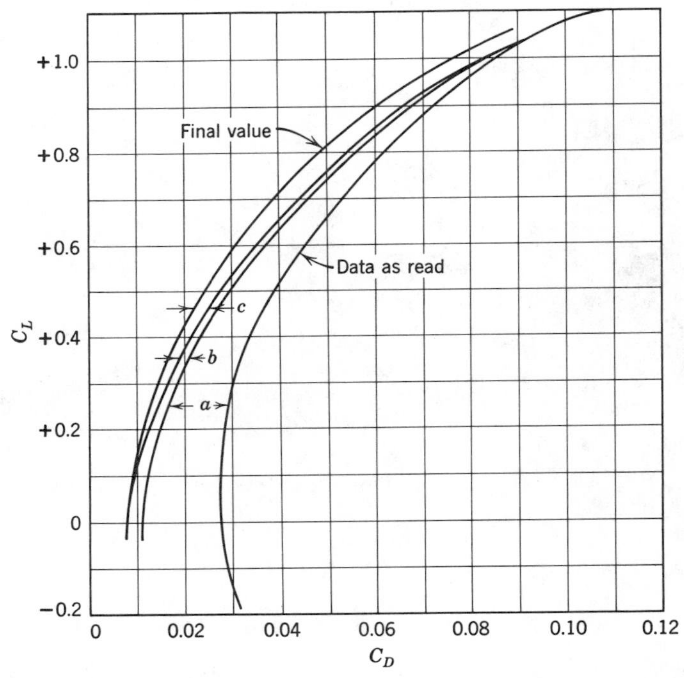

FIGURE 4.47. Corrections for a wing-alone test. (*a*) Sting and support tare and interference. (*b*) Drag of exposed incidence strut. (*c*) Alignment.

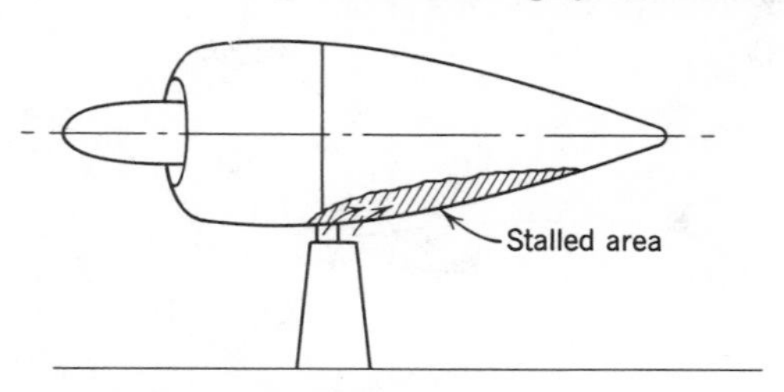

FIGURE 4.48. Effect of bleeding.

underside of the model and produce results that are not only wrong but also unsteady and difficult to evaluate. It is, therefore, frequently advantageous to terminate the windshields well below the model and let the test be subjected to added but well-efined tare drag (see Fig. 4.48) and to provide seals to stop the flow.

Subsonic testing of military aircraft are often done over angle of attack ranges from 0° to 90°, and missiles from 0° to 180°. When the model is similar to a slender body and is mounted on a swept strut parallel to the body axis and freestream, the strut will be similar to a splitter plate behind a cylinder. A splitter plate behind a cylinder can inhibit the vortex formation and reduce the wake pressure. When this occurs with a wind tunnel model, there is an improper flow simulation. The effect of the splitter plate on a cylinder is a function of the plate size and location relative to the cylinder and Reynolds number (Ref. 4.5).

The problem of properly taking and evaluating support tares and interferences for models at a high angle of attack is difficult. The models are generally mounted on stings with internal balances making the use of mirror images difficult. If the model is supported by auxiliary supports, there can be mutual interferences between the support, the model, and the normal sting. Auxiliary supports may require a second internal balance, adding possible complexity to the model. And finally, the complexity caused by the large range of angles that must be covered.

4.18. PROFILE DRAG BY THE MOMENTUM METHOD

It should be noted that a balance is not always required in a wind tunnel. The drag may be obtained by comparing the momentum in the air ahead of the model with the momentum behind the model, and the lift may be found by integration of the pressures on the tunnel walls. These artifices are most generally employed in airfoil section research in a two-dimensional tunnel.

It has been generally accepted that spanwise integration for profile drag is not necessary since the flow is two dimensional, other than a region near the wall. Several tunnels, however, have noted a spanwise variation of the profile drag coefficient when measured with wake rakes behind the airfoil.

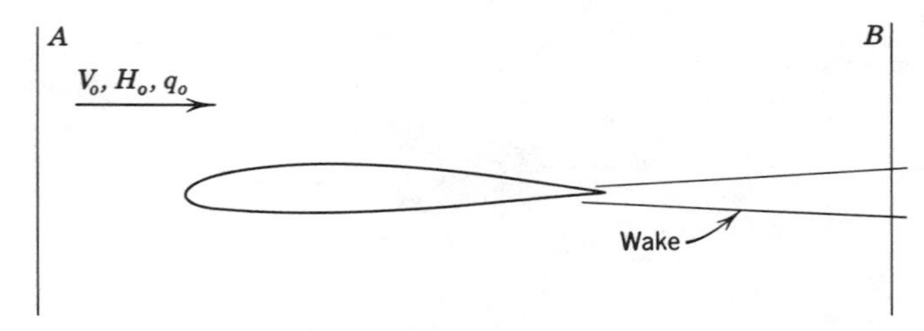

FIGURE 4.49

These spanwise variations are repeatable, but can change with model configuration. In Ref. 4.6, Fig. 11, the variation in profile drag for a four-element airfoil is shown and in Ref. 4.7 and Fig. 4 from Ref. 4.8 similar variations are shown in profile drag at four Reynolds numbers from 1×10^6 to 3×10^6. The spanwise variation of profile drag decreased with increasing Reynolds number, but the variation is significant at all Reynolds numbers.

This problem is an area where further investigation appears warranted. Because the profile drag varies with span, how does one measure it? Obviously a measurement at a single spanwise station is suspect, and thus some average value should be determined. Should this be a mean or a weighted average? Furthermore, Mueller and Jansen in Ref. 4.7 have shown that at low Reynolds numbers based on chord (below 100,000) there are large-scale vortices in the wake similar to a cylinder. Do these vortices persist at higher Reynolds numbers, and thus make the use of wake surveys questionable due to rotational losses? It is well known that the wake survey is not valid in a stalled airfoil or where separation is present. In Ref. 4.6 Brune et al. state that there was good agreement of both streamwise velocities and turbulent intensity at a given chord location, and different spanwise locations, which implies two-dimensional flow. Nonetheless, the profile drag varied with span. Based on Refs. 4.7 and 4.8 is true two-dimensional flow a function of Reynolds number or does it really exist? Or is there some three-dimensional distortion between the airfoil's trailing edge and location where the wake survey is made? At present none of these answers is known.

The basic theory of the wake survey measurement is the following. Consider the flow past an airfoil (Fig. 4.49). It may be seen that the part of the air that passes over the model suffers a loss of momentum, and this loss is shown by and equal to the profile drag of the airfoil, or

$$D = \frac{\text{Mass}}{\text{Sec}} \times \text{Change in velocity}$$

$$D = \iint \rho V \, da (V_0 - V)$$

where D = drag, V_0 = initial airspeed (at A), V = final airspeed in the wake (at B), and da = small area of the wake perpendicular to airstream. Hence

$$D = \iint(\rho V V_0\, da - \rho V^2\, da)$$

and

$$c_{d0} = 2 \iint\left(\frac{V}{V_0}\frac{da}{S} - \frac{V^2}{V_0^2}\frac{da}{S}\right)$$

Also

$$V_0 = \sqrt{2q_0/\rho}$$

and

$$V = \sqrt{2q/\rho}$$

Therefore

$$c_{d0} = 2 \iint\left(\sqrt{\frac{q}{q_0}} - \frac{q}{q_0}\right)\frac{da}{S} \tag{4.22}$$

For a unit section of the airfoil, $S = c \times 1$, and the area da is equal to $dy \times 1$, where y is measured perpendicular to the plane of the wing. Finally,

$$c_{d0} = 2 \int\left(\sqrt{\frac{q}{q_0}} - \frac{q}{q_0}\right)\frac{dy}{c} \tag{4.23}$$

From Bernoulli's equation

$$H_0 - p_0 = \tfrac{1}{2}\rho V_0^2 = q_0$$

and

$$H - p = \tfrac{1}{2}\rho V^2 = q$$

where H and H_0 = total head in wake and freestream, respectively, and p and p_0 = static head in wake and freestream, respectively. Hence we have

$$c_{d0} = 2 \int\left(\sqrt{\frac{H - p}{H_0 - p_0}} - \frac{H - p}{H_0 - p_0}\right)\frac{dy}{c} \tag{4.24}$$

The ordinary pitot-static tube reads $(H - p)$ directly, but practical difficulties usually prevent the construction of a bank of them. The customary method of obtaining values for Eq. (4.23) is to use a wake survey rake (Fig. 3.12). This is simply a bank of total head tubes spaced about a tube diameter apart with the total head orifice about one chord ahead of the rake body. The tubes are individually connected in order to the tubes of a multiple manometer or a scanivalve; and, since only the ratio q/q_0 is needed, the readings are independent of the specific gravity of the fluid in the manometer and its angle.

A traversing pitot tube can be used rather than a wake rake and is often used in conjunction with a data system. The data system measures the pitot-static position as well as the pressures. It is useful to have the probe displacement and the total head plotted on a CRT or X-Y plotter to ensure that the whole wake is traversed. If such an output is not available, care must be taken to ensure that the traverse is long enough to encompass the width of the wake. Also, integrating wake rakes can be used.

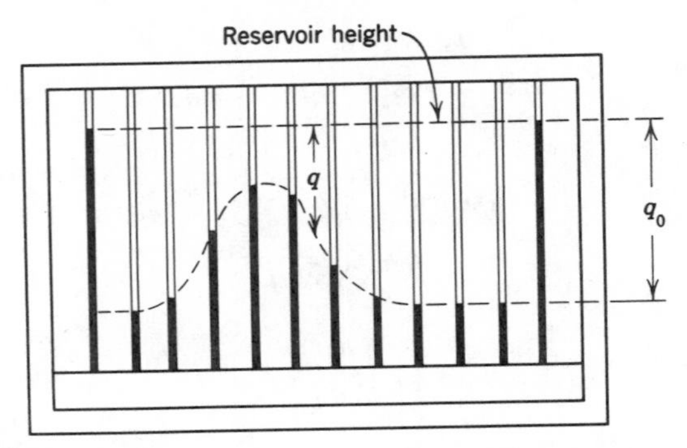

FIGURE 4.50. The wake as it appears on the multiple manometer.

The manometer will appear as in Fig. 4.50. In actual practice many more readings will be available through shimming up the rake in small increments. A small amount of "splash" outside the wake proper may also appear, caused probably by the total pressure gradient present in the tunnel. The engineer must fair the curve according to his experience.

The constant readings of the outside tubes indicate that they are out of the wake and hence may be used to determine q_0. Note that q_0 should be used from the wake rake reading, not from the tunnel q at the model location, because the longitudinal velocity gradient in the tunnel invalidates q calibrations made upstream. The other tubes read the values of q corresponding to their position on the rake.

It will be seen that the proper values of q can be obtained only if the rake is situated far enough behind the wing so that the wake has returned to tunnel static pressure, since a difference in static pressure across the wake will void the values for q. A solution to this problem is to locate the rake at least 0.7 chord behind the trailing edge of the wing. At this position the rake will be approximately at tunnel static pressure. A second solution is to equip the rake with static orifices, the usual practice being to employ three, one at each end and one in the middle, which are averaged. Since the measurement of freestream static pressure close to a body is a difficult thing at best, extreme caution must be exercised in locating the static holes. A satisfactory procedure is to locate them out of the plane of the rake body as in Fig. 3.12 and calibrate them with a standard pitot-static tube, adjusting the tip length of each static tube until true static is read. If the tunnel is not at atmospheric static pressure normally, reference tubes on the multiple manometer or scanivalve should be connected to tunnel static pressure.

It is a tedious job to measure the pressure in each tube, divide it by q_0, take the square root, and perform the other measures necessary for the calculation of Eq. (4.23). Since the ratio q/q_0 is close to 1.0 if the rake is fairly well downstream, the assumption that $\sqrt{q/q_0} = 0.5 + q/(2q_0)$ is valid. Substituted into Eq. (4.23) it yields

$$c_{d0} = \frac{Y_w}{c} - \frac{1}{q_0 c}\int q\, dy \tag{4.25}$$

where Y_w = wake width.

Equation (4.25) makes possible the direct integration of the wake survey data as received, greatly reducing the time necessary for calculation.

The wake survey cannot be used to measure the drag of stalled airfoils or of airfoils with flaps down. Under these conditions a large part of the drag is caused by rotational losses and does not appear as a drop in linear momentum.

If practical reasons prohibit the location of the rake far enough downstream so that the wake has not yet reached tunnel static pressure, additional corrections are necessary (Ref. 4.9), and if tests are made at large Mach numbers, still further changes are required.

It has been found that a round total head tube will not read the true pressure at its centerline if it is located in a region where the pressure is varying from one side of the tube to the other. An allowance may be made for this (Ref. 4.9), or the total head tubes may each be flattened at the tip. The latter procedure is recommended, although the usual correction for the lateral pressure variation is quite small (Refs. 4.10, 4.11).

The momentum method of measuring drag is basically a measurement of the variation of velocity through the wake. These measurements can also be made with a laser, hot wire, or thin-film probe (Sections 3.6 and 3.7). These devices have the same limitations as a pitot tube: they cannot measure momentum loss caused by separation or fluid rotation.

4.19. LIFT AND DRAG BY PRESSURE DISTRIBUTIONS

Still a third method exists whereby the lift and drag may be measured: the integration of the static pressures over the wing. For these tests the airfoil is equipped with many flush orifices, each individually connected to a tube of a multiple manometer or scanivalve. For lift determinations the pressures are plotted perpendicular to the chord, yielding the normal force coefficient C_N. When plotted parallel to the chord, they give the chord force coefficient, C_C. The approximate C_L may be found from

$$C_L = C_N \cos \alpha \tag{4.26}$$

The actual static-pressure distribution over a wing is shown in Fig. 4.51A. The same pressure distribution plotted normal to the chord for the determination of normal force is shown in Fig. 4.51B, and parallel to the chord for chord force determination in Fig. 4.51C. Several of the pressure readings are labeled so that their relative positions may be followed in the various plots.

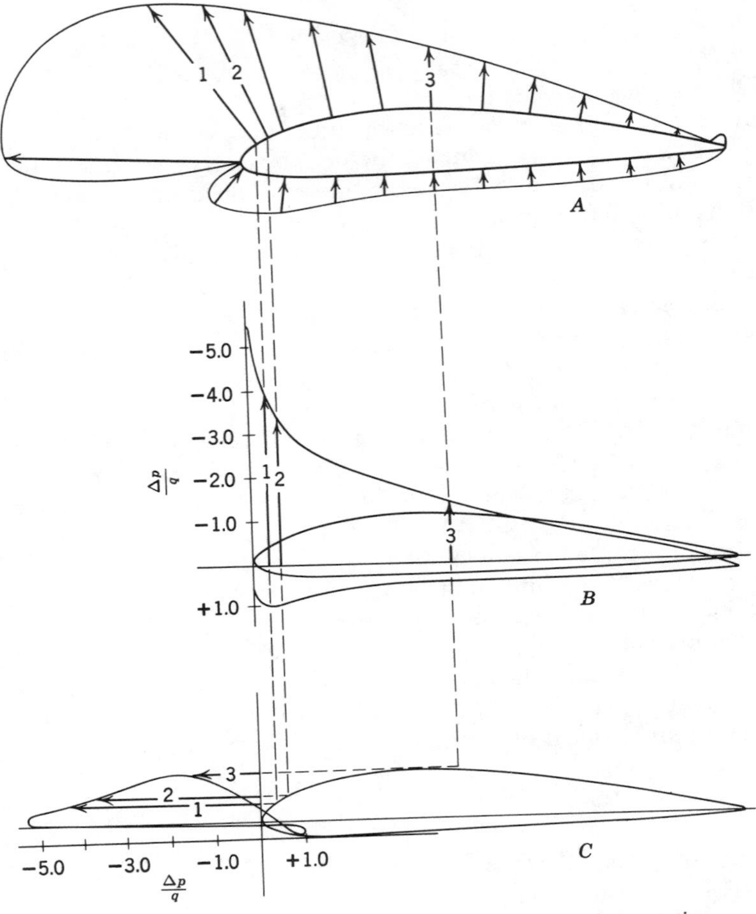

FIGURE 4.51. The actual pressure distribution and its presentation.

The growth of the pressure distribution with angle of attack is shown for a typical airfoil in Fig. 4.52; in this figure may also be found a partial answer to the oft-repeated question "Which lifts more, the upper or the lower surface?" At zero lift, both surfaces have both positive and negative lift. With increasing angle of attack the upper surface increases in proportion until it finally is lifting about 70% of the total.

Many interesting observations may be made from pressure distributions. These include:

1. The location of the minimum pressure point and its strength.
2. The load that the skin is to withstand and its distribution.
3. The location of the point of maximum velocity and its value. This follows from item 1.

4. The location of the maximum pressure point and its strength.
5. The probable type of boundary layer flow and its extent.
6. The center of pressure location.
7. The critical Mach number. This follows from item 3.

It may be worthwhile to note that a stagnation point on the trailing edge of an airfoil occurs, even theoretically, only when the airfoil does not have a cusp trailing edge. The pressure distribution may also be measured by use of a scanivalve and data system.

For students in aerospace engineering the measurement of the lift curve and drag polar by pressures recorded on manometers and photographed is an illustrative experiment. Students can see the change in width of the wake for all angles of attack as the α increases, and the change in pressure distribution on the wing (similar to Fig. 4.52) as the manometers plot them in real time.

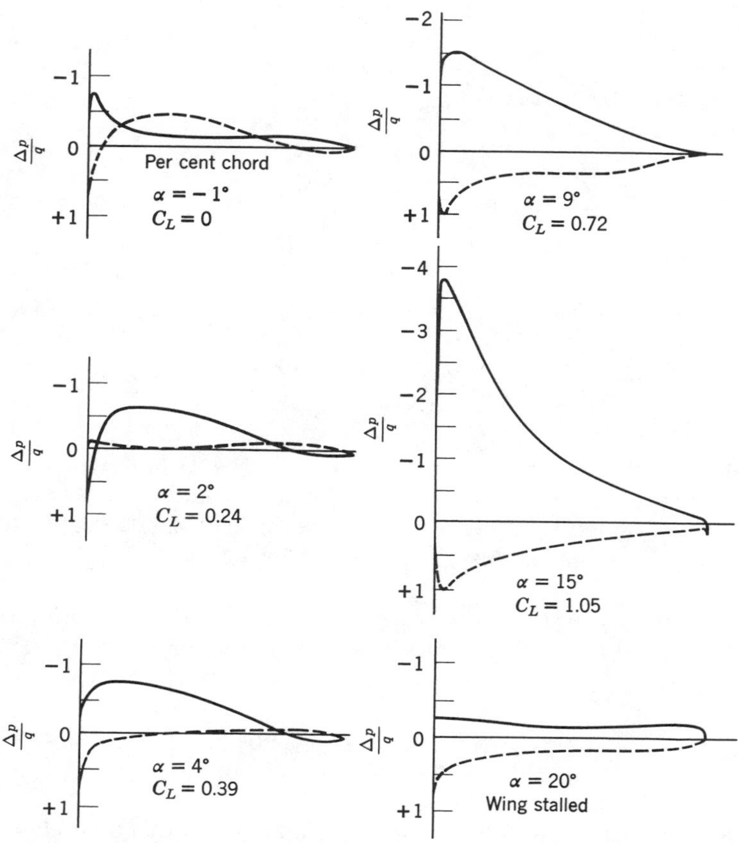

FIGURE 4.52. Growth of static-pressure distribution with angle of attack.

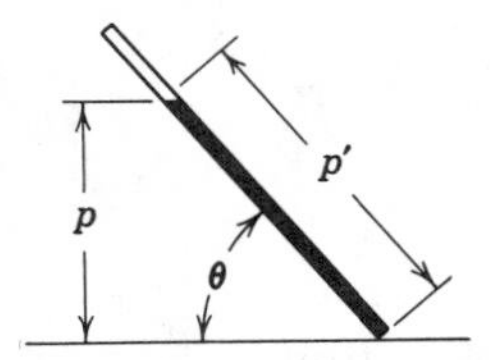

FIGURE 4.53

Two-dimensional tunnels often use ceiling and floor pressures to obtain lift and moment data.

Pressure distributions are usually plotted as follows:

Pressures are read with a multiple manometer that may or may not be inclined. The true head in any one tube, p, equals ($p' \sin \theta$) × (sp. gr. of liquid), where p' is the measured head as defined in Fig. 4.53. The normal force

$$N = -\int \Delta p \, dS$$

where $\Delta p = p_u - p_l$, p_u = pressure on upper surface, p_l = pressure on lower surface, and S = wing area.

For a unit depth of span,

$$N = -\int \Delta p \, dc$$

where c = wing chord.

By definition,
$$N = (\rho/2) S V^2 C_N$$

and hence
$$C_N = \frac{N}{qc} = -\frac{1}{c} \int \frac{\Delta p}{q} \, dc \tag{4.27}$$

It follows that the pressures may be plotted in units of dynamic pressure against their respective locations on the chord. Furthermore, the area under such a curve divided by the chord is the normal force coefficient, and the moment of area about the leading edge divided by the area is the center of pressure.

When a trailing edge flap is lowered, it is customary to show the flap pressures normal to the flap chord in its down position (see Fig. 4.54). For finding the total C_N due to the main wing and flap we have

$$C_N = C_{N\,\text{wing}} + C_{N\,\text{flap}} \cos \delta_F \tag{4.28}$$

where δ_F = flap angle.

It should be noted that, though good agreement between C_N and C_L can be obtained, the drag measured by the pressure distribution,

$$C_{D\,\text{pressure}} = C_C \cos \alpha + C_N \sin \alpha, \tag{4.29}$$

does not include skin friction or induced drag.

The point is sometimes raised that a fallacy is involved in plotting the pressures that act normal to the curved surface of the wing as though they were normal to the chord. Actually there is no error. A simple analogy is observable in the pressure that is acting radially in a pipe but whose force trying to split the pipe is the pressure times the section area made by a plane that contains a diameter.

The mathematical explanation is as follows. Consider a small element of surface ds, which is subjected to a static pressure p acting normal to it. The total force on the element is $p\ ds$, directed along p, and the component of this force normal to the chord line is $p\ ds \cos \alpha$. (See Fig. 4.55.)

But $ds \cos \alpha = dc$, where dc is a short length of the chord, so that the total force normal to the wing chord line is $N = \int \Delta p\ dc$.

It will be noted in Fig. 4.52 that a maximum stagnation pressure of $\Delta p/q = +1.0$ is usually developed near the leading edge of a wing. This may be accepted as the rule for section tests, but swept-back panels will show less than $\Delta p/q = +1.0$ at all stations except at the plane of symmetry.

The pitching moment can also be obtained from the chordwise pressures by use of

$$C_{M_{\text{LE}}} = \frac{1}{C} \int \frac{\Delta P}{q} (x) dc \tag{4.30}$$

where x = distance from leading edge to the ΔP.

See Chapter 6 on data reduction of pressure data with blockage corrections for dynamic pressure, and wall corrections for α.

Since a laser can measure velocity parallel to the crossing light beams (Section 3.7), it is also possible to measure the two-dimensional lift coefficient with a laser. This can be done on a three-dimensional wing, if desired,

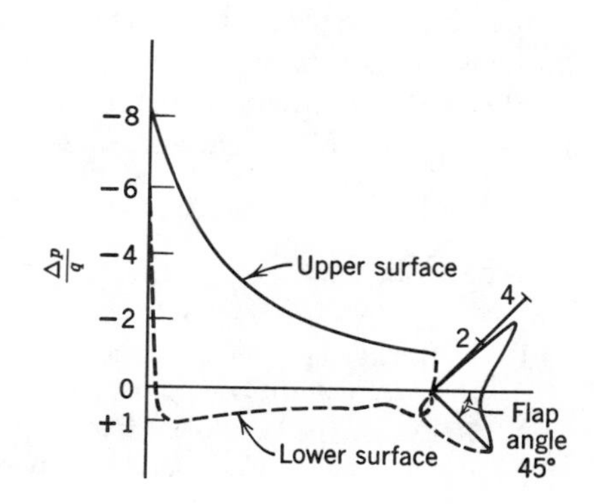

FIGURE 4.54

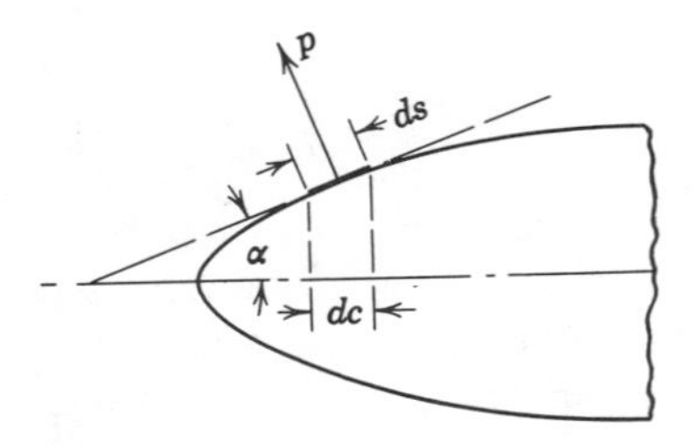

FIGURE 4.55

by measuring the velocity around a closed path that encircles the wing (a line integral). The circulation Γ is

$$\Gamma = \oint_A^B V \cdot ds \tag{4.31}$$

Since the laser can measure the dot product, the values of the circulation can be obtained. The Kutta–Joukowski theorem states that

$$L/b = \rho V\Gamma \tag{4.32}$$

Thus the lift per unit span or the two-dimensional lift can be determined and easily converted to a two-dimensional lift coefficient. This method gives no information about the chordwise distribution of pressure.

This method can, however, be used to determine the distribution to the spanwise lift distribution of a wing caused by the tunnel walls in a three-dimensional test. The tunnel measured c_l can be compared to the calculated values by a vortex latice or other methods. Such information might be desired in the following circumstances. As stated in Chapter 6 the maximum model span should be about 0.8 times the tunnel width. Over their life, many aircraft are stretched and their gross weight increases, as does the wing loading. At some point it may be necessary to increase the wing area. This can be accomplished by extending the wing tips. If the original model's wing span was close to maximum, the span will now be too large. The tunnel walls induce an upwash at the wing (Fig. 6.24) which distorts the model span load, and can distort the wing's stall near the tip. The method outlined can be used to check this distortion. If the tip stall pattern is greatly distorted, it may be necessary to build a new model.

REFERENCES

4.1. L. B. Gratzer, Optimum Design of Flexure Pivots, Report 390, University of Washington Aeronautical Laboratory, Seattle, Washington, 1953.

4.2. B. Ewald, The Development of Electron Beam Welded One-Piece Strain Gage Balances, Paper 78-803, AIAA 10th Aerodynamic Testing Conference, 1978, pp. 253–257.

4.3. T. M. Curry, Multi-Component Force Data Reduction Equation, *Proceedings of Technical Committee on Strain Gages,* 3rd SESA International Congress on Experimental Stress Analysis, May 1971.

4.4. M. Dubois, Calibration of Six Component Dynomometric Balances, *Proceedings of Round Table Discussion on Measurement of Force and Mass,* 7th IMECO Conference, London, 1976, pp. 40–58.

4.5. T. N. Mouch and R. C. Nelson, The Influence of Aerodynamic Interference on High Angle of Attack Wind Tunnel Testing, Paper 78-827, AIAA 10th Aerodynamic Testing Conference, 1978, pp. 426–432.

4.6. G. W. Brune, D. A. Sikavi, E. T. Tran, and R. P. Doerzbacher, Boundary Layer Instrumentation for High-Lift Airfoil Models, Paper 82-0592, AIAA 12th Aerodynamic Testing Conference, 1982, p. 161.

4.7. T. J. Mueller and B. S. Jansen Jr., Aerodynamic Measurements at Low Reynolds Number, Paper 82-0598, AIAA 12th Aerodynamic Testing Conference, 1982.

4.8. D. Althaus, Drag Measurement on Airfoils, OSTIV Congress, Paderborn, West Germany, 1981.

4.9. B. Melville Jones, Measurement of Profile Drag by the Pitot-Traverse Method, *R&M* 1688, 1936.

4.10. J. Bicknell, Determination of the Profile Drag of an Airplane Wing in Flight at High Reynolds Numbers, *TR* 667, 1939.

4.11. A. Silverstein, A Simple Method for Determining Wing Profile Drag in Flight, *JAS,* May 1940.

CHAPTER 5

Testing Procedure

There is very little sense in testing an aircraft component or a complete model unless the data are going to be used. Accordingly, we could say that the discussion of each testing procedure should be followed by a few words on the degree to which the data may be trusted. This plan has been decided against, however, and the use of wind tunnel data in general has been treated in a single chapter (Chapter 7). This permits a more direct approach when extrapolation to full scale is under consideration.

Before we begin to discuss testing procedure, however, it seems useful to discuss the building of the models for testing, planning the tests, and obtaining a tunnel.

5.1. MODEL DESIGN AND CONSTRUCTION

The type and the construction of the wind tunnel model are dictated by the tunnel in which it is to be tested and the type of test to be made. After the obvious and paramount necessity of extreme accuracy, accessibility and maintenance are next in importance. Working conditions in a wind tunnel are at best very trying. The temperature may vary from 30°F in winter to 140°F in summer. The model is usually so placed that accessibility is at a premium, and repair facilities may or may not be available. All these factors demand that changes be as simple as possible, and that the model with all its parts and additions be thoroughly tested outside the tunnel before tests are commenced.

In general, models made of laminated mahogany will be adequately strong without steel or aluminum reinforcement for tests up to 100 mph. Above that, and to about 300 mph, wood or various epoxy models with steel load members are satisfactory; for the higher speeds metal is needed. These speed criteria are, of course, very rough and general. A very thin model might easily require solid steel construction although testing is to be at only 100 mph. The criterion for model strength is deflection rather than yield load limits, as great rigidity is desirable. Although the practice varies with tunnels, for low-speed tests either four factors based on yield strength or five factors based on ultimate strength tend to be used for margins of safety. It is advisable to provide metal beams for any control surfaces in order to maintain the best accuracy of the hinge alignment. All parts of wood models that must be removed and added must be attached with machine screws. This usually required that a threaded metal block be bonded into one part of the model and the other part have a metal block for the screw head. Wood screws are intended to be used only once. Care must be taken in design and construction to ensure that the parts fit together with a method such as alignment pins to ensure proper and repeatable alignment.

The only serious design changes between model and full scale include (1) provision of variable horizontal and sometimes the vertical tail incidence whether the real airplane will have them or not, and (2) omission of miniature parts such as pitot tubes, etc., whose Reynolds number would be too low at model scale.

A wing fitting of the type shown in Fig. 5.1 is in general use. It provides an attachment for a dummy bayonet. Two sealing blocks are needed for both upper and lower surfaces, one with a slot to allow a strut to pass, and one solid to be used when the dummy system is not employed. (Recall that during the process of testing the model may be mounted both normal and inverted, with and without the image system.) The second set of blocks may be omitted and the slot sealed with tape, clay, tunnel wax, or epoxy filling compounds. The material used depends on the desired surface condition. Tape may not follow the contour, clay and wax are soft, and the epoxy filler can be difficult to remove.

A satisfactory material for wind tunnel models to be used up to about 100 mph is well-seasoned Honduras mahogany. This wood is easy to work, glues well, takes all types of finish, and is little subject to warping. A second choice is walnut, also a nice wood to work but likely to have curly grain and hence more difficult to work to close tolerances. Holly has high tearing resistance and is excellent for trailing edges. Some of the softer woods, plastic, plaster, and metals can be used; each has its characteristic advantages and disadvantages for construction and maintenance.

The mahogany form block should be composed of laminations from $\frac{3}{8}$ to $\frac{3}{4}$ in. wide. They should be cut from larger pieces and have alternate strips turned end to end so that any warping tendency will be resisted. The block should be glued according to standard practices; that is, glue should be fresh, pressure adequate, and drying time sufficient (1–5 days). When practical, 2 or 3 weeks' seasoning is desirable.

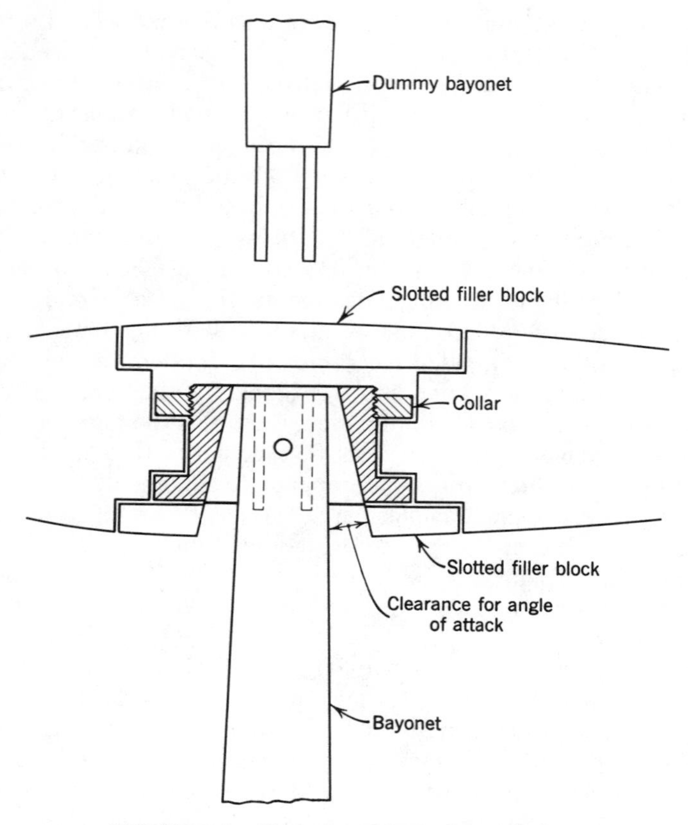

FIGURE 5.1. Typical model attachment fitting.

Some of the new plastics, epoxy resins, and fiberglass are often very useful in model building. Where weight is a problem (as on flutter models) fiberglass can be used for external stores, slipper tanks, etc. Even the low-melting-point alloys can be used in many ways. For example, fillets can be made of Cerrobend, which works easily and takes a good polish.

Basically, the choice of model construction methods is a function of the available shop facilities and skills. For projects built by students, wood is an excellent choice, while for a large corporation that has computer-aided design available, the loft lines from the CAD can be transformed into a set of instructions for numerically controlled machine tools (Fig. 5.2). In this case, depending on loads, steel or aluminum is ideal for lifting surfaces and flow-through jet nacelles. The fuselage can be built on a steel–aluminum spar with aluminum bulkheads that are threaded to take thick (0.125–0.250 in.) fiberglass–epoxy pieces of shell that form the body.

Wings and tails can also be built of large flat plates (spars) of aluminum–steel to which epoxy or wood are bonded and contoured. For sharp trailing

FIGURE 5.2. Contouring a swept wing with a numerically controlled milling machine. (Photograph courtesy of Boeing Aerodynamic Laboratories.)

edges and for plastic contour material the leading edge should also possibly be metal. In any case, the model designer must design the model to fit the capabilities of the people and shop that will produce the model.

A simple way to build a model to be used for configuration trade-off studies is to build the load-carrying capability into the model out of metal. Styrofoam is bonded to these metal spars and cut to contour with a hot wire. The styrofoam is then covered with thin fiberglass or Mylar sheets (about 2 mil) to provide a smooth contour. The fiberglass and styrofoam are not considered to be load bearing as in sandwich construction used for radio-controlled models. In general, these are one-shot models and do not lend themselves to major modifications—but their cost is relatively low.

Usually models have many component parts, each to be made separately. When models are made of wood, it is advisable to make each part oversize by 0.0625 or 0.125 in. and allow some additional seasoning. This permits any strains in the laminated block to relieve themselves by warping.

The last fraction of an inch is then worked down to female templates by files, scrapers, and sandpaper. The day should be so planned that time remains to spray at least one coat of clear lacquer after the piece is done to seal it and prevent warping due to changes in its moisture content.

Hinged surfaces present a problem for the model designer largely because the smallest hinge possible to construct is far too large in scale. In view of the hopelessness of accurate reproduction the designer just does the best he or she can.

Several basic types of hinges are in general use. The first is simply a set of brackets for each angle setting. This is a slow method with respect to model changes, and it slows model construction as well. A serious factor is that if test results indicate during the program that additional angles are required, the new brackets usually require shop work above the level of tunnel engineers and tie up the tunnel during fitting as well.

A second method is to furnish the surfaces with hinges and to use an exterior sector with drilled holes for each angle setting. This is a good positive method that at least yields the same setting for repeat points, and it is only the work of a moment if additional settings become necessary. One must assume, of course, that such a sector in the "breeze" will not cause appreciable trouble. However, brackets for control surfaces give a repeatable angle setting. Brackets, often with fairings, are about the only way to attach leading edge slats, slots, and trailing edge slotted flaps, especially if they have a Fowler-type action.

A third method is to furnish hinges and an interior setting lock. We have had too many of these slip, and have broken too many, to be enthusiastic about this type except when it is designed by someone with much experience. The engineer using this type of hinge should check the setting at the *end* of each run.

A fourth method for metal control surfaces is to use hinges near the tip. The inboard end has a reamed hole that slides over a large pin. For each angle a hole is drilled and reamed through the control surface and the large pin. The holes are spaced spanwise. The surface angle is then held with a straight or roll pin.

Model makers differ in their opinions of suitable finishes. For wood models, the choice is usually between three types, all quick drying: shellac, clear lacquer, or pigmented lacquer. Shellac and the clear lacquers seem to yield a slightly thinner coat (0.002–0.005 in.) than the pigmented lacquer (0.003–0.008 in.). On the other hand, many model makers believe that the smoothest finish is obtained only with the pigmented lacquer. Regardless of the choice, an adjustment in the templates should be made to allow room for the finish. Finishes put on and sanded back to zero thickness are not believed to offer sufficient protection and moisture seal.

For metal or plastic models, either sprayed enamel or lacquer can be used. Usually a primer coat is applied first. This is necessary for a metal model to obtain proper bond, and then pigmented finish coats are added. Generally a flat finish rather than high gloss is preferred to avoid highlights in both configuration and flow visualization photographs. Some of the tough, and somewhat thick, aircraft epoxy paints should be avoided. They can only be removed by sanding in many cases, and often a model must be refinished when it is used for more than one test. It is amazing how chipped and pocked a model becomes during a test. One source of the nicks are razor blades used to cut tape for sealing joints or applying tufts on the model surface.

A model for a 7 × 10 ft tunnel should have the wing contour accurate to 0.005 in. to the true contour, and the fuselage to within 0.01 in. No perceptible ridges or joints should be permitted. If a metal model is to be employed, a surface finish of rms 10 (25 microinches) may be attained by using No. 600 wet emery paper. If this discussion of model accuracy and finish seems nonsensical in the light of later (Section 7.2) additions of roughness to trip the boundary layer, it is not quite as bad as it seems. The drag of a turbulent boundary layer on a smooth surface is both different and more repeatable than that on a questionable surface, and the smoothness behind intentional roughness thus makes sense. Dimensions may be harder to justify, except to note the large difference in pitching moment associated with changes in camber or camber shape. In attempts at extended laminar flow, the need for accuracy rapidly makes itself evident.

Air passages, radiator openings, and cooling entrances may be simulated by an indenture of the entry without any completed flow passages. Such passages if completed could have Reynolds numbers too low for satisfactory testing. A parallel situation exists for all small excrescences: aerials, bomb racks, pitot-static tubes, and the like. They too would show such scale effect that their true effect could not be measured, and hence they are left off. Quite often jet engine nacelles are of the flow-through type. In some cases the nacelles are equipped with four or more static orifices. The nacelles are calibrated so that the internal aerodynamic forces can be calculated from the static pressures and then subtracted from the force data; or a nacelle wake rake is used to obtain nacelle forces (Section 5.13).

Pressure models for tests of the type described in Sections 4.18, 4.19 and 5.20 (see Fig. 5.3) require additional care in design and construction. Usually the pressure taps on the wing are located at 0, 1.25, 2.5, 5.0, 10.0, 15.0, 20.0, 30.0, 40, 50, 60, 70, 80, 95, and 100% of the chord on both upper and lower surface at several spanwise stations. This obviously necessitates a large number of tubes, which should be brought out from the model under circumstances least influencing the flow. First, however, let us consider the design of the pressure orifices themselves.

If static-pressure orifices are kept small (say about $\frac{1}{32}$ in. in diameter), negligible difference is found between drilling them perpendicular to the surface or pendicular to the chord. But it is certain that they must be absolutely flush. A metal tube in a wood model has a tendency to form a slight ridge as the softer wood is filed down about it. Some designers use metal strips at the section where the orifices are to be, thus avoiding the difficulties of filing dissimilar materials. An artifice practiced by the Canadian National Research Council seems a satisfactory arrangement. This embraces a solid transparent plastic plug leading down to the buried metal tubes. After the airfoil is shaped, holes are drilled down to the tubes through the plastic.

FIGURE 5.3. Model with scanivalves and tubing for surface pressures in aft body and high-pressure air lines for nacelles in forward body. (Photograph courtesy of Boeing Aerodynamic Laboratories.)

Since the plastic offers filing characteristics similar to those of wood, remarkable smoothness is attained. If the wings of the model are exceedingly thin, it is sometimes advantageous to put the upper surface orifices on one wing and those for the lower surface on the other. Wings of metal and epoxy over metal usually have milled slots cut into the wing. Thin-wall stainless tubing is then laid in the slots and brought to the model surface. The slots are then filled with epoxy and tubes and epoxy returned to contour. On steel wings soft solder can be used to fill slots. Often it is desirable to cut a large chordwise slot for tubes about 0.5 in. from the chord line of the pressure taps. Then the individual tubes are brought from thin slots to the larger slot. This large slot then intersects a spanwise slot that is used to get the tubes into the fuselage.

When sanding or filing surfaces with pressure orifices, care must be taken to avoid plugging the tubes. One way this can be accomplished is to blow dry compressed air through all of the tubes when sanding or filing.

Though many satisfactory pressure models have been built using copper tubing, scanivalve or manometer fill time can usually be saved by going to annealed stainless-steel tubing since, for a given (and usually critical) outside diameter, its inside diameter is a maximum. The stainless tubing is less likely to kink than copper, but it is harder to solder without an acid treatment.

Manometer fill times, which run as much as 2 min under some circumstances, may be estimated with good accuracy from the data in Refs. 5.1 and 5.2.

If manometers are used, the leads from the model are most easily made of plastic tubing, which has little tendency to kink or leak and has good bending qualities. A sort of multiple tube is available that furnishes 10 leads about $\frac{1}{32}$ in. in diameter in a thin flat strip $\frac{1}{16}$ in. thick by 1 in. wide. It is hard to envision a more compact arrangement.

Since automatic data-acquisition equipment have become generally available, two trends have occurred with pressure or loads models. First, the number of pressure ports has increased. Second, pressures are recorded by the use of stepping scanivalves and transducers. Since the electrical cables required for a scanivalve are many orders of magnitude smaller than plastic tubing, the massive problem of getting tubes out of the model is gone. The use of scanivalves also reduces tunnel set-up time, because large numbers of pressure ports do not have to be connected, leak checked and phased out to a manometer after the model is installed. With scanivalves these tasks can be done prior to tunnel entry, because the transducers measure pressures from a reference pressure, quite often a pressure tube is brought into the model and is then manifolded to the transducers. This tube can also be used to apply a ΔP to the transducer at the start of each day or shift to check the calibration of the transducers. Often the first and last port of the scanivalve are connected to the same source. If the first and last port have the same pressure reading, one is sure that the valve has not skipped or hung up on any of its steps.

When speed rather than accuracy is the major factor, pressure models can be made from the solid type by belting the model parallel to the airstream with strips of the multiple plastic tubing mentioned above. Holes drilled in the tubing at the selected stations enable pressures to be read at one chordwise station per tube. Actually, of course, the presence of the flat tubing alters the true contours and hence also alters the pressure distribution about the model. The error so introduced is often surprisingly small. But extreme care must be taken to ensure that the tubing belt does not pull up from the trailing edge as it leaves the model. Airfoil pressure distributions are very sensitive to camber at the trailing edge.

Though perhaps not a wind tunnel design criterion per se, the fact remains that wind tunnel models have to be moved about, and, depending on their size, this may become a ticklish problem. Most tunnels have a lifting crane for moving the model into the test section, and in turn whenever possible model designers should provide an attachment near the center of gravity of the model. Some tunnels provide canvas sandbags for supporting the model when it is resting on the floor or a table.

Some sort of modeling-clay-type material is frequently needed for filling cracks, covering screw holes, and making minor contour changes. Children's modeling clay or, better, sculptors' Plastalina No. 4, will suffice for low-speed work. When extra strength or high temperatures must be considered, a stiff wax made according to the formula given below will be found excellent. The acetone and pyroxylin putties are also very good although they require a few minutes' drying time. Automotive car body fillers made of a two-part epoxy are excellent for filling holes, and so on. They cure rapidly and are easy to shape. If the filler covers a screw or bolt, paper or tape should be put into the hole first. It is a difficult job to dig filler out of the head of a screw. A heated Allen wrench can be used to clean the filler from the head socket, but it is best not to fill the head socket. These fillers do not work too well for removable fillets as they tend to be brittle and will break very easily when the cross section is thin.

Tunnel Wax Formula

Beeswax	About 80% by weight
Venice turpentine	About 20% by weight
Powdered rosin	About $\frac{1}{2}$% by weight

Bring turpentine to a boil; add the rosin, and stir. Add the beeswax in small chunks and allow to melt. Stir thoroughly.

Remove from heat and pour into trays for cooling.

5.2. PLANNING THE TEST

A wind tunnel test should be run only if (1) some new knowledge is desired and (2) the test as planned has a reasonable chance of obtaining the knowledge sought with the necessary accuracy. In view of the cost of models and tunnel time, it should be determined that the "new knowledge" does not already exist. In many cases—too many cases—a good library search could have saved both time and money.

It is hard to write specific rules for setting up a test and taking data, since there are many types of tests. However, the following procedures do stand as accepted and good. If some seem obvious to the experienced engineer, we hasten to add that at one time or another we have seen all the rules stated below completely disregarded.

1. Check all calibration curves of new equipment before, during, and after a test. Always calibrate for the full-load range, and always use a number of loads—not a single load and the assumption that the calibration is linear.

2. Take enough points so that the loss of any one point will not hurt the fairing of the curve.

3. Always repeat the wind-off zero and the wind-on first point at the end of the run. Have acceptable balance "drift" limits set up before the program starts (0.1 or 0.2% of maximum reading is a reasonable drift allowance).

4. Take points on base runs at every degree plus 0.5° readings at the stall or other points of interest. Take routine runs with 2° readings and 1° increments at the stall.

5. Check all models against their templates and check the templates. Do not hesitate to cancel a program and pay the cancellation fee if the model is not within acceptable limits. If you do, the *new* shop foreman will have future models right.

6. Be very careful when you shorten a program by omitting "irrelevant" components. For instance, changes that primarily affect the pitching moment only might lead one to read the pitching balance. This omission would make it impossible to plot the data completely since the angle of attack correction is affected by the lift, which must also be known, and, indeed, if later a change to a new center of gravity seems desirable the drag values must be obtained. Similarly reading less than six components in the interest of saving time on yaw runs can also lead to serious work-up troubles of the same nature. This problem is not as serious with automatic data systems. Many tunnels have the policy to take all six components as required by balance interaction equations, thus avoiding the hassle caused by these "time savers."

7. Plan model variations of wide enough scope to bracket needed data so that interpolation rather than extrapolation is in order.

8. Whenever possible find out how others do the type of test you contemplate and profit from their experience.

9. Be clear in all instruction and data presentation. Never use the word "pressure" when "static pressure" or "total pressure" might be confused. Always use a subscript for pitching and yawing moments to indicate the axis about which they are measured, and specify the desired center of gravity location. Also, if multiple entries are made with a model, make sure that the statement "the same as the last test" is really true before you use it.

5.3. TUNNEL OCCUPANCY PROCEDURE

Each tunnel has a somewhat different procedure for use, and no exact rules can be written to cover them all. Nevertheless, a description of a typical tunnel procurement may be useful in giving a general familiarity with the system.

Most of the large wind tunnels are scheduled about 6 months in advance, and hence an inventor seeking to prove some new idea may be very disappointed in the delay he or she may be subjected to. Aircraft companies avoid this problem by regularly scheduling tests of 100 hours' duration every few weeks. Then, as their time approaches, they select from needed tests the one upon which the greatest urgency rests. As a testing time approaches, the following procedure is usual.

1. About 2 months before a test, the tunnel is informed of the tunnel configuration desired: external balance, swept strut, two-dimensional test section, etc. If the desired setup does not meld with the other programs scheduled about that time, a shift of a week or so may be necessary to avoid excessive tunnel changes.

2. Three weeks before the test complete model drawings, stress analysis, desired tunnel operating conditions, and a preliminary run list are sent to the tunnel manager.

3. Two weeks before the test a meeting is arranged between the engineers who will supervise the test for both the tunnel and the airplane company. At this meeting any points not apparent in the pretest information may be ironed out. Agreement on the special equipment needed is reached: pressure measurements (number of tubes and expected pressure ranges), cameras both still and motion picture. All constants required for data reduction should be provided, including model reference dimensions such as wing area, span, mean aerodynamic chord, transfer dimensions for center of gravity positions desired, and reference for angle of attack. The realistic definition of desired accuracy should include forces and moments, angle settings, pressures, and model location; the acceptance of reduced accuracy where it actually is sufficient may result in considerable savings in time and money. A list of plots needed and the form of the tables of data are presented for both tab data and magnetic tape. Agreement is reached on a date for this presentation of preliminary data and for the final report.

4. One week before the test the company representatives arrive with the model and commence as much of the setup and weight tares as feasible outside the tunnel. There may still remain some questions about the program, which must be settled before running. The representatives must do a certain amount of legwork to ascertain that all the items previously agreed upon have actually been accomplished.

5. During the last day before the test, company representatives remain on 1-h alert, ready to move into the tunnel and start their test.

5.4. GENERAL TESTING PROCEDURE

Depending on the innovations incorporated and the terms of the development contract or program, a new model airplane may require from one to six models (or more) and up to six different wind tunnels. A typical program is as follows:

After the preliminary layout of the proposed new airplane has been made, a "complete" model is designed and constructed. This first* model, usually of 6–16% scale, is a breakdown model; that is, the different configurations of

* High-speed tests for transonic airplanes may precede the low-speed tests.

the airplane may be built up progressively through additions to the wing alone, making possible the evaluation of the relative effect of each component. Testing this model requires measurement of all six forces and moments: lift, drag, and side force, and rolling, yawing, and pitching moments. The important criteria of maximum lift (stalling speed), minimum drag (high speed), and static stability are evaluated. The breakdown model aids in determining the exterior configuration of the airplane so that the specialized models can be designed. (See Fig. 5.4.)

The breakdown model is useful for much more than simply satisfying the aerodynamicist's curiosity about the contributions of each component. One instance comes to mind where much of the breakdown was omitted ("We can only fly the whole airplane"). When the performance fell far short of that predicted, and after the tunnel and tunnel crew had been duly excoriated, a breakdown model disclosed that the horizontal jet pod supports were lifting "nearly as much as the wing," and in the opposite direction!

The second model (after the first breakdown model there is no specific order for the additional ones) may be a small-scale spin model for determining the spin-recovery characteristics in a spin tunnel. Here the model is put into a tailspin in the vertical airstream of the tunnel, and a remotely operated

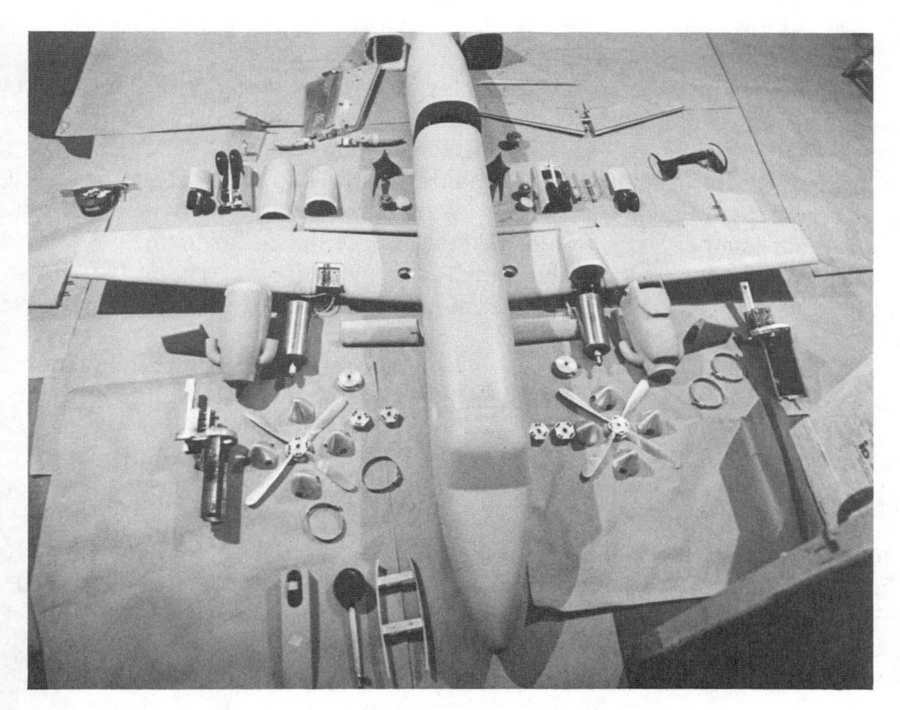

FIGURE 5.4. Breakdown of wind tunnel model of Beech 1900 turbo-prop airliner. (Photograph courtesy of Beech Aircraft Corporation.)

mechanism moves the control surfaces as desired to bring the plane out of the spin. Motion pictures of the recovery can be examined to see whether the procedure is satisfactory (see Section 5.17).

A third model, also light and fragile, may be flown in a free-flight tunnel such as the Langley 30 × 60 ft where motion pictures record its stability and maneuverability.

A fourth model, so constructed that its rigidity is related to the full-scale airplane, may be tested for critical flutter speed (see Section 5.18).

If the design appears satisfactory or can be made so after these preliminary tests, larger models of component parts may be tested. Aileron panels and tail surfaces to perhaps 40% scale may be tested, and nacelles to a similar scale for propeller aircraft may be investigated for cooling and drag.

Compressibility effects are investigated with high-speed models in high-speed tunnels. Sometimes additional section tests of the airfoils to be used are made in a two-dimensional tunnel, and if the design is entirely untested, pressure distributions over the flap, flap vanes, etc., may be taken to determine loads for the structural design (see Sections 4.19 and 5.20).

Finally, for smaller aircraft, when the first actual airplane is finished, it can be tested in a full-scale tunnel for aerodynamic "clean-up" changes. Here also the manufacturing irregularities can be examined and improvements suggested. Military airplanes can be subjected to simulated battle damage so that studies can be made of possible catastrophic effects.

The cost of such a program is not small, of course. Yet, compared to the cost of building the actual airplane, testing it, and changing it, the model-testing cost becomes minor indeed. Rarely would a single concern have the entire facilities required for a complete testing program. The customary solution is for the complete model and the control surface panels to be tested in the company's own wind tunnel, leaving the spin, stability, flutter, and high Mach number work to tunnels specially designed for them.

The basic parameters of the tunnel that need direct attention before the wind is turned on include considerations of the average angle of flow, the average q, and the balance loads.

The average angle of flow need not be considered before the full test of a three-dimensional wing. As shown in Section 4.16 it is accurately determined by the model normal and inverted tests of the alignment determination. A two-dimensional model should also be run normal and inverted. However, when inversion is not to be employed for any of a number of reasons, or when it is actually impossible, as for a panel model, the procedure outlined in Section 4.16 may be followed for finding the average angle for a given model. A rough check may be made from the first run by comparing the expected and obtained angles of zero lift. Indeed, particularly for the panels, so much advantage accrues in the analysis of later data from assuming the models to be absolutely accurate and hence making expected and obtained angles of zero lift identical that this is the usual procedure.

The average dynamic pressure must be calculated for each model of different planform by a method like that for obtaining the average angle. That is, the product of local q (from the dynamic pressure survey of the test section) and the model chord at the same station is plotted against the model span. If the area under the qc versus span curve is then divided by the total wing area, the resulting quotient is the average dynamic pressure. This average dynamic pressure is *not* used to find the various coefficients until it is increased by the blocking factor obtained from Section 6.11.

Last, but not least, a check of expected loads should be made to ascertain that ample provision has been made to run the entire program at one speed. Changing the tunnel speed during a program adds one more effect to the data unless, of course, Reynolds number effects are required.

1. After the first run has been made, it should be checked thoroughly against expected results. If possible, the setup should be arranged so that the first run is simple enough for comparison with previous tests. Items to be checked include $\alpha_{Z.L.}$, $dC_L/d\alpha$, $C_{L\max}$, $C_{D0\min}$, and C_{m0}.

2. Determine the testing accuracy by:

(a) Running a test twice without any change in it at all. This tests the reproducible accuracy of the balance and the speed control.

(b) Resetting and repeating a run made previously after there have been several intervening runs. This determines the reproducible accuracy of setting the flaps, tabs, etc., as well as the accuracy of the balance and speed control.

3. Keep a running plot of all data as they come out. Any uncertain points can be substantiated immediately by taking readings at small increments above and below the uncertain ones. Many tunnels have the capability of plotting the data on a point by point basis as they are acquired. These data may or may not have all of the required data reduction steps; however, they show any bad data points, unexpected results, and usually give Δ's that can be used to plan the next run.

4. Occasionally repeat a basic run. This will indicate any gradual model warpage or other alterations occurring with time. It must be realized that the surface condition is gradually degraded in a long test and repeats of baseline configuration monitors this effect.

5. Always repeat the first reading at the end of each run. This will indicate any control surface slippage, etc. Inspect the model frequently, checking all control settings, and wherever possible make angular measurements of controls, etc., with the inevitable slack taken out in the loaded direction.

Besides checking the control settings one must check that the setting quadrants are right. An example comes to mind of a model whose stabilizer quadrant markings were an even 10° off. Fortunately the model had been run before and the wide difference in trim conditions started a search for trouble. One can imagine much difficulty arising later if a test pilot is told to set the stabilizers 10° wrong for the first flight!

6. Make every data sheet self-contained. Avoid using expressions such as "Same as Run 6," for this necessitates looking up Run 6. Every data sheet must contain the model designation, configuration, test speed, date, and tunnel temperature and pressure. Further data, such as effective Reynolds number and model dimensions, are valuable. In tunnels with computerized data systems there are no data sheets. Thus all pertinent information must be on the run log.

7. Keep an accurate log of everything that happens. When analyzing the data, the exact point at which changes were made may be of paramount importance.

8. Always keep a run list in chronological order; never assign a number like "Run 3b" just because the later run is a check of Run 3.

The size and design of the tunnel determine the size of the model that can be accommodated and, sometimes, other important criteria such as model weight or power arrangements. Occasionally a gasoline engine can be operated in the tunnel (Ref. 5.3). The tunnel itself also determines the complexity of the model and hence its cost. The cost of a model is a function of both size and complexity. A simple wing for student experiments built by the students will be minimal, usually just material cost. The cost of a complete low-speed tunnel model shows in both the model designer's time and shop time. Models with complex leading and trailing edge flaps to be tested at several angles, gaps, and overlaps, are very expensive. The same is true of a powered or jet lift V/STOL model. A good policy is to try to keep the model as simple as possible with careful consideration of its intended use. For example, incorporation of movable control surfaces early in the preliminary design of an aircraft may not be necessary, but the model should be designed so that these features may be added as required in later tests. There is a strong tendency to grossly underestimate the time required to build a complete airplane model at the first attempt. The amount of remotely operated model equipment is a function of the number of tests to be made and the type of tunnel.

5.5. TESTING THREE-DIMENSIONAL WINGS

The first wind tunnel tests ever made were concerned with the behavior of wings (Fig. 5.5). The wing is still the most critical item in the success of any aircraft. The high-speed digital computer has given the aerodynamicist a powerful tool in wing design. Airfoils can be designed to yield a desired chordwise pressure distribution. Vortex lattice methods can determine the distribution of loads along the span. The variables in wing design are airfoil section, taper, twist, and sweep. The wing is usually designed for minimum drag at cruise lift coefficients. To obtain take-off and landing performance the camber is increased to obtain large values of maximum lift and accept-

FIGURE 5.5. Three-dimensional wing being tested with image system in. Omission of the image pitch strut reduces the total drag of the system with little error in data. (Courtesy Wichita State University.)

able stall patterns by use of leading edge flaps, slots, droop, and trailing edge flaps. This large number of variables requires extensive wind tunnel tests to obtain an optimum wing.

Some aerodynamicists prefer to run the wing and fuselage rather than the wing alone. The argument for this choice is based on the possible effect of the fuselage on the wing stall pattern and the fact that the aircraft will fly with a fuselage.

Tests of a wing should be run at as high a Reynolds number as possible to aid extrapolation to flight Reynolds number. The achievement of large test Reynolds numbers is constrained by tunnel size and aircraft size. As a rule of thumb, the model span should be less than 0.8 of the tunnel width. This constraint on model span determines the maximum model scale. The wing aspect ratio will then determine the model chord. The Reynolds number may be increased by increasing the velocity of the test, but test velocity may be constrained by both the balance load limits and the fact that the lift curve slope is a function of Mach number. To see the latter as a first approximation using the Prandtl–Glauert correction to a two-dimensional lift curve slope:

$$a_0 = \frac{a_{0_{M=0}}}{\sqrt{1 - M^2}} \tag{5.1}$$

A test Mach number of 0.20 will yield a change in the incompressible lift curve slope of 2%. Therefore, tests in which the maximum lift is desired should be run at speeds near the full-scale take-off or landing speed.

As tare and interference drag can be as large as the wing minimum drag, flaps up, the tare and interference must be carefully evaluated.

In general the aerodynamicist or student will be most interested in the $dC_L/d\alpha$, $\alpha_{Z.L.}$, $C_{L\,max}$ and how stall develops, $C_{D0\,min}$, e, location of aerodynamic center, C_{mac}, and possibly the center of pressure. The following comments will give a student or tunnel engineer some feel for these items.

(a) $dC_L/d\alpha$. The theoretical lift curve slope for a nonswept wing based on Glauert's solution for the span load is, for wings with an $AR > 5.0$,

$$a = \frac{a_0}{1 + (57.3a_0/\pi AR)(1 + \tau)} \tag{5.2}$$

Glauert's solution takes into account twist (both geometric and aerodynamic), chord distribution along the span, and downwash by using a series of horseshoe vortices, the bound vortices on the quarter chord. The factor τ is a small positive number that increases the induced angle of attack over the minimum value for a wing with an elliptic distribution of lift. For a first approximation τ is often taken as zero.

For swept wings the lift curve slope decreases roughly with the cosine of the sweep angle of the quarter chord. From NACA TN 2495 the following equation, which agrees with experimental data, may be used:

$$a = 0.95 \left(\frac{a_0}{1 + (57.3a_0/\pi AR)}\right) \sqrt{\cos \Lambda c/4} \tag{5.3}$$

Quite often swept wings, especially those with leading and trailing edge flaps, do not have a linear lift curve.

There are more elaborate methods for predicting the lift curve slope and wing performance using vortex lattice methods that will distribute the load both spanwise and chordwise, but they are beyond the scope of this book. The equations given here are intended to be used to check wind tunnel results to determine if they are approximately correct. If results of a more elaborate analysis are available, they should be used.

(b) $C_{L\,max}$. The maximum lift coefficient for airfoils varies from 0.6 for very thin profiles to about 1.7 for highly cambered thick profiles. In general it increases with Reynolds number (see Section 7.4.) The wing maximum lift coefficient usually runs from 85 to 90% of the airfoil values, and is never more than the airfoil values. Swept wings show a loss considerably more than the above values.

Besides determining the value of $C_{L\,max}$ the shape of the stall portion of the stall region is important. In general, it is desired to have a gentle stall. This means that the curve should be gently curved near maximum value without abrupt drops in values of C_L just past the maximum value.

The location of the start of the stall is also important. It is desirable that the stall start inboard of the ailerons so that lateral control can be maintained. A stall that starts at the root of the wing may cause excessive tail buffet. To study the stall pattern, flow visualization techniques such as oil flow and tufts described in Section 3.10 are used.

(c) $\alpha_{Z.L.}$. The angle of zero lift in degrees is roughly equal to the amount of camber in percent for airfoils and untwisted wings of constant section. It requires a considerable amount of calculation to determine $\alpha_{Z.L.}$ for a twisted wing.

(d) $C_{D0\,\min}$. The minimum coefficient of drag decreases with increasing Reynolds number (see Section 7.3) and usually has a value between 0.0050 and 0.0085 in the tunnel after the tare and interference have been subtracted. The values given may be higher for wings equipped with lower surface trailing edge fairings for flap tracks (Fowler action flaps) or hard points for external stores.

The value of $C_{D,P\,\min}$ is sensitive to the transition point between laminar and turbulent flow. The location of transition can be determined by the flow visualization methods of Section 3.10. Often it is desired to fix the transition by boundary layer trips. These are at various locations on the model and they will increase $C_{D,P\,\min}$. The effect of trip strips on $C_{D,P}$ can be determined (see Section 7.2).

(e) The C_L versus C_D curve or drag polar below the start of flow separation is often approximated by Eq. (5.4). This curve fit is used for two purposes. First, it can simplify performance calculations in cruise and climb. Second, by dividing the drag into parasite and induced parts the effect of configuration changes can be more easily monitored.

$$C_D = C_{DP_e} + \frac{C_L^2}{\pi ARe} \tag{5.4}$$

$C_{D,P,e}$ is the equivalent parasite drag coefficient and e is called Oswald's efficiency factor. For a wing only it accounts for a drag increase due to nonelliptic span loading. For a whole model it accounts for the change in span loading caused by the fuselage, nacelles, etc., and the increase in tail drag with C_L.

The values are determined by a plot of C_L^2 versus C_D (Fig. 5.6). Over a range of C_L approximately between 0.2 and 0.8–1.0 the curve is linear. At low values of C_L it is curved, and at large values of C_L separation causes the curve to deviate from a linear relationship. Often the curve is slightly nonlinear for swept wings and judgment will have to be used. The value of $C_{D,P,e}$ is less than $C_{D,P\,\min}$.

Equation (5.4) is of the form

$$C_D = C_{DP_e} + KC_L^2$$

where

$$K = \frac{1}{\pi ARe} \tag{5.5}$$

Thus the slope K is used to determine e and the intercept on the C_D axis to determine $C_{D,P,e}$.

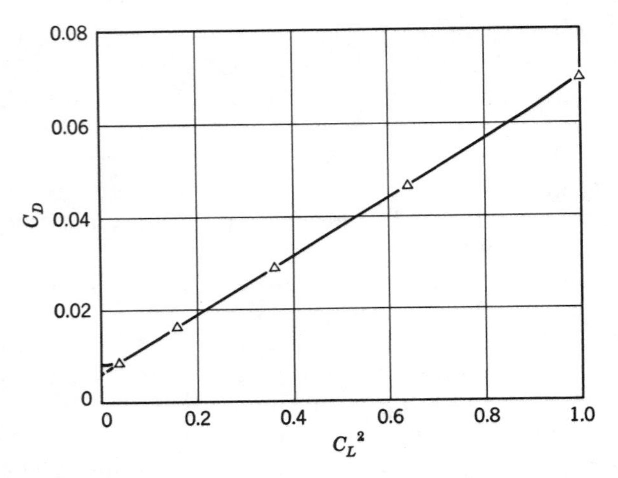

FIGURE 5.6. C_D versus C_L^2 for an NACA 23012 wing.

Often during wind tunnel tests of either wings or complete aircraft it is desirable to find the effect of configuration changes on the level of parasite drag. This is done by setting $e = 1.0$ in Eq. (5.4) and then calculating the induced drag coefficient. This value is subtracted from the measured C_D to yield an approximate C_{D0}, which is plotted versus C_L.

Example 5.1. Values of C_L and C_D are given in Table 5.1. Find Oswald's efficiency factor e if $AR = 7.0$.

From a plot of C_D versus C_L^2 the slope $K = 0.05173$. $C_{D,P,e} = 0.0081$, $C_{D,P\,\min} = 0.0090$, $e = 0.879$.

(f) Location of the aerodynamic center may be computed as follows.* Consider a wing mounted so that the axis of rotation is at some point behind and below the probable location of the aerodynamic center (Fig. 5.7). (The aerodynamic center is defined as the point about which the moment coefficient is constant.) Let the distance along the chord from the trunnion to the aerodynamic center be x, and let the distance above the trunnion be y. Both x and y are measured in fractions of the MAC.†

* A second method of calculating the location of the aerodynamic center is given in the appendix of NACA *TR* 627.

† The mean aerodynamic chord may be found from either

$$\text{MAC} = \frac{2}{3}\left(C_T + C_R - \frac{C_T C_R}{C_T + C_R}\right)$$

where C_T = wing tip chord and C_R = wing root chord for straight tapered wings, or

$$\text{MAC} = \frac{2}{S}\int_0^{b/2} c^2\, dy$$

for other planforms.

TABLE 5.1

α	C_L	C_D
−3.93	0.0688	0.0090
−1.82	0.2411	0.0111
+0.28	0.4309	0.0177
2.39	0.6123	0.0277
4.48	0.7932	0.0406
6.57	0.9712	0.0583
8.65	1.1296	0.0777
10.74	1.2665	0.0987
12.80	1.3734	0.1244

It will be seen that

$$M_{\text{ac}} = M_{\text{tr}} - xc(L\cos\alpha + D\sin\alpha) - yc(D\cos\alpha - L\sin\alpha) \quad (5.6)$$

where M_{tr} = the moment measured about the mounting trunnion. Hence

$$C_{m\text{ac}} = C_{m\text{tr}} - x(C_L\cos\alpha + C_D\sin\alpha) - y(C_D\cos\alpha - C_L\sin\alpha) \quad (5.7)$$

Applying the condition that $C_{m\text{ac}}$ does not vary with C_L we get

$$\begin{aligned}\frac{dC_{m\text{ac}}}{dC_L} = 0 = {} & \frac{dC_{m\text{tr}}}{dC_L} \\ & - \left[\left(1 + C_D\frac{d\alpha}{dC_L}\right)\cos\alpha + \left(\frac{dC_D}{dC_L} - C_L\frac{d\alpha}{dC_L}\right)\sin\alpha\right]x \\ & - \left[\left(\frac{dC_D}{dC_L} - C_L\frac{d\alpha}{dC_L}\right)\cos\alpha - \left(1 + C_D\frac{d\alpha}{dC_L}\right)\sin\alpha\right]y \quad (5.8)\end{aligned}$$

The data may easily be used to find C_L, C_D, α, and also the slopes $dC_{m\text{tr}}/dC_L$ and $d\alpha/dC_L$ since they are straight lines. The determination of dC_D/dC_L is difficult, for it is a curve.

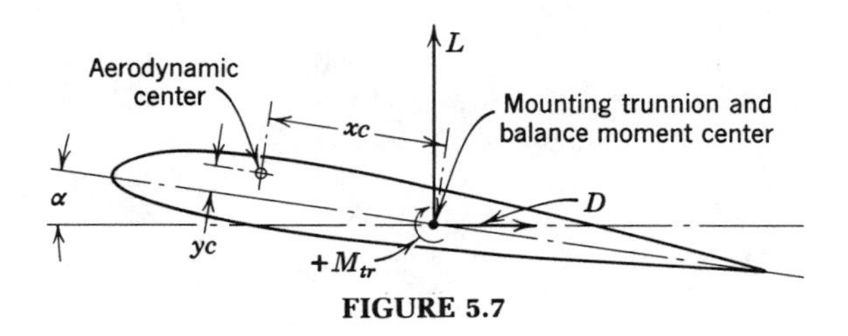

FIGURE 5.7

If the wing efficiency factor has been determined, dC_D/dC_L may be found directly from

$$C_D = C_{D0\ \min} + \frac{C_L^2}{\pi ARe}$$

$$\frac{dC_D}{dC_L} = \frac{2C_L}{\pi ARe} \tag{5.9}$$

If information for the above equation is not available, the slope of the drag curve at the proper point may be obtained by the familiar mirror method. In this method a small hand mirror is set directly on the plotted curve and adjusted until the reflected curve appears as a smooth continuation of the original. Under these conditions the plane of the mirror will be perpendicular to the drag curve at the selected C_L, and the drag curve slope may then be computed.

Equation (5.8), having two unknowns, requires the substitution of two points and then the simultaneous solution of the resulting equations. The approximation of measuring dC_D/dC_L may be eliminated for one of these points by selecting for the point the angle at which C_D is a minimum. At this point, obviously, $dC_D/dC_L = 0$.

Example 5.2. Find the aerodynamic center of an airfoil whose tests yield the data in Table 5.2. The mounting trunnion is at the 49% chord point and 17.9% below the chord line.

Plots of the data yield $dC_{m\text{tr}}/dC = 0.254$.

$$dC_L/d\alpha = 0.1025 \text{ per degree}$$
$$= 5.873 \text{ per radian}$$

At $\alpha = 2.3°$, $C_L = 0.34$, $C_{D0\,\min} = 0.0087$, and $dC_D/dC_L = 0$.

At $\alpha = 8.8°$, $C_L = 1.00$, $C_D = 0.0210$, and by the mirror method $dC_D/dC_L = 0.048$.

TABLE 5.2

α	C_L	C_D	$C_{m\text{tr}}$
−2	−0.086	0.0120	−0.023
0	0.111	0.0095	0.024
2	0.326	0.0087	0.079
4	0.531	0.0096	0.131
6	0.737	0.0138	0.183
8	0.943	0.0195	0.231
10	1.118	0.0267	0.281
12	1.260	0.0369	0.317

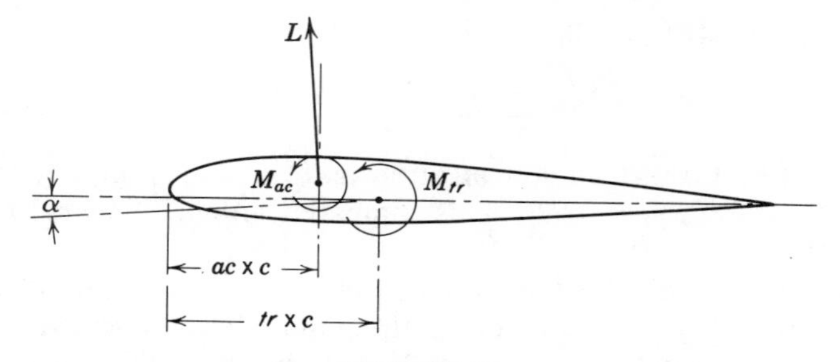

FIGURE 5.8

Substituting into Eq. (5.8), the equations simplify to

$$\begin{aligned} 0.99916x + 0.01765y &= 0.254 \\ 0.97300x + 0.06216y &= 0.254 \\ x = 0.2516 &= 0.252 \\ y = 0.1475 &= 0.148 \end{aligned}$$

The aerodynamic center is $(0.490 - 0.252) = 0.238$ or 23.8% of chord or 1.2% chord ahead of the quarter chord, and $(-0.179 + 0.148) = -3.1\%$ below the chord line.

In order to save time in locating the aerodynamic center, the assumption is sometimes made that the moment is due entirely to the lift and that the aerodynamic center is on the chord line. Since the lift and drag act through the aerodynamic center, the moment about the trunnion is (Fig. 5.8)

$$M_{\mathrm{tr}} = M_{\mathrm{ac}} + L(\mathrm{tr} - \mathrm{ac})c \qquad (5.10)$$

where M_{ac} = moment about the aerodynamic center, and tr = chordwise location of the balance trunnion.

Rewriting Eq. (5.10) in coefficient form, we have

$$C_{m\mathrm{tr}} = C_{m\mathrm{ac}} + C_L(tr - \mathrm{ac}) \qquad (5.11)$$

and differentiating and transposing ($dC_{m\mathrm{ac}}/dC_L = 0$),

$$\mathrm{ac} = \mathrm{tr} - dC_{m\mathrm{tr}}/dC_L \qquad (5.12)$$

The aerodynamic center is theoretically a small amount behind the quarter chord. In practice, it is found ahead of the quarter chord for the older profiles and behind for the new profiles.

Example 5.2a. Calculate the location of the aerodynamic center for the data of Example 5.2, using Eq. (5.11).

1. From a plot of C_{mtr} versus C_L,

$$dC_{mtr}/dC_L = 0.254$$

2. Substituting the trunnion location and dC_{mtr}/dC_L in Eq. (5.12), we have ac = 0.49 − 0.254 = 0.236. This compares with 0.238 by the method of Eq. (5.8).

Equation (5.11) indicates that when $C_L = 0$, $C_{mac} = C_{mtr}$. In other words, the value of the moment coefficient at the point where the curve strikes the C_m axis is approximately the value of C_{mac}. Rather than call it that, the usual practice is to label the above intersection C_{m0}.

(g) After the location of the aerodynamic center has been obtained the moment coefficient about it may be found from

$$C_{mac} = C_{mtr} - x(C_L \cos \alpha + C_D \sin \alpha) - y(C_D \cos \alpha - C_L \sin \alpha) \quad (5.7)$$

The value of C_{mac} varies with the amount and shape of the camber line. For wings using symmetrical airfoils, it is zero; for nonswept wings, the value is −0.10 or less. For swept wings it becomes larger and more negative, and is a function of sweep and aspect ratio. For wings with flaps on the trailing edge the values of −1.0 can be exceeded. It should be noted that large negative values of C_{mac} can often reduce the airplane $C_{L\max}$ at some center of gravity positions due to large down loads on the tail.

Example 5.3. Calculate the C_{mac} for Example 5.2.

Substituting each point into Eq. (5.7) we have the data shown in Table 5.3.

It is not unusual to find some small spread in the values of C_{mac}, although strictly speaking, the definition states that it must be constant. The small nonlinearity in the values of C_{mac} is due to the vertical displacement terms of the ac in Eq. (5.7). This can easily be verified by plotting the x and y displacement terms versus C_L or α.

TABLE 5.3

α	C_{mac}
−2	−0.0026
0	−0.0053
2	−0.0026
4	0.0016
6	0.0076
8	0.0118
10	0.0291
12	0.0383

It is a surprising fact that the location of the aerodynamic center is practically unchanged by flaps. The explanation lies in the manner in which the moment is generated:

$$C_{m\,\text{total}} = C_{m\,\text{due to changing}\,\alpha} + C_{m\,\text{due to camber}}$$

As indicated by theory, the C_m due to changing α is constant about the quarter chord. The C_m due to camber is a constant about the half chord. Hence adding camber in the form of flaps merely increases the value of $C_{m\text{ac}}$ without changing the location of the aerodynamic center as determined by changing α.

(h) The center of pressure is defined as "that point on the chord of an airfoil through which the resultant force acts." Though its usefulness has declined with the introduction of the concept of the aerodynamic center, it must occasionally be determined from force tests. The procedure is as follows:

The forces measured appear as a lift force L, a drag force D, and a moment about the mounting trunnion M_{tr} (Fig. 5.9). At the point through which the resultant force acts, the moment vanishes. Hence

$$M_{\text{CP}} = 0 = M_{\text{tr}} + L(p \cos \alpha) + D(p \sin \alpha)$$

where p is the distance from the trunnion to the center of pressure, positive to the rear of the trunnion.

We then have

$$-C_{m\text{tr}} = C_L(p/c) \cos \alpha + C_D(p/c) \sin \alpha = 0$$

and

$$\frac{p}{c} = \frac{-C_{m\text{tr}}}{C_L \cos \alpha + C_D \sin \alpha} \tag{5.13}$$

The location of the center of pressure from the wing leading edge is then

$$\text{CP} = (p/c) + (\text{tr}/c) \tag{5.14}$$

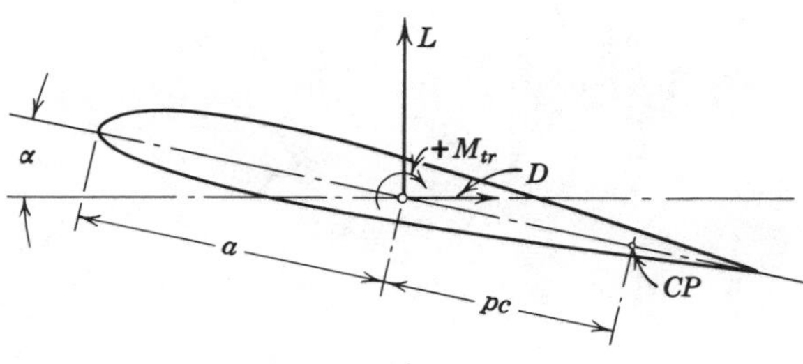

FIGURE 5.9

Example 5.4. A wing is mounted with the trunnion at the 23.5% chord point. At $\alpha = 7.3$, we have $C_L = 0.6$, $C_D = 0.0320$, and $C_{mtr} = -0.011$. Find the center of pressure.

$$\frac{p}{c} = \frac{-C_{mtr}}{C_L \cos \alpha + C_D \sin \alpha} = \frac{0.011}{(0.6)(0.9919) + (0.0320)(0.127)}$$

$$= 0.0184$$

$$\text{CP} = 0.0184 + 0.235 = 0.253 = 25.3\%$$

5.6. TESTING TWO-DIMENSIONAL WINGS

Alterations to airfoil sections are frequently investigated in two-dimensional tunnels wherein a short constant chord section of a wing completely spans the jet width (Fig. 5.10), simulating infinite aspect ratio. The jet is usually $2\frac{1}{2}$–4 times higher than it is wide. In some tunnels of this type the drag is read by the momentum survey method, and the lift by the pressure on the tunnel walls (Sections 4.18 and 4.19). The pitching moment may also be read from the pressure on the tunnel walls, but in many cases the wing is mounted on trunnions and the moment is read with a simple beam balance. The proportionately large drag of two endplates prohibits accurate drag measurements by the usual force tests and complicates the endplate seal.

FIGURE 5.10. Airfoil model in two-dimensional tunnel. (Photograph courtesy of NASA.)

Since the models customarily used in two-dimensional tunnels are larger in proportion to jet size than others, corrections for constriction, buoyancy, and camber must be considered (Chapter 6). Excessive errors due to wall effect will arise in $c_{l\,\max}$ if the model chord exceeds $0.4h$ and in $c_{d0\,\min}$ if it exceeds $0.7h$, where h is the tunnel height.

The information obtained from two-dimensional tests will be reducible to section coefficients c_l, c_{d0}, and $c_{m\frac{1}{4}}$. These coefficients (unlike wing coefficients C_L, C_D, and $C_{m\frac{1}{4}}$, which are an average of conditions including varying Reynolds number and effective angles of attack across the span) consider only a section under constant load and hence constant effective angle of attack. It is customary to consider the minimum profile drag coefficient C_{D0} as equivalent to c_{d0} when both are at the same Reynolds number. Likewise, it is assumed that $C_{mac} = c_{mac}$.

The location of the aerodynamic center and the center of pressure may be calculated as discussed in Section 5.5 (f) and (h). The lift coefficients c_l and C_L may also be considered equal except at their maximum, where the spanwise lift distribution usually results in a diminution in lift of at least 10–15% for straight wings. Expressed symbolically,

$$C_{L\,\max} = 0.90c_{l\,\max} \text{ (approximately)} \tag{5.15}$$

Two-dimensional testing is used to obtain data for aerodynamic comparison of various airfoil shapes. Comparisons would include $c_{l\,\max}$ and the shape of the lift curve at maximum lift, minimum drag, and the variation of c_d and c_{mac} with c_l or α. The lift curve slope usually decreases from the 2D value to a wing 3D, and is slightly affected by Reynolds number and increases with Mach number for Mach numbers less than 1. There is some question about the validity of two-dimensional drag data (see Section 4.18). The reduction of data to coefficient form is covered in Sections 4.18 and 4.19 when pressure measurements are used, and for the effect of tunnel walls in Sections 6.2–6.7.

5.7. TESTING COMPONENT PARTS OF AN AIRPLANE

The testing of large-scale models of part of an airplane offers many advantages provided that the data can be properly applied to the airplane. Nacelles, tail surfaces, dive brakes, and ailerons are parts belonging in this group.

To take a concrete example, models of 8-ft span are about the maximum usually tested in a 10-ft tunnel. If the full-scale aircraft has a wing span of 80 ft the largest model that can be tested will be $\frac{1}{10}$ scale. This reduction in size makes it nearly impossible to reproduce small items accurately; their Reynolds number will be very small; and it will be exceedingly difficult to measure the hinge moments of control surfaces. A 30% model of the vertical tail could be tested as well as a 30% aileron model, but, if such large-scale panel models are to be employed, they must, of course, be tested under flow conditions that simulate those on the complete airplane. Mounting the panel like a complete wing (Fig. 4.22) permits an endflow about the inboard tip that does not actually exist on the real airplane. Such flow may easily invalidate the test results, and unfortunately even the addition of an endplate will not provide sufficient sealing to produce complete flow conditions. It also can be difficult to account for the forces and moments of the endplates when balance data are taken.

Two arrangements are satisfactory: mounting the panel on a turntable or mounting it with a small gap (less than 0.005 span) between its inboard end and the tunnel floor (Figs. 4.21 and 5.11). Either of these arrangements will seal the inboard end of the panel and subject it to nearly the same flow conditions as would occur on the actual aircraft. Usually the effective aspect ratio then developed will be about $0.95(2b)^2/(2S_p)$, where b is the panel span and S_p is the panel area.

An important and sometimes insoluble problem may arise for aileron panels from sweptback wings. Here the flow over the aileron is affected very strongly by the remainder of the wing and cannot be simulated by any simple reflection plane. It is therefore suggested that a thorough study of spanwise loadings be made and adequate correlation ensured before attempting panel tests of such ailerons. A full half-span model is satisfactory.

A minor point in panel testing of the type shown in Figs. 4.21 and 5.13 is that, though the hinge moments should be reduced to coefficient form by using the tunnel dynamic pressure, the force coefficients should be based on tunnel dynamic pressure corrected to allow for the diminished velocity in the boundary layer. The corrected q is usually about 99% of the centerline q. If the boundary layer velocity profile is known, it can be used to adjust the q.

Minimum confusion in applying the test data will result if the panel is selected so that positive tunnel directions are positive airplane directions. Thus a left panel should be mounted on a right wall,* and a right panel on a left wall. Floor-mounted models should be left airplane panels.

In the selection of the panel span the inboard juncture of aileron or flaps should not occur at the floor but a few inches above it.

Owing to asymmetry, much of the panel data will require that the loads from the panel with flap or aileron zero be subtracted from those with surface deflected in order to determine the contribution of the control. Obviously, since the surface zero data are to be repeatedly used, only very good data should be used for the base run, but good data are not always obtainable. For instance, the deflection of the surface will cause a change of lift and hence a change in wall correction angle. The proper basic coefficient should be obtained from curves of surface zero data plotted against corrected angle of attack. The problem is further confused by the fact that the surface de-

* Not upside-down on a left wall.

FIGURE 5.11. Reflection-plane tail test. (Courtesy General Dynamics—Convair.)

flected may allow greater angles of attack than the surface zero run and no basic data can exist for the entire range.

Details of the calculations necessary for referring panel tests of ailerons, rudders, and elevators to the complete aircraft are given in the following sections.

5.8. TESTING CONTROLS: AILERON PANELS

The airplane designer is interested in three items connected with proposed ailerons. One, the rolling moment, he or she is seeking. The other two, the yawing moment and the hinge moment, are the price that must be paid. In addition, the tunnel engineer is concerned with referring the data to the complete airplane. The problem first apparent is that the panel model will not have the same lift curve slope as the complete wing. Actually, however, this is of small import as long as the roll due to a given aileron deflection is at a known wing lift coefficient. This, it will be shown, requires that the complete wing be previously tested and its lift curve known and available. The procedure for then referring the aileron panel angles of attack to the complete wing is as follows:

1. From the estimated performance of the airplane, note the speeds that correspond to various important flight conditions such as minimum control speed, climb speed, etc. Calculate the complete wing lift coefficients that correspond to these speeds.

2. Plot the span loading curve, which is usually known before the test (for a nonswept wing, Schrenk's method can be used for an approximation). The total area under this curve, A_T, divided by the total wing area, S_T, is a measure of the wing lift coefficient. Likewise, the area of that part of span loading curve above the panel span A_p, when divided by the area of the wing panel S_p, is a measure of the panel lift coefficient. The ratio of these two ratios, then, is the ratio of the complete wing lift coefficient C_{Lw} to the panel lift coefficient C_{Lp}. That is,

$$\frac{C_{Lw}}{C_{Lp}} = \frac{A_T/S_T}{A_p/S_p} \tag{5.16}$$

Equation (5.16) may be used to find the panel lift coefficients that correspond to the selected wing lift coefficients.

3. Test the panel with aileron zero and obtain a plot of C_{Lp} versus α_u, where α_u is the angle of attack uncorrected for tunnel wall effect.

4. From 3, read the uncorrected angles that should be set to read the desired panel lift coefficients. The tunnel operator may then set these uncorrected angles with arbitrary aileron deflections and maintain proper panel model to complete model correlation.

The span loading as required above may be simply found by a method proposed by Schrenk (Ref. 5.4), or by one proposed by Pearson (Ref. 5.5). Schrenk's method for untwisted wings without flaps is as follows:

1. Plot the wing chord against the span. If the airfoil section varies, then a weighted chord

$$c_1 = ca_0/\bar{a}_0 \tag{5.17}$$

should be used. The term with the bar is the average section slope given by

$$\bar{a}_0 = \frac{2}{S}\int_0^{b/2} a_0 c\, dy \tag{5.18}$$

2. On the same graph, plot a quarter ellipse whose area is equal to half the wing area, and whose span equals the wing span.

3. The span loading will be represented by a line midway between 1 and 2. (See Fig. 5.12).

The data for an aileron test require some special consideration for three reasons: the structural problems due to cable stretch and wing twist; the simulation of a symmetrical case as a result of the carry-over arising from the wall reflection (see Section 6.28); and the doubled force increments arising from the same source. Cable stretch and wing twist are problems that must be evaluated by the structural engineers for a particular airplane. Assuming that the carry-over is small, one is fully justified in using the measure force data (corrected for wall effects and blockage) for computing the rolling and yawing moments about the aircraft centerline. But if one is interested in the complete wing lift, drag, and pitching moment, the measured data, including as they do the full reflection, are too large. In other words a 50-lb lift increment ($\Delta C_L = 0.1$, say) for one aileron down is not a 100-lb increment ($\Delta C_L = 0.1$) when the whole wing is considered, since the image aileron should not be down. Thus, letting a subscript d mean aileron down, and o mean control neutral, wing data for an asymmetrical model may be found from

$$C_L = \frac{C_{Lo} + C_{Ld}}{2} \tag{5.19}$$

$$C_D = \frac{C_{Do} + C_{Dd}}{2} \tag{5.20}$$

$$C_m = \frac{C_{mo} + C_{md}}{2} \tag{5.21}$$

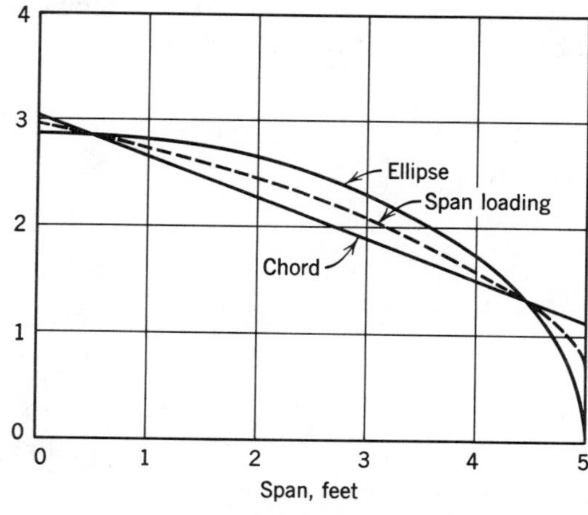

FIGURE 5.12

The roll and yaw data may be treated as outlined below.

In testing one panel of a wing, it is seen that a yawing and rolling moment about the aircraft centerline is produced that in actuality would be canceled by the panel on the other side. Thus it is necessary to subtract the moments due to the panel with aileron zero from the moments with the aileron deflected. The subtraction also acts to remove the tare effects of the turntable, the net result being the yawing and rolling moments due to the deflection of the aileron only. The only requirement is that the test conditions must simulate the proper spanwise loading over the aileron.

In working up the data, the coefficients must be corrected to complete wing areas and spans so that the results will be usable. The definitions are as follows (the subscript p indicates "panel"):

$C_{Lp} = \frac{L_p}{qS_p}$ panel lift coefficient,

$C_{D'p} = \frac{D'}{qS_p}$ panel drag coefficient including mounting plate drag,

$C_{np} = \frac{\mathrm{YM}_p}{qS_w b_w}$ panel yawing moment coefficient,

$C_{lp} = \frac{\mathrm{RM}_p}{qS_w b_w}$ panel rolling moment coefficient about balance rolling axis, based on wing area and span.

To get the rolling moment at the aircraft centerline we have (Fig. 5.13)

$$\mathrm{RM}_a = \mathrm{RM}_p + La'$$

$$C_{l'a} = C_{lp} + C_{Lp}\frac{S_p}{S_w}\frac{a'}{b_w} \tag{5.22}$$

where a' is the distance from the balance rolling moment axis to airplane centerline (the subscript a indicates "aircraft").

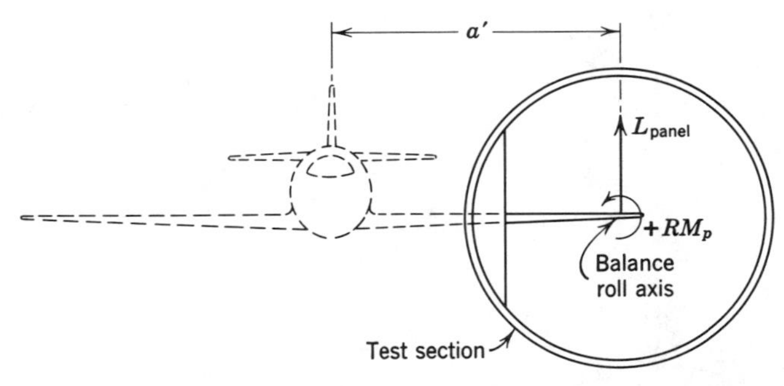

FIGURE 5.13. Panel rolling moment and its relation to the rest of the airplane. (Front view.)

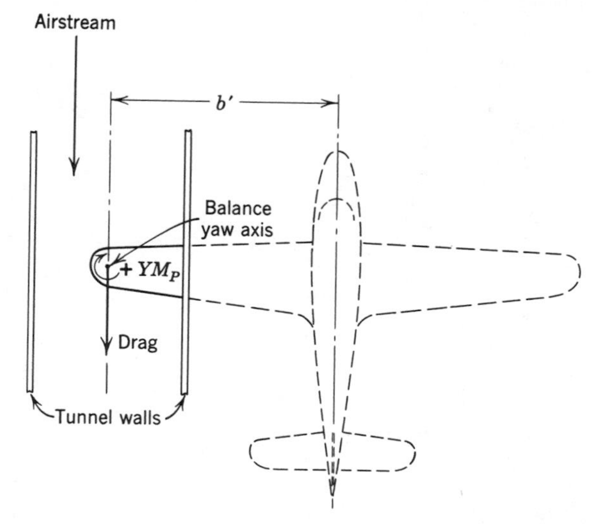

FIGURE 5.14. Panel yawing moment and its relation to the rest of the airplane. (Plan view.)

However, this represents the moment of one panel plus the aileron about the aircraft centerline; to find the part due to the aileron only, we subtract the rolling moment coefficient of the panel with aileron zero, C_{l0}. Hence the rolling moment coefficient of one aileron about the airplane centerline is*

$$C_l = C_{lp} + C_{Lp} \frac{S_p}{S_w} \frac{a'}{b_w} - C_{l0} \tag{5.23}$$

By a similar process, the yawing moment coefficient due to one aileron is (Fig. 5.14)*

$$C_n = C_{np} - C_{D'p} \frac{S_p}{S_w} \frac{b'}{b_w} - C_{n0} \tag{5.24}$$

where $C_{np} = YM_p/qS_w b_w$; b' = distance of balance yaw axis to aircraft; and C_{n0} = yawing moment coefficient of one panel about aircraft centerline, aileron zero.

For structural purposes, or a check of the spanwise load calculations, the lateral center of pressure may be found by dividing the semi-span rolling moment about the airplane axis of symmetry by the lift.

Although not directly apparent, the rolling moment coefficient also determines the helix angle (see Fig. 5.15).

* Panel rolling (C_{lp}) and yawing moment (C_{np}) coefficients; do not confuse with damping-in-roll $dC_l/d(pb/2V)$ or damping-in-yaw $dC_n/d(rb/2V)$.

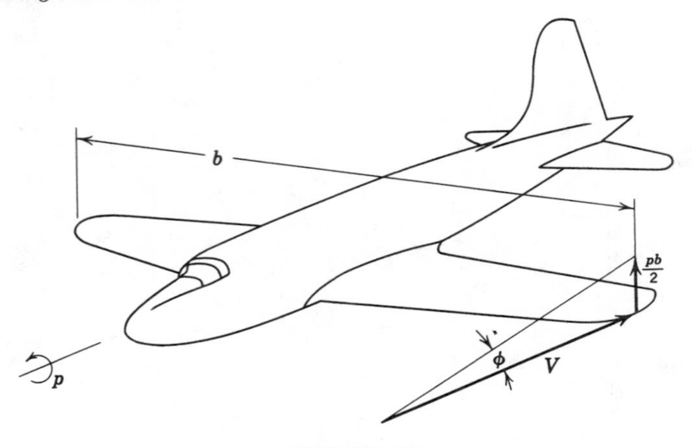

FIGURE 5.15

At a given rate of steady roll the rolling moment is opposed by an equal and opposite damping moment. In the usual symbols, the rolling moment

$$\text{RM} = (\rho/2)SV^2C_l b = \text{DM}$$

where DM = damping moment. Dividing through by the helix angle, $pb/2V$ (p = rolling velocity, radians per second),

$$\text{DM} = \frac{pb}{2V}\frac{\rho}{2}SV^2 b\frac{C_l}{pb/2V} = \frac{\rho}{2}SV\,p\frac{b^2}{2}\frac{dC_l}{d(pb/2V)}$$

The term $d(C_l)/d(pb/2V)$ is called the damping-in-roll coefficient (frequently written C_{lp}) and is a function of wing taper and aspect ratio, both of which are constant for a given airplane. For unswept wings, values of C_{lp} may be found in Fig. 5.16. The helix angle is then*

$$\frac{pb}{2V} = \frac{C_l}{C_{lp}} \tag{5.25}$$

and the rolling velocity is

$$p = \frac{2V}{b}\frac{C_l}{C_{lp}} \tag{5.26}$$

The maximum rolling velocity is usually limited by stick force considerations rather than by the airplane's actual ability to roll. Power-driven con-

* Equation (5.25) applies only to roll without yaw or sideslip and can be misleading at high angles of attack, where adverse yaw in flight may be appreciable.

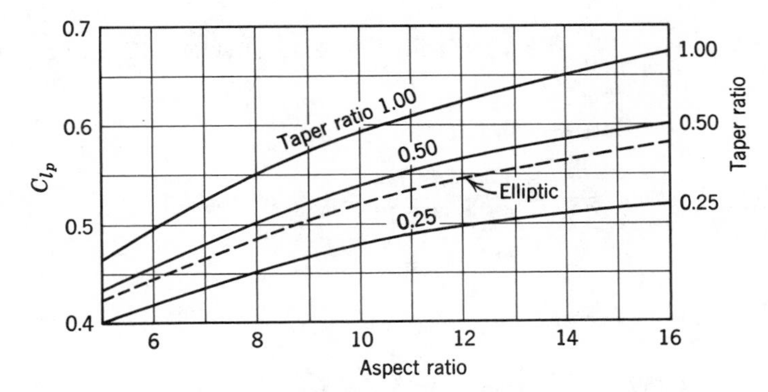

FIGURE 5.16. C_{lp} for various values of aspect ratio and taper.

trols as well as aerodynamic balances to decrease the hinge moments are used to increase the rolling velocity; however, wing twist will decrease rolling velocity.

Example 5.5. Data from a test run at $q = 25.6$ lb/sq. ft on a wing panel of 12 sq. ft yields $C_L = 0.789$, $C_D = 0.0599$ *at* $\alpha = 5.44°$, left aileron down 15°. (Data have been corrected for tunnel wall effect.) $C_{lp} = -0.0102$, $C_{np} = -0.0010$. Find the rolling and yawing moments about the aircraft centerline if the distance from the balance roll axis to the aircraft centerline is 8 ft. The rolling moment for aileron zero is $C_{l0} = 0.0191$, and the yawing moment is $C_{n0} = -0.0028$. $S_w = 70$ sq. ft, $b_w = 25$ ft.

$$C_l = C_{lp} + C_{Lp}\frac{Sp}{S_w}\frac{a'}{b_w} - C_{lp0}$$

$$= -0.0102 + 0.789\frac{12}{70}\frac{8}{25} - 0.0191$$

$$= -0.0102 + 0.0433 - 0.0191 = 0.0140$$

$$C_n = C_{np} + C_{D'p}\frac{S_p}{S_w}\frac{b'}{b_w} - C_{np0}$$

$$= -0.0010 - 0.0599\frac{12}{70}\frac{8}{25} + 0.0028$$

$$= -0.0015$$

Example 5.6. Assume that sufficient control force exists to develop the above C_l at 150 mph. Calculate the helix angle and rate of roll. The wing taper ratio is 2 : 1 ($\lambda = 0.50$), and the model is 40% scale.

1. $AR = \dfrac{b^2}{S} = \dfrac{25^2}{70} = 8.93$. From Fig. 5.16 at $AR = 8.93$ and $\lambda = 0.50$,

$$C_{lp} = 0.520$$

$$\frac{pb}{2V} = \frac{C_l}{C_{lp}} = \frac{0.014}{0.520} = 0.0269 \text{ radian}$$

2. $$p = \frac{pb}{2V}\frac{2V}{b}57.3$$

$$= 0.0269\,\frac{(2)(150)(1.47)}{25/0.40}\,57.3$$

$$= 10.9 \text{ deg/s}$$

Typical data that might be expected from tests of a aileron are shown in Figs. 5.17, 5.18, and 5.19. The roll versus yaw plot is particularly useful when figuring ratios for differential ailerons.

Perhaps at this point it would be fitting to discuss the lateral axis more fully in order to explain the reaction obtained before and during a roll. Consider the case when the ailerons are deflected, but no roll has yet had

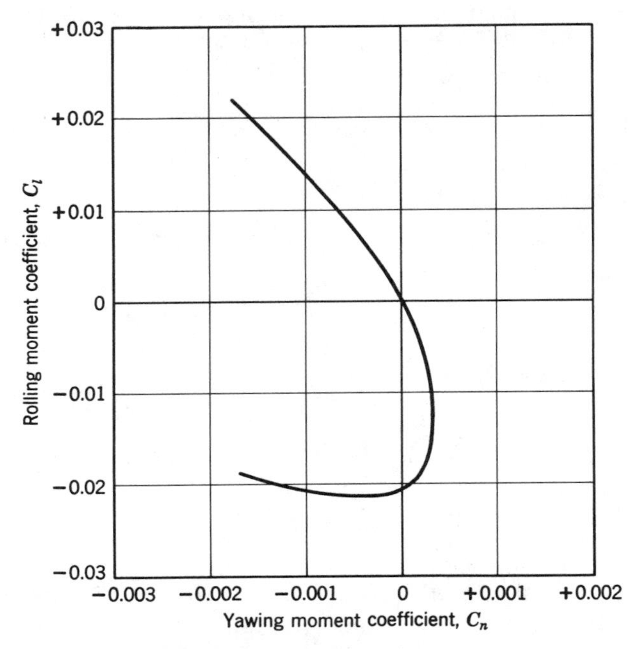

FIGURE 5.17. Aileron adverse yaw.

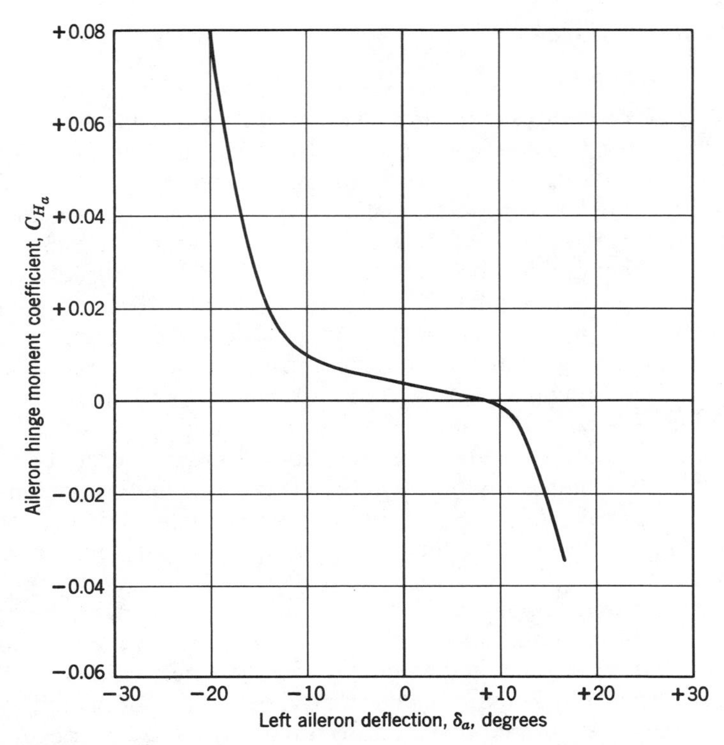

FIGURE 5.18. Hinge moment coefficient C_{Ha} versus aileron deflection δ_a.

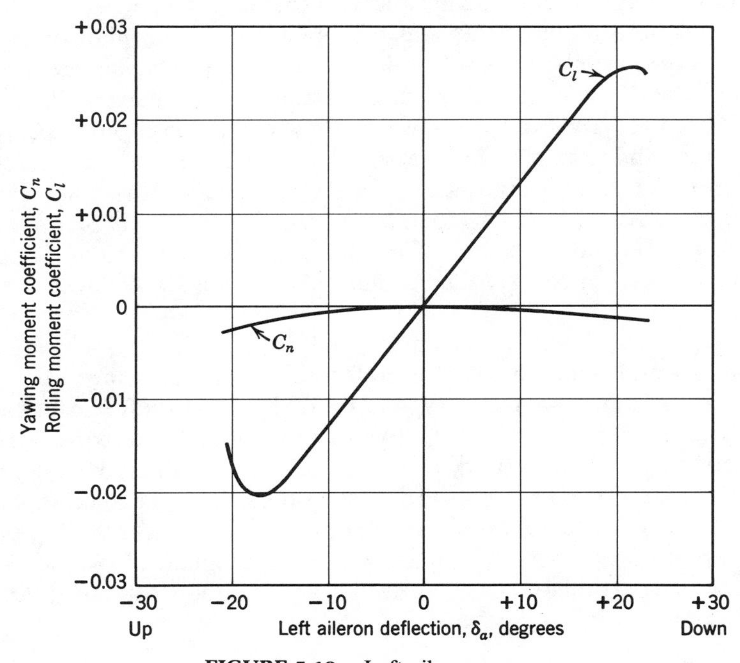

FIGURE 5.19. Left aileron power.

time to develop, or perhaps an asymmetrical span loading is being resisted. The down aileron creates more lift and induced drag and usually more profile drag, whereas the opposite effect is noted for the up aileron. The net result is a yawing moment opposite in sign to the rolling moment: left yaw for right aileron. Since nose left yaw is negative and right roll positive, the yaw is adverse when it has an opposite sign to the roll. Many methods are suggested for balancing this adverse yaw, among them an artificially increased profile drag of the raised aileron, which tends to pull the lowered wing into the turn. Such a profile-drag increase can be obtained either by a special aileron design or by gearing the aileron controls so that the raised aileron has a greater deflection than the depressed one. The latter system is referred to as "differential ailerons." Differential ailerons reduce adverse yaw but are usually accompanied by an overall reduction in maximum rate of roll.

The designer, therefore, notes not only the maximum amount of roll ($C_{l\,\max}$) from graphs such as Fig. 5.19 but also examines data such as those shown in Fig. 5.17 to observe the amount of adverse yaw, a minimum being desirable.

Now we come to a point *important* to the tunnel engineer. Referring to Fig. 5.17 again, we note that when the curve appears in the first and third quadrants, favorable yaw is indicated, but it will actually exist only when the airplane does not roll. When rolling occurs, the direction of the relative wind over each wing is so altered that a strong adverse yaw is developed, and the results of the static tunnel test may be entirely erroneous.

Under most conditions, the air loads on the ailerons oppose their deflection, producing a moment that must be supplied by the pilot or by some outside means. Methods employed to help the pilots include powered "boosters" and mass and aerodynamic balance. The mass balances may only balance the weight of the surface or they may be arranged to provide an inertia force while the ship is rotating. The aerodynamic balances control surface areas of various cross-sectional shapes ahead of the hinge and area disposition such as horns, shielded horns, and internal, medium nose, and sharp nose aerodynamic balance. They may also include various devices aft of the hinge line, such as balance tabs, spring tabs, and beveled trailing edges.

Aerodynamic balances are simpler and lighter than mass balances or power boost and hence are to be preferred as long as icing need not be considered for general-aviation-type aircraft. Large aircraft and fighters almost always need power boost. Unfortunately, most aerodynamic balances are effective for only a portion of the aileron travel, as may be seen by the extent of the decreased slope in Fig. 5.18. A measure of balance superiority is then the range of decreased hinge moments as well as the slope of the balanced part of the curve. Complete aileron data must, of course, include the effect of the other aileron as well as the amount of the differential decided upon.

The aileron hinge moment coefficient C_{Ha} is defined by

$$C_{Ha} = \frac{\text{HM}}{qS_a c_a} \tag{5.27}$$

where HM = aileron hinge moment, positive when it aids control deflection; S_a = area of aileron aft of hinge line; and c_a = average chord of aileron aft of hinge line.

Another definition of the hinge moment is based on the use of the root mean square chord aft of the hinge as follows:

$$C_{Ha} = \frac{\text{HM}}{q\bar{c}_f^2 b_f} \tag{5.28}$$

where $\bar{c}_f$ = root mean square chord flap or aileron aft of the hinge, and b_f = flap or aileron span.

The quantity $\bar{c}_f^2 b_f$ is most easily obtained by integrating the area under the curve of local flap chord squared against flap span.

Figure 5.20 shows a hinge moment calibration setup.

The variance of definitions again demonstrates the necessity of clear and complete definitions of every item in the wind tunnel report.

Complete consideration of hinge moments must include the gearing ratio of the controls. Depending on the purpose of the aircraft, various limiting conditions are imposed by both the military and FAA.

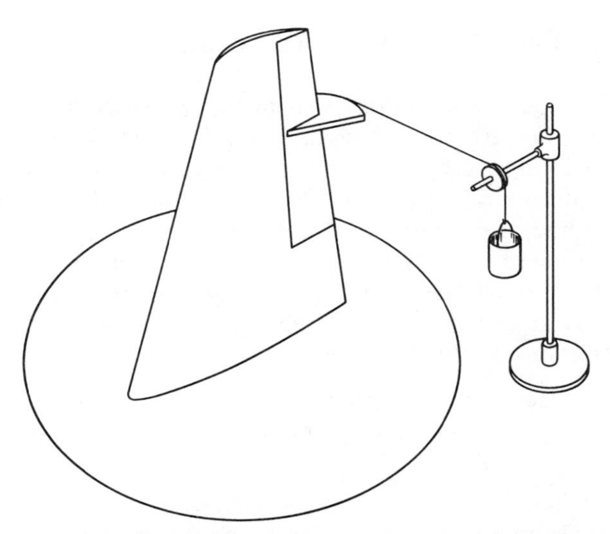

FIGURE 5.20. Setup for calibrating hinge moment strain gages on a panel model. For quick checks a good spring scale is convenient.

Example 5.7. An airplane has the following specifications:

$W = 40{,}000$ lb,
$S = 755$ sq. ft,
$S_{ail} = 17.43$ sq. ft,
Wheel radius $= 0.625$ ft,
Aileron differential $= 1:1$,
At $+10°$ aileron,
$C_{Ha} = -0.0089$,
At $-10°$ aileron, $C_{Ha} = 0.0188$,
$b = 71.5$ ft,
Aileron chord aft of hinge $= 1.426$ ft,

$$\frac{\text{Wheel throw}}{\text{Aileron deflection}} = 3.33.$$

Calculate the wheel force (one hand) necessary to deflect the ailerons 10° at 262 mph indicated airspeed.

For the total hinge moment coefficient resulting from both ailerons, we have

$$C_{Ha} = 0.0188 - (-0.0089) = 0.0277$$

$$M_{ail} = qS_a c_a C_{Ha} = (175.2)(17.43)(1.426)(0.0277)$$

$$= 120.5 \text{ lb-ft}$$

$$\text{Wheel moment} = \frac{120.5}{3.33} = 36.2 \text{ lb-ft}$$

$$\text{Wheel force} = \frac{36.2}{0.625} = 58.0 \text{ lb}$$

5.9. TESTING CONTROLS: RUDDERS

The rudder produces a side force that in turn produces a yawing moment about the center of gravity of the airplane. This may or may not produce yaw, for the lateral loading may be asymmetrical and the rudder employed only to maintain a straight course. Some drag, a small moment about the quarter-chord line of the tail itself, some roll, and a rudder hinge moment will also be created. The fact that the drag moment is stabilizing is no argument in favor of a large vertical tail drag, since, in maintaining a straight course with asymmetrical loading, drag is decidedly harmful.

The designer of a vertical tail seeks:

1. A large side force with minimum drag.
2. The steepest slope to the side force curve so that small yaw produces large stabilizing forces.
3. The smallest hinge moment consistent with positive control feel. The maximum pedal force is defined by FAA and the military.
4. Proper rudder balance so that under no conditions will the pilot be

unable to return the rudder to neutral, and preferably it should not even tend to overbalance.

5. A zero trail angle* so that control-free stability is the same as control-fixed stability. It will be seen that the zero trail angle permits smaller pedal forces and rudder movements to return a yawed aircraft to zero.

6. The largest yawing moments about the airplane's center of gravity. This moment is due almost entirely to the side force. The proportions due to the yawing moment of the vertical tail about its own quarter chord and the yawing moment of the vertical tail surface due to its drag are quite small but not always insignificant.

The rudder calculations, unlike the aileron panel tests, will require the absolute value of the drag coefficient. [See Eq. (5.29).] This is not easily obtained with a panel test, but, in view of the small contribution of the drag effect, an approximation may be made by reading the section drags with a wake survey rake at stations along the vertical tail with the rudder angle zero, and by summing them to get the total drag coefficient. This may be subtracted from the minimum drag as read by the balance to get the drag of the endplate. The method as outlined makes the very questionable assumption that the tare drag is unaffected by rudder angle, which is justified only by the peculiar conditions of this setup in which tare accuracy is not vital.

The sign convention for control surfaces and tabs is as follows: A positive angular deflection yields a positive force on the main surface. A positive hinge moment produces a positive deflection.

The rudder setup for finding n (and hence C_n) is shown in Fig. 5.21. The contributing parts are (1) the moment due to the vertical tail side forces, (2) the moment due to the vertical tail drag, (3) the moment due to the vertical tail moment about its own quarter chord.

In symbols these factors become

$$\begin{aligned} n_{cg} &= (n_v)_v - Y_v l_v \cos\psi - D_v l_v \sin\psi \\ &= qS_v\bar{c}_v C_{n\frac{1}{4}v} - qS_v C_{Yv}\cos\psi\cdot l_v - qS_v C_{Dv}\sin\psi\cdot l_v \end{aligned}$$

$$C_{ncg} = \frac{S_v\bar{c}_v}{Sb}C_{n\frac{1}{4}v} - \frac{S_v l_V}{Sb}C_{Yv}\cos\psi - \frac{S_v l_v}{bS}C_{Dv}\sin\psi \tag{5.29}$$

where n_{cg} = yawing moment of vertical tail about center of gravity,
S_v = vertical tail area,
$C_{n\frac{1}{4}v}$ = vertical tail moment coefficient about its own quarter chord,
$\bar{c}_v$ = mean aerodynamic chord of vertical tail,
S = wing area,

* For zero trail angle the rudder is so balanced that it remains at a zero deflection even when the airplane is yawed.

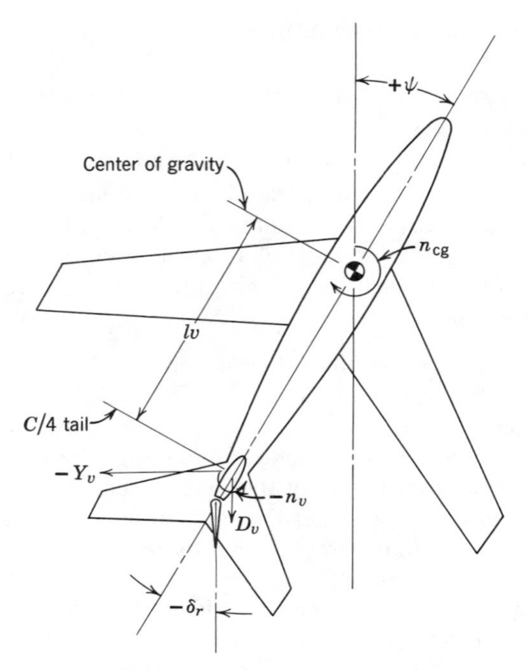

FIGURE 5.21. Plan view of yawed airplane.

b = wing span,
l_v = tail length (distance from tail quarter chord to center of gravity),
$C_{Yv} = Y_v/qS_v$,
ψ = angle of yaw,
$(n_v)_v$ = moment of vertical tail about its own quarter chord,
D_v = drag of vertical tail.

Sometimes it is desired to estimate the tail contribution to yawing moment; in this case some simplification of Eq. (5.29) is possible. The tail drag term and the tail moment about its own quarter chord are small, and for moderate yaw angles cos $\psi \approx 1.0$. Letting $S_v l_v/Sb$ equal the tail volume coefficient, $\overline{V}_v$, and putting $C_{Yv} = C_{Lv} = a_v\psi$, we have

$$C_{ncg} = -a_v\psi\overline{V}_v,$$

where

$$a_v = dC_{Lv}/d\alpha_v$$

This equation does not include the sidewash effect discussed in the next paragraph; it also assumes that the ratio of dynamic pressures q_v/q is unity.

It was mentioned in Section 5.8 that span loading must be considered in order to apply data properly from an aileron panel test to the complete airplane. In Section 5.10 attention is drawn to the proper method of applying the data from an isolated horizontal tail. The vertical tail is less affected by the remainder of the airplane, but some sidewash does exist when the airplane is yawed. Hence 15° yaw by no means results in a vertical tail angle of attack of 15°. Proper evaluation of the sidewash can be made by equipping the complete model with a vertical tail whose incidence is variable, and by going through a procedure similar to that outlined for the horizontal tail in Section 5.11.

Example 5.8. A vertical tail model whose area is 12 sq. ft is tested at 100 mph. The model MAC = 3.0 ft. The actual airplane, of which the model is 40% scale, has a wing area of 750 sq. ft and a span of 78 ft. The tail length is 30 ft. Find the tail yawing moment coefficient about the center of gravity if $C_{Yv} = 0.794$, $C_{Dv} = 0.0991$, $C_{n\frac{1}{4}v} = -0.1067$ for $\psi = 6°$. The rudder is deflected 10°.

$$C_{ncg} = \frac{S_v}{S_w}\frac{(\text{MAC})_v}{b_w} C_{n\frac{1}{4}v} - \frac{S_v}{S_w}\frac{l_v}{b_w} C_{Yv}\cos\psi - \frac{S_v}{S_w}\frac{l_v}{b_w} C_{Dv}\sin\psi$$

$$= \frac{12}{(0.40)^2(750)}\frac{3}{(0.4)78}(-0.1067) - \frac{12(0.4)(30)}{(0.40)^2(750)(0.4)(78)}0.794(0.9945)$$

$$- \frac{12}{(0.40)^2(750)}\frac{0.4(30)}{0.4(78)}(0.0991)(0.1045)$$

$$= -0.001025 - 0.0304 - 0.00398$$

$$= -0.0354$$

5.10. TESTING CONTROLS: ELEVATORS

The elevators may be also tested by the panel mounting method. With this arrangement, one-half the horizontal tail is usually mounted as shown in Fig. 5.11 and the results are doubled to get the data for the entire tail.

It will be noted that for airplanes of conventional dimensions, the pitching moment of the horizontal tail about its own quarter chord and the pitching moment about the airplane center of gravity produced by the horizontal tail drag are negligible when compared to the moment produced by the tail lifting force. Hence it will probably be necessary only to measure the lift of the panel model along with the elevator hinge moments to evaluate the desired qualities. Occasionally it will be desirable to compare two different methods of trimming to determine which has less drag for a given lift. Then, of course, drag measurements will be necessary.

The tail lift curve slope as determined from the panel model may require adjustment in order to apply the test results (Fig. 5.22). For example, sup-

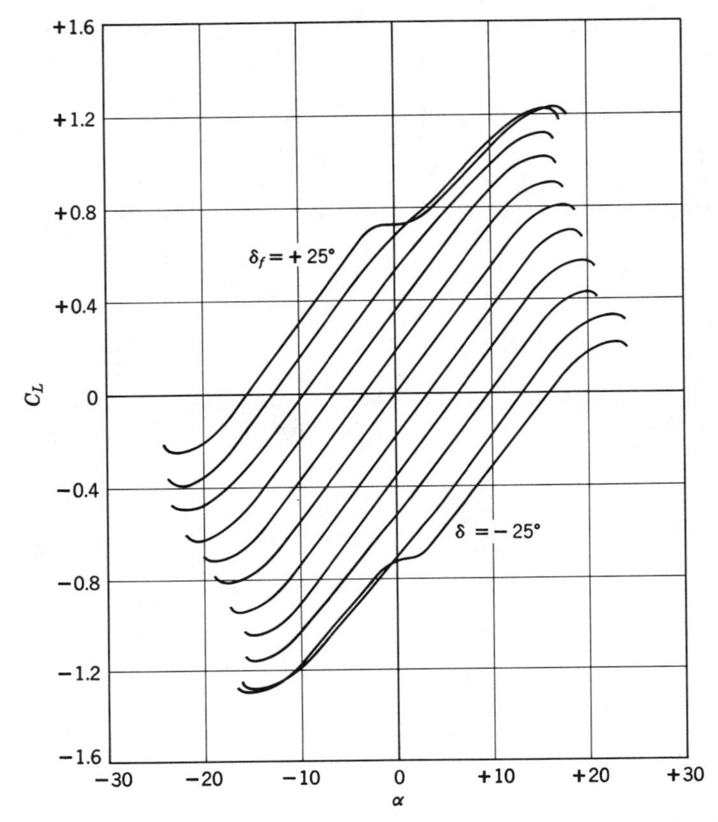

FIGURE 5.22. Typical panel lift curves. The breaks in the curves for large flap deflections occur when the flap stalls.

pose that the complete model has been tested at a constant angle of attack with varying settings of the stabilizer. The pitching moment about the airplane center of gravity due to the horizontal tail,

$$M_t = -l_t q_t S_t C_{Lt} \tag{5.30}$$

may be measured, and with the known tail area S_t and tail length l_t the value of C_{Lt} may be determined. In these calculations it is probably better to use $q_t = q_{\text{freestream}}$ than the very questionable $q_t = 0.8 q_{\text{freestream}}$ sometimes arbitrarily employed. From the calculated C_{Lt} and known stabilizer angles the slope of the tail lift curve on the airplane $(dC_{Lt}/d\alpha_t)$ may be established. It then remains to diminish the panel lift curve slope by the factor $(dC_{Lt}/d\alpha_t)_{\text{airplane}}$ divided by $(dC_{Lt}/d\alpha_t)_{\text{panel}}$.

The procedure followed to align the hinge moment data to the airplane may be traced through Fig. 5.23.

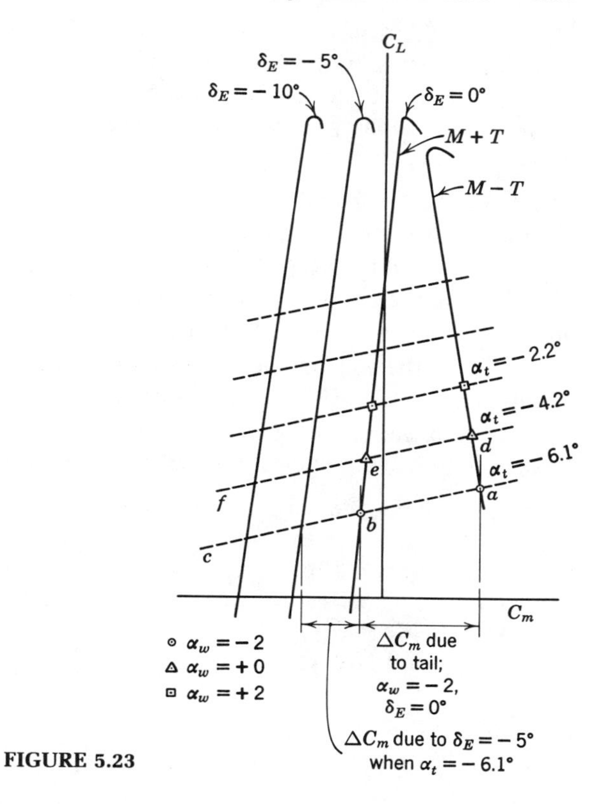

FIGURE 5.23

Let us suppose that the power-off moment curves of our example airplane are as shown in Fig. 5.23 for the model with tail and without tail, some center of gravity location* being specified. If lines *abc*, *def*, etc., are drawn between points of equal angle of attack values, the difference between the model plus tail ($M + T$) and $M - T$ curves is the contribution of the horizontal tail with elevator zero at the specified α_W. The moment due to the tail $\Delta C_{mt} = q_t(C_{Lt})S_t l_t$ and values of C_{Lt} are readily determined. From the previously prepared tail lift curve we find the tail angles of attack that correspond to the C_{Lt} values, and so label the lines *abc*, *def*, etc., as $\alpha_t = -6.1, -4.2$, etc. This procedure furnishes the relation between the panel tests and the complete airplane, since values of hinge moment and tail lift coefficient for various angles of attack are available from the panel tests.

To carry this chart to completion, the values of ΔC_m for various elevator deflections are calculated from the C_{Lt} values corresponding to the tail angles

* If another location is desired for the center of gravity, the curves may be rotated about zero lift by the relation

$$\Delta \frac{dC_m}{dC_L} = \frac{\% \text{ MAC change}}{100} = \Delta \text{cg}$$

of attack, and curves of constant elevator deflections may be drawn in. The complete chart may then be used to read the amount of elevator needed to trim at any C_L or the amount of moment available for maneuvering with a specified stick load. The maneuvering investigation requires the hinge moment data and the mechanical advantage, as follows:

1. Assume the chart to indicate that an elevator deflection of $-15°$ is required to trim at $C_L = 1.9$ and $\alpha_{tail} = +12$. From the airplane geometry and $C_L = 1.9$, the value of q may be found from $q = W/SC_L$.
2. From the chart of hinge moment versus α_{tail} we read (say) for $\alpha = +12$ and $\delta_e = -15°$ a $C_{He} = 0.0640$.
3. From the airplane geometry and q (step 1) the elevator hinge moment is calculated from $HM_e = qS_e c_e \cdot C_{He}$.
4. From the curve of mechanical advantage versus δ_e, and the known linkage lengths, the stick force may then be found.

In this manner the various flight conditions may be investigated and desirable balance changes evaluated. Note the limit to forward center of gravity travel is a function of elevator power, which is a function of elevator deflection and size. The elevator power must be sufficient to balance the airplane at maximum lift ($C_{m,CG} = 0$). The critical condition is gear and landing flaps down in ground effect.

5.11. TESTING COMPLETE MODELS

The six-component test of a complete model is the most difficult of all wind tunnel tests. More variables are under consideration than in other tests, for one thing, and the individual tests are more complicated, for another. For instance, each run requires *three* additional runs to evaluate the tare, interference, and alignment. (See Sections 4.16 and 4.17.)

Because tare and interference runs are extremely time consuming they are often replaced with either a set of support tares taken from a calibration model or not applied at all to the data. When no tares are applied, the data are compared with a base airplane configuration and are incremented from the baseline data. This is usually done only when the firm conducting the test has had a lot of experience with a particular tunnel, model-mounting system, and flight test.

Although tests of a complete model are difficult they are the most common type of testing. Usually there are several wind tunnel entries of the model as the airplane design develops.

Tests to determine optimum leading and trailing edge flap configurations and angles can be quite long owing to the large number of variables, which include pitch up on swept wings. The effect of control surface deflections both in pitch and yaw are checked not only for cruise but take-off and landing flap settings, in and out of ground effect. The effect of power for prop-driven airplanes or high bypass turbine engines merely adds to the test complexity with various power settings and engine out tests for multiengined aircraft required.

The data are usually corrected for the effect of tunnel walls, tunnel upflow, and tunnel blockage (Chapter 6). The use of trip strips on the model is discussed in Section 7.2. Power lift testing limits for V/STOL type aircraft is discussed in Section 6.33.

Before a brief discussion of the required runs for the model of a general-aviation-type aircraft is given, a brief discussion of how a test is run is in order.

Before a wind tunnel test is made with the model, the angle of attack is usually calibrated. This is done by the use of an inclinometer. The model reference plane is set to zero angle of attack (fuselage reference line or chord plane of the mean aerodynamic chord are usually used as references). Then the model is pitched through the desired angle-of-attack range and the relations between the alpha indicated and measured or geometric is determined. The next calibration is a weight tare. The model center of gravity does not coincide with the balance moment center and thus there is a change in pitching moment measured with angle of attack. This must be subtracted from the data as it is not an aerodynamic moment. In yaw there is both a rolling and pitching moment. If a sting balance is used, the model is rolled rather than yawed, and if the model CG is on the model centerline, there is no weight tare. Many weight tares will be taken during a test since they are a function of the model center of gravity location and hence model configuration.

In a low-speed wind tunnel with unpowered models there is usually one wind-on run made per model configuration. If the model is powered, several runs at different power settings may be made per model configuration without stopping the tunnel. Because of one run per setup, it is desirable to have a method of displaying the test results as they are obtained. The best form of data visibility is that of plots (Fig. 5.24). For pitch runs C_L-α, C_L-C_D, C_L-C_m and sometimes C_l-α are required. For yaw runs C_n-ψ, C_l-ψ, C_Y-ψ, or sometimes side slip angle β is used instead of yaw angle ψ.

These plots serve three purposes:

1. They will show any points that do not fit the curves (so-called bad points). When these occur, the test point is repeated (a check point). Note: the check point should not be approached from a stalled condition owing to the possibility of hysteresis in stalled regions. If the check point fits the curve, it replaces the initial point. If the check point repeats the initial point, additional test points are then taken to define the curve in this region.

2. Sometimes in the stall region the angle increments are too large to define the curve. When this happens additional points are taken to define the curves.

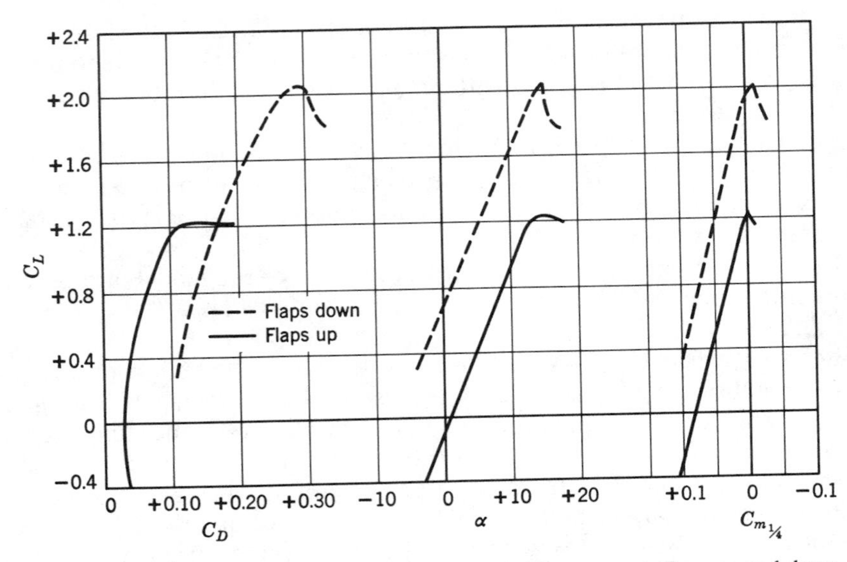

FIGURE 5.24. Typical curves of lift, drag, and pitching moment, flaps up and down.

3. The plots are often used to make comparisons between model configurations, and to check values and slopes.

The plots can be obtained in two ways. First, for small student-type tunnels without computerized data systems, the uncorrected (raw) balance data in engineering units can be plotted by hand. This task argues for an external balance of the pyramid type (Section 4.6) where the forces and moments can be obtained without excessive calculations, rather than platform- or sting-type balances.

For tunnels equipped with computerized data systems where the data are both recorded and reduced by a computer, the type of balance is not critical. As it is desirable to see the data in coefficient form as close to real time as possible, the speed of the data-reduction process in the computer will determine the number of corrections applied to the data. Balance corrections, if small, blockage, and mounting tare corrections may be omitted to save time. Angle calibrations, weight tares, and wall corrections should be applied. In most cases the wall corrections will be approximate wall corrections. The wall corrections to α, C_D, and C_m are all based on wing-alone or tail-off lift coefficients (Section 6.30). The pitching moment correction is also a function of the location of the wing wake system relative to the horizontal tail and is measured during the test. It is not always practicable, owing to the time required, to change the model to make the tail-off and the two runs required for pitching moment correction. In this case the wall corrections can be based on tail-on lift coefficients, but such data should only be used for quick-look data during the test. To avoid using the quick-look plots for careful data analysis, the plots can be marked "Raw data—do not use for final analysis." This minimizes the improper use of data that have been reduced by approximate methods, in this case wall corrections, which can lead in some components to significant errors (Section 6.30). The data can be presented on either *X-Y* plotters or CRTs. If CRTs are used, it is useful to also have hard copies. Straight-line fairing is adequate for these plots and helps to identify bad data points.

The use of a computer to carry out data analysis during a test is quite common. The computer is often used to determine slopes, values, or an increment or Δ in values or slopes. These data analyses must be carried out on fully corrected data. During a test, extreme care must be taken in making such analyses. The data used for such analyses should be plotted to ensure that there are no bad points or other unpleasant surprises in the data used. These plots should use the same fairing routine that will be used for the final plots. This is especially critical when the computer is asked to take increments or Δ's between two runs. When slopes are required, care must be taken in the routine used and the range of values over which the slope is required. As an example, if one is trying to find $C_{D,P,e}$ and Oswald's efficiency factor e (Example 5.1) and one uses a linear regression curve fit over the C_L range from 0° to 3° or 4° past stall, one will not get the correct values for either term. The values of $C_{D,P,e}$ and e are based on the linear portion of the C_L^2-C_D curve, and not the whole curve. Thus the computer must be given the lower and upper limits of C_L to be used.

The other problem with using a computer to analyze data during a test is that it often is not the quickest way to obtain the answers. If plots of the fully corrected data that were made to check for bad data points are available, the values for increments usually can be read from plots faster than by the use of a computer. The same is true for slopes. Assume that the lift curve slope is required and zero lift is not at $\alpha = 0$. The slope can easily be read by the use of two drafting triangles. Place one triangle with an edge on the linear portion of the C_L-α curve. Place the second triangle against the first so that the first triangle can be translated to put the slope of the C_L-α curve through $C_L = \alpha = 0$. Read C_L at $\alpha = 10°$ and the slope is $C_L/10$. The same technique can be used to obtain the linear portion of a C_L-C_m curve at a different center of gravity position than on the plotted data. The slope will change directly with the position of the CG (see Section 5.10). Calculate the slope for the new CG position. Set one triangle edge to pass through $C_m = C_L = 0$ and set the same edge to pass through the value of the slope at $C_L = 1.0$. With the second triangle slide the first to pass through the C_{m0} value (this is not a function of CG position) and draw the new linear curve. The new curve can be computed by using Eq. (5.6) with $C_{m,\text{TR}}$ replaced by $C_{m,\text{CG}}$ and X and Y, the distances from the original CG to the desired CG. Or, Eq. (5.10) can be used with the trunnion replaced by the CG. It should be noted that this is only exactly true when dealing with a wing. When dealing with a complete airplane, the tail contribution is slightly affected by CG location through the tail volume coefficient, which is a function of the distance from the CG to the ac of the tail surface. But this usually has a minor effect on the results and it often is ignored during a quick analysis.

If there is going to be a lot of similar type analysis made on tunnel data during the test, then it would pay to develop a set of canned computer programs that are easy to use for this purpose. But even with these programs available the engineer has to think of what he or she is trying to do and assure himself or herself that the program will yield the correct desired results. Unfortunately, the computer is not concerned with whether or not the results are correct.

A list of the runs usually employed for an unpowered model is provided in Table 5.4. The numerical values in this table are, of course, only approximate, exact values being dictated by the particular design in question. Additional runs would be needed to check fillets, alternative tail surfaces, ground effect, etc.

Attention is drawn to items in Table 5.4 marked "correlation." In many instances correlation runs are added to evaluate separate effects of configurations that would never be flown. For example, a two-engine model usually has a run made without nacelles. The data from this run compared with those from the run with them in place aid in identifying the effect of the nacelles on the airplane's efficiency, drag, and lift. After several models have been tested the usual effects of a "good" nacelle become known, and, when a "poor" one turns up, it is so identified and attention is directed toward improving it. A standard procedure is to list the important performance parameters in tabular form, noting the change in each as each component is added to the wing. Studies made of such tables can be informative indeed.

Comments on the customary curves and information desired follow.

The Lift Curve, Flaps Up

Items of interest on the flap-up lift curve include the value of $C_{L\,\max}$ for determining flap-up stalling speed and minimum radius of turn, the shape of the curve at the stall (it should be moderately smooth, but may not be), the angle of zero lift, the slope of the lift curve $dC_L/d\alpha$, and the value of negative $C_{L\,\max}$. At the Reynolds numbers usually encountered in the wind tunnel, $C_{L\,\max}$ will be from 0.6 to 1.7, $dC_L/d\alpha$ for nonswept wings from Eq. (5.2), swept wings from Eq. (5.3). Construction of the power-off trim lift curve is shown in Fig. 5.25. A similar method can be used for trim drag.

The Lift Curve, Flaps Down

This curve will have very nearly the same slope as the flap-up curve and the same location of the aerodynamic center. The value of flap-down $C_{L\,\max}$ is important for determining the increment due to the flap $\Delta C_{L\,\max}$, for this

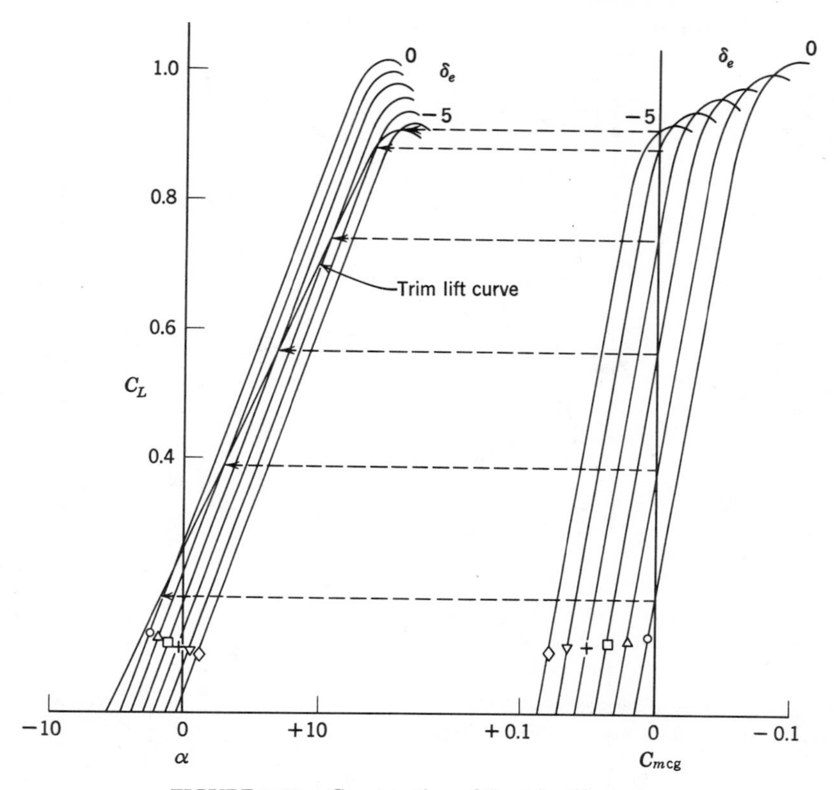

FIGURE 5.25. Construction of the trim lift curve.

apparently does not change with Reynolds number and may be used to determine full-scale $C_{L\,\max}$, flaps down (see Section 7.4), which is needed for landing and take-off runs. The value of $C_{L\,\max}$, flaps down, will vary from 1.2 to 3.5, the higher value being sought after but rarely attained. Trailing edge flaps increase α for zero lift, increase the $C_{L\,\max}$ and reduce the α for stall. They often will change the shape of the C_L-α curve near stall. Leading edge flaps, slats, slots, and Krueger flaps tend to extend the lift curve to higher α stall and greater values of $C_{L\,\max}$. The basic purpose of flaps is to reduce the wing area through increased $C_{L\,\max}$ and thus reduce the parasite drag in cruise. To be usable the $C_{L\,\max}$ must be for trimmed flight. Thus, flap systems that generate large negative pitching moments may require a large tail to develop adequate down loads for trim, which reduces the total $C_{L\,\max}$ (wing and tail).

The angle of 0.9 $C_{L\,\max}$ is of interest for landing-gear-length considerations. It will probably be from $\frac{1}{2}°$ to 3° less for flaps down than for flaps up if the flaps cover the inboard wing area, and 5–8° less if they cover the entire span. Again the sharpness of the stall is of interest, since large lift coeffi-

TABLE 5.4. Test Program of Unpowered Unswept Model[a]

W = wing
B = fuselage
H = horizontal tail
"Tare" = dummies in
"Correlation" refers to data accumulated to assist in laying out new designs.

V = vertical tail
G = gear
F = flaps
Polar plot = C_L vs α, C_D, and C_m from $\alpha_{Z.L.}$ through stall;
Polar run = L, D, M, from $\alpha_{Z.L.}$ through stall

Runs	Model Configuration	Data Sought	Run Consists of
1–4	W	Tare, interference, and alignment; final polar plot	Polars, model normal and inverted, dummies in, dummies out
5–6	W	Wing lateral stability for future correlation	Yaw ±30°, at C_L = 0.3 and 1.0
7–10	WB	Tare, interference, and alignment, final polar plot	Polars, model normal and inverted, dummies in, dummies out
11–12	WB	Wing and body lateral stability for future correlation	Yaw ±30° at C_L = 0.3 and 1.0
13	WBH	Polar plot, effect of horizontal tail	Polar
14–15	WBH	Lateral stability for evaluating vertical tail and correlation	Yaw ±30° at C_L = 0.3 and 1.0
16	$WBHV$	Polar plot; effect of vertical tail	Polar
17–18	$WBHV$	Directional stability	Yaw ±30° at C_L = 0.3 and 1.0

19–23	*WBHV*	Tailsetting and downwash	Polar with tail incidence −4, −3, −2, −1, 1
24–25	*WBHV*	Effect of yaw on static longitudinal stability	Polar with $\psi = 5°, 10°$
26–43	*WBHV*	Rudder equilibrium and power	Yaw ±30° at $C_L = 0.3$ and 1.0; rudder, 2°, 5°, 10°, −2°, −5°, −10°, −15°, −20°, −25°
44–63	*WBHV*	Aileron power	Yaw ±30° at $C_L = 0.3$ and 1.0 with aileron −25°, −20°, −15°, −10°, −5°, 5°, 10°, 15°, 20°, 25°
64–72	*WBHV*	Elevator power	Polars with elevators from −25° to 15°, $\psi = 0$
73–89	*WBHVF*	Effect of flaps on elevator and trim	Polars with elevators from −25° to 15°, $\psi = 0$, for take-off and landing flaps
90–95	*WBHVF*	Effect of flaps on lateral stability	Yaw ±30° at $\alpha = 3°$ and 10°, for take-off and landing flaps
96–104	*WBHVF*	Effect of flaps on lateral control	Polars with flaps 55°, $\psi = 0$, ailerons at −25° to 20°
105	*WBHFG*	Effect of gear down	Polars with landing flaps, $\psi = 0$ model inverted to reduce interference between supports and gear, and vertical tail removed to avoid physical interference with tail strut
106	*WBHFG*	Effect of gear down on lateral stability	Yaw ±30° with $\alpha = 3°$ and 10°, flaps 55°; model inverted

[a] An airplane with sweep requires more stall studies than are shown in the table, particularly for longitudinal, lateral, and directional stability.
Note: Alpha calibration and weight tares are not shown.

cients that are perilously close to a violent stall cannot safely be utilized to their full value. There is usually little need to take the flap-down lift curve as low as the angle of zero lift. The stall should be read in very small steps so that its shape is accurately determined. To increase the span of the wing affected by flaps and hence the maximum lift the ailerons are sometimes drooped (Fig. 5.26). Spoilers on top of the wing can be used for roll control, to decrease lift usually in landing approach, and, when on the ground, to increase drag and kill lift, putting more weight on the wheels to increase the effective use of the brakes. If such devices are used, the number of runs in a wind tunnel increases as their effectiveness is determined.

Here are some hints about testing for $C_{L\,max}$:

1. If possible, plan $C_{L\,max}$ tests at flight Mach number since for many airfoils the local velocity over the leading edge becomes sonic when $M_{freestream} = 0.1$ or 0.2.

2. Low aspect ratio wings with sweep have a leading edge vortex and are not usually sensitive to Reynolds number unless the leading edge radius is large. $RN = 2{,}000{,}000$ is usually adequate.

3. High aspect ratio wings are quite sensitive to Reynolds number, although not above $RN = 5{,}000{,}000$.

4. Leading edge slats are usually insensitive to Reynolds number, but

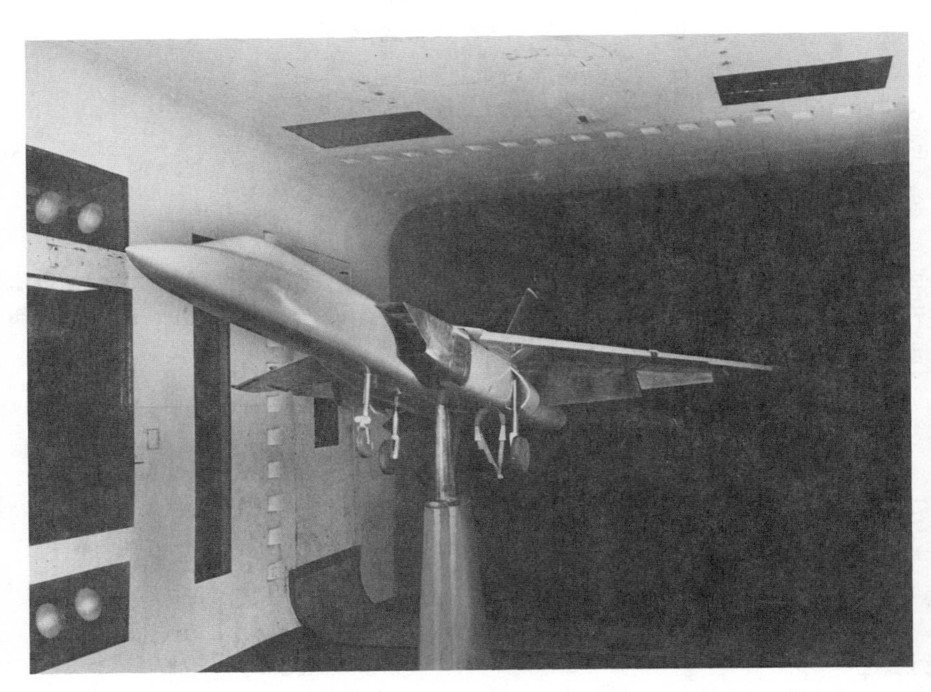

FIGURE 5.26. Testing for $C_{L\,max}$. (Courtesy North American Aviation Corp.)

TABLE 5.5. Lift Curve Data from Tests of a Twin-Jet Fighter

Configuration	$\alpha_{Z.L.}$	$\Delta\alpha_{Z.L.}$	$C_{L\alpha}$	ΔC_L	$C_{L\,max}$	$\Delta C_{L\,max}$
W	−0.3	—	0.0612	—	1.115	—
WB	+0.2	0.5	0.0650	0.038	1.115	0.0
WBC	+1.0	0.8	0.0650	0.0	1.115	0.0
$WBCHV$	+0.9	−0.1	0.0709	0.059	1.197	+0.082
$WBCHVF_{20}$[a]	−1.2	−2.1	0.0700	−0.009	1.34	+0.143
$WBCHVF_{50}$	−4.2	−3.0	0.0700	0.0	1.37	+0.03
$WBCHVG$[b]	+0.9	0.0	0.0700	0.0	1.197	0.0
$WBCHVGF_{20}$	−1.2	−2.1	0.0700	0.0	1.34	+0.143
$WBCHVGF_{50}$	−4.2	−3.0	0.0700	0.0	1.37	+0.03

[a] Nose flap 30° with all T.E. flap deflections.
[b] Compared with *WBCHV*.

although slats reduce the effect of Reynolds number on $C_{L\,max}$, a substantial variation of pitching moment at the stall may be found.

5. In studying $C_{L\,max}$, the model should be as close to trim as possible. On low aspect ratio models with short tails, the tail effectiveness varies with angle of attack as the local dynamic pressure changes.
6. The addition of nacelles usually reduces $C_{L\,max}$.
7. For swept wing models attention must also be paid to pitch-up (reversal of slope of C_m curve) in the C_L-C_m curve as it can limit the usable C_L.

Data from tests of a model of a twin-jet penetration fighter are presented in Table 5.5. As is shown, it is customary to list both the fundamental data and the progressive increments, the increments drawing attention to both good and bad items more directly than the total numbers. The data as shown are for an untrimmed model, and the increase of lift curve and maximum lift with the addition of the horizontal tail will become a decrease when trim is considered. A pair of nacelles usually reduces the lift curve slope about 0.02. The angle for 0.9 $C_{L\,max}$ (which is of interest in designing the landing gear) should not be taken too seriously until ground board tests are completed.

The Drag Curves, Flaps Up and Down

The designer is particularly interested in $C_{D\,min}$ because of its effect on aircraft performance. To ensure accuracy in this range the readings should be made every degree near $C_{D\,min}$. Aircraft $C_{D\,min}$ will vary widely with the type of airplane and wing loading, a value of 0.0120 being not unreasonable for a clean fighter.

The airplane drag coefficient, C_D, at $C_{L\,max}$ is needed for takeoff and landing calculations. Varying widely, depending on type of airplane and amount of flap, this coefficient may range from 0.1000 to 0.5000.

The shape of the drag curve is important for climb and cruising, a minimum change with C_L being desirable.

Drag data for a twin-jet fighter are shown in Table 5.6. Of interest here is the decrease of effective induced drag when flaps are down. The increments of drag due to gear, flaps, tip tanks, etc. are presented on a basis of wing area for consideration on the particular airplane at hand. It is also common to see them quoted on their own frontal area so that their losses may be compared from airplane to airplane.

The Pitching Moment Curve

The slope of the pitching moment curve must be negative for stability, of course, although definite values for the desired slope have not yet been agreed upon. An excellent discussion of wind tunnel procedure for finding the critical stability and control characteristics for single- and twin-engine propeller-driven aircraft can be found in Ref. 5.6.

Quite often in reducing data the moments are reduced about a nominal CG location (usually near 25% MAC) and at desired forward and aft CG limits. The forward and aft CG limits usually allow for the destabilizing effects of power and possibly free controls depending on the aircraft's control system. It is often necessary to test for power effects on propeller aircraft with low power loadings (Section 5.12).

Sometimes the stability is stated in terms of the added rearward travel possible without instability. This might be 0.1 MAC, meaning that the aircraft will still be stable if the center of gravity is moved one-tenth chord aft of the normal rearward location. It should further be stated whether this is for a control-free or control-fixed condition.

TABLE 5.6. Drag Data from Tests of a Twin-Jet Fighter-Type Aircraft

Configuration	$C_{D0} + kC_L^2$	ΔC_{D0}	e
W	$0.0049 + 0.0933C_L^2$	—	0.87
WB[a]	$0.0111 + 0.0982C_L^2$	0.0062	0.85
$WBHV$	$0.0131 + 0.0982C_L^2$	0.0020	0.85
$WBHVT$[b]	$0.0148 + 0.0923C_L^2$	0.0017	0.875
$WBHVG$	$0.0305 + 0.0982C_L^2$	0.0174[c]	0.85
$WBHVGF_{20}$	$0.058 + 0.0577C_L^2$	0.0275	1.105
$WBHVGF_{50}$	$0.1130 + 0.0292C_L^2$	0.055	1.555

[a] Includes canopy.

[b] Tip tanks.

[c] Compared with *WBHV*.

In the case of many modern high-performance aircraft equipped with *unit horizontal tail* or *stabilators* that are power operated, the forward center of gravity location is not as restricted. In this case the stick force is not a function of center of gravity, since it is artificial. Also there is no distinction between stick-fixed and stick-free stability because of the irreversible control system for which no "floating" is possible.

In general, the pitching moment curves are used to determine if the airplane has static stability throughout the desired center of gravity range at all flight conditions (cruise, take off, landing). The airplane's controls also must be able to trim the aircraft under the same conditions. The aft CG limit is a function of stability (slope) and determines the tail area and the forward CG limit is a function of control, $C_{m,\mathrm{CG}} = 0$, usually in landing configuration near the ground, and determines the elevator size and travel. For swept wing aircraft special attention is paid to pitch up as it will limit the usable C_L. Static stability does not guarantee dynamic stability, but usually it is a prerequisite. This constraint may not be as critical for an aircraft using sensors and a computer to operate in a fly-by-wire mode.

The lift, drag, and moment data are usually presented on one sheet (see Fig. 5.24). The reversal of the moment positive and negative values makes the moment curve appear "normal" when viewed with the page on end.

Longitudinal stability data for two airplanes are listed in Table 5.7. Of interest is the customary destabilizing effect of the fuselage and of course the large stabilizing effect of the horizontal tail. For performance calculations trimmed values of C_L and C_D are often desired. These can be constructed as described in Fig. 5.25. Another approach that can be used for a tunnel with a computer data system is as follows: The horizontal tail can be pitched remotely with the tail angle measured. Then, using real-time data reduction, pitch the tail until the C_m about the desired CG location in zero, and then take the data.

Stabilizer effectiveness ($dC_m/d\delta_s$) is obtained by holding α_{wing} constant and varying the tail incidence and computing the resulting pitching moment data. This slope should be evaluated early in the program, since it is needed for the tail-on wall corrections (Sections 6.21 and 6.30).

The Elevator or Stabilizer Power Curve

The plot of ΔC_m against elevator deflection or stabilizer incidence is made at several values of the lift coefficient. It indicates the amount of elevator or stabilizer deflection needed to produce a certain moment coefficient (Fig. 5.27). Usually the plot is nearly a straight line from $+15°$ to $-20°$ deflection with a slope of about -0.02 for elevators and about twice that for the complete tail.

A further study of the elevator may be made from a plot of $C_{m,\mathrm{CG}}$ versus C_L for several elevator angles (Fig. 5.28). The intersections of the curves with the axis indicate trim condition. This plot may also be made against α. The

TABLE 5.7. Longitudinal Stability Data for Two Propeller-Driven Airplanes, Power Off

	Single-Engine Airplane					Twin-Engine Airplane		
Configuration	C_{m0}	ΔC_{m0}	$C_{L\text{ trim}}$	$\frac{dC_m}{dC_L}$	AC	C_{m0}	ΔC_{m0}	$\frac{dC_m}{dC_L}$
W	−0.021	—	0.50	0.042	25.8	−0.007	—	0.017
WB	−0.036	−0.015	0.40	0.091	20.9	−0.015	−0.008	0.025
WBH	0.062	0.098	0.53	−0.120	42.0	−0.020[a]	−0.005[a]	0.085[a]
$WBHV$	0.032	−0.029	0.25	−0.130	43.0	0.024	0.044	−0.102
$WBHVF_{30}$	0.102	0.070	0.86	−0.118	41.8	—	—	—
$WBHVF_{55}$	0.167	0.065	1.22	−0.138	43.8	—	—	—
	Center of gravity at 30.0% MAC, 0.4% above MAC.					Center of gravity at 30% MAC, on MAC.		

[a] WBN, not WBH.

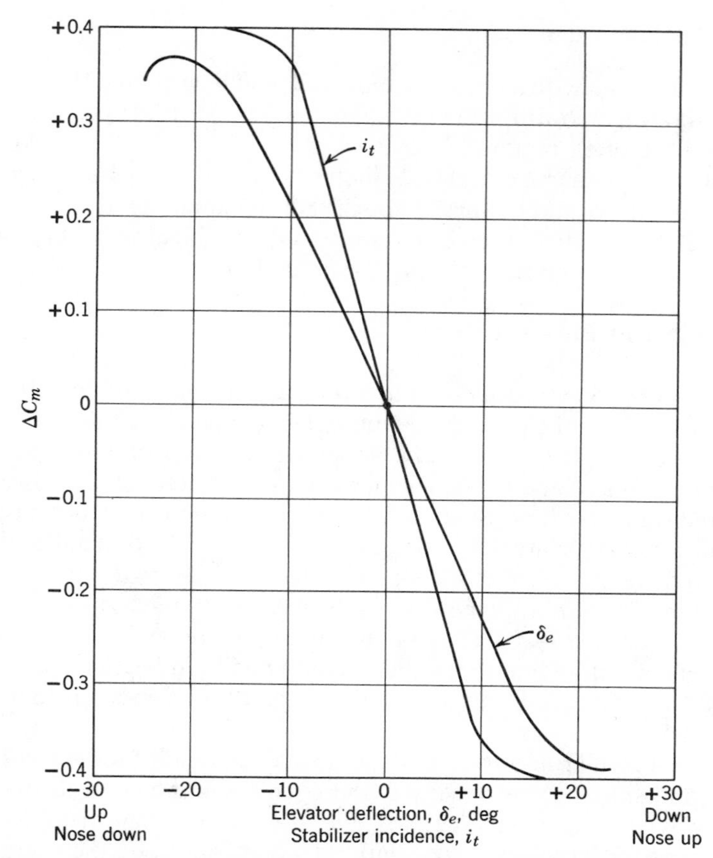

FIGURE 5.27. Typical plot of change in moment coefficient with elevator and stabilizer deflection.

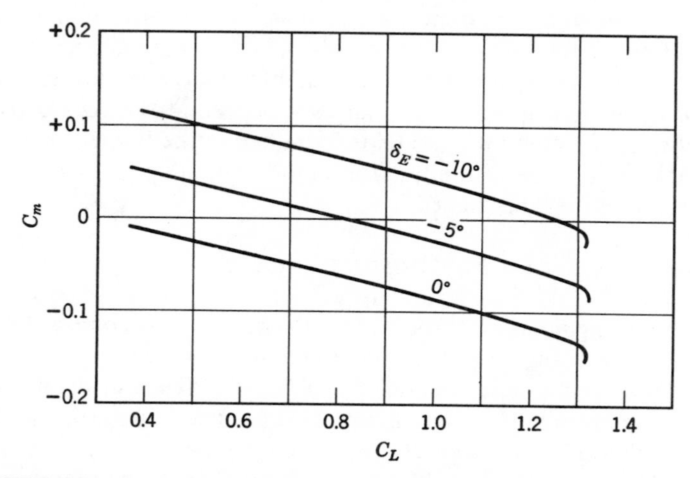

FIGURE 5.28. Typical plot of $C_{m,\text{CG}}$ versus C_L for several elevator deflections.

plot for an airplane with a unit horizontal tail would be similar. For propeller-driven aircraft, runs with propellers, windmilling or at idle power, flaps and gear down are often required.

It is also necessary to determine whether the elevator balance is sufficient to keep the control forces small enough that maximum load factors may be developed. The control force criterion is usually critical in landing or accelerated flight, flaps and gear up, props windmilling.

The Aileron Power Curves

The aileron criteria are usually determined at zero yaw and may be considered from the plots of C_l versus C_n (Fig. 5.17), C_l and C_n versus δ_a (Fig. 5.19), and C_{Ha} versus δ_a (Fig. 5.18). The important qualities of good ailerons are high rolling moment and low hinge moments—the latter can also be obtained from panel models. (See Section 5.8.) The maximum rate of roll and the maximum helix angle are determined from $C_{l\,\max}$ (see Section 5.8; Ref. 5.6), a $C_{l\,\max}$ of 0.03 being satisfactory for one aileron.

Tests required to obtain the above information embrace runs at the proper angles of attack, $\psi = 0$. The rolling moments are measured for various aileron deflections and the helix angles computed as in Section 5.8. (See runs 44–63 of Table 5.4.) A discussion of aileron hinge moments is also presented in Section 5.8.

When taking aileron or rudder data, a small amount of rolling and yawing moment and side force is usually found even when the controls are neutral and there is zero yaw. This delta is due to either asymmetrical flow in the tunnel or model asymmetry, and both the appearance and the usefulness of the data are improved if aileron and rudder moments at zero yaw are subtracted from the data.

It is a nice refinement to run aileron tests with the horizontal tail off, for two reasons. The first (and minor one) is that it saves effort in data reduction since tunnel wall effect on the horizontal tail is then nonexistent. The second is that when the ailerons are deflected in flight the airplane normally rolls and the inboard aileron trailing vortices are swept away from the horizontal tail by the helix angle. When the model is immobile in the tunnel, these vortices stream back quite close to the horizontal tail and induce a loading on it that does not occur in rolling flight.

Rudder Power and Equilibrium Curves

In most single-engine aircraft the rudder is not a critical component. It must furnish adequate control on the ground and in the air, but no criteria similar to "rate of roll" or "pounds of stick force per g" have been established. The problem of the propeller-driven high-powered single-engine aircraft becomes difficult under the high-power low-airspeed (wave-off) condition. Here it is not unusual to require full rudder to overcome torque to maintain straight flight. The criteria become more those of hinge moments (Sections 5.8 and 5.9) than those usually obtainable from the complete model. Particular attention must be paid to avoiding overbalance at high rudder deflections.

The modern high-performance multiengine airplane must possess sufficient directional stability to prevent it from reaching excessive angles of yaw or developing rudder forces that tend to keep the plane yawing. Furthermore, it must also be able to be balanced directionally at the best climb speed with asymmetric power (one engine out, the other at full power for a two-engine airplane) as specified by the FAA and the military.

The most critical condition for the criterion of decreasing rudder pedal force occurs at high thrust coefficient, flaps and gear down, at large angles of sideslip. For propeller aircraft depending on propeller rotations the sideslip angle may be critical to the right or left, and test runs must be made accordingly.

The asymmetric power* condition requires yaw runs at the attitude corresponding to 1.2 V_{stall} gear down, flaps at take-off setting, take-off power on right engine, left engine windmilling.

Usually the rudder information is grouped into two curves. The first, rudder equilibrium, is a plot of rudder deflection against angle of yaw, or in other words δ_r for $C_n = 0$. This need be taken for yaw in only one direction, for it will be similar in the other owing to symmetry. The slope $d\psi/d\delta_r$ can be about -1.2 for maneuverable airplanes on down to -0.5 for the more stable types (Fig. 5.29).

The second curve, rudder power, is a plot of C_n versus δ_r. A slope of $dC_n/d\delta_r = -0.001$ is reasonable, but varies with airplane specifications; one usual criterion is "one degree slip per degree rudder deflecton;" that is,

$$\frac{dC_n}{d\delta_r} \Big/ \frac{dC_n}{d\psi} = \frac{d\psi}{d\delta_r} = 1.00$$

Again the curve need be plotted only for either plus or minus rudder (Fig. 5.30).

The effect of yaw on the characteristics of an airplane is shown in Fig. 5.31.

The Amount of Lateral Stability as Compared with the Amount of Directional Stability

Information about the roll axis is needed to determine whether sufficient dihedral is incorporated in the design to provide lateral stability at the most critical condition. This will be, for most propeller aircraft, the approach with flaps down and power on, where power and flaps combine to reduce the dihedral effect. The ailerons should be free if possible.

* This is *the* critical condition for multiengine types.

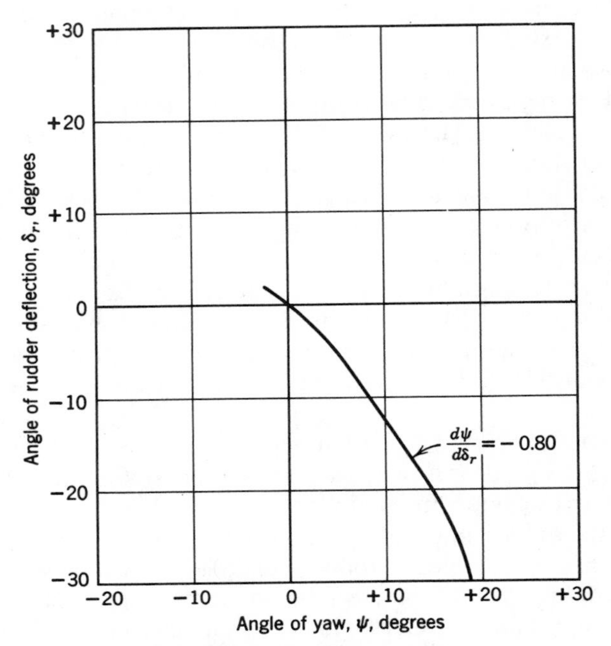

FIGURE 5.29. Rudder equilibrium.

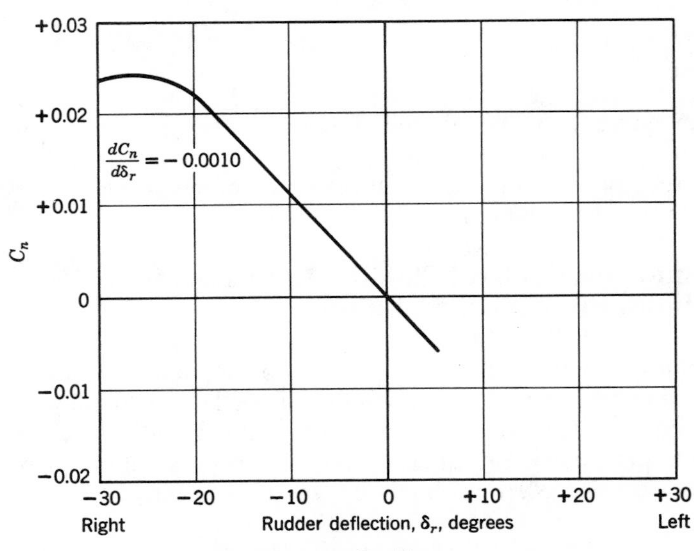

FIGURE 5.30. Rudder power.

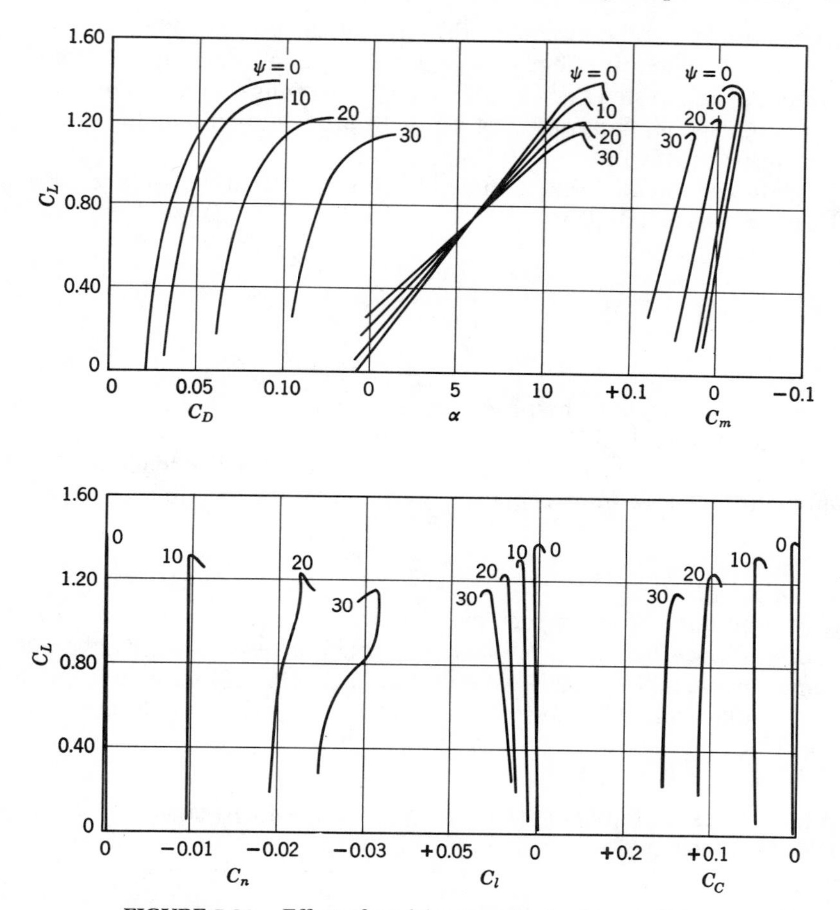

FIGURE 5.31. Effect of model yaw on basic characteristics.

The tests for lateral stability embrace yaw runs at the approach attitude, flaps and gear down, and 50% normal power for propeller aircraft. (See runs 90–95 in Table 5.4 for gear-up data; add runs with gear down.) The angle of attack for the approach should be chosen on the basis of tunnel $C_{L\,\max}$ (used to get 1.2 V_{stall}), but the thrust coefficient should be based on full-scale conditions for propeller aircraft.

Too much lateral stability for a given amount of directional stability results in an objectionable motion called a *Dutch roll*. Too little lateral stability for a given amount of directional stability results in spiral instability. However, the advantages in general control and handling characteristics are so great with a relatively large vertical tail that some spiral instability is acceptable. Hence dihedral investigations are usually more concerned with avoiding Dutch roll than escaping spiral instability.

For most propeller airplanes the critical condition will occur at high speed, where dihedral effect will be a maximum and directional stability a minimum owing to small power effects. The test runs therefore embrace yaw runs at high speed, flaps and gear up, with propeller windmilling or at high-speed thrust coefficient.

Tests have indicated that a value of roll to yaw that will give what pilots call satisfactory stability is

$$\frac{dC_l/d\psi}{dC_n/d\psi} \cong -0.8$$

See also Table 5.8.

A very rough idea of the proper distribution of dihedral and fin area may be obtained from Fig. 5.32, which is an adaptation from Fig. 4 of Ref. 5.2 and from Fig. 5.33. The value of γ in Fig. 5.32 is for a lightly loaded high-wing monoplane; for low-wing airplanes, γ should be replaced by γ_L, where

$$\gamma_L = \gamma + 3°$$

The 3° difference between high and low wing configurations is due to fuselage cross-flow effects. The effective dihedral is always wanted and it may be obtained by recalling that a value of $dC_l/d\psi = 0.0002$ is equivalent to 1° effective dihedral. Wing sweep also can have a pronounced effect on the

TABLE 5.8. Direction Stability for a Single-Engined Airplane, Low and High Angles of Attack

Configuration	$C_{n\psi}$	$\Delta C_{n\psi}$	$C_{l\psi}$	$\Delta C_{l\psi}$	dC_l/dC_n
W	−0.00012[a]	—	0.00056	—	—
	−0.00014	—	0.00037	—	—
WB	0.00118	0.00130	0.00058	0.00002	—
	0.00082	0.00096	0.00037	0	—
WBH	0.00063	−0.00055	0.00070	0.00012	—
	0.00027	−0.00055	0.00087	0.00050	—
WBHV	−0.00165	−0.00228	0.00120	0.00050	−0.727
	−0.00186	−0.00213	0.00087	0	−0.467
$WBHVF_{45}$	−0.00230	—	0.00040	—	−0.174
	−0.00250	—	0.00012	—	−0.480
$WBHVF_{55}$	−0.00230	—	0.00040	—	−0.174
	−0.00250	—	0.00	—	0

[a] The upper value is for $\alpha = 2.5°$; the lower, for $\alpha = 11.1°$.

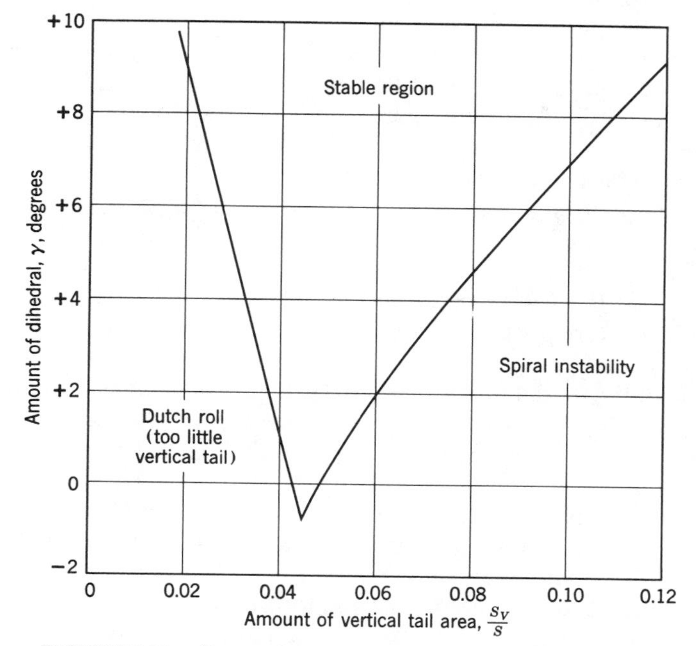

FIGURE 5.32. Proper dihedral for various amounts of fin area.

effective dihedral; highly sweptback wings display too much dihedral at moderate lift coefficients and are normally constructed with no geometric dihedral.

Since geometric wing dihedral alone does not indicate the overall effective dihedral which includes vertical tail, wingtip shape, sidewash, and power effects for propeller aircraft, a more quantitative indication of the

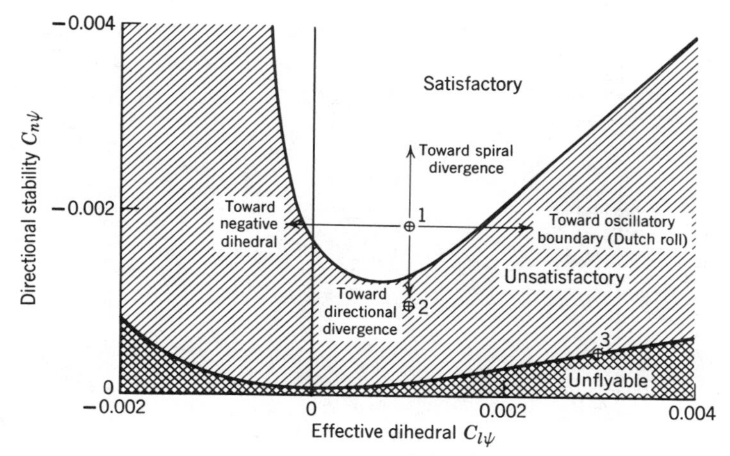

FIGURE 5.33. Free flight tunnel results showing good and bad combinations of $C_{n\psi}$ and $C_{l\psi}$.

effect of combinations of directional stability and effective dihedral is shown in Fig. 5.33. As shown excessive dihedral coupled with weak weathercocking leads to the oscillatory or Dutch roll boundary. Excessive directional stability leads to the spiral divergence boundary. The complete wind tunnel model yields the total stability and dihedral including all interference effects.

Usually lateral wind tunnel data is plotted as C_L, C_D, C_m, C_n, C_l, C_C, or C_Y versus ψ (yaw angle) or β (side slip angle) (Fig. 5.34). The generally preferred axis system is the stability axis system.

As discussed in Chapter 4, most external balances measure forces and moments about the wind axes, an axis system aligned with the tunnel centerline, while internal balances use body axes, which pitch, roll, and yaw with the airplane model. In lateral stability it is desirable to use stability axes that yaw with the model but do not pitch. Section 6.30 gives the equations to transfer from wind axes to body or stability axes, and for body to wind axes.

When yaw runs are presented on the stability axis the cross wind force is replaced with the side force C_Y and the drag C_D is along the airplane centerline. No particular slope or values to C_Y are required. The only use of C_Y is to calculate the side force for asymmetrical flight and hence the necessary angle of bank to counteract said side force with a tangent component of lift. The side force needed to overcome the torque reaction at low speed while maintaining straight flight may also be evaluated for propeller-powered aircraft.

Tailsetting and Average Downwash Angle

To avoid the drag of cruising with elevators deflected, and the loss of maximum ΔC_m due to elevator if partial elevator is needed for trim, it is usually desirable to set the stabilizer incidence so that the aircraft is trimmed at

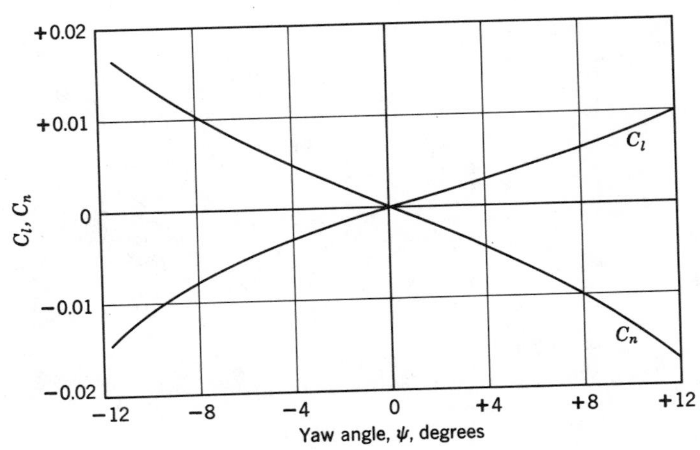

FIGURE 5.34. Typical yaw characteristics.

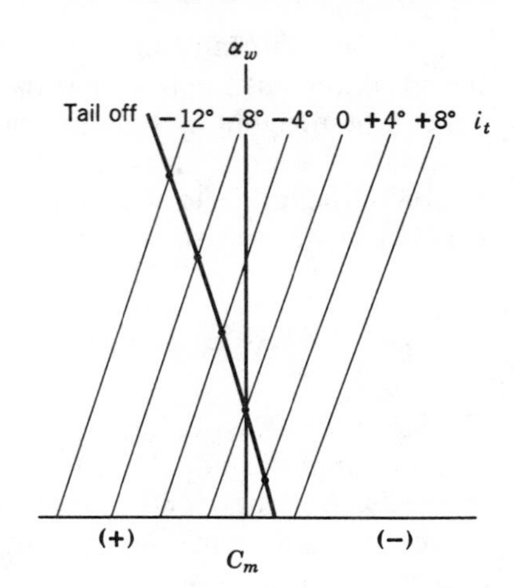

FIGURE 5.35

cruising with $\delta_e = 0$. For stability considerations as well as correlation for future designs, it is necessary to know the angle of downwash at the tail for each wing angle of attack. The procedure is as follows:

1. Run the model with the horizontal tail removed, obtaining a tail-off stability curve similar to that shown in Fig. 5.35.
2. Next run the model with the horizontal tail on, using tail incidence, i_t, angles of, say, −8°, −4°, 0°, 4°, 8°. Curves as indicated in Fig. 5.35 will be obtained.

Now the intersections of the horizontal tail-on curve with the tail-off curve are points where, for a given wing angle of attack α_w, the tail-on pitching moment equals the tail-off pitching moment; that is, the tail is at zero lift, and hence

$$\alpha_t = \alpha_w + i_t - \epsilon_w = 0 \tag{5.31}$$

where ϵ_w = downwash angle at the tail; α_t = tail angle of attack.

Since α_w and i_t are known for the points of intersection, ϵ_w may be determined from Eq. (5.31), and a plot of ϵ_w against α_w or C_L may be made. This plot and the usual effect of flaps on downwash are shown in Fig. 5.36. Not infrequently the curve of ϵ_w against α_w is a straight line.

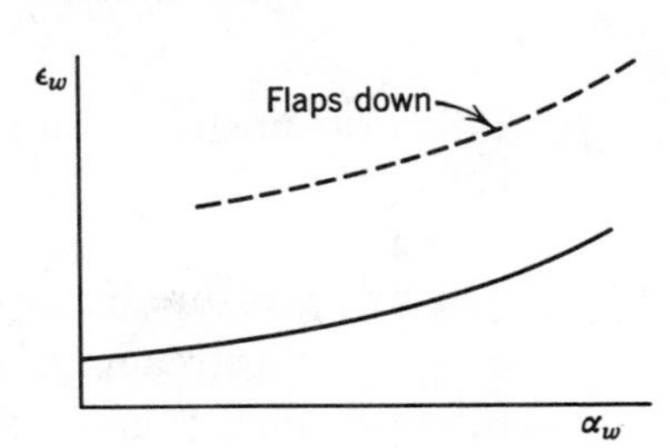

FIGURE 5.36

Methods shortcutting the above lengthy procedure have been devised based on the assumption that the wing downwash is zero at zero lift. However, when a wing is twisted, the total lift is zero, but it is not zero at all spanwise stations. Thus, the wing is usually producing a downwash in the region of the tail. The short cut is only true for a nontwisted wing. In general, it is better not to use such methods as they eventually lead to errors.

5.12. POWER EFFECTS, PROPELLER AIRCRAFT

For an aircraft with high power loadings (gross weight/horsepower) and conventional design, the effect of power can be estimated by well-known methods. But the trend in executive-type aircraft to turbo props and low-power loadings and higher wing loading leads to power effects that are difficult to predict—but they must be accounted for.

The effect of the propeller on stability and control can be broken down into direct and indirect effects. The direct effects are:

1. Pitching and yawing moments arising from the thrust line not passing through the center of gravity.
2. The propeller normal force in the plane of rotation will produce pitching or yawing moments. This contribution can be large even at zero thrust coefficients and adversely effects longitudinal and lateral stability.
3. The torque reaction to the propeller.
4. For multiengined aircraft there is a yawing and rolling moment for the engine-out condition. This usually is the critical design condition for the rudder.

In general the direct effects are amenable to analysis, as they involve a force (thrust) and moment arm. However, it can be difficult to obtain data for the normal force variation with flow angle in some cases.

The indirect effects are caused by the interaction of the slipstream with other parts of the aircraft. Determining the indirect effects accurately is very difficult by analytical methods. The indirect effects are sensitive to the airplane configuration and can be broken down into the following broad categories:

1. Effect of slipstream on the moments of the wing, nacelles, and fuselage.
2. Effect of slipstream on the wing's lift coefficient due to higher local dynamic pressure over portions of the wing.
3. Effect of slipstream on downwash, and cross flow at the tail.
4. Effect of slipstream on the dynamic pressure at the tail.

In general, the effect of the slipstream on the fuselage and nacelle moments is small compared to other power effects and analytically difficult to analyze. The slipstream effect on the wing's pitching moment with the flaps down can, however, be large. The same is true for the wing's lift, both with the flaps up and down, which affects stalling speeds power on and off. The partial immersion of the wing in the slipstream will alter the downwash and thus change the horizontal tail's angle of attack. The normal force at the propeller will also change the downwash at the tail. The rotational component of the slipstream will change the angle of attack across the horizontal tail, as well as the vertical tail. The critical condition is at high power and low speed. And, finally, the increased velocity of the slipstream will change the tail's contribution to stability. Just how much the downwash due to the propeller, rotation effect and velocity increase affect the tail's contribution is a function of how much or how little of the tail is immersed in the slipstream. In general, the propeller downwash is destabilizing even at zero thrust. It also should be noted that the displacement of the slipstream in side slip results in one flap being immersed in the slipstream to a greater extent than the other, which reduces the dihedral effect, and this effect is maximized at low speeds and with high thrust. From the foregoing discussion, it can be seen that the effect of power tends to be destabilizing both longitudinally and laterally, with the critical conditions generally occuring at high thrusts and low speeds with the flaps down. For multiengined aircraft, the magnitude of the power effects are also a function of the propeller's rotation, that is, both right handed or one right and one left, etc.

The Effect of Power

To illustrate the effect of power, let us look at a single engine tractor with a wing loading of 39 psf and a power loading of 7 lb per horsepower. The thrust line of this airplane was very close to the center of gravity. As can be seen from Fig. 5.37, there is a change in the longitudinal stability between power off and 100% power, both with flaps up and down. It is interesting to note that with the flaps up there is about an 80% reduction in the moment curve slope from power off to 100% power and a 50% reduction with the flaps down. Figure 5.38 shows the effect of power on the lateral characteristics with the flaps down. The effect of the partial immersion of the flaps in the slipstream can be seen in the right half of the figure. These figures shown that the effect of power can be fairly dramatic in a high-performance airplane.

Fortunately, the above effects may be evaluated in the wind tunnel, and the aircraft may be revised or the pilot forewarned.

The simulation of the propeller slipstream for a constant-speed propeller requires matching both the axial and rotational velocity ratios. To match these ratios over the entire range of lift coefficients would require an adjustable pitch propeller. However, a satisfactory approximation of the slipstream may be accomplished with a single setting of a fixed pitch model

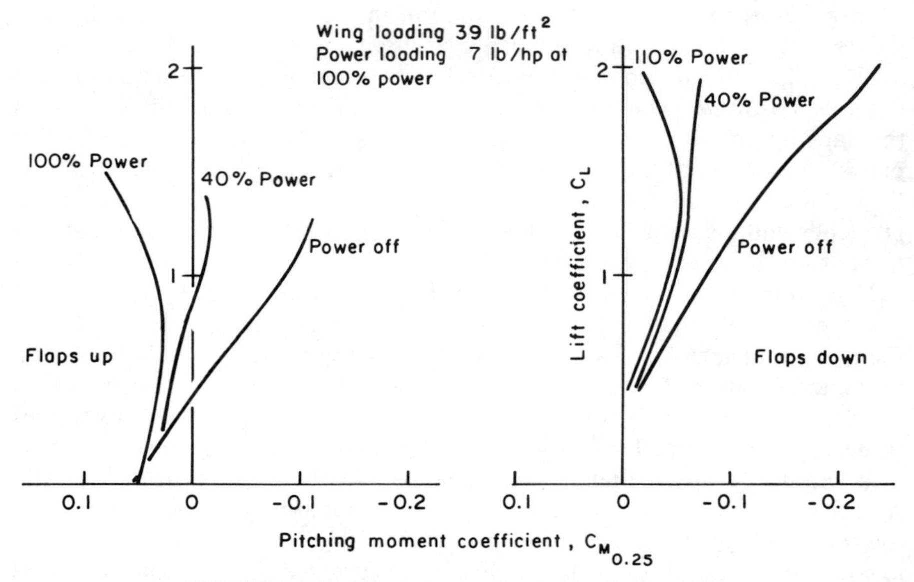

FIGURE 5.37. Effect of power on longitudinal stability.

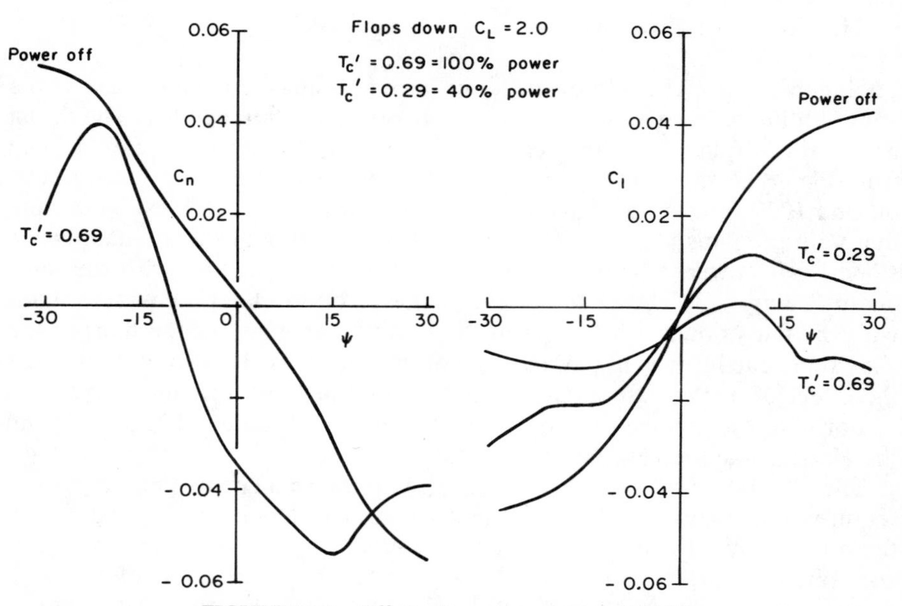

FIGURE 5.38. Effect of power on lateral stability.

propeller over a large part of the lift coefficient range where power effects are of interest. From momentum theory, it can be found that the axial velocity ratio can be matched with a propeller of scale diameter. The rotational velocity can be matched by using a geometrically similar propeller operating at the proper advance ratio. Typical fixed pitch model propellers have aluminum or steel blades with steel hubs, and allow the blades to be set over a wide range of pitch angles.

The relationship between the model and full scale propeller are as follows:

$$T_c = \frac{T}{\rho V^2 d^2} \tag{5.32}$$

$$Q_c = \frac{Q}{\rho V^2 d^3} \tag{5.33}$$

where T and Q are thrust and torque, respectively, and d is the propeller diameter. Then using subscript S for full-scale airplane and subscript M for model, we have for similarity

$$\frac{V_S}{n_S d_S} = \frac{V_M}{n_M d_M} \tag{5.34}$$

Also

$$T_{cM} = \frac{T_M}{\rho V_M^2 d_M^2} \qquad \text{and} \qquad T_{cS} = \frac{T_S}{\rho V_S^2 d_S^2}$$

Dividing, we have

$$\frac{T_{cM}}{T_{cS}} = \frac{T_M}{\rho V_M^2 d_M^2} \frac{\rho V_S^2 d_S^2}{T_S}$$

Substituting from Eq. (5.34) and clearing, we obtain

$$\frac{T_{cM}}{T_{cS}} = \frac{T_M}{T_S} \frac{n_S^2 d_S^4}{n_M^2 d_M^4}$$

Now it can also be shown that the thrust

$$T_S = \rho n_S^2 d_S^4 C_{TS} \qquad \text{and} \qquad T_M = \rho n_M^2 d_M^4 C_{TM}$$

and that, for a given V/nd, if the two propellers are geometrically similar, $C_{TS} = C_{TM}$ (this omits scale effect). Hence

$$\frac{T_M}{T_S} = \frac{n_M^2 d_M^4}{n_S^2 d_S^4} \qquad \text{and} \qquad \frac{T_{cM}}{T_{cS}} = 1$$

or, if the model is tested at $T_{cM} = T_{cS}$, similarity of thrust will be preserved. In a similar manner, Q_{cM} should equal Q_{cS}.

The motors used to drive the propeller are usually of small diameter and high-power output. They are generally water-cooled, variable-frequency, alternating-current motors. These motors operate at frequencies up to 400 Hz with maximum voltages of 240 or 480 V. Maximum rpm varies between 12,000 and 24,000 rpm. Motors have either a two- or four-pole tachometer. The rpm is best measured by counting the tach frequency and dividing by 60 or 30 to get the rpm. The motors also have one or two built-in thermocouples to monitor their temperature. One limitation on the choice of a motor is the physical size of the model into which it must fit. Another consideration in selecting a motor is that usually maximum power can only be held for 1 or 2 min before overheating, thus the motor should be selected to avoid continuous operation near maximum power. This problem can be reduced by operating at a lower wind tunnel velocity.

A variable-speed alternator is used to supply power to the motors. It must have a variable-frequency range commensurate with the motor rpm requirements. Since the motors require a specified volts/cycle input to minimize current and thus heating, the power source must match the motor's specified volts/cycle range.

The wire bundle providing power and water to the model motors should be brought into the model on a lower surface. The bundle should have some sort of fairing around it to make the model support tares more repeatable.

Selection of the motor and test velocity are dependent on each other. Knowing full-scale flight conditions and power setting to be simulated, model power may be calculated with the following equation:

$$P_m = P\left(\frac{\sigma_M}{\sigma}\right)\lambda^2\gamma^3 \tag{5.35}$$

where λ = model scale, γ = model test velocity/full scale velocity, and σ = density ratio. Knowing the model scale, a plot can be constructed of required power versus lift coefficient (Fig. 5.39) for a range of dynamic pressures. Lines of constant rpm may also be plotted. Limits on rpm are defined by the motor characteristics and considerations of tip losses if the model propeller tip speed is higher than full-scale tip speed.

Model Propeller Calibration

With the tunnel test velocity determined, the selection of an appropriate blade angle can be accomplished. This required the following full-scale information (Fig. 5.40):

1. A plot of T_c' versus C_L for the power setting to be simulated. $T_c' = C_{D\ \text{power-off}} - C_{D\ \text{power-on}}$.

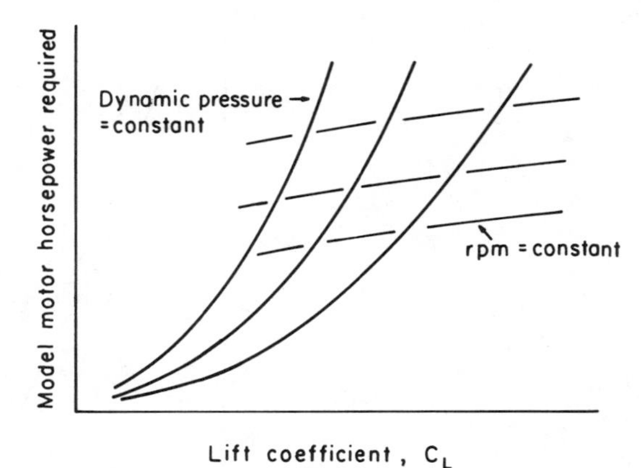

FIGURE 5.39. Model motor power requirements.

2. A plot of J versus C_L; J = advance ratio = V/nd, n = rpm.
3. A plot of β versus C_L.
4. The range of lift coefficients that are to be simulated.

Then, with the model in the tunnel at the minimum drag point, runs are made at several values of β, which cover the range of full-scale β's over the desired C_L range. A run is also made with the propeller off and the difference in drag between power off and on determines T_c'. Thus, for each β selected run through the rpm range of the motor and record drag, test-section static temperature, and pressure for each motor rpm. From these data (Fig. 5.41):

1. Calculate T_c' versus rpm for each β and plot.
2. Calculate T_c' versus J for each β and plot.

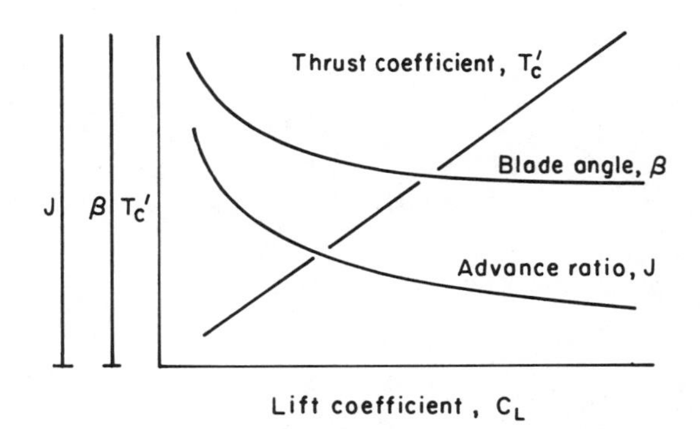

FIGURE 5.40. Full-scale airplane propeller characteristics.

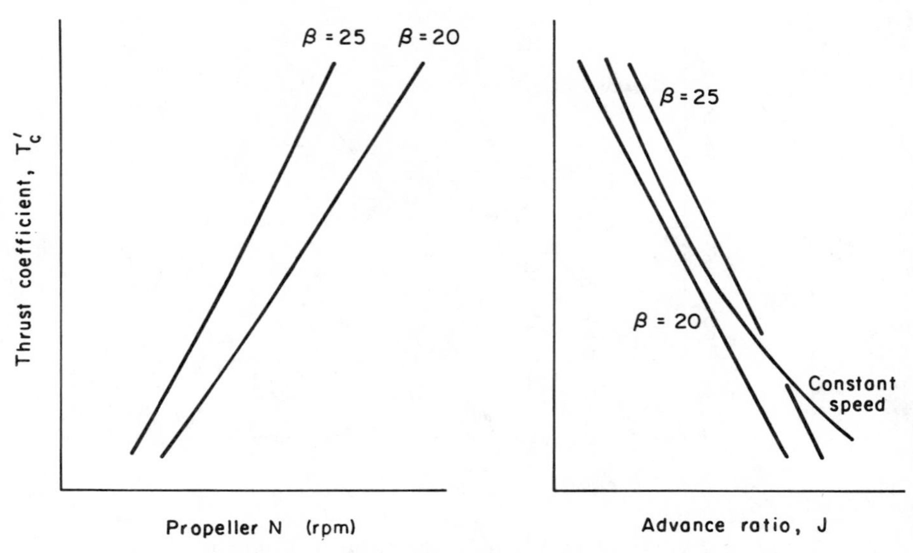

FIGURE 5.41. Model propeller blade angle selection.

To match the full scale T_c' versus C_L relationship plot J versus C_L for each β on the plot that has the full-scale J versus C_L curve (Fig. 5.42). As plotted, all blade angles match the full-scale axial velocity ratio. To match rotational velocity ratios select the blade angle that best matches the full-scale curve. If the rotational effect is to be matched at both extremes of the C_L range, it will be necessary to use two blade angles during the tests. Now the required β or β's for the model propeller as well as the relationship between T_c' and rpm for those β's are known. If the static temperature and pressure vary from values used in calibration, the test rpms are adjusted to hold the advance ratio at the same value used in calibration.

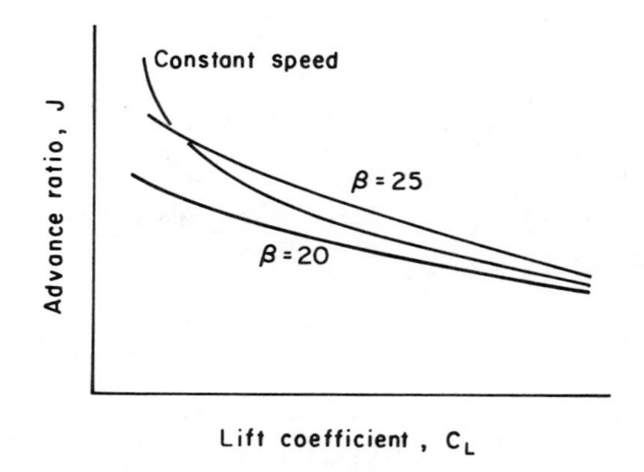

FIGURE 5.42. Model propeller final selected blade angle.

Actual Methods of Power Testing

Two different methods are commonly used for power testing. The first is to test at a constant thrust coefficient, that is, constant rpm. The T_c' versus rpm plot (Fig. 5.40) is used to select the proper rpm. A series of runs are made at different T_c' values. The data may be used directly for yaw runs and calculations involving no airspeed charges, such as short period motions. The data may be cross-plotted to match the full-scale airplane T_c' versus C_L relationship for information when the airplane velocity does change. The second method involves testing at constant power. The rpm is varied to match the T_c' versus C_L relationship of the full-scale airplane. A plot of model lift versus rpm (Fig. 5.43) is required and is obtained from the T_c' versus C_L and T_c' versus rpm plots (Fig. 5.40).

For each configuration which changes the zero lift angle, an operator's plot of rpm versus α indicated is required (Fig. 5.43). The model is pitched through the α range and at each angle of attack the rpm is varied until the trimmed lift and rpm match a point on the lift versus rpm plot. The trimmed lift can be calculated from the balance pitching moment and lift readings. Additional runs with the same configuration but with different tail angles, etc., can use the same operator's plot. The data generated can be used directly in calculations where the airplane velocity is variable. For calculations involving other T_c' versus C_L relationships, additional constant power runs at different powers must be made to allow cross-plotting the data. As can be seen, the second method is more time consuming than the first. In practice, a combination of the two methods is often used. Various T_c' = constant runs are made which allow the construction of different T_c' versus

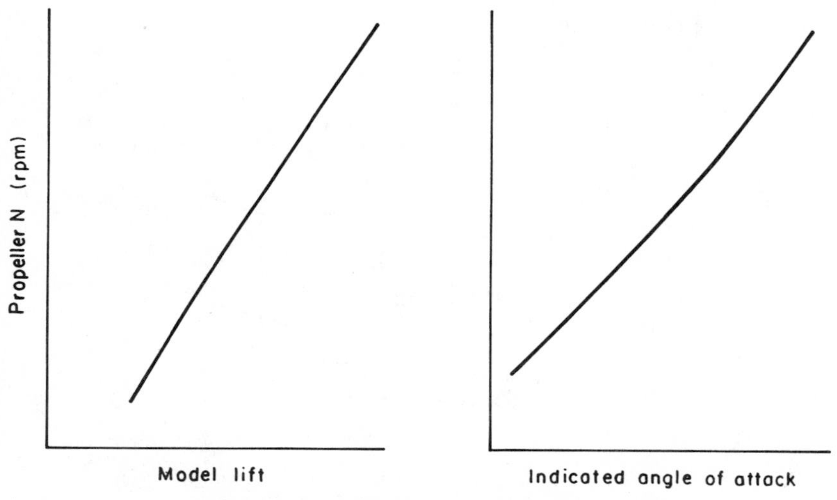

FIGURE 5.43. Model power operating charts.

C_L relationships. Constant power runs at the maximum engine power may be made to give a closer definition of the curves than would be obtained by cross-plotting the T'_c = constant data. It should be noted that if the airplane power is increased, the constant power runs yield no data that can be extrapolated for the new engine.

If it is desired to match both the axial velocities or T'_c and rotational velocities Q_c, the model propeller must be exactly geometrically similar to the full scale and the test run at the same J's. If there is room in the model, the motor and propeller can be mounted on a strain gage balance that will measure the torque and yield Q_c. If a torque balance cannot be installed in the model, the following procedure can be used.

Set the motor in a dynamometer and obtain curves for bhp for various values of rpm and input kW. The results will yield a plot similar to Fig. 5.44. When making the calibration, monitor the motor temperature to not exceed limits, nor stall the motor as there is great risk of burning up the windings. It also should be noted that most ac voltmeters and ammeters are intended for 60 Hz. They may not be accurate at other frequencies.

As outlined before, when the model is in the tunnel to calibrate T'_c versus rpm also measure Q_c from the torque balance or the $kW_{in.}$. Then for several values of β plot T'_c versus Q_c (Fig. 5.45). The curve for the actual propeller is put on the same plot. From this plot select a model β that most closely matches the full-scale propeller. If the full-scale propeller is a constant-speed propeller, it may be necessary to pick two model β's to match high and low speeds or J's as before.

The test is then run as described before in the constant-power method. Either Q_c from the torque balance or $kW_{in.}$ should be recorded with the data. If desired, constant T'_c runs can be made for cross-plotting.

Table 5.9 lists some of the power and sizes of the electric motors used for this purpose (Figures 5.46 and 5.47).

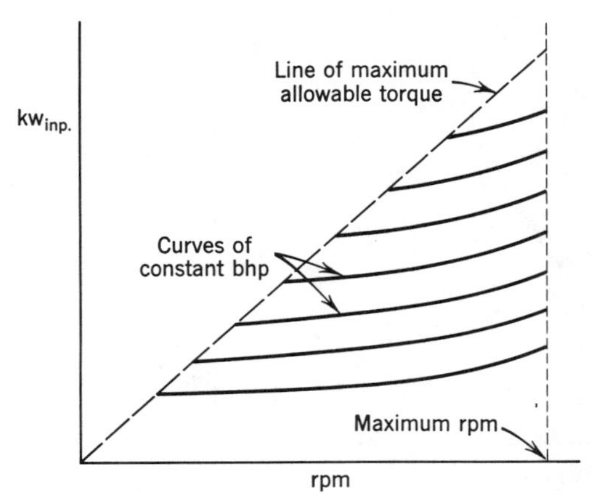

FIGURE 5.44

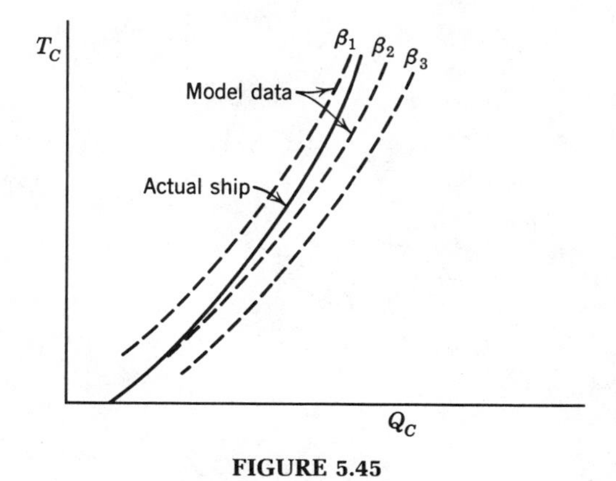

FIGURE 5.45

Because of the small size of hydraulic motors and the high rpm they develop, they are now being used in many instances for V/STOL propeller drive systems. Typically, pressures of from 600 to 5000 psi are needed to meet the power requirements. Some performance curves for a typical hydraulic motor are shown in Fig. 5.48. Also available are some air-driven turbines that can be used to drive propellers. When considering such a device, it should be borne in mind that the required rpm for propeller windmilling cases is usually much less than the free wheeling rpm of the propeller. Thus, the turbine must be capable of holding such rpms.

The problem of "jumping the balance" with water lines, power leads, etc., is quite simple as once the desired water flow rate is established (and not changed) the load put on the balance is constant and not a function of the model power. The problem of jumping the balance with hydraulic or com-

TABLE 5.9. Dimensions of Some Wind-Tunnel Model Motors

hp	Diameter, in.	Length, in.	rpm
6.4	2.16	12.00	12,000
9	2.2	7.5	27,000
20	3.2	7.0	18,000
35	4	10.0	18,000
52	4	17	11,500
75	4.5	12	18,000
130	8	16	5,400
150	7.5	14	8,000
200	10	33	5,000
1000	28	38	2,100

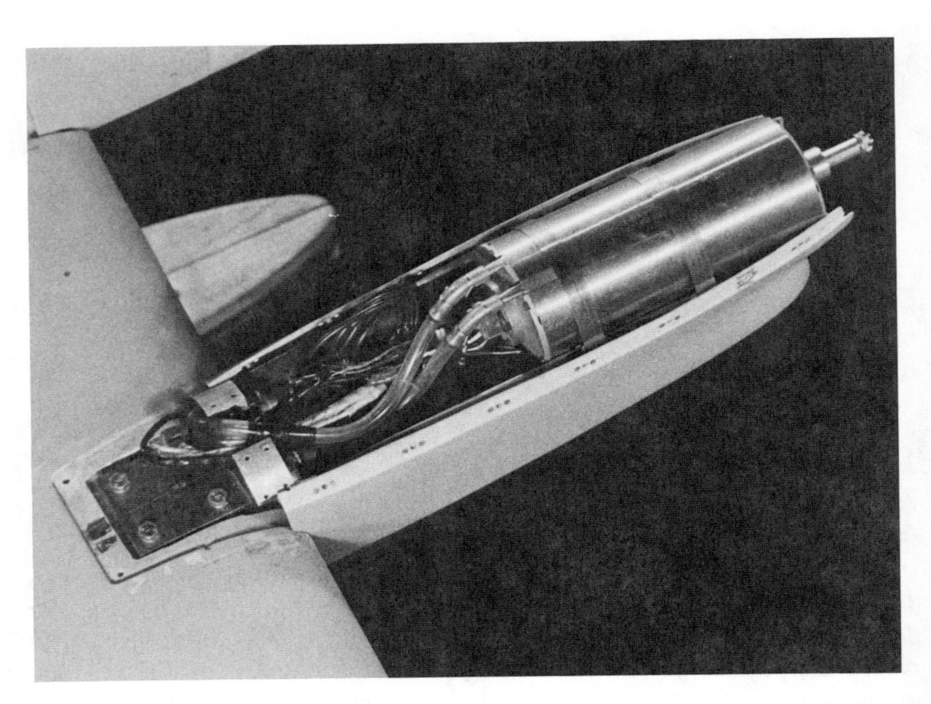

FIGURE 5.46. Water-cooled ac motor in nacelle. (Photograph courtesy of Beech Aircraft Corporation.)

FIGURE 5.47. A powered model. The power and water leads leave the model between the mounting forks and enter the balance fairing through streamline tubing. (Photograph courtesy of Beech Aircraft Corporation.)

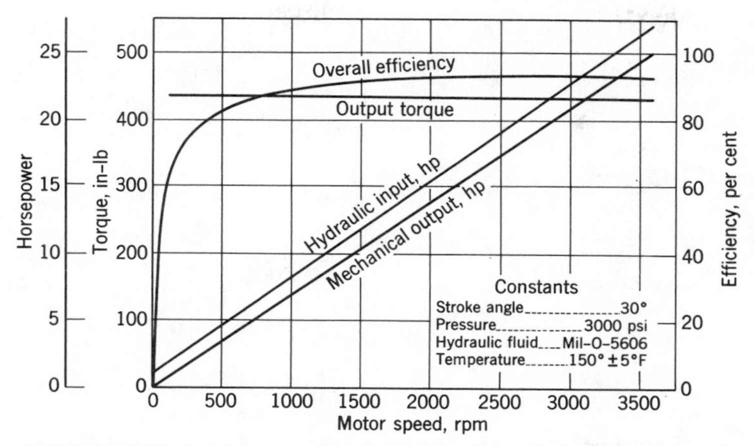

FIGURE 5.48. Hydraulic motor characteristic performance curves (intermittent hp = 34.9 at 4850 rpm). (Courtesy Vickers Inc.)

pressed air is not that simple since both the pressure and mass flow will vary with the required power, thus balance loads also vary in both forces and moments. These balance loads as functions of pressure and mass flow must be determined and subtracted from the data. See Section 5.13 for a possible method of jumping the balance with hydraulic pressure lines.

5.13. POWER EFFECTS OF JET AIRCRAFT

The need for power-on tests is far less acute for a jet-engine airplane than for a propeller-driven one. The effect of the thrust moment is easily calculable, and there is no large slipstream of high rotation which strikes the fuselage and tail with a wide variety of effects. Indeed, the sting mounting usually employed helps simulate the jet stream for the single-engine airplane. However, as the engine bypass ratios increase, the nacelle inlet lip normal force may become more important for engines mounted close to the wing. This is destabilizing and may require power testing in take off and landing configurations.

There are two methods of simulating a jet engine in power-off testing. The first is to fair in the inlet and exhaust with smooth fairings. With engines mounted close, but external to the body, or close to the wing, this approach may significantly distort the flow over the nacelle and adjacent areas. The same may be true on aircraft with engines buried in the fuselage with inlets near the leading edge of the wing root. The method may be acceptable for single engines on a sting mount with inlet in the fuselage nose.

The second approach is to use flow-through nacelles, which can be of two types. The first is a simple flow-through nacelle. The internal drag is either ignored or it is estimated or measured and subtracted from the data. The

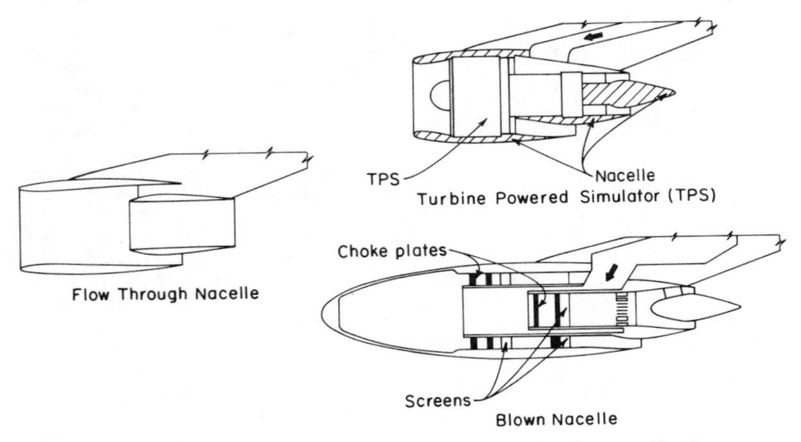

FIGURE 5.49. Three ways to simulate jet-engine nacelles.

second type has an internal cowl or plug that is used to provide a correct inlet condition corresponding to a specific flight condition. The internal drag of these can also be estimated or measured and subtracted from the data (Fig. 5.49).

A blowing nacelle is used when a thrust-producing jet is required. The inlet is covered by a faired dome that is used for a high-pressure plenum. By the use of choke plates and screens the flow is more uniformly distributed as it is exhausted from the nacelle. Often a blowing nacelle has two independently measured and regulated flows to represent both the fan air and primary air (Fig. 5.49).

The use of a flow-through nacelle for inlet flow simulation and the fan cowl geometry effects, and blown nacelles for the jet effect will simulate the engine airframe interactions successfully if the following three assumptions can be satisfied:

1. There must be no coupling in the near-field flow between the inlet and exhaust flow fields.
2. The ram condition plume of the blown nacelle must accurately represent the plume exiting from the flow-through nacelle.
3. The potential flow field that is displaced by the domed inlet must not significantly alter the adjacent aerodynamic flow field.

The difficulty of always meeting these three assumptions can be avoided by using a turbine-powered simulator (TPS), Figs. 5.49 and 5.50. These provide a practical model equivalent of a real engine in a real airplane. The TPS uses high-pressure air to drive a turbine that drives a fan stage that compresses the inlet air. The inlet air is exhausted through a fan nozzle and the turbine air through a primary nozzle. The TPS will simulate 80–90% of

FIGURE 5.50. A sting-mounted model near the tunnel floor with two turbine-powered simulators (TPS). The sting support is capable of vertical translation. (Photograph courtesy of DNW.)

the inlet flow, the pressure ratios of the fan and the core jets, the fan temperature, but the core temperature is very low. The error in core temperature does not significantly affect an accurate representation of the thrust and exhaust flow. This is because in high bypass engines the core flow is a small part of the total flow and it is surrounded by the large fan flow. To accurately duplicate the full-scale engine–airframe the TPS must be calibrated.

The simulation of jet engines puts two requirements on the wind tunnel facility: a large air supply and some method of jumping either an external or internal balance with this air. Since the blown nacelles or similar ejectors to simulate an engine exhaust usually require large masses of air when compared to a TPS, they will define the air-supply system. Furthermore, to keep the size of the air lines into the model reasonable, as required by support tares, the air must be delivered to the model at high pressure. The mass flow is, of course, also a function of the tunnel and hence the model size. For tunnels of the size used for developmental testing, this leads to an air supply that can deliver on the order of 20 lb/s at 1000 psig. or more.

There are many schemes in existence for getting air of this mass flow and pressure across balances. These same methods could be adapted for hydraulic fluid if a hydraulic motor is used to drive a propeller. When faced with this problem for an external balance, the easiest solution appears to be the use of a loop of high-pressure hose.

This is not a correct approach, as can be seen when one realizes that most dial-type pressure gages use a Bourdon tube. This is a curved tube closed at one end, used to measure pressure by gearing the pointer to the tube and using the tendency of the tube to straighten out under pressure to drive the pointer. A curved piece of hydraulic tubing or hose will do the same thing and load the tunnel balance when the curve is between the balance and a connection to the tunnel structure.

Ideally the air should pass along the vertical centerline of an external balance so that it can enter a large high-pressure plenum in the fuselage. From the model plenum the air flow is controlled by small electrically driven ball valves or their equivalent as it is routed to its desired use (Fig. 5.3). For an internal balance the air will usually pass through the sting and then through the internal balance to a plenum. One note of caution: high-pressure piping falls under piping or boiler codes. This means that all welds usually must be made by a certified welder; welds may or may not require x-ray tests and pipes and welds must pass hydrostatic tests before they can be used.

One highly successful method jumping an external balance is to use an L-shaped air line. Starting from the compressor side, at some point near the balance the pipe is firmly attached to the structure of the building. From this point the bottom of the L runs toward the balance. Very close to the ground connection there is a gimbal made out of X flexures. Within the gimbal there is a bellows with a liner (to prevent the air from vibrating the bellows). The pipe then makes a 90° bend up to the balance. In this leg there are two gimbals with bellows. The lengths of the pipe are critical as the gimbals should have very little deflection when the system is at rest. With adequate pipe length any small motions due to the heating–cooling of pipe, etc., will be taken up in the three gimbals (Fig. 5.51). The effect of pressure in the pipe can be calibrated by caping the test section end and pressurizing the pipe. Note: this will also require a valve to release the pressure. The effect of mass flow can be achieved by the use of a zero thrust nozzle. This is T-shaped pipe with two calibrated sharp edge orifice plates at the end or the top of the T. This is attached to the air line at the model trunnion. The mass flow can be changed by changing orifice plates.

To jump an internal balance with compressed air requires the balance to be designed for this purpose. The air is delivered through a sting system that is nonmetric. The six-component balance has a central air duct that matches the sting. The air duct has opposing bellow seals and is sealed at the model end. Holes in the circumference pass the air to a chamber that encloses the bellows and from this chamber to the model plenum. The balance is calibrated with the balance pressurized at the expected running pressures to account for small interactions and sensitivity changes due to the air pressure. Momentum tares are evaluated by calibration with a zero thrust nozzle (Fig. 5.52).

What is required to get high-pressure air to the model has been outlined. But before we turn to equipment to calibrate the jet engine simulators a word of caution about high-pressure air systems. Air or any other gas in large quantities at high pressure represents large amounts of stored energy, and it must be treated with respect. The air controls in the model and most of the other controls in the system are remotely operated by either electrical or pneumatic methods.

To protect people working on the model there must be some interlock system to prevent the model from inadvertently being charged with high-

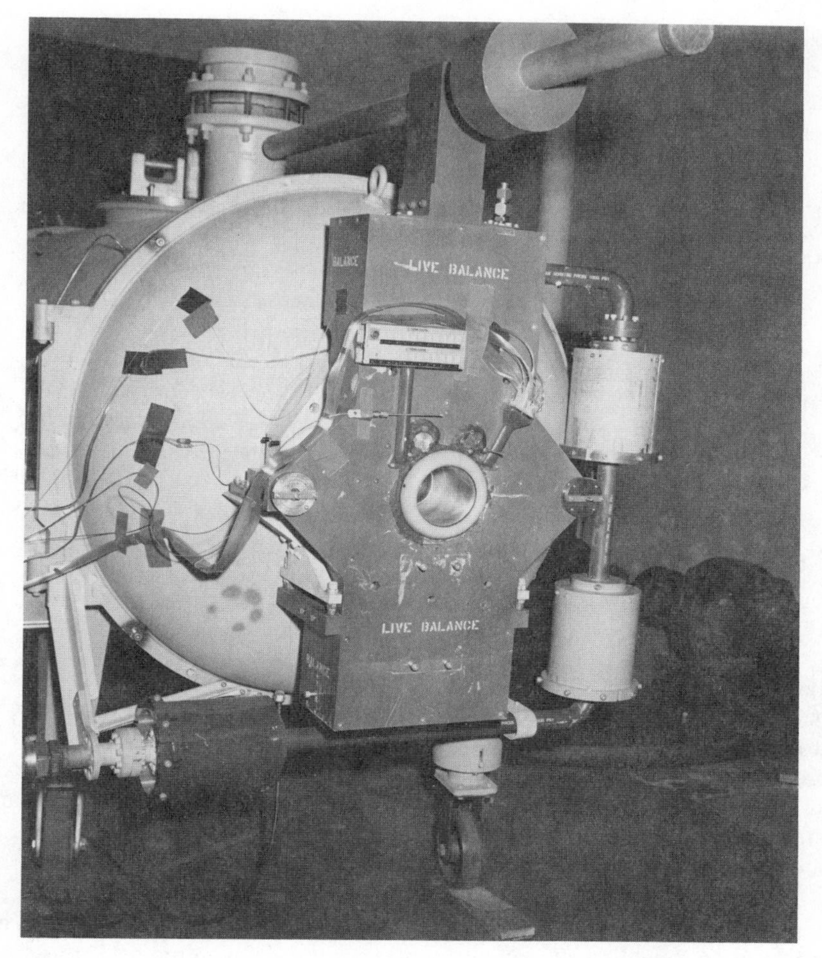

FIGURE 5.51. Calibration stand for model jet engines. Two six-component balances are at the top and bottom with the nacelle inlet in the center. The high-pressure air enters from the lower left through a shielded gimbal. Two other gimbals are in the vertical leg of the L. (Photograph courtesy of Boeing Aerodynamic Laboratories.)

pressure air. Any place in the system where air can be trapped must have bleed valves and a pressure gage to ensure that there is no high-pressure air when disassembling. An example is the piping across the balance when checking pressure tares. When removing flanges, each bolt should be slightly backed off one after the other. Never completely remove each bolt one at a time. People have been killed when removing a cap on a high-pressure pipe when pressurized. The load on the cap can exceed the strength of the bolt and the cap blows off. When going from a pipe designed for high pressure to one designed for a lower pressure through a pressure-reducing

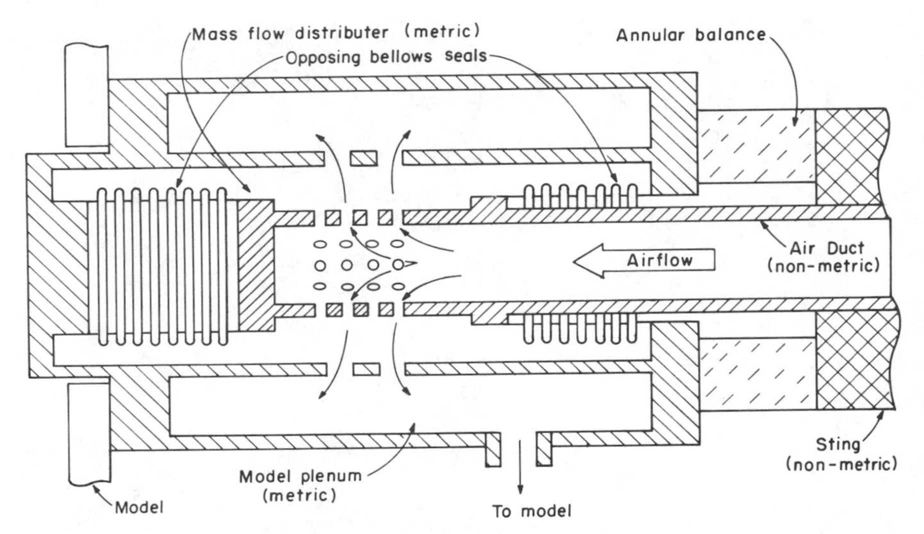

FIGURE 5.52. Schematic of jumping air across an internal (sting) balance. The forces on the two bellows cancel out, leaving small air line tares.

valve, blow out or rupture disks are required to protect the lower-pressure pipe in case of a valve failure. Any high-pressure-air system should be designed by an engineering firm conversant and experienced in high-pressure piping. Do not try to cheat the safety system such as replacing a rupture disk that continually ruptures with one of a higher rating, nor should the safety system be bypassed. Also, extreme care must be taken with positive displacement pumps to prevent "dead heading" the pump. The pressure will build up at an alarming rate until something comes unzipped. Do not clean any pipe connection with any petroleum product or an in-line explosion may occur due to dieseling. Treat all high-pressure-air systems as if they are a loaded and fused bomb—because they are.

The problems of getting the air across the balance and into the model have been discussed and now the requirements for calibrating a flow-through nacelle, a blown nacelle, or a TPS will be discussed.

If the requirements for the nacelle drag in a flow-through, or the thrust of a blown nacelle or TPS, are not stringent, they can be obtained by momentum methods using a wake rake. This method is theoretically correct, but it is difficult to get the correct momentum when integrating the rake output owing to distorted velocity profiles.

A better way to calibrate these devices is to use a special calibration stand that will cover the full range of flight operations (Fig. 5.51). One such device is a chamber 4 ft in diameter by 12 ft long. At the forward end two six-component balances support a force-balance assembly to which either a TPS, blown, or flow-through nacelle is mounted. The inlet of the TPS or flow-through nacelle is open to the ambient pressure of the room, and the exhaust confined to the chamber. A bellows seals the air passage around the nacelle and force due to pressure is canceled by compensating bellows. Thus the balance measures the nacelle thrust. Air is jumped across the balance by the same method used for the tunnel external balance. The air flow required to drive the blowing nacelle or TPS is measured by either a calibrated single critical flow venturi (CFV)—or a set of CFVs that operate in parallel in any combination desired, called multiple critical venturis (MCV). These have throat areas ranging in proportion to 1, 2, 4, 8, and two throats at 16. This gives a total of 47 effective venturi sizes. When the throat is at sonic velocity, the CFV only requires one pressure and temperature measurement to obtain the mass flow. Venturis built to the dimensions given in ASME standards can have errors in mass flow of up to 0.5% owing to manufacturing tolerances and finishes. Therefore, the CFV's should be calibrated and their calibrations should be traceable to primary air flow standards of the Colorado Engineering Experimental Station, Inc. or equivalent. A TPS requires one CFV or MCV while a blown nacelle with both fan and core flow requires two in which to measure the mass flow.

The inside of the calibration chamber is filled with screens to break up the exhaust jet and prevent recirculation and entrapment around the jet. They also diffuse and mix the flow before exiting at a low-pressure MCV at the rear of the chamber. The low-pressure MCV measures the air flow exiting the chamber, and controls the nacelle pressure ratio. Two air ejectors exhaust the flow and maintain the low pressure at the MCV at sonic throat conditions.

For a TPS the fan air flow is the difference between the high-pressure air flow into the nacelle and the low-pressure MCV at the end of the chamber. For a description of the problems of model testing of powered nacelles see References 5.7, 5.8, and 5.9.

Flow-through nacelles are calibrated for internal drag by using the difference between ideal thrust and the measured thrust. Flow-through nacelles often have trip strips just inside the inlet lip. The location of the trip strip is often determined by flow visualization to ensure the desired turbulent boundary layer within the duct. The nacelle has four or more internal static ports. The average of these pressures is used to determine the mass flow via the nacelle calibration when running the test. From the test mass flow the nacelle drag is determined via the nacelle calibration.

5.14. TESTING FUSELAGES, NACELLES, AND BODIES OF REVOLUTION

Tests of fuselages alone are rarely made, for the interference effect of the wing on the fuselage is of such prime importance and magnitude that tests of a fuselage without the presence of the wing are of very questionable value. When a fuselage alone is tested the buoyancy effects (see Section 6.9) are usually important. Blockage corrections are commonly moderate to large.

Nacelle tests are much more valuable than fuselage tests because usually entirely different items are being investigated. Generally the nacelle tests are concerned only with cooling pressure drops and cooling drags and not with the total nacelle drag, which would be largely dependent on the wing–nacelle interference.

If a power-driven propeller is to be utilized in the set-up, careful consideration should be given to the control and measurement of the rpm. For nacelles simulating modern high-power units, the loss of a single revolution per minute can represent a large thrust decrement and, in turn, can invalidate any drag measurements that may be made. It is usually advantageous to fix the rpm by means of a synchronous driving motor and to vary the tunnel speed and propeller blade angle to get various flight conditions. Such an arrangement corresponds to the customary constant-speed setup of most airplanes.

The usual nacelle is of such dimensions that buoyancy, constriction, and propeller corrections are important. For clarity, let us assume that a model is to be tested at 100 mph. The constriction effect of the closed jet increases the velocity over the model so that the results are similar to those encountered in free air at a slightly higher speed, say 102 mph. The effect on the propeller is opposite, however, yielding the results expected in free air at 96 mph. It is therefore necessary to increase the tunnel speed to approximately 104 mph, at which time the propeller slipstream is the same as in free air at 100 mph. The propeller coefficients are then based on 100 mph.

The buoyancy effect is assumed to be the same as expected at 100 mph without the propeller.

Drag coefficients for a nacelle may be based either on nacelle frontal area or other reference area. The choice should be clearly stated. The quantity of cooling air per second Q is usually defined by

$$Q = KS\sqrt{2\Delta p/\rho} \tag{5.36}$$

where K = engine conductivity; S = nacelle or engine frontal area, sq. ft; Δp = baffle pressure drop, lb/sq. ft; ρ = air density, slugs/ft^3.

Bodies of revolution or fuselages are best tested on their sides, using a single strut and yawing the model to simulate angles of attack. This procedure both reduces the tare and interference and makes their determination easier. For accurate angle determination the torsional deflection of the strut under torque load should be calibrated and corrected in the data work-up.

5.15. TESTING PROPELLERS

Propellers are frequently investigated in wind tunnels either alone or in conjunction with a fuselage or nacelle. If an entire model is tested, the propeller diameter will be small compared to the tunnel jet diameter, and the corrections to be described will become small also. For tests in which pro-

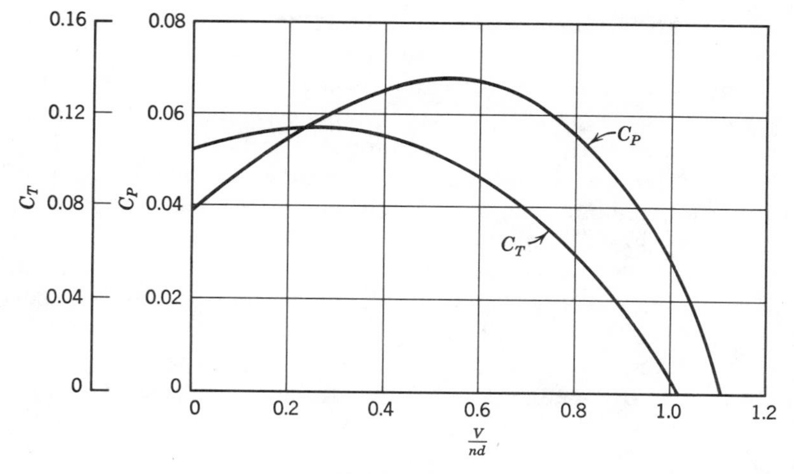

FIGURE 5.53

peller characteristics are to be determined, the propeller diameter may well be 60% of the jet diameter, and the corrections are not only large but unfortunately not as accurate as those for wings. An approach to the problem has been made by Glauert; boundary corrections for propellers may be found in Section 6.32.

Numerous coefficients have been advanced for presenting propeller data, each type perhaps being advantageous for particular applications. One of the more popular forms is

$$C_P = \frac{P}{\rho n^3 d^5} = \text{Power coefficient} \tag{5.37}$$

$$C_T = \frac{T}{\rho n^2 d^4} = \text{Thrust coefficient} \tag{5.38}$$

where P = power input, ft-lb/sec; n = rps; d = propeller diameter, ft; T = thrust, lb. The "variable" of propeller testing is usually the advance ratio, $J = V/nd$, where V is in feet per second and n in revolutions per second. A form using more conventional quantities is $J = 88\ V/Nd$, where V is in miles per hour and N in revolutions per minute. It will be seen that the two are numerically identical.

Plots of C_P and C_T against J for a typical propeller are shown in Fig. 5.53.

5.16. TESTING FOR CAVITY RESONANCE

One of the newer problems that besets the modern high-speed aircraft is cavity resonance, a high-intensity vibration of wheel wells, bomb bays, or cockpits that arises when their covers are removed and the high-speed airstream moves by (and into) the opening.

This phenomenon has been located by means of tunnel tests, and overcome by means of Helmholtz resonators, tuned chambers which are opened into the offending cavity. The procedure for the tunnel test is to open the various cavities one at a time and to pick up their natural frequencies with a pressure pickup fed into a scope or recording oscillograph. The resonance, if any, will occur close to the same speed at which it will occur on the airplane, but at a frequency increased by the scale factor. If space is available for Helmholtz resonators they may be tried; if not, scoops or lips may be added to the cavities intuitively until the intensity is down.

A second approach, if the natural frequency has already been determined by flight test, is to mount the model on strain gages selected so that their spring constant and the mass of the model result in the natural frequency value already known. Decreased amplitudes of vibration then indicate the success of the cure being considered.

5.17. TESTING FOR SPIN CHARACTERISTICS AND SPIN RECOVERY

The study of spin and spin recovery is covered in three broad areas. These are (1) entry into stall and loss of control, (2) post stall entry into a spin, and (3) spin and recovery.

A complete spin program will require models and tests in each of these areas. However, in many cases it may only be necessary to demonstrate that it is possible to consistently recover from a spin. Thus spin and recovery tests are often made first in a spin tunnel.

In Ref. 5.10 an outline of a design procedure is given for predicting spins. It is not simple and requires several types of wind tunnel tests other than spin tunnel tests. These include static six-component full model tests at $C_{L\,\max}$ and beyond at various side slip angles accounting for Reynolds number effects at stall and post-stall regions.

There are two types of dynamic tests made using balances. The first is a forced oscillation test with the model using an internal balance mounted on an oscillating sting (Ref. 5.11). The results from this test yields combined derivatives that are used to analyze the airplane's motion in stall departure. The second dynamic test uses a rotary balance. The Langley spin tunnel's rotary balance (Fig. 5.54) allows the model to be tested through a $\pm 15°$ sideslip and 8°–90° angle of attack range. Usually the balance moment center and the desired aircraft center of gravity coincide. The balance moment center is either on the spin axis or at a desired offset from the same axis. The model can be rotated either right or left up to 90 rpm. By use of the tunnel airspeed and rotational speed steady spins can be simulated. The rotary balance is equipped with an on-line data system. Tares in the form of inertial

FIGURE 5.54. Beech Model 76 on rotary balance in NASA Langley 20-ft vertical spin tunnel. Model is 0.18 : 1 scale. (Photograph courtesy of Beech Aircraft Corporation.)

forces and moments of the model at different attitudes and rotational speeds at zero tunnel speed are taken and subtracted from the wind-on test data. Tares are taken by surrounding the model with a covered bird cage structure, which encloses the model but does not touch it. This allows the air immediately surrounding the model to rotate with the model. The results of the rotary balance tests are used in inertial/aerodynamic computer programs based on Ref. 5.12 to produce plots that are used to predict possible spin modes. Component buildup testing can be performed on the rotary balance to show the influence of the various components on spin characteristics (Ref. 5.13). It also should be noted that the rotary balance when not rotating can be used to get static data at high angles of attack. The spin tunnel is not the only tunnel equipped with a rotary balance.

The Langley 30 × 60 tunnel is used with a free flight model for studies through stall and loss of control. This facility has a mini computer that is used to simulate the flight control laws and provide control inputs along with the pilot's (Section 1.3). Due to model support cables, departure from stall into a spin cannot be simulated in this facility.

NASA uses a free flight model dropped from a helicopter to study spin entry characteristics. These tests predict the airplane's susceptibility to spin entry and the dominate spin modes. Radio controlled model airplanes are also used for similar data (Ref. 5.15). A good radio controlled model program can cost as much as the rotary balance and spin tunnel program.

To study the spin and recovery characteristics the spin tunnel is used. The dynamically scaled model is hand launched into the tunnel at various pitch attitudes with pre-rotation (Fig. 5.55). The tunnel air speed is adjusted to balance the model's sink rate, thus holding the model level at the viewing window. The spin and recovery are photographed with a 16 mm movie camera and also video taped. The film is used to get the angle of attack, bank angle, spin rate, and turns for recovery. The tunnel speed yields the rate of descent. The recovery from the spin is initiated by remote control to set the model's aerodynamic controls to a predetermined position. The models are often built out of balsa wood or thin fiberglass, because both weight and moments of inertial requirements are critical.

Spin tunnel tests can determine (1) spin modes and recovery characteristics, (2) effects of mass distribution and center of gravity, (3) the effect of external stores, (4) the type and size of required spin recovery chute, and (5) exit trajectory of air crew. It is necessary to determine both the spin and spin recovery in both right and left spins for all combinations of rudder, elevator, and ailerons. This will require a large number of tests as indicated in Ref. 5.14, where approximately 500 spins were made in one aircraft study.

FIGURE 5.55. Beech Model 76 being launched in NASA Langley 20-ft vertical spin tunnel. Model scale is 0.07 : 1. (Photograph courtesy of Beech Aircraft Corporation.)

The data are in the form of film and observation notes. The film can be analyzed in a cross-hair-equipped film viewing machine using a protractor to yield both fuselage and spin axis angles to at least $\pm 1°$. The fuselage angle (θ) is measured from the horizontal and is negative nose down. The spin axis angle (ϕ) is measured from the horizontal and is positive left wing up. The angle of attack is

$$\alpha = 90° - (-\theta) \tag{5.39}$$

The number of turns for recovery from the film can be determined to one quarter turn. The spin rate (Ω) can be determined to $\pm 2\%$.

The scale factors between the model and full scale where M is the model, A is the full-scale aircraft, and N is the model scale (i.e., $\frac{1}{16}$ scale, etc.) are

$$\text{Length} \quad L_M = L_A N \tag{5.40}$$

$$\text{Area} \quad S_M = S_A N^2 \tag{5.41}$$

$$\text{Weight} \quad W_M = W_A N^3 \left(\frac{M}{A}\right) \tag{5.42}$$

$$\text{Moment of inertia} \quad I_M = I_A N^5 \left(\frac{\rho M}{\rho A}\right) \tag{5.43}$$

$$\text{Velocity} \quad V_M = V_A \sqrt{N} \tag{5.44}$$

$$\text{Spin rate} \quad \Omega_M = \Omega_A / \sqrt{N} \tag{5.45}$$

Because the spin tunnel only simulates developed spins and recovery, it yields no information on the spin susceptibility of the aircraft. Also, if the model shows two spin modes, it is almost impossible to predict which mode, if either, will be predominate on the airplane in the spin tunnel.

For similtude and scaling requirements, especially for dynamic models, see Ref. 5.16.

5.18. DYNAMIC AEROELASTIC TESTING

Today's design trend toward thinner wings equipped with external stores, engine pods, and the like has aggravated the already existing problem of structural deformation. Since the deflection of the aircraft structure can change the dynamic behavior and the flutter characteristics, the testing of flexible models in the wind tunnel is necessary to determine these "elastic" effects.

Essentially, two basic types of aeroelastic models have evolved, the dynamic stability model, and the flutter model. As pointed out in Ref. 5.17, in dynamic stability tests one is interested in a dynamic behavior dominated by

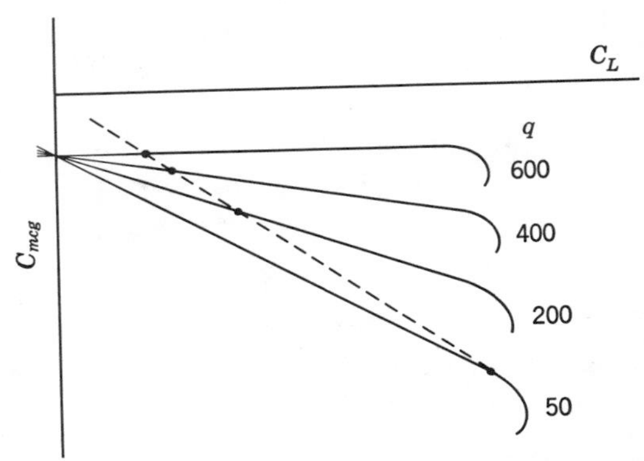

FIGURE 5.56. Effect of a swept wing aircraft elastic deformation on longitudinal static stability.

rigid body modes of motion. On the other hand, for flutter tests, one is interested in the elastic modes (or, really, a dynamic instability brought on by the elasticity of the structure).

Elastic models may be used in steady-state testing to determine the effect of deflection on static stability by mounting the model on the balance support system. It is important, however, to duplicate to scale the deformations of the full-scale vehicle and to be sure that the support system allows the model to deform elastically. Figure 5.56 illustrates the effect of deflection on the static stability curve; the dashed curve is the effective stability curve, since the aircraft does not fly at constant q of over the C_L range.

Most model suspension rigs allow freedom in pitch, vertical translation, and yaw with limited roll, and fore and aft freedom. Such a system will not allow a determination of the long-period or phugoid motion because the tunnel operates at constant speed whereas wide changes in velocity occur during the phugoid on the full-scale vehicle.

The model may be "kicked" into a displaced altitude and released; the ensuing motion defines the short-period oscillation which occurs at constant speed. Camera studies, recording oscillographs, or similar instrumentation may be used to record the motion. Such quantities as the period of the motion, the time to damp to one-half amplitude, and the number of oscillations to damp to one-half amplitude may be determined.

Flutter testing seems to be more common for low-speed tunnels than testing for dynamic stability. In flutter testing, one looks for

1. Critical flutter speed.
2. Flutter frequency.
3. Flutter mode (symmetrical or antisymmetrical).
4. Fuselage coupling.
5. Wing-empennage interaction.

FIGURE 5.57. Boeing 747/Space Shuttle flutter model on vertical rod mount. (Photograph courtesy of Boeing Aerodynamic Laboratories.)

In most cases where complete models are tested, it is common practice to mount the model on a vertical rod. This system (see Fig. 5.57) allows fairly free motion in pitch, roll, yaw, and vertical translation. Fore and aft and spanwise motions are considerably restrained, however.

A second method is to constrain the model by a cable bridle that holds the model centered in the tunnel (Fig. 5.58).

Component testing of wing panels (aileron flutter) and empennage models is frequently undertaken to obtain individual flutter characteristics of the particular component. Figure 5.59 shows a large-scale empennage model mounted in the tunnel.

It is interesting to note that the high-speed camera is one of the most useful devices in flutter testing because it will indicate the flutter mode shape; accelerometers, strain gages, and so on mounted on the wing spars or other parts of the main structure will record flutter frequency.

The actual testing procedure is to approach the expected critical flutter speed slowly. At each new speed setting, the model can be excited by means of a "jerk wire." Tracings of instrumentation output can be monitored on a suitable chart recorder, FM tape, and so on. As critical speed is approached, the damping time increases and can easily be detected on the recorder. Finally, when the recording pen indicates divergence, the cameras are turned on and the run is terminated by cutting the tunnel. Some tunnels have a device called a "q stopper" which drops speed rapidly and thus reduces

FIGURE 5.58. Rockwell B-1 flutter model on cable mount. (Photograph courtesy of North American Aircraft Operations of Rockwell International.)

FIGURE 5.59. DC-10 tail flutter model with double hinged rudder. (Photograph courtesy of NASA Langley and McDonnell–Douglas Corporation.)

the model amplitudes. This consists of two splitter plates on the ceiling and floor. The splitter plates are equipped with spring-loaded flaps with snubbers at the end of their travel. When the tunnel is cut during model flutter, the flaps are automatically deployed toward the tunnel centerline. This reduces the speed by increasing the drag in the test section near the model and forces some of the high-energy air to pass between the splitter plates and the ceiling and floor.

There is always a risk during a flutter test of getting such a severe model flutter that the model is either partially or totally destroyed. The loss of a model or parts can also damage the wind tunnel propellers.

In flutter tests the tunnel is expected to provide the true air speed at which data were taken as the tunnel was cut. This requires that the section static temperature and pressure be recorded as well as the dynamic pressure. It is most convenient to run the tunnel during a flutter test by an aircraft air speed meter rather than q, as the flutter speed is an air speed and dynamics engineers are used to this term.

The primary purpose of a flutter test is to ensure that the airplane will not encounter flutter within its flight envelope, at all possible loadings. The loading requirement means that in aircraft with wet wings the distribution of fuel must be considered. On some military aircraft with soft wings, external stores must also be considered. The structural stiffness of the airplane will determine the loads that must be considered.

Data from flutter tests are presented as the true air speed for flutter against altitude. One may also spot on the design speed of the aircraft which, of course, should be less than the flutter speed at any particular altitude.

Flutter Model Design and Scaling

In constructing flutter models, the proper scaling of model characteristics is important. Since the model geometric scale ratio affects other parameters, it is usually fixed by a consideration of wind tunnel limitations. The maximum model span that the tunnel can accommodate should not exceed 0.8 of the tunnel width (without excessive blocking or wall interference); this sets the ratio b_M/b_A (Ref. 5.17). The subscripts refer to model and full scale airplane. The quantity b can be any linear dimension, although wing span appears more convenient to use. The model scale stiffness, mass distribution, bending, and torsional stiffness should be duplicated. The requirement on mass is that the mass or weight ratio should be:

$$\frac{(m/\pi\rho b^2)_M}{(m/\pi\rho b^2)_A} = 1 \tag{5.46}$$

$$\frac{m_M}{m_A} = \frac{\rho_M b_M^2}{\rho_A b_A^2} \tag{5.47}$$

where m = mass per foot
ρ_M = tunnel operating density
ρ_A = flight density of full-scale aircraft
The total mass or weight ratio then becomes

$$\frac{M_M}{M_A} = \frac{\rho_M}{\rho_A}\left(\frac{b_M}{b_A}\right)^3 \tag{5.48}$$

The frequency ratio that should be preserved is

$$\frac{\left(\dfrac{V}{b\omega}\right)_M}{\left(\dfrac{V}{b\omega}\right)_A} = 1 \tag{5.49}$$

or

$$\left(\frac{V}{b\omega}\right)_M = \left(\frac{V}{b\omega}\right)_A$$

Other relations important for flutter models are

1. The velocity ratio

$$\frac{V_M}{V_A} = \left(\frac{b_M}{b_A}\right)^{\frac{1}{2}} \tag{5.50}$$

2. The "static" moment scale ratio

$$\frac{S_M}{S_A} = \frac{\rho_M}{\rho_A}\left(\frac{b_M}{b_A}\right)^4 \tag{5.51}$$

3. The weight moment of inertia ratio

$$\frac{I_M}{I_A} = \frac{\rho_M}{\rho_A}\left(\frac{b_M}{b_A}\right)^5 \tag{5.52}$$

4. The stiffness ratio

$$\frac{\text{Model stiffness}}{\text{Airplane stiffness}} = \frac{\rho_M}{\rho_A}\left(\frac{V_M}{V_A}\right)^2\left(\frac{b_M}{b_A}\right)^4 \tag{5.53}$$

Although the foregoing ratios are used in model design, the completed model is given vibration tests to determine the true frequency.

Since actual-scale reproduction of the airplane structure is not practical, the model designer seeks a simplified structure that will give the right bending and torsional stiffness. For low-speed models, a single spar as shown in

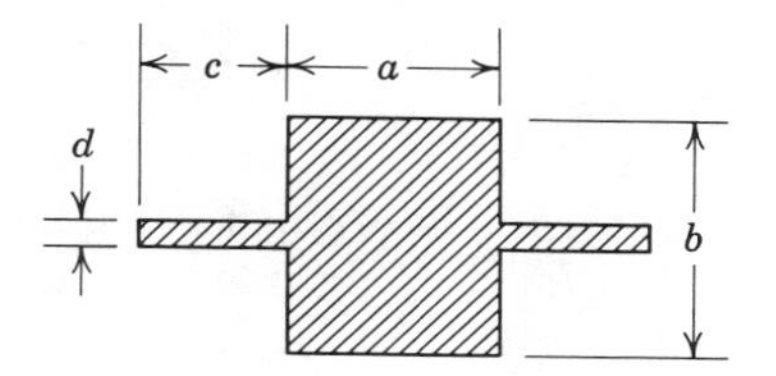

FIGURE 5.60. Typical spar cross section for low-speed flutter model. Dimensions a and b determined by I_h and J required. Dimension c determined by I_f required. Dimension d set at $0.1b$. I_h and I_f are moments of inertia in vertical and fore and aft bending.

Fig. 5.60 can frequently be used to obtain the moments of inertia for both vertical and fore and aft bending. It is customary to have all the stiffness in the spar with the covering providing only the aerodynamic shape. To accomplish this, the wing, fuselage, or tail external shape is made in sections. The gaps between the sections can be covered with thin rubber dental dam, or better, filled with a thin soft foam rubber. The filler should not increase the stiffness of the structure especially aerodynamic surfaces such as the wing and tail.

After the spar is fabricated, the stiffness can be checked by static deflection tests. For these tests, the spar is mounted as a cantilever beam. An accelerometer that can measure angles is attached to the spar which is loaded, and the slope of the elastic curve is determined by the accelerometer.

Now from beam deflection analysis the deflection is known to be

$$\frac{d^2y}{dx^2} = \frac{M}{EI} \tag{5.54}$$

or

$$\frac{M}{EI} = \frac{d\left(\frac{dy}{dx}\right)}{dx}$$

with

$$\frac{dy}{dx} = \Theta$$

Then

$$\frac{M}{EI}\,dx = d\Theta$$

and since the spar stations are finite distances apart the equation

$$_n\Delta\Theta_{(n+1)} = \int_n^{n+1} \frac{M}{EI}\,dx \tag{5.55}$$

applies. Therefore to obtain the average value of M/EI between stations it is only necessary to subtract successive values of the measured deflections to get

$$_n\Delta\Theta_{(n+1)} = \Theta_{(n+1)} - \Theta_n \tag{5.56}$$

Since $\Delta\Theta$ is equal to the area of the M/EI curve between stations, the average value of M/EI is obtained by dividing $\Delta\Theta$ by the distance between stations. Finally, $1/EI$ is obtained by dividing out the known applied bending moment. The values for bending stiffness in the other plane and for the torsional stiffness can be found by the same procedure. All quantities should be compared to the airplane data to check the exactness of the model spar design.

The low-speed flutter model usually does not have either the right scaled weight (gravitional force) or the right deflection ratio. The ratio of gravitational force to aerodynamic force is

$$\frac{m}{\rho b^2} = \frac{g}{V^2/b} \tag{5.57}$$

The ratio $g/(V^2/b)$ is seldom scaled properly, so that some additional vertical force must be applied if the model is to fly at the proper lift coefficient. As for deflections, the ratio of the deflection (due to a scaled load) of an aeroelastic model to that of the full-scale airplane should be

$$\frac{\delta_A}{\delta_M}\frac{b_M}{b_A} = 1.0 \tag{5.58}$$

However, for most flutter models this unit value is seldom achieved, but fortunately the product in Eq. (5.58) may go as high as 3.0 without introducing any appreciable error. Some typical values of the various ratios introduced in this section are listed in Table 5.10. Steps in the construction of a flutter model are shown in Fig. 5.61.

TABLE 5.10. Typical Flutter Model Scale Ratios for a Four-Engine Turbofan Cargo Plane

Ratio	Symbol	Numerical Value
Geometric	b_M/b_A	1/24[a]
Density	ρ_M/ρ_A	1.44[b]
Velocity	V_M/V_A	1/7.5
Frequency	$\dfrac{(V/b\omega)_M}{(V/b\omega)_A}$	3.20
Deflection	$\dfrac{\delta_A/\delta_M}{b_A/b_M}$	2.34
Weight	W_M/W_A	1/9,600
Static moment	S_M/S_A	1/230,400
Weight moment of inertia	I_M/I_A	1/5,529,600
Stiffness	EI_M/EI_A or GJ_M/GJ_A	1/12,960,000

[a] Dictated by size of tunnel to be used.
[b] Dictated by tunnel and flight conditions.

FIGURE 5.61. A nearly completed flutter model. The dark lines are joints rather than glue locations. (Courtesy McDonnell Aircraft Corp.)

5.19. TESTING WINDMILL GENERATORS

The need for a power source to operate when a jet-engine airplane has a high-altitude flame-out has reactivated the interest shown many years ago in wind-driven generators or hydraulic pumps, and not infrequently the tunnel engineer is called upon to evaluate a particular generator by an operational test. When this is so, the special precautions that are paid to models such as rotors, propellers, and the like, whose possibility of failure is higher than that of rigid models, should be applied.

Corrections to the data from a windmill test are subject to boundary corrections as outlined in Chapter 6, specifically wake blockage, and propeller corrections with a negative sign. However, the emergency power generator is so small relative to the tunnel and the interest in very accurate data as compared to proof testing is so slight that corrections may be neglected. This is not true of windmills intended to be used for ground power generation (see Section 9.9).

It is of interest to look into the mechanism of a windmill from the theoretical side in order to gain an understanding of how it works. Of course the device takes energy from the air, but surprisingly the total stream energy is not available to the windmill. In words, the slowing of the stream makes a portion of the air go around the windmill instead of through it, and a point may be reached beyond which an attempt to take more energy from the stream is fruitless.

Looking at the problem from a momentum standpoint, we find that if the velocity at the windmill is $V(1 - a)$, the final velocity will be $V(1 - 2a)$, and letting the windmill radius be R, we have for the power out

$$\begin{aligned} P_o &= V(1 - a)[\rho\pi R^2 V(1 - a) \cdot V - \rho\pi R^2 V(1 - a) \cdot V(1 - 2a)] \\ &= 2\pi\rho R^2 V^3 a(1 - a)^2 \end{aligned} \tag{5.59}$$

Differentiating and solving, we find the maximum power out occurs when $a = \frac{1}{3}$. Substituting this value, and comparing the maximum power out with the total in a freestream of the same radius P_s, we have

$$\frac{P_o}{P_s} = \frac{0.296\rho\pi R^2 V^3}{0.5\rho\pi R^2 V^3} = 0.594$$

or even with no-drag blades the windmill can hope for only 59.4% of the stream energy. A good figure in estimating windmill sizes seems to be about one-half that theoretically available, or roughly 30% of the total stream energy. During testing, stalled blades should be avoided, for when they unstall a runaway may occur. This can be avoided by use of a governor to control the rpm. As the windmill is often connected to a generator, some method such as a resistor bank must be provided to dissipate the output.

5.20. TESTING FOR LOADS

Local loads on landing gear doors, bomb bay doors, flaps, nacelles, and so on, are best obtained by the use of strain gage balances designed for the expected loads. When the distribution of air loads are desired on the wing, fuselage, and tail, a special model is built as described in Section 5.1.

The data are usually presented in the form of a pressure coefficient versus model location. These data can then be integrated to obtain, for example, local lift coefficients on a wing or tail and plotted versus span for span load. In most cases the purpose of the test is to obtain aerodynamic loads for structural design (Sections 4.18 and 4.19).

The use of pressure transducers and scanivalves makes the data reduction quite easy (see Section 3.8). The transducer measures a ΔP from a reference pressure to a point on the model directly. The millivolt output of the transducer through the transducer's calibration is changed to engineering units (psi or psf) and corrected, if necessary, for the location of the reference source and divided by the true dynamic pressure. The test dynamic pressure is the true dynamic pressure when corrected for wake and solid blockage (see Section 6.11). The sample time for pressure transducers is usually quite short compared to balance data, which often are averaged over a longer time. If both balance data and transducer data are taken simultaneously, and the balance data are used to calculate solid blockage, care should be taken to ensure that balance data agree with the standard balance data. This problem usually occurs when only a few steps are used on the scanivalves.

Although the data reduction for a pressure test is quite simple, the huge amount of data can have a dramatic effect on the computer time required to reduce the data. Often 800–1000 pressure ports are measured at each test point. This large mass of data also makes it nearly impossible to monitor the test data in any sort of real-time mode (see Section 3.8). Even if the data could be reduced and plotted in real time, no loads engineer could look at all of the plots and ask the tunnel operator for a check point. Thus, large load tests cannot be completely monitored in a real-time mode.

Invariably, in a large pressure test there will be plugged and leaking ports. These problems are supposed to be caught in the model checkout, but some will still show up in the final data. Because of the bad port problems there is a tendency to label any port that appears to fall outside the expected curve as a bad port. This can be dangerous, especially on leading edge flaps. In one case on a leading edge flap several ports at the same chordwise location very close to the leading edge showed very large negative C_p's. When looking at one chordwise pressure station, one might assume that the port was bad. But it is hard to believe that at the same percent chord at several spanwise stations the same model port is bad in exactly the same way.

5.21. TESTING LOW-ASPECT-RATIO WINGS

The advantages of using low aspect ratios for supersonic airplanes are quite impressive, and not infrequently the tunnel engineer finds himself or herself testing such configurations for low-speed characteristics.

Low-aspect-ratio lift curves may look quite different from those at high aspect ratio. Below $AR = 2.0$ the curve is usually concave upward (Figs. 5.62 and 5.63). The lift curve slope at zero lift may be approximated by

$$dC_L/d\alpha = 0.008 + 0.018(AR) \quad \text{(per degree)} \tag{5.60a}$$

below AR = 3.0. Between AR = 3.0 and 5.0 the following equation may be used:

$$\frac{dC_L}{d\alpha} = 0.1 \frac{AR}{AR + 2} \tag{5.60b}$$

For greater aspect ratios, Eq. (5.2) should be employed.

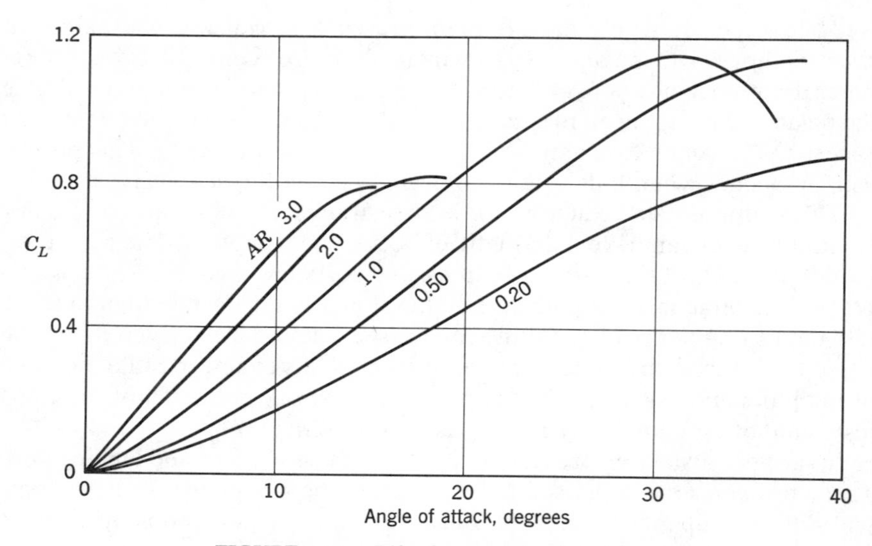

FIGURE 5.62. Lift of rectangular flat plates.

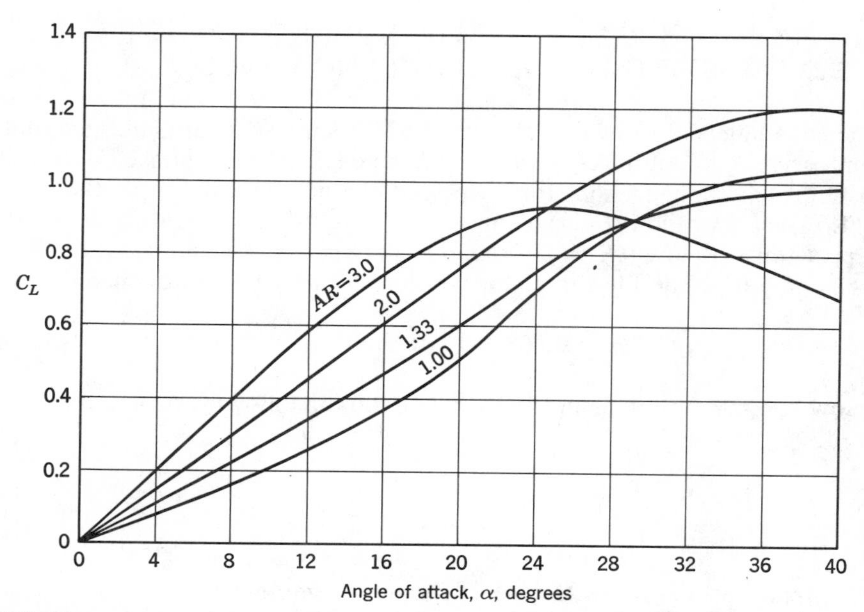

FIGURE 5.63. Lift coefficients for delta wings of various aspect ratios, NACA 0012 profile.

5.22. TESTING ENGINES

The actual operation of piston or jet engines in a wind tunnel for development reasons is a very specialized type of test possible in only a very few wind tunnels. Of the two, the piston engines present less of a tunnel problem, since their exhaust is smaller in quantity than that from a jet engine. The jet engine requires a huge scavenging system sometimes using half as much power as the tunnel itself in order to keep contamination low. This problem, incidentally, has an interesting facet in high-speed work, where the presence of rather small amounts of exhaust changes the values of γ and hence confuses the operating Mach number. In some low-speed tunnels the air exchangers can handle the exhaust problem.

If an engine is run in the tunnel with minimal air exchange, extreme care must be taken to protect personnel from high concentrations of carbon monoxide. After the test run the tunnel should be opened up and run to vent the CO before personnel enter the test section.

5.23. JETTISON TESTS

It is often necessary to determine the satisfactory release characteristics of tip tanks, underwing stores, bombs, or other devices. Although it is most direct to state that we will design the model and test to duplicate the ratio of inertia forces to gravity forces (Froude number), it is probably more instructive to go through the mental gymnastics of a hypothetical case, as follows:

Assume a store 16 ft long and a model 1.6 ft long. Further assume that whenever the full-scale store falls a length, it is pulled back half a length by aerodynamic drag, and pitches 10°. The linear acceleration is hence 16 ft/s^2, and the rotational acceleration 20 deg/s^2. Obviously we would like the model to pitch 10° while it is pulled back half a length also, so that the trajectory is similar to the full-scale condition.

The first thing we note is that while the full-scale store takes 1 s to fall a length and pitch 10°, the model must do the same in 0.316 s. Since half the model length is 0.8 ft, the linear acceleration needed for the model is again 16 ft/s^2. But the angular acceleration turns out to be 200 deg/s^2, or, in other words, the angular acceleration is increased by the scale factor λ (which is equal to l_{FS}/l_M, see below).

The aerodynamic force which produces the linear displacement is proportional to the body area and hence decreases as λ^2, and if we follow the dimensionally sound procedure of reducing the model weight by λ^3, the linear acceleration will be increased by λ. We get around this by reducing the test speed by $\sqrt{\lambda}$.

The torque is largely due to the force on the fin area (down by λ^2), the dynamic pressure (down by λ), and the length of the lever arm (down by λ).

In order to get λ times the full-scale pitch acceleration, we must reduce the model moment of inertia by λ^5.*

Hence we have (using W for weight, I for moment of inertia, l for typical length, and subscripts M and FS for model and full scale)

$$W_M = W_{FS} \frac{\rho_M}{\rho_{FS}} \left(\frac{l_M}{l_{FS}}\right)^3 \tag{5.61}$$

$$I_M = L_{FS} \left(\frac{l_M}{l_{FS}}\right)^5 \tag{5.62}$$

$$V_M = V_{FS} \left(\frac{l_M}{l_{FS}}\right)^{\frac{1}{2}} \tag{5.63}$$

Poor releases (wild pitching or hitting the airplane with the store) are almost invariably cured by jettison guns, and may be cured by store tilt, flaps on the

* Note that we more or less arbitrarily reduced the full-scale weight by λ^3. If we had used λ^2 and let $V_M = V_{FS}$, the moment of inertia would have come down by λ^4. This type of "heavy scaling" is useful at high Mach number and is discussed in Ref. 5.18.

FIGURE 5.64. Multiple-flash pictures of the release and separation of a bomb shape. In multiple-flash pictures the static items (the airplane fuselage and the catch-net in the above photograph) will always appear brighter than the moving model, since their image is reinforced by each flash. (Courtesy Sandia Corp.)

airplane fuselage near the store fins, flaps on the store-mounting pylon, or toed-in stores.

Drop data may be presented as moving pictures or multiple-flash stills (Fig. 5.64) or they may be reduced and plotted as in Fig. 5.65. When cameras are used, both side and top cameras are needed, and usually extra windows must be added to the tunnel.

A more elaborate approach to the separation problem that avoids the difficulty of matching model and full-scale moments of inertia is briefly as follows: the store model is mounted on an internal balance on a sting. The balance output in terms of angles, forces, and moments is fed into a computer. The computer, using the equations of motion, calculates the motion of

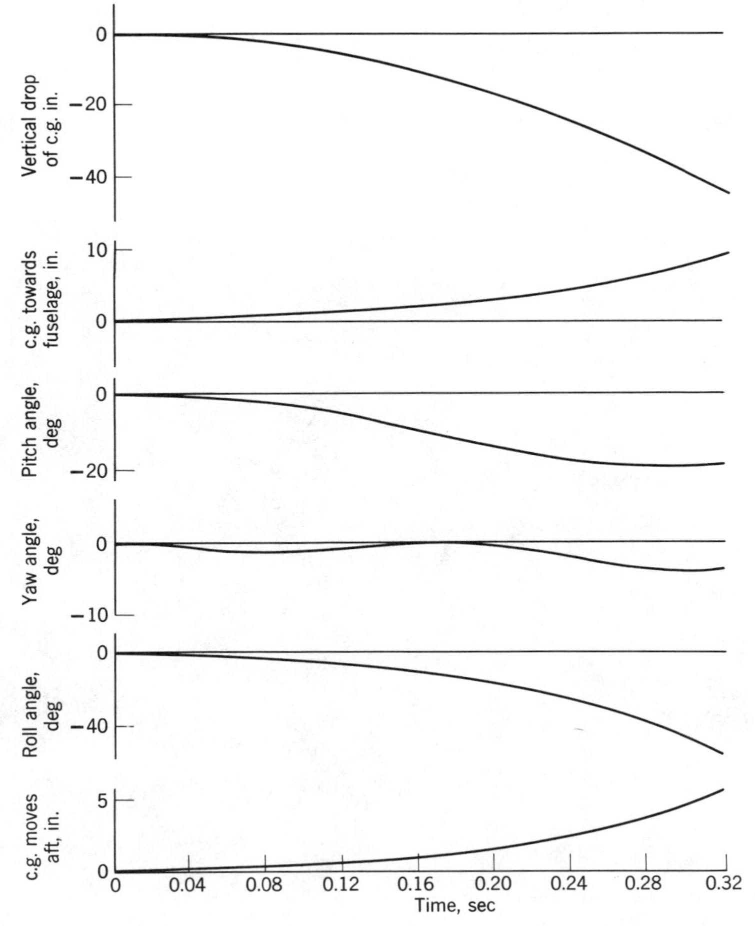

FIGURE 5.65. Presentation of store drop data. Other configurations could be plotted on the same sheet to aid in selecting the best configuration. Stores do not necessarily drop at the acceleration of gravity.

the store and directs the sting to move a short distance. This process is repeated in short steps and yields the path of the store as it leaves the aircraft. This method does not require the safety net with its large drag penalty, nor does it run the risk of the break up of the model and possible damage to the tunnel's propellers.

5.24. PARABRAKE TESTING

The use of drag chutes to provide ''air braking'' is quite common now, and the wind tunnel may be used to determine the drag characteristics of such devices. The chute may be packed in the model tail section and opened remotely during the test. Figure 5.66 shows a drag chute deployed during a tunnel test of a turboprop cargo-troop transport plane. Troublesome oscillations of the chute occurred during this test program, probably as a result of too short a bridle and too short suspension lines on the chute. Wake from the airplane could also have been contributory.

FIGURE 5.66. Model parabrake deployed behind a cargo transport model. The trailing loop from the windshield contains wires to the solenoid-operated chute compartment doors. (Courtesy Lockheed Georgia Co.)

5.25. TESTING STORES

The word "stores" is a catch-all for all types of jettisonable fuel tanks, bomb shapes, and other capacity devices. In general, they are designed for minimum drag while on the aircraft, and adequate stability after jettison. Tests run with the stores alone largely seek solutions for the stability problems; their cargo is typically of uniform density and their centers of gravity correspondingly from 40 to 55% of their length from their nose.

Necessary wall corrections for the tests include wake and solid blockage. The lift is ordinarily so small that downwash corrections are not needed.

Tests consist of angle of attack runs up to very large angles (say 50° perhaps 60°). These sorts of angles in flight hurt the impact circle of probable error and are to be avoided, but stability must still exist if a bad pitch-off produces them. The shape should be tested with the fins in both the + and × positions.

The best drag data possible are not good enough for serious bomb work, nor is it possible to correlate tunnel data with flight, simply because the flying bomb is always rolling and sometimes pitching. Tare and interference tests of a store shape are illustrated in Fig. 5.67.

A few words of *free-spinning* rocket or bomb models might be added here in order to reduce heart failure among wind tunnel engineers. A little thought will reveal that with a model at high angle of attack the downward moving fin

FIGURE 5.67. Testing using a single or double support strut. Both the single and double support struts are quite weak in torsion (yaw), and any models tested with them should not have natural frequencies close to those of the tunnel. Seven-by-ten tunnels of the type shown often have low-speed oscillations around 1 cps, and lateral restraint wires may have to be added if the model natural frequency is in that range. The model shown is installed with an image system. (Courtesy Sandia Corporation.)

will stall first, and the lift on the upward moving fin then becomes a "driver." The result can be a great and sudden increase in rpm—in one instance an increase from 300 to 2000 rpm in a very few seconds. Don't leave, just shut down the tunnel.

5.26. TESTING REENTRY LANDING CRAFT

Low-speed tests of reentry landers are made to determine the performance, stability, and control during the approach and landing of the spacecraft. The low-speed test program for these vehicles is the same as for an unpowered airplane.

5.27. V/STOL TESTING

Aircraft can be characterized by their mode of take off and landing. There are large numbers of aircraft that use conventional take off and landing (CTOL) techniques. These aircraft require relatively long runways. The exception to the long runway for CTOL's is operation off an aircraft carrier. A second mode is vertical take off and landing (VTOL), such as helicopters and the Harrier AV-8. The third mode is short take off and landing (STOL). The required length of runway can vary from 500 to 2500 ft depending on the size and weight of the aircraft. The STOL field length is the hardest to define and perhaps, as one wag put it, "the STOL field length is the length required by our aircraft." The VTOL and STOL type are often called V/STOL aircraft.

Over the years there have been a multitude of V/STOL vehicle configurations proposed, studied, and built.

Helicopters

The helicopter is the most successful and widely used V/STOL vehicle, both commercially and militarily. It can take off vertically, or with a short ground run if heavily loaded; it can hover; and it can maneuver in any direction. Outside of the military, the helicopter tends to be used more for industrial purposes rather than as public transport, such as airlines, owing to its high operating costs. The high-speed performance of a helicopter tends to be limited by rotor tip losses due to compressibility and retreating blade stall, plus the ability of the rotor to produce propulsive thrust at high speed. Various schemes have been proposed to overcome the helicopter's speed limits, such as wings to unload the rotor at high speed, additional sources of thrust other than the rotor, folding and storing rotors, etc. (Figures 5.68 and 5.69).

FIGURE 5.68. Four-blade helicopter rotor in U.S. Army 7 × 10 ft tunnel at NASA Ames. (Photograph courtesy of U.S. Army Aeromechanics Laboratory, AVRADCOM.)

FIGURE 5.69. Two-blade helicopter rotor in DNW open test section for aeroacoustic testing. (Photograph courtesy of U.S. Army Aeromechanics Laboratory, AVRADCOM.)

Tilt Wing

This aircraft obtains the high lift for vertical take off by rotating the wing, engines, and propeller—rotor through 90°. The propeller is in a horizontal plane for take off and landing and a vertical plane for forward flight. As a large portion of the wing is immersed in the slipstream the wing is not stalled through the major portion of the flight regime (Fig. 5.70).

Vectored Thrust

Aircraft in this category rotate the thrust producer. It would include both jet powered, where the whole engine is rotated, or the exhaust jet which is rotated through a pivoting nozzle, tilt propellers (rotors), and tilting ducted fans (propellers). See Figures 5.71 and 5.72.

Vectored Slipstream

The propeller slipstream is deflected downward by the use of the wing flaps. When the aircraft is powered by jet engines, this method of powered lift is often called either upper surface blowing or lower surface blowing. In upper surface blowing the engine's jet blows across the upper surface of the wing and flap. The flow follows the wing-flap due to the Coanda effect and the high energy in the boundary layer prevents separation. The wing also reduces the noise on the ground. On lower surface blowing the engine exhaust is below the wing and impinges on the lowered flaps to deflect the flow (Fig. 5.73).

FIGURE 5.70. XV-142 tilt wing model above ground plane in NASA Ames 40 × 80 ft tunnel. (Photograph courtesy of NASA Ames.)

FIGURE 5.71. AV-8B Harrier vector jet thrust V/STOL fighter in NASA Ames 40 × 80 ft tunnel. (Photograph courtesy of NASA Ames.)

Jet Flaps

High-pressure air is ducted along the wing span and is blown over the wing or parts of the wing in several ways. At the trailing edge of the wing the air is either blown over the upper surface of the flap using the Coanda effect over a curved surface at the trailing edge, or the jet nozzle is built to deflect the jet wake. In either case the thin jet is turned downward. When the flap is blown, the high-energy air delays separation. The air can also be ejected at the leading edge to delay separation of the wing, and this can be used alone or in combination with the two trailing edge blowing methods. When the amount of air on a blown or jet flap is greater than that required to prevent separation, additional circulation lift is produced, which is greater than that predicted by either jet reaction or potential flow.

FIGURE 5.72. Flutter model of tilt rotor V/STOL on rod mount in NASA Langley 16 × 16 ft Transonic Dynamics Tunnel. (Photograph courtesy of NASA Langley and Bell Helicopter.)

FIGURE 5.73. Vector slipstream by upper surface blowing of jet engine exhausts in NASA Ames 40 × 80 ft tunnel. (Photograph courtesy of NASA Ames.)

FIGURE 5.74. Fan-in-wing model in NASA Ames 40 × 80 ft tunnel. (Photograph courtesy of NASA Ames.)

Fan in Wing

A large fan is buried within the wing airfoil contour. In hover the wing acts as a duct, improving the static thrust of the fan. In forward flight at low speed the fan aids the wing lift. The fan is primarily used for vertical lift and transition, and jet engines or other propulsion methods are used for thrust in forward flight (Fig. 5.74).

Miscellaneous Systems

Vertical-take-off aircraft have been built that are tail sitters. These aircraft take off and land in a vertical attitude and use either jets or propellers both to lift the aircraft and to power it in horizontal flight.

The autogyro uses an unpowered rotor (same as a helicopter in autorotation) to provide lift. The thrust for forward flight was usually supplied by a piston engine propeller combination.

The above list of types of V/STOL aircraft is not intended to be all inclusive, and aircraft using combinations of powered lift systems have been proposed.

At low flight velocities used in V/STOL operations, the wing lift can be produced in three ways:

1. Basic lift of a wing or unpowered lift. This lift is due to circulation and is independent of thrust.
2. Lift due to deflected thrust by any of several methods as described previously. This lift varies linearly with thrust.
3. Additional circulation lift due to either jet exhaust or a propeller slipstream moving over the wing. This lift is a function of the increased velocity, and the increase in the effective chord of the flap used to deflect the air downward. The increase in effective flap chord is due to high-speed air being approximately parallel and in the same plane as the flap. This lift varies in a nonlinear manner with thrust, as the rate of increase in lift decreases with increasing thrust.

The basic concept in all V/STOL aircraft is to create lift by using power to produce a downward directed momentum. For the purpose of discussion of wind tunnel tests of V/STOL aircraft, the powered lift can be divided into two broad categories. The first is a distributed power lift such as produced by a helicopter rotor or a blown flap. The second is point power lift similar to the vectored thrust from a jet engine.

Testing Philosophy

Testing of V/STOL vehicles, including isolated rotors, is the same as power testing of a conventional aircraft. If a developmental test is being carried out, the powered lift is matched to a desired flight condition. Often several data points are taken near the desired point to allow the data to be cross plotted to get the desired data. This requires that the test schedule be carefully determined in advance. If the purpose of the test is to obtain basic data, then a different procedure is followed. Usually the purpose of these research tests is to determine the effect of one parameter on the other parameters. As an example, for a pure jet flap on a wing one might desire the effect of varying the jet momentum coefficient on lift, drag, and pitching moment from zero lift to maximum lift. Thus pitch runs would be made at various momentum coefficients.

V/STOL Instrumentation

As can be seen from the description of the various methods of producing powered lift, the simulation will be accomplished in the wind tunnel by two general methods. The first is by rotational devices such as a helicopter rotor, propeller, ducted fan, or fan in wing. The second is usually by compressed air to simulate jets and blown flaps. The rotational devices can be powered by electric motors (see Section 5.12), hydraulic motors, or air motors. The rotation devices can also be used to simulate a jet exhaust similar to a turbine power simulator. When the motor power is an electric motor, the problem of jumping from ground to the balance is relatively easy even for a water-cooled motor. However, when either compressed air or hydraulic power is used, the problem of jumping the balance is more complex as there can be loads applied to the balance that vary with both pressure and mass flow (see Section 5.13).

Depending on the model, V/STOL models require much more instrumentation within the model than a standard nonpowered force model.

Measuring RPM

If the model has rotating machinery, such as a helicopter rotor, lift fan, propeller, etc., the rpm must be controlled to match tip speed ratios on rotors, advance ratio on propellers, and thrust on lift fans. There are two types of optical systems that can be used. The first is a transmissive sensor; this is basically a light on one side of a disk with a hole in it and a light detector on the other side of the disk. For one hole in the disk you would get one pulse per revolution. The second is a reflective sensor where a light shines on the shaft that has a painted mark that reflects light to the detector, yielding a one per revolution pulse for a single mark. Both of the above units may have to be shielded from ambient light. A third method is an ac generator where a magnet is rotated with the shaft inside a coil producing a sine wave whose frequency varies with shaft rpm. All three of these methods can use a counter that will measure pulses or frequency per second, which can be converted to rpm. The output of the counter must be visual for the operator. Many counters have binary outputs that can be adapted for input to a data system. A fourth way to measure rpm is to use a dc generator whose output voltage is proportional to an rpm which can be calibrated. This is the least accurate. A digital voltmeter could be used for visual output and data system input, or the voltage itself could be used.

The rpm of rotors and propellers can be checked by the use of commercial strobotachometers. These use a xenon flash tube that is flashed at various frequencies, which are set on the dial in terms of rpm. When a marked blade is rotating at the same rpm, it is stopped by the light. These units must be used with care, for if the desired rpm is 2000, the blade will appear to stop also at 4000 and 1000, or even multiples of the blade speed. This problem is further complicated when multiple blades are in use. This makes the use of a strobotachometer difficult and precludes its use as a device to set rpm.

Compressed Air

When compressed air is used for simulating vectored thrust, etc., the mass flow must be known and controlled. This usually requires calibration of the nozzle by one of two methods. The first is to calibrate the nozzle in a calibration facility so that the thrust, mass flow, etc., are known by measuring pressure and temperature for desired thrusts. The second method is to use a pitot rake to calibrate the thrust. This is a simpler method, but not as accurate as the first (see Section 5.13).

Often jet flaps and lift fans in wings are calibrated on a balance, sometimes the one used during the test. For the jet pressure and temperature are used; the fan uses rpm (Fig. 5.75).

All of these calibrations yield a static thrust versus measurement that can be used during the test to set operating points.

If multiple air-driven devices are used, it may be desired or required to control each device independently. This can be accomplished by use of small motor-driven ball valves, placed between the model's high-pressure-air plenum and the device. Each device must have pressure and temperature measurements to control its mass flow.

Rotors

Often on tests of rotors both the lead–lag angle and the flapping angle versus azimuth position are required. These are measured by strain gage beams calibrated for angle versus strain. When testing rotors and propellers, stresses or moments on the blades may be desired, and these too are usually measured by strain gages. Such measurements give insight into blade twist and vibration frequencies when operating. As these measurements are taken on rotating devices the signals are transmitted through slip rings (see Section 3.16). Much, if not all, of the data from rotating blades is needed as a function of blade position, which requires a continuous trace of the data signal. The most convenient method of acquiring the data is then on an FM tape deck. An event marker can be put on one of the tape channels to be used to determine the blade azimuth position (Fig. 5.76).

Vectored Thrust

In the testing of V/STOL vectored thrust systems, when the lift is produced by jet engines, it may be desired to separate the wing lift from the thrust lift to determine the interference between the component parts. This necessitates the model being designed so that the loads on the engine can be measured separately from those of the model. For example, the first, often an external, balance can be used to measure the power and aerodynamic forces and moments while the model is attached to the power section by a second balance that measures the aerodynamic forces and moments. The difference between the two balances is the power effects plus the interferences. The other approach is to have the first balance measure the aerodynamic forces and moments plus interferences and the second balance measure the power effects. The second method may be more difficult because of the tares on the air lines due to pressure and mass flow.

Balances for rotors, especially, must be carefully designed or selected to avoid resonances between the balance and the rotor. The exciting frequen-

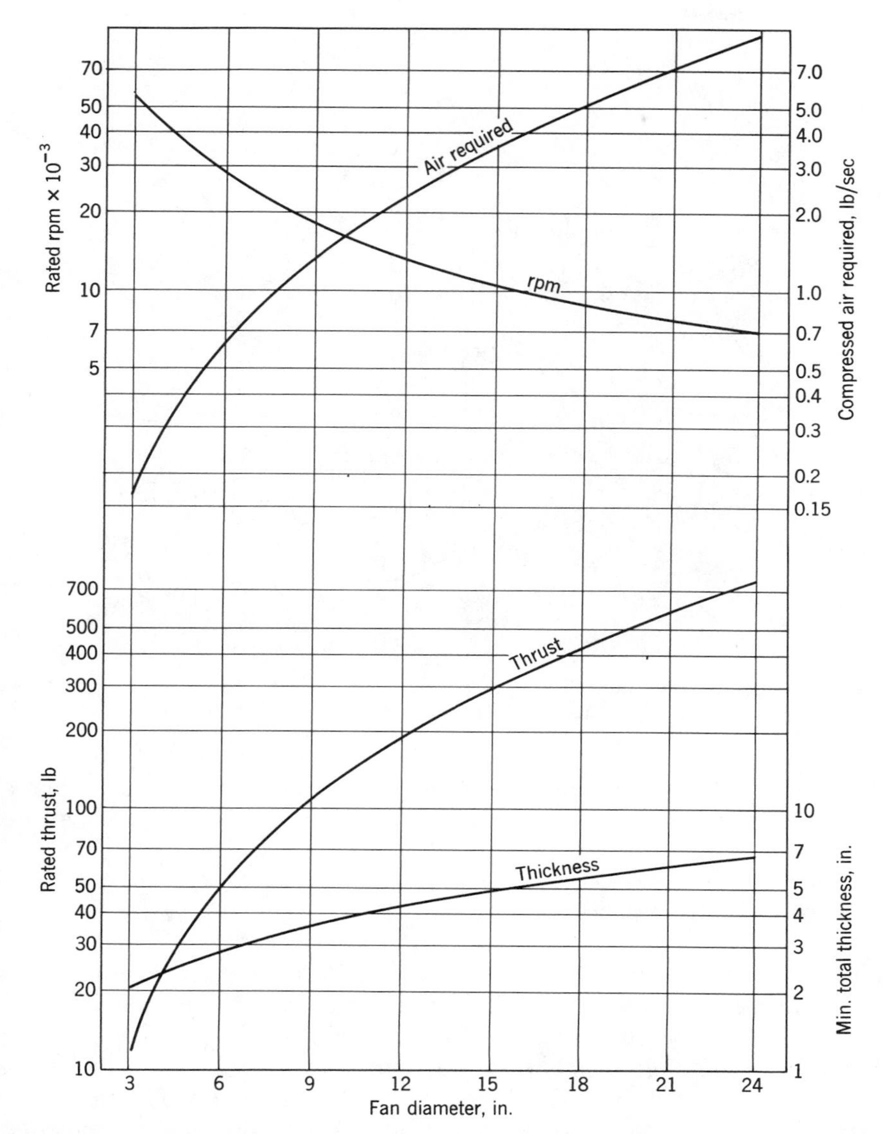

FIGURE 5.75. Typical performance curves for a reaction-type jet fan. Thrust and rpm performance as shown above can be obtained from *impulse-type jet fans* of smaller thickness, lower weight, and somewhat higher air consumption. (Courtesy Tech. Development, Inc.)

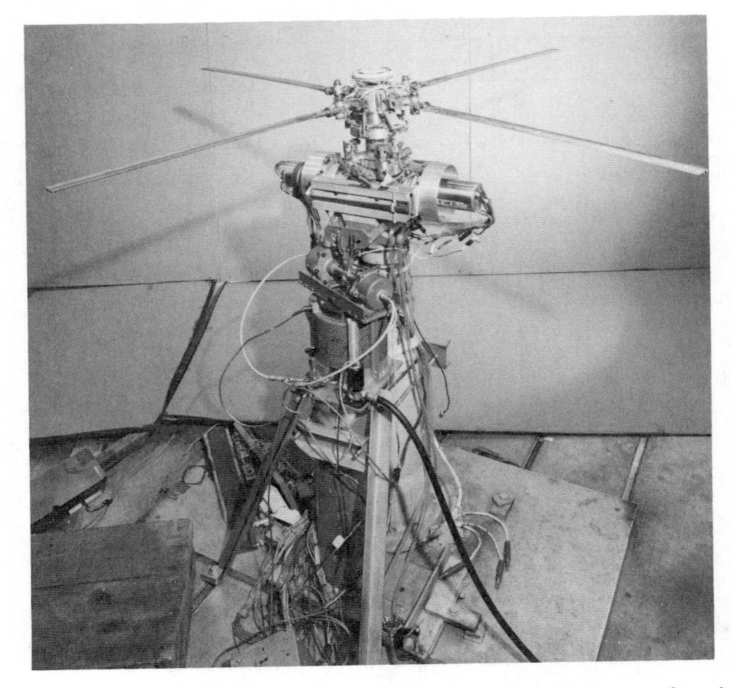

FIGURE 5.76. Four-bladed rotor model showing complexity of this type of model. (Photograph courtesy of U.S. Army Aeromechanics Laboratory, AVRADCOM.)

cies from the rotor are multiples of the rotor speed. For existing external balances it is necessary to avoid operating in regions of balance-rotor resonance.

Tare and Interference

Tare and interference measurements for V/STOL models are the same as for conventional models with the following exceptions. If the model uses compressed air, the pressure and mass flow tares must be evaluated (see Section 5.13). The second problem is with helicopter rotors, because it is difficult to use an image system from the ceiling. Generally, for rotors only the tare of the model supports themselves is evaluated. This method neglects the effect of the rotor wake on the support system. The tares are taken with the rotor blades removed, and the forces, moments, and driving torque are recorded at the various air speeds. If the rotor is mounted to a fuselage, the tares and interference of the fuselage is handled the same as any model.

Model Sizing

The determination of model size for a given tunnel for a V/STOL model that must be tested in the transition flight range is complicated. As discussed in

Section 6.33 there is a lower speed test limit that is a function of the model area and tunnel cross-sectional area for models with distributed lift. This limit requires a small model in a big tunnel. A small model at low velocity means low Reynolds numbers. If care is not taken, the Reynolds number may get in the range of 200,000 or less. At this low Reynolds number the aerodynamic properties of conventional airfoils may be quite different than at $RN = 6 \times 10^6$ (see Section 8.4). It should be noted that the full-scale aircraft also may be operating at low Reynolds numbers in transition. Because of the Reynolds number and difficulty in model construction, most of the tunnels built in the 1960s for V/STOL are large.

There is no known easy way out of the dilemma of small models relative to tunnel size. If the model is not to be tested toward the hover end of transition, the size can be increased. But for models with large downwash angles and large wing span to tunnel width ratios, the distribution of the tunnel interference may become nonuniform. Heyson in Ref. 6.43 has a large number of charts for tunnels of different geometry that can be used to estimate the wall corrections as applied to model design. There is the additional risk of building a large model that is not planned to be used near hover. Since models are expensive, sooner or later the model will be pushed closer to hover, ignoring the risk of nonvalid data, which will be followed by requests to the wind tunnel to both explain the abnormal data and to correct it.

The Rotor Model

The design of a model rotor presents some difficulties not encountered with the usual wind tunnel model of an airplane. The hub and hinge design and construction can usually be worked out in a satisfactory manner, but some inherent difficulties arise with the rotor blade representation. For one thing, it is common practice in rotor design to have the blade statically balanced about its quarter-chord line. Such a balance rules out the homogeneous blade and requires either a built-up blade or a solid wood one with a metal leading edge. For most model sizes the built-up blade is not practical, both because of the small size of the skin, ribs, etc., and because of the exaggerated effect of the skin wrinkles due to the scale of the model. The wood blade works well, however, and the metal leading edge is convenient to use as a tie-in to the metal hub. Mass balances for achieving static balance may be built into each blade tip, with, of course, a secure locking device.

However, the actual blade twists during flight, and when it is possible, usually when larger models are employed, a built-up model will be used—especially designed so that its dynamic characteristics match those of the full-scale craft, and realistic aeroelastic deformations and vibratory stresses are obtained.

The performance of a rotor is helped aerodynamically by root cutaway, inverse taper, and twist, and the model designer may be expected to produce such designs despite their difficulty.

The model should be equipped with adequate flat surfaces for leveling and angle measurements, some type of hinge lock to be used during balancing, and an ample supply of spare parts. Blade angles should be measured with their slack taken out in the direction of low pitch, since they will be so held during operation by the centrifugal torque that develops (see Ref. 5.19).

Hinged Rotor Operation

There are certain operational procedures that must be followed with rotors equipped with flapping hinges operating in a horizontal plane.

1. The rotor is brought up to operating speed with the tunnel off. As the rotor starts to rotate, there is very little centrifugal force on the blade. If the tunnel is running, the advancing blade will flap up to very large angles, owing to its large lift, and the blade will not track, leading to relatively large oscillating loads.
2. When the rotor is at the desired rpm, the tunnel is brought up to desired speed.
3. When shutting down at the end of a run, the tunnel is brought down to zero or a very low speed and then the rotor rpm is reduced.

When large blade angles are used, this procedure may not be possible owing to either limits on the power of the rotor drive or blade strength being marginal to carry the large static thrust or flow recirculation between the rotor and tunnel floor. For these conditions it may be possible to tilt the rotor forward, start the tunnel, and gradually increase the speed. The rotor will autorotate and, as the rpm builds up, rotor power may be added. Extreme care must be taken when operating in an autorotation mode to avoid excessive rotor rpm. When testing at large collective pitch angles with the rotor shaft tilted aft to the flow, the rotor can also enter an autorotating mode. This can be detected by a reduction in rotor torque, with the torque sign changing as the rotor begins to autorotate. Again, extreme care must be observed when the rotor is powered by an electric motor to avoid large increases in rpm due to the rotor driving the motor. The increase in rpm is usually very dramatic if such an increase occurs; control can be obtained by cutting the tunnel and pitching the rotor forward.

Flow Visualization for Rotors—Propellers

Small tufts can be attached to the blades; the centrifugal force does not seem to seriously affect the tufts. The minitufts in Section 3.10, which are fluorescent and viewed under ultraviolet light, are ideal for this application. When an event-marked signal (used to determine blade position) is available, it can be used to trigger the flash stopping the blade at the desired azimuth position.

A smoke generator can also be used to visualize the flow through the rotor–propeller disk. This is preferable to the use of balsa or other types of dust, which present monumental cleanup jobs. In these type flow visualizations the smoke–dust is often illuminated by a strong light entering the test section through a slit (Fig. 3.28).

REFERENCES

5.1. A. R. Sinclair and A. Warner Robins, A Method for the Determination of the Time Lag in Pressure-Measuring Systems Incorporating Capillaries, *TN* 2793, 1952.

5.2. Time Lags Due to Compressible-Poiseuille Flow Resistance in Pressure-Measuring Systems, *NOL Memo* 10,677, 1950.

5.3. B. Lockspeiser, Ventilation of 24 ft Wind Tunnel, *R&M* 1372, 1930.

5.4. O. Schrenk, A. Simple Approximation Method for Obtaining the Spanwise Lift Distribution, *TM* 948, 1940.

5.5. H. A. Pearson, Span Load Distribution for Tapered Wings with Partial-Span Flaps, *TR* 585, 1937.

5.6. H. J. Goett, Tunnel Procedure for Determining Critical Stability Procedure, *TR* 781, 1943.

5.7. E. H. Fromm, The Boeing Flight Simulation Chamber for Static Calibrations of Engine Simulators, Paper, 45th Meeting of the Supersonic Tunnel Association, 1976.

5.8. E. H. Fromm, Wind Tunnel Testing of Integrated Aerodynamic and Propulsion Effects, Paper, 48th Meeting of the Supersonic Tunnel Association, 1977.

5.9. M. Harper, The Propulsion Simulator Calibration Laboratory at Ames Research Center, AIAA Paper 82-0574, 1982.

5.10. J. R. Chambers, Overview of Stall/Spin Technology, Paper 80-1580, AIAA Atmospheric Flight Mechanics Conference, 1980.

5.11. K. J. Orlik-Ruckemann, Techniques for Dynamic Stability Testing in Wind Tunnels, *Dynamic Stability Parameters,* AGARD CR-235, May 1978, pp. 1-1–1-24.

5.12. W. Bihrle Jr. and B. Barnhart, Spin Prediction Techniques, *J. Aircraft,* **20,** 97–101, February, 1983.

5.13. W. Bihrle Jr. and J. S. Bowman Jr., Influence of Wing, Fuselage, and Tail Design on Rotational Flow Aerodynamics Beyond Maximum Lift, *J. Aircraft,* **18,** 920–925, November 1981.

5.14. R. R. Tumlinson, M. L. Holcomb, and V. D. Gregg, Spin Research on a Twin-Engine Aircraft, Paper 81-1667, AIAA Aircraft Systems and Technology Conference, 1981.

5.15. M. L. Holcomb, The Beech Model 77 "Skipper" Spin Program, Paper 79-1835, AIAA Aircraft Systems and Technology Meeting, 1979.

5.16. C. H. Wolowicz, J. S. Bowman Jr., and W. P. Gilbert, Similitude Requirements and Scaling Relationships as Applied to Model Testing, *NASA Technical Paper* 1435, August 1979.

5.17. R. L. Bisplinghoff, H. Ashley, and R. L. Halfman, *Aeroelasticity,* Addison-Wesley Publishing Co., Cambridge, MA, 1957.

5.18. J. F. Reed and W. H. Curry, A Wind Tunnel Investigation of the Supersonic Characteristics of Three Low-Fineness-Ratio Stores Internally Carried in a Simulated F-105 Bomb Bay, Sandia Corporation SCTM30-56-51, 1956.

5.19. W. C. Nelson, *Airplane Propeller Principles,* Wiley, New York, 1944, p. 60.

CHAPTER 6

Wind Tunnel Boundary Corrections

The conditions under which a model is tested in a wind tunnel are not the same as those in free air. There is no difference traceable to having the model still and the air moving instead of vice versa,* but the longitudinal static pressure gradient usually present in the test section and the open or closed jet boundaries in most cases produce extraneous forces that must be subtracted out. These may be summarized as follows:

The variation of static pressure along the test section produces a drag force known as "horizontal buoyancy." It is usually small and in the drag direction in closed test sections, and negligible in open jets, where in some cases it becomes thrust.

The presence of the lateral boundaries produces:

1. A lateral constraint to the flow pattern about a body, known as "solid blockage." In a closed wind tunnel solid blockage is the same as an increase of dynamic pressure, increasing all forces and moments at a given angle of attack. It is usually negligible with an open test section, since the airstream is then free to expand in a normal manner.

2. A lateral constraint to the flow pattern about the wake known as "wake blockage." This effect increases with an increase of wake size (drag), and in a closed test section increases the drag of the model. Wake blockage is usually negligible with an open test section, since the airstream is then free to expand in a normal manner.

* With the sole exception of ground board testing.

3. An alteration to the local angle of attack along the span. In a closed test section the angles of attack near the wingtips of a model with large span are increased excessively, making the tip stall start early. The effect of an open jet is just the opposite (tips unstalled), and in both cases the effect is diminished to the point of negligibility by keeping model span less than 0.8 tunnel width.

4. An alteration to the normal curvature of the flow about a wing so that the wing moment coefficient, wing lift, and angle of attack are increased in a closed wind tunnel, decreased with an open jet.

5. An alteration to the normal downwash so that the measured lift and drag are in error. The closed jet makes the lift too large and the drag too small at a given geometric angle of attack. An open jet has just the opposite effect.

6. An alteration to the normal downwash behind the wing so that the measured tailsetting and static stability are in error. In a closed jet the model has too much stability and an excessively high wake location, the opposite being noted in an open jet. This stability effect is large.

7. An alteration to the normal flow pattern so that hinge moments are too large in a closed test section, too small in an open one.

8. An alteration to the normal flow about an asymmetrically loaded wing such that the boundary effects become asymmetric and the observed rolling and yawing moments are in error.

9. An alteration to the flow about a thrusting propeller such that a given thrust occurs at a speed *lower* than it would in free air when a closed jet is employed. The effect is opposite when the propeller is braking. Wake effects such as these are negligible when a free jet is employed.

Fortunately, it is a rare test indeed when all the above corrections must be applied. We should note, however, that the additional effects resulting from the customary failings of wind tunnels—angularity of flow, local variations in velocity, tare, and interference—are extraneous to the basic wall corrections discussed in this chapter, and it is assumed that the errors due to these effects have already been removed before wall effects are considered. Methods governing their removal have been discussed in Chapters 3 and 4.

Since the manner in which the two- and three-dimensional walls affect the model and are simulated is quite different, they will be individually considered in the sections to come.

6.1. THE METHOD OF IMAGES

It is well known that the flow pattern about a wing may be closely simulated mathematically by replacing it with a vortex system composed of a lifting line vortex and a pair of trailing vortices. For detailed near-flow-field simulation the lifting line approximation can be replaced by a vortex lattice. Similarly, a solid body may be represented by a source–sink system, and a wake by a source. Thus the entire airplane or component may be simulated by "artificial" means to almost any degree of accuracy desired, depending, of course, on the complexity of calculations that can be tolerated. Fortunately, a very simple first-order setup usually suffices.

Students of fluid theory are well acquainted with the simulation of a boundary near a source, sink, doublet, or vortex by the addition of a second source, sink, doublet, or vortex "behind" the boundary to be represented. Solid boundaries are formed by the addition of an image system which produces a zero streamline matching the solid boundary. An open boundary requires an image system that produces a zero velocity potential line which matches the boundary in question. After the image system is established, its effect on the model is the same as that of the boundary it represents.

We may see how to make up an image system by considering the following case for a vortex in a closed rectangular tunnel; and we note herewith that it is usually necessary to satisfy the conditions only in the plane of the lifting line. A three-dimensional image system is necessary to get only the boundary induced upwash aft of the wing, the streamline curvature effect, or the corrections for a wing with a lot of sweepback.

Consider a single vortex A which we wish to contain within the solid walls 1 and 2 (Fig. 6.1). To simulate wall 1, we need a vortex B of sign opposite to that of A, and for wall 2, a vortex C of the same sign as B. Now, however,

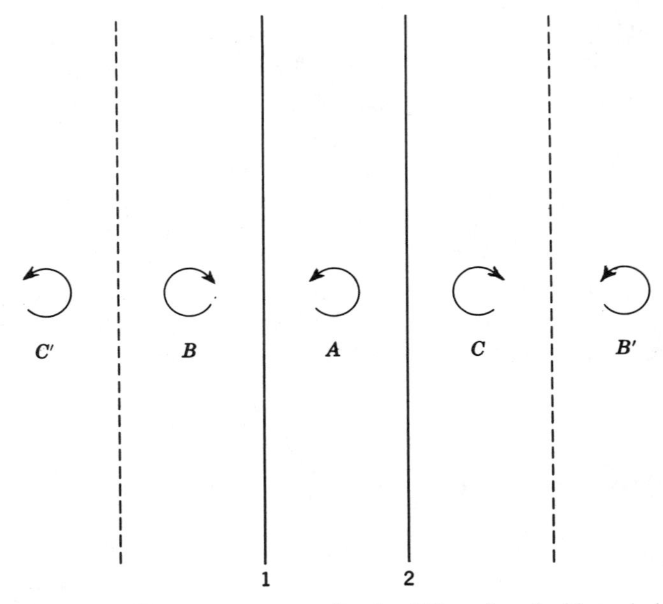

FIGURE 6.1. Vortex arrangement for simulation of vertical boundaries.

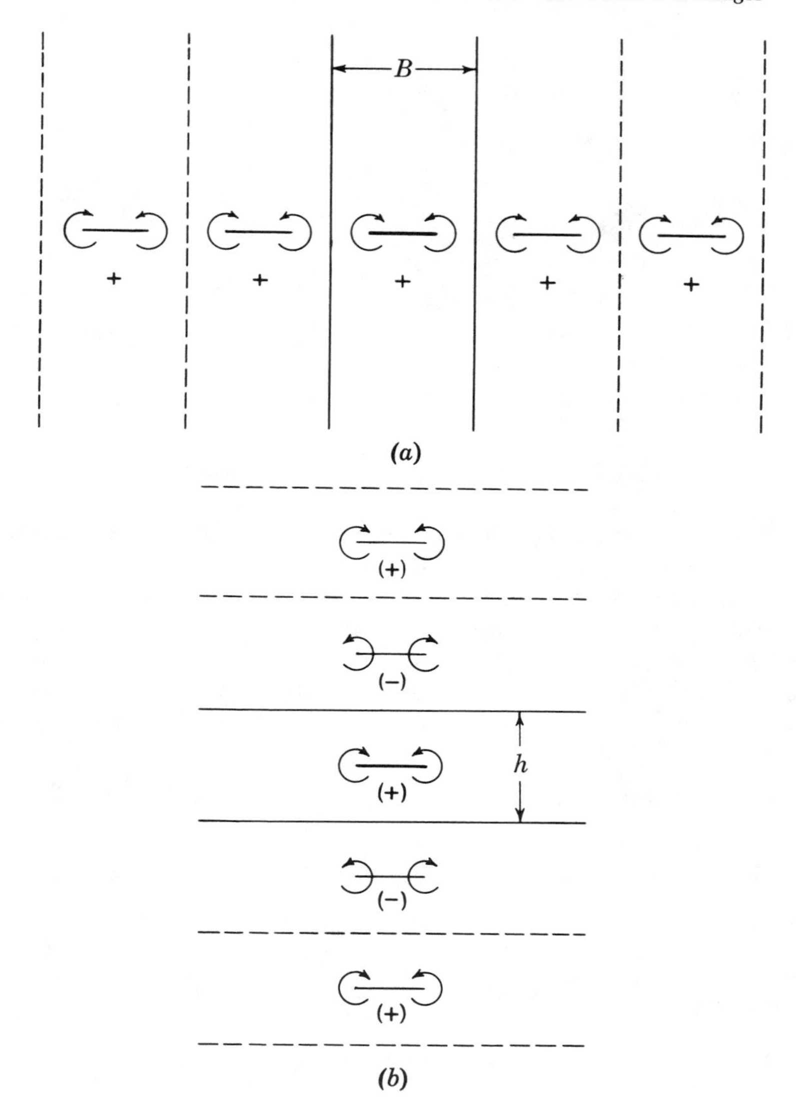

FIGURE 6.2. Vortex pair arrangement for simulation of (*a*) vertical boundaries and (*b*) horizontal boundaries.

vortex B needs a vortex B' to balance it from wall 2, and vortex C needs a vortex C' to balance it from wall 1, and so on out to infinity with vortices of alternating sign.

The containment of a wing or vortex pair similarly becomes that shown in Fig. 6.2*a* for vertical walls and in Fig. 6.2*b* for horizontal walls.

The image system for a closed rectangular test section thus becomes a doubly infinite system of vortices. Figure 6.3 shows the image system

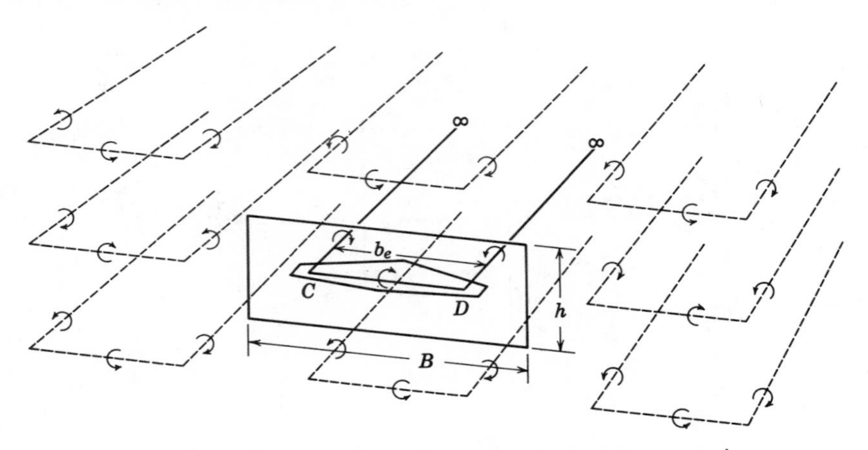

FIGURE 6.3. Image system for a closed rectangular test section.

needed for a wing in a closed rectangular tunnel when the three-dimensional quantities are required.

We may through elementary vortex theory and logic develop the form that corrections for boundary-induced upwash will take for an arbitrarily shaped test section. The only mathematical tools needed are the expression for the induced velocity w due to a vortex of strength Γ at a distance r. This is for a semiinfinite vortex, that is, starting at lifting line and trailing to infinity in one direction.

$$w = \Gamma/4\pi r \tag{6.1}$$

The relation between lift and circulation for a uniformly loaded wing

$$\Gamma = (SV/2b)C_L \tag{6.2}$$

Combining the two we get

$$w = (SV/8\pi rb)C_L \tag{6.3}$$

Now r represents the vortex spacing in the image system, which we may express as some constant times a tunnel dimension, say the tunnel height h, and the model wing span may be expressed in terms of the tunnel width B. The induced angle at the centerline of the test section is then

$$\Delta(\Delta\alpha_i) = \frac{w}{V} = \frac{S}{8\pi k(b/B)(hB)} C_L$$

for any one image, and, summing the whole field, setting $B/8\pi kb = \delta$, and noting that hB is the test-section area C, we have

$$\Delta\alpha_i = \delta(S/C)C_{LW}(57.3) \tag{6.4}$$

for the complete system. This equation is in degrees, the subscript W denotes that the correction is based on the wing lift.

It develops that δ is completely determined by the shape of the test section, the size of the model relative to the test section, the type of spanwise load distribution over the model, and whether or not the model is on the centerline of the test section. Equation (6.4) is hence a general form useful for all wings and test sections as long as the wing is small (less than $0.8B$) relative to the test section so that the upwash at the tunnel centerline may be taken as the average upwash.

Since the induced drag coefficient may be written as

$$C_{Di} = \alpha_i C_L$$

where α_i = induced angle, the change in induced drag caused by the boundary induced upwash becomes

$$\Delta C_{Di} = \Delta\alpha_i C_{LW} = \delta(S/C)C_{LW}^2 \tag{6.5}$$

Equation (6.5) is also a general form. The manner in which the downwash affects larger models, and how it must be handled for the special cases of asymmetrical loadings, is covered in later pages.

We will turn now to image systems and other corrections for two-dimensional testing. Later, the corrections for three-dimensional tests will be covered in detail.

6.2. WALL CORRECTIONS FOR TWO-DIMENSIONAL TESTING

In order to study effects primarily concerned with airfoil sections, it is customary to build models of constant chord which completely span the test section from wall to wall.* The trailing vortices are then practically eliminated, and the image system for a *small* wing consists of a vertical row of vortices (having alternately plus and minus signs) above and below the model. Usually, however, when two-dimensional tests are made, the models are made of large chord to obtain the highest Reynolds number possible, and the wing must be represented by a distribution of vortices instead of a single one. The effect of the floor and ceiling of the tunnel is to restrain the naturally free air curvature of the flow so that the model acts like one with extra camber (Section 6.6).

* There is no occasion in this type of test to have the model off the tunnel centerline, and the corrections mentioned in this section will cover only the symmetric cases.

The effects of the walls on the model thickness and wake are subject to solid and wake blockage, and buoyancy if the tunnel has a longitudinal static pressure gradient. These effects will be considered separately.

The wall corrections for two-dimensional testing have been discussed by Allen and Vincenti in Ref. 6.1, and it seems logical to follow their treatment in general. Since the trailing vortices that escape in the boundary layer are quite weak, no downwash corrections are needed.

Often the end plates in the wall are equipped with various schemes to remove or thin the boundary layer and to avoid separation for models with high lift flaps. Terminating the flaps short of the wall will produce a pair of shed vortices and destroys the concept of a two-dimensional wing with a uniform span load.

6.3. BUOYANCY (TWO DIMENSIONS)

Almost all wind tunnels with closed throats have a variation in static pressure along the axis of the test section resulting from the thickening of the boundary layer as it progresses toward the exit cone and to the resultant effective diminution of the jet area. Some tunnels have slightly expanding test sections to minimize this effect. It follows that the pressure is usually progressively more negative as the exit cone is approached, and there is hence a tendency for the model to be "drawn" downstream.

Glauert finds that the magnitude of the gradient may be expressed as a nondimensional factor k defined by

$$\frac{dp}{dl} = -k\frac{(\rho/2)V^2}{B}$$

where l = jet length, ft; p = pressure, lb/sq. ft; B = jet width, ft. An ideal tool for this measurement is a long static tube (see Section 3.2). The factor k is from 0.016 to 0.040 for a closed square jet of width B, but should be experimentally measured for a given tunnel.

The amount of "horizontal buoyancy" is usually insignificant for wings, but for fuselages and nacelles it is larger and becomes important. Corrections may be calculated as follows:

Suppose that the static pressure variation along a jet is as shown in Fig. 6.4 and that the model to be tested has the cross-section area S as shown in Fig. 6.5. It will be seen that the variation of static pressure from, say, station 2 to station 3 is $p_2 - p_3$ and that this pressure differential acts on the average area $(S_2 + S_3)/2$. The resulting force for that segment of the fuselage is therefore

$$\Delta D_B = (p_2 - p_3)\left(\frac{S_2 + S_3}{2}\right)$$

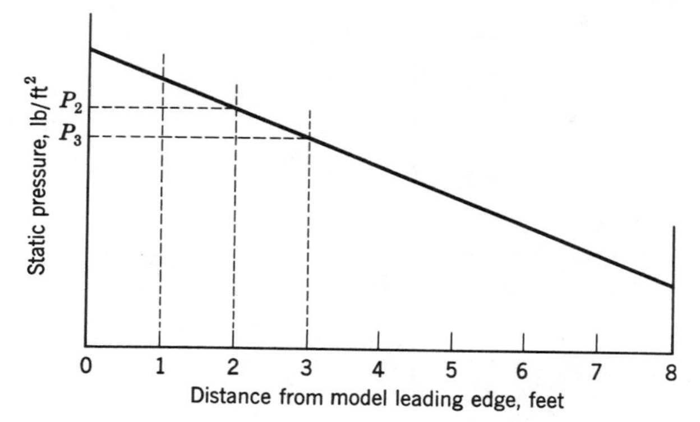

FIGURE 6.4. Static pressure gradient.

This equation is most simply solved by plotting local static pressure against body section area, the buoyancy then becoming the area under the curve.*

For the case where the longitudinal static pressure gradient is a straight line the equation becomes

$$\Delta D_B = -\Sigma\ S_x(dp/dl)\ dl$$

where S_x = fuselage cross-section area at station x; l = distance from fuselage nose; dp/dl = slope of longitudinal static pressure curve.

Since $\Sigma\ S_x\ dl$ = body volume, we have

$$\Delta D_B = -(dp/dl)\ \text{(body volume)} \tag{6.6}$$

* Or by plotting the local static pressure coefficient against body section area divided by wing area to get the buoyancy drag coefficient directly.

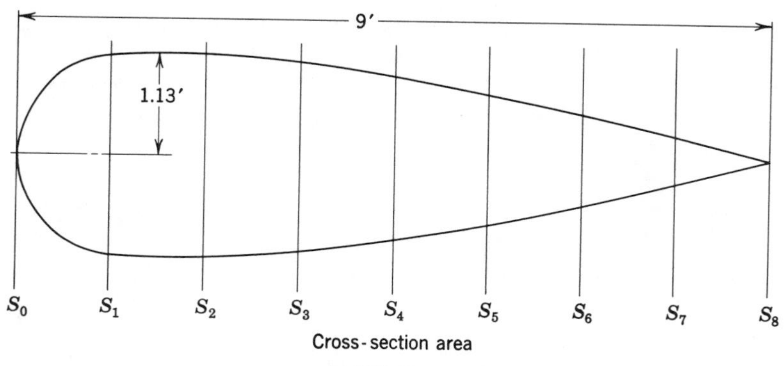

FIGURE 6.5

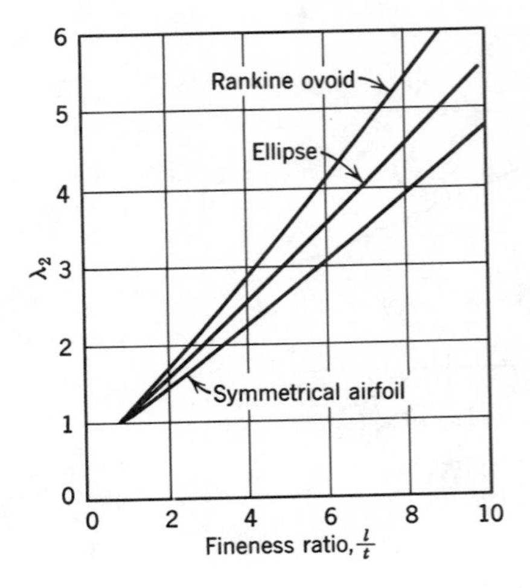

FIGURE 6.6

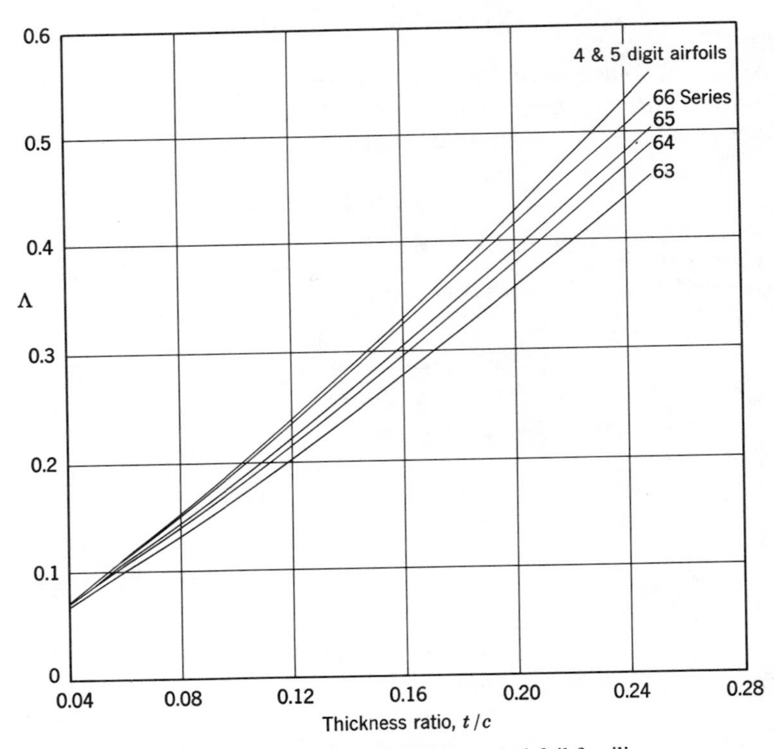

FIGURE 6.7. Values of Λ for several airfoil families.

Now the existence of a falling static pressure gradient implies that the test section is getting effectively smaller; in other words, the streamlines are being squeezed by the contracting tube. Adding the squeezing effect to the pressure-gradient effect, Glauert (Ref. 6.2) found that the total drag increment (for a two-dimensional body) is

$$\Delta D_B = -\frac{\pi}{2}\lambda_2 t^2 \frac{dp}{dl} \text{ lb per ft of span}$$

where t = body thickness and λ_2 = body shape factor from Fig. 6.6.

Allen and Vincenti in Ref. 6.1 replace $\lambda_2 t^2$ by $\frac{1}{4}\Lambda c^2$ and hence get

$$\Delta D_B = -\frac{\pi}{8}\Lambda c^2 \frac{dp}{dl} = -\frac{6h^2}{\pi}\Lambda\sigma \frac{dp}{dl} \tag{6.7}$$

where h = tunnel height, c = model chord, and $\sigma = (\pi^2/48)(c/h)^2$; and

$$\Lambda = \frac{16}{\pi}\int_0^1 \frac{y}{c}\left[(1-P)\left(1+\frac{dy}{dx}\right)\right]^{\frac{1}{2}} d\frac{x}{c} \tag{6.8}$$

The terms x and y are the airfoil coordinates, c its chord, and P its no-camber, symmetrical, pressure distribution.

Values of Λ for a number of airfoils are in Fig. 6.7; more are available in Ref. 6.1, or by direct integration of the above equation. Reference 6.3 may be consulted if the integration is used. Application of the buoyancy correction is covered in example 6.1.

6.4. SOLID BLOCKAGE (TWO DIMENSIONS)

The presence of a model in the test section reduces the area through which the air must flow, and hence by continuity and Bernoulli's equation increases the velocity of the air as it flows over the model. This increase of velocity, which may be considered constant over the model for customary model sizes, is called "solid blockage." Its effect is a function of model thickness, thickness distribution, and model size, and is independent of the camber. The solid-blockage velocity increment at the model is much less (about one-fourth) than the increment one obtains from the direct area reduction, since it is the streamlines far away from the model that are most displaced. The average velocity in the lateral plane of the model *is* proportionately increased, of course.

To understand the mathematical approach, consider solid blockage for a right circular cylinder in a two-dimensional tunnel. The cylinder, which may be simulated by a doublet of strength $\mu = 2\pi V a^2$ (Ref. 6.4, p. 46), where a = cylinder radius, is "contained" by an infinite vertical series of doublets of the same strength as the one simulating the model. The axial velocity due to the first doublet (Ref. 6.4, p. 73) is

$$\Delta V = \mu/2\pi h^2$$

so that

$$\Delta V/V_u = a^2/h^2$$

where V_u is uncorrected velocity.

Since the velocity produced by a doublet varies inversely with the square of the distance from the doublet, the doubly infinite doublet series may be summed as

$$\epsilon_{sb} = \left(\frac{\Delta V}{V_u}\right)_{\text{total}} = 2\sum_{1}^{\infty}\frac{1}{n^2}\frac{a^2}{h^2}$$

$$\epsilon_{sb} = (\pi^2/3)(a^2/h^2)$$

It is seen that a large 2-ft-diameter cylinder in a tunnel 10 ft high would act as though the clear jet speed were increased by 3.3%.

Now, the blockage due to a given airfoil of thickness t may be represented as that due to an "equivalent" cylinder of diameter $t(\lambda_2)^{1/2}$, and with this approach the solid blockage for any two-dimensional body may be found from simple doublet summation. Glauert in Ref. 6.2 wrote the solid-blockage velocity increment as

$$\epsilon_{sb} = \frac{\pi^2}{3}\frac{\lambda_2}{4}\frac{t^2}{h^2} = 0.822\lambda_2\frac{t^2}{h^2} \tag{6.9}$$

Values of λ_2 may be found in Fig. 6.6. For an open jet the constant becomes -0.411.

Allen and Vincenti in Ref. 6.1 rewrite Eq. (6.9) by introducing σ as in Eq. (6.8), and $\Lambda = 4\lambda_2 t^2/c^2$. Their result is then

$$\epsilon_{sb} = \Lambda\sigma \tag{6.10}$$

where Λ and σ have the same values as in Eq. (6.7). The manner of using this increment will be held until a later time.

A simpler form* for the solid-blockage correction for two-dimensional tunnels has been given by Thom in Ref. 6.5. It has the merit of showing the parameters upon which the correction depends a little more clearly than Eq. (6.10). Thom's solid-blockage correction is

$$\epsilon_{sb} = \frac{K_1\,(\text{model volume})}{C^{3/2}} \tag{6.11}$$

* Corrections for very large models are given in Ref. 6.6 and Section 6.11.

where $K_1 = 0.74$ for a wing spanning the tunnel breadth and 0.52 for one spanning the tunnel height. (A good approximation for airfoil model volume is 0.7 × model thickness × model chord × model span.)

The term C above is the tunnel test section area, which, if a little greater accuracy is desired, may be properly taken as the geometric area less the boundary layer displacement thickness taken around the perimeter. Usually, the approximation that the displacement thickness is one-sixth of the boundary layer thickness works well, since it is inevitable that the boundary layer be turbulent. (For the laminar case, as a matter of interest, one-third would be a good approximation for the displacement thickness.) It is not possible to give an approximate value for the boundary layer thickness in a wind tunnel, since it varies with roughness, cracks, leaks, Reynolds number, and Mach number. Thus, the thickness will have to be measured using a boundary layer mouse, hot wire, total probe, etc. (see Sections 3.5, 3.6, and 3.7).

The boundary layer displacement thickness, when it is desired, may be figured from

$$\delta^* = \int_0^Y \frac{u}{V_0}\, dy$$

where u = local velocity in boundary layer at a height y above the surface, Y = boundary layer thickness, and V_0 = freestream velocity.

Several wind tunnels of the 7 ft by 10 ft general size seem to have boundary layer displacement thicknesses of about $\frac{1}{2}$–$\frac{3}{4}$ in.

A fundamental source of error in the above solid-blockage method is the simulation of the body by a doublet or a doublet system. This error may be circumvented if the pressure at the tunnel wall is measured model in and model out, the resultant velocity increment computed, and the image system theory used to compute the ratio between blockage at the wall and at the tunnel centerline. For many tunnels the difficulty is in having walls free from glass windows, doors, etc., so that valid pressure measurements can be obtained. For a two-dimensional tunnel the velocity increment at the tunnel ceiling or floor is three times that at the tunnel centerline. See Section 6.11 for use of this method for corrections due to very large blockage.

6.5. WAKE BLOCKAGE (TWO DIMENSIONS)

Any real body without suction-type boundary layer control will have a wake behind it, and this wake will have a mean velocity lower than the freestream. According to the law of continuity, the velocity outside the wake in a closed tunnel must be higher than freestream in order that a constant volume of fluid may pass through the test section (Fig. 6.8). The higher velocity in the

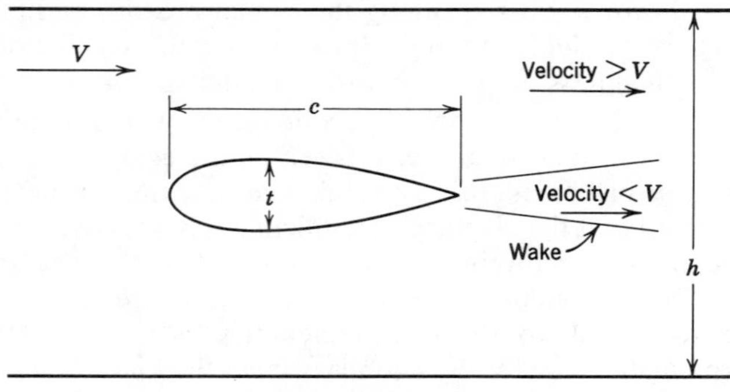

FIGURE 6.8

mainstream has, by Bernoulli's principle, a lowered pressure, and this lowered pressure, arising as the boundary layer (which later becomes the wake) grows on the model, puts the model in a pressure gradient and results in a velocity increment at the model.

To compute this wake effect we must first mathematically simulate the wake and the tunnel boundaries. The wake simulation is fairly simple. In the two-dimensional case a line source at the wing trailing edge emitting, say "blue" fluid will result in a "blue" region similar to a wake. Since the only drag existent is represented by this wake, the proper quantity Q to be emitted may be determined by

$$D = \rho Q V$$

In order to preserve continuity, a sink of the same absolute strength should be added far downstream.

The simulated wake may be contained within the floor and ceiling by an infinite vertical row of source–sink combinations. The image sources produce no axial velocity at the model, but the image sinks will induce a horizontal velocity in the amount

$$\Delta V = \frac{Q/h}{2}$$

where h = spacing between sources.

The factor $\frac{1}{2}$ arises since half of the sink effect will be upstream and the other half downstream. Thus, an incremental velocity is produced at the model by the walls which should be added to the tunnel-clear results to allow for "wake blockage." A useful form of the above statement is

$$\epsilon_{\text{wb}} = \frac{\Delta V}{V_u} = \tau c_{du}$$

where
$$\tau = \frac{c/h}{4}$$

Thom's paper yields the same relation for two-dimensional wake blockage.

Maskell (Section 6.11) has examined the effect of the flow outside the wake and how its higher speed results in a reduced pressure over the rearward portion of the model. For the two-dimensional case this equals the wake image effect and he suggests that the correction be

$$\epsilon_{\text{wb}} = \frac{\Delta V}{V_u} = \frac{c/h}{2} c_{du} \tag{6.12}$$

The wake gradient effect, from Eq. (67) of Ref. 6.1 is

$$\Delta C_{d\,\text{wb}} = \Lambda\sigma \tag{6.13}$$

and is usually quite small.

Wake blockage may be neglected for the rare case of a two-dimensional test section with open top and bottom.

6.6. STREAMLINE CURVATURE (TWO DIMENSIONS)

The presence of ceiling and floor prevents the normal curvature of the free air that occurs about any lifting body, and—relative to the straightened flow—the body appears to have more camber (around 1% for customary sizes) than it actually has. Accordingly, the airfoil in a closed wind tunnel has too much lift and moment about the quarter chord* at a given angle of attack, and, indeed, the angle of attack is too large as well. This effect is not limited to cambered airfoils, since, using the vortex analogy, any lifting body produces a general curvature in the airstream.

We may gain an insight into the streamline curvature effect, and calculate values as well, by assuming that the airfoil in question is small and may be approximated by a single vortex at its quarter-chord point. The image system necessary to contain this vortex between floor and ceiling consists of a vertical row of vortices above and below the real vortex. The image system extends to infinity both above and below and has alternating signs according to the logic of Section 6.1. Let us start by considering the first image pair as shown in Fig. 6.9. It is apparent that they induce no horizontal velocity since the horizontal components cancel, but, as will also be seen, the vertical components add.

* The moment about the half chord is independent of the camber and has no curvature effect.

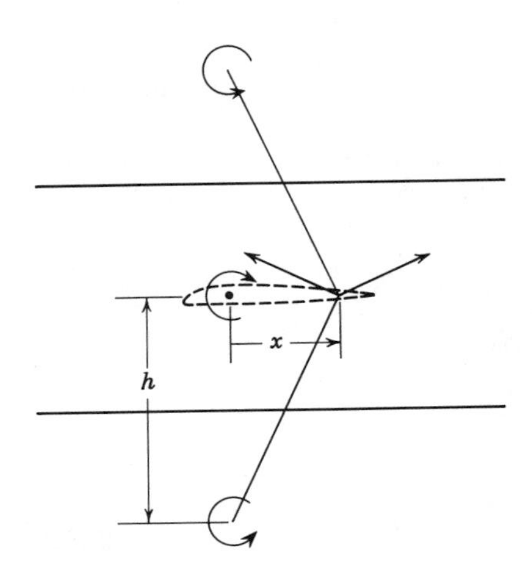

FIGURE 6.9

From simple vortex theory, the vertical velocity at a distance x from the lifting line will be

$$w_v = \frac{\Gamma}{2\pi} \frac{1}{\sqrt{h^2 + x^2}} \frac{x}{\sqrt{h^2 + x^2}} = \frac{\Gamma}{2\pi} \frac{x}{h^2 + x^2}$$

Substitution of reasonable values for x and h into the above equation reveals that the boundary-induced upwash *angle* varies almost linearly along the chord, and hence the stream curvature is essentially circular.

The chordwise load for an airfoil with circular camber may be considered to be a flat plate loading plus an elliptically shaped loading.* The magnitude of the flat plate load is determined from the product of the slope of the lift curve (2π per radian) and the boundary-induced increase in the angle of the tangent at the half-chord point because for circular camber the curve at this point is parallel to the line connecting the ends of the camberline. The load is properly computed as an angle of attack correction.

The elliptical loading is determined by the product of the slope of the lift curve and the angular difference between the zero lift line (i.e., the slope of the curve at the three-quarter chord point) and the chord line (the angle at the half chord). The lift, pitching-moment, and hinge-moment corrections are due to this elliptic component of the load.

* See p. 110 of Ref. 6.4.

Considering the flat plate loading first, we find that the upwash induced at the half chord by the two images closest to the real airfoil is

$$w_v = 2\frac{\Gamma}{2\pi}\frac{c/4}{h^2 + (c/4)^2}$$

Since $\Gamma = cc_lV/2$, the angular correction needed for the nearest images becomes

$$\Delta\alpha = \frac{w_v}{V} = \frac{1}{8\pi}\frac{c^2}{h^2 + (c/4)^2}c_l$$

Assuming that $(c/4)^2$ is small compared to h^2, and again using

$$\sigma = \frac{\pi^2}{48}\left(\frac{c}{h}\right)^2$$

we get

$$\Delta\alpha = \left(\frac{6\sigma}{\pi^2}\right)c_l$$

The second pair of vortices being twice as far away will be roughly one-fourth as effective, and the third pair one-ninth, so that for the images above and below the real wing we have

$$\begin{aligned}\Delta\alpha_{sc} &= \frac{6\sigma}{\pi^3}\left(1 - \tfrac{1}{4} + \tfrac{1}{9} - \tfrac{1}{16}\cdots\right)c_l \\ &= \frac{6}{\pi^3}\frac{\pi^2}{12}\sigma c_l = \frac{1}{2\pi}\sigma c_l\end{aligned}$$

since the alternating series shown above equals $\pi^2/12$ when summed to infinity. The additive lift correction is

$$\begin{aligned}\Delta c_{l\,sc} &= -2\pi(\tfrac{1}{2}\pi)\sigma c_l \\ &= -\sigma c_l\end{aligned} \tag{6.14}$$

and the additive moment correction is

$$\Delta c_{m\frac{1}{4}\,sc} = \frac{-\sigma}{4}\Delta c_{l\,sc} \tag{6.15}$$

Allen and Vincenti in Ref. 6.1 spread the vorticity out along the airfoil chord instead of concentrating it at the quarter chord. The lift and moment values of the simple analysis remain the same, but the angle of attack correction becomes

$$\Delta\alpha_{sc} = \frac{57.3\sigma}{2\pi}(c_{lu} + 4c_{m\frac{1}{4}u}) \tag{6.16}$$

If the chord is kept less than 0.7 tunnel height (and it usually is), wall effects on the *distribution* of lift may be neglected.

Since there is no drag in theoretical two-dimensional flow, there is no streamline curvature correction for drag.

6.7. SUMMARY OF TWO-DIMENSIONAL BOUNDARY CORRECTIONS

The complete low-speed wall effects for two-dimensional wind tunnel testing are summed below for ease in use. The data with the subscript u are uncorrected data based on clear stream q, with the exception of drag which must have the buoyancy due to a longitudinal static-pressure gradient removed before final correcting.

Velocity [from Eqs. (6.10) and (6.12)]:

$$V = V_u(1 + \epsilon) \tag{6.17}$$

where $\epsilon = \epsilon_{sb} + \epsilon_{wb}$.

Dynamic pressure [from expanding Eq. (6.17), and dropping higher-order terms] is

$$q = q_u(1 + 2\epsilon) \tag{6.18}$$

Reynolds number [from Eq. (6.17)] is

$$R = R_u(1 + \epsilon) \tag{6.19}$$

For α, c_l, and $c_{m\frac{1}{4}}$ [from Eqs. (6.14), (6.15), (6.16)] we have

$$\alpha = \alpha_u + \frac{57.3\sigma}{2\pi}(c_{lu} + 4c_{m\frac{1}{4}u}) \tag{6.20}$$

$$c_l = c_{lu}(1 - \sigma - 2\epsilon) \tag{6.21}$$

$$c_{m\frac{1}{4}} = c_{m\frac{1}{4}u}(1 - 2\epsilon) + \frac{\sigma c_l}{4} \tag{6.22}$$

For c_{d0} (from the dynamic pressure effect plus the wake gradient term) we get

$$c_{d0} = c_{d0u}(1 - 3\epsilon_{sb} - 2\epsilon_{wb}) \tag{6.23}$$

For the above,

$$\sigma = \frac{\pi^2}{48}\left(\frac{c}{h}\right)^2$$

The case of the free two-dimensional jet (floor and ceiling off, but wingtip walls in place) requires an additional factor that accounts for the downward deflection of the airstream as follows (both flow curvature and downwash deflection are included):

$$\Delta\alpha = -\left[\frac{1}{4}\left(\frac{c}{h}\right)c_l + \frac{\pi}{24}\left(\frac{c}{h}\right)^2(c_l)\right](57.3) \tag{6.24}$$

$$\Delta c_{d0} = -\frac{1}{4}\left(\frac{c}{h}\right)c_l^2 \tag{6.25}$$

$$\Delta c_{m\frac{1}{4}} = -\frac{\pi^2}{96}\left(\frac{c}{h}\right)^2 c_l \tag{6.26}$$

These values should be added to the observed data.

It is noted that a drag correction is present here, and further that these corrections are extremely large. Since the jet is free to expand, blockage corrections are not necessary.

The case where a wing completely spans a free jet without lateral restraining walls is not of much value in practice. Such a setup is rarely used except in small tunnels for preliminary tests. The spillage around the wingtip makes the wing area less effective, so that the coefficients as obtained should be increased. In this case the two-dimensional airfoil is in reality a very low aspect ratio wing. One test (unpublished) indicates that for $c/h = 0.2$, the measured lift was 18% low.

Example 6.1. Find the corrected data for the following two-dimensional test:

Model 65-209 airfoil; test speed 100 mph; test section 2 ft by 7 ft; model chord 2.5 ft, standard sea-level air; $\alpha_u = 4.0°$; lift 61.30 lb; drag 7.54 lb; moment about quarter chord -7.98 ft-lb; tunnel longitudinal static pressure gradient -0.02 lb/sq. ft/ft. Neglect area reduction by boundary layer.

From Fig. 6.7, $\Lambda = 0.163$, and from Eq. (6.7) and following, $\sigma = 0.0262$ and the buoyancy is

$$\Delta D_B = -\frac{6(7)^2}{\pi}(0.163)(0.0262)(-0.02)$$

$$= 0.008 \text{ lb}$$

The uncorrected coefficients are

$$c_{lu} = \frac{61.30}{25.58 \times 5.0} = 0.48$$

$$c_{d0u} = \frac{7.54 - 0.008}{25.58 \times 5.0} = 0.00589$$

$$c_{m\frac{1}{4}u} = \frac{-7.98}{25.58 \times 5.0 \times 2.5} = -0.025$$

The corrected coefficients are

$$\alpha = 4.0° + \frac{(57.3)(0.0262)}{2\pi}[0.48 + 4(-0.025)] = 4.09°$$

$$c_l = 0.48[1 - 0.0262 - 2(0.163)(0.0262) - 2(0.0893)(0.00589)]$$
$$= 0.472$$

$$c_{d0} = 0.00589[1 - 3(0.163)(0.0262) - 2(0.0893)(0.00589)]$$
$$= 0.00577$$

$$c_{m\frac{1}{4}} = (-0.025)[1 - 2(0.163)(0.0262) - 2(0.0893)(0.00589)]$$
$$+ (0.0262)(0.472/4)$$
$$= -0.0216$$

6.8. EXPERIMENTAL VERIFICATION OF TWO-DIMENSIONAL WALL CORRECTIONS

By testing models of several sizes at the same Reynolds number, data were obtained that have yielded an excellent check on the wall corrections presented. These (from Ref. 6.1) are shown in Fig. 6.10, uncorrected, and Fig. 6.11, corrected. It is seen that the method given brings the data into agreement.

6.9. BUOYANCY (THREE DIMENSIONS)

The philosophy behind the buoyancy correction has been covered in Section 6.3. For the three-dimensional case the total correction (pressure gradient and streamline squeezing effect) has been given by Glauert in Ref. 6.2 as

$$\Delta D_B = -\frac{\pi}{4}\lambda_3 t^3 \frac{dp}{dl} \tag{6.27}$$

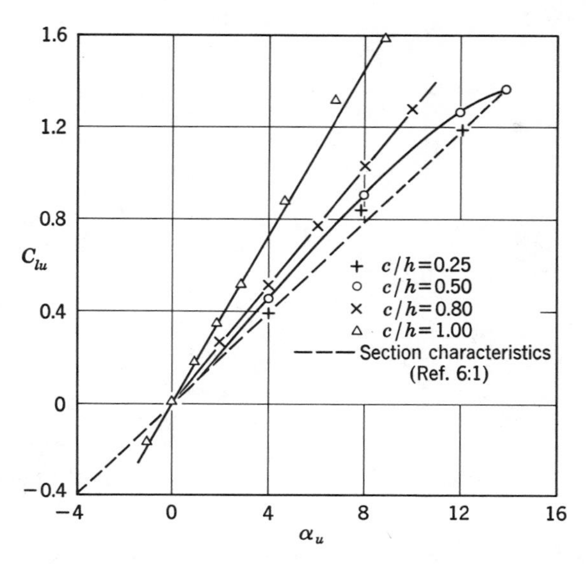

FIGURE 6.10. Lift characteristics for NACA 0012 airfoil section uncorrected for tunnel-wall interference.

where λ_3 = body-shape factor for three-dimensional bodies from Fig. 6.12 and t = body maximum thickness.

Example 6.2. Calculate the drag due to buoyancy for the model of Fig. 6.5 when tested in a closed round tunnel of 9 ft diameter at 100 mph. The static pressure gradient is −0.026 lb/sq. ft/ft.

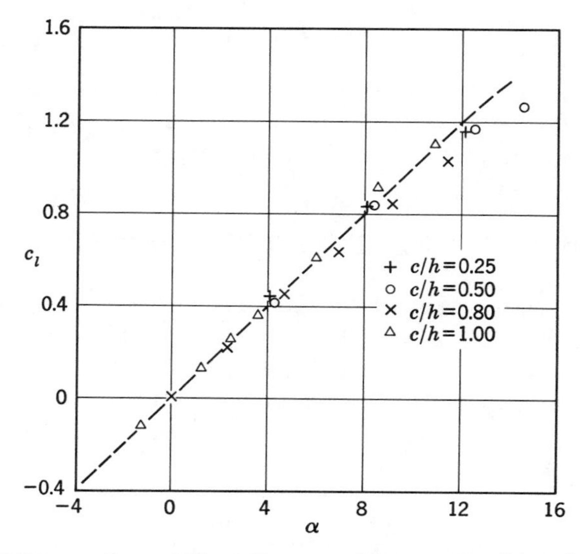

FIGURE 6.11. Data of Fig. 6.12 corrected for tunnel-wall interference.

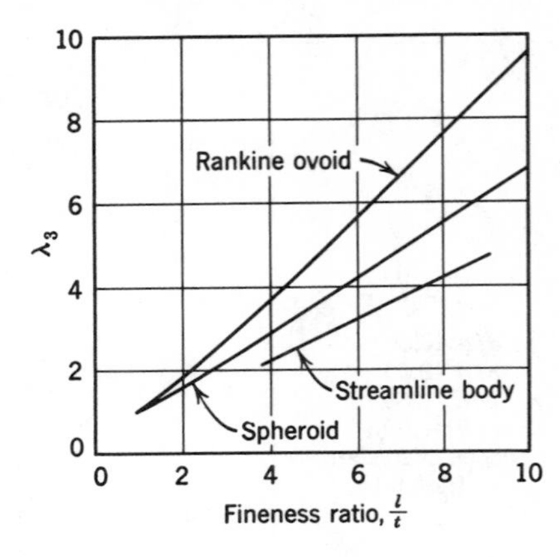

FIGURE 6.12. Values of λ_3.

1. The volume of the body is 16.62 ft^3. Neglecting the virtual mass

$$\Delta D_B = -\frac{dp}{dl}\text{(volume)} = (0.026)(16.62) = 0.43 \text{ lb}$$

2. From Fig. 6.12 for an $l/t = 3.98$, $\lambda_3 = 2.2$ (estimated), $t = 2.26$ ft

$$\Delta D_B = -\frac{\pi}{4}\lambda_3 t^3 \frac{dp}{dl}$$

$$= -\frac{\pi}{4}(2.2)(2.26)^3(-0.026)$$

$$= 0.519 \text{ lb}$$

As seen from Example 6.2, neglecting the virtual mass may change the buoyancy drag as much as 20%, but this in turn would be about 1% of the total model drag for models of ordinary dimensions.

6.10. SOLID BLOCKAGE (THREE DIMENSIONS)

The solid-blockage corrections for three-dimensional flow follow the same philosophy described in Section 6.4 for two dimensions. According to Herriot (Ref. 6.7), the body is again represented by a source–sink distribution, and contained in the tunnel walls by an infinite distribution of images.

Summing the effect of the images, we have for the solid-blockage velocity effect for a wing

$$\epsilon_{\text{sb}\,W} = \frac{\Delta V}{V_u} = \frac{K_1\tau_1\,\text{(wing volume)}}{C^{\frac{3}{2}}} \tag{6.28}$$

where K_1 is the body-shape factor from Fig. 6.13, and τ_1 is a factor depending on the tunnel test section shape, and model span to tunnel width ratio, from Fig. 6.14.

For bodies of revolution a similar approach results in

$$\epsilon_{\text{sb}\,B} = \frac{\Delta V}{V_u} = \frac{K_3\tau_1 \text{ (body volume)}}{C^{\frac{3}{2}}} \tag{6.29}$$

where K_3 is the body-shape factor from Fig. 6.13; τ_1 is a factor depending on the tunnel test section shape, and model span (*assumed zero*) to tunnel width ratio from Fig. 6.14.

Thom's short-form equation for solid blockage for a three-dimensional body is

$$\frac{\Delta V_{\text{sb}}}{V_u} = \epsilon_{\text{sb}} = \frac{K \text{ (model volume)}}{C^{\frac{3}{2}}} \tag{6.30}$$

where $K = 0.90$ for a three-dimensional wing and 0.96 for a body of revolution. (A good approximation for the volume of a streamline body of revolution is $0.45ld^2$, where l = length and d = maximum diameter.)

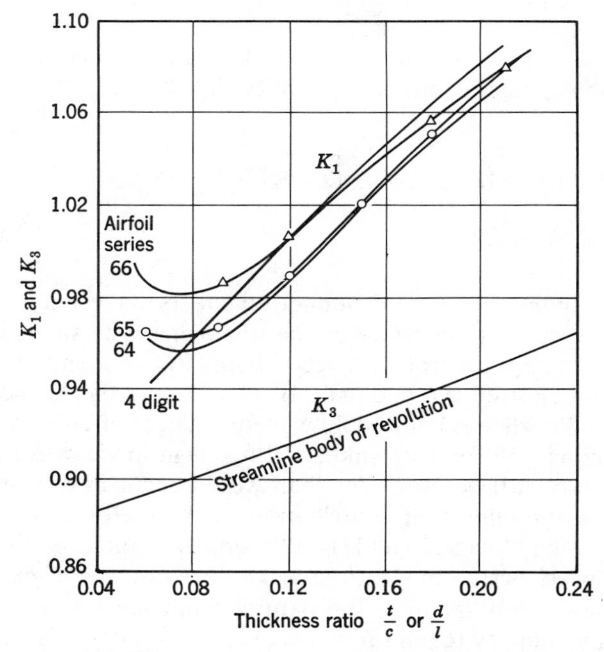

FIGURE 6.13. Values of K_1 and K_3 for a number of bodies.

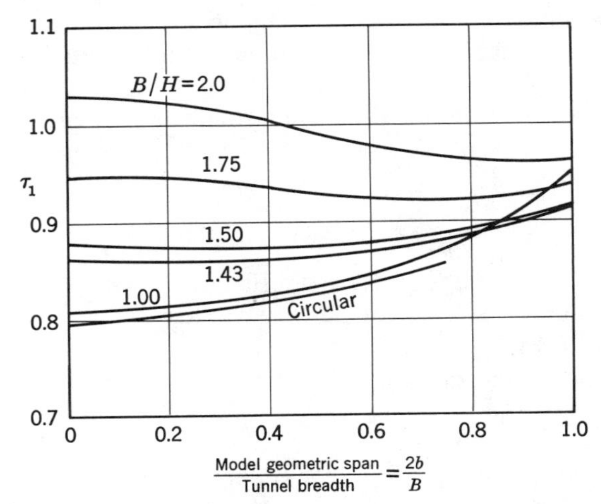

FIGURE 6.14. Values of τ_1 for a number of tunnel types. Use $b/B = 0$ for bodies of revolution.

Solid blockage for a wing–body combination is simply the sum of each component as determined from the above relations.

The velocity ratio method (see Section 6.4) yields the result that the velocity increment at the wall of a round closed wind tunnel is 2.2 times that at the tunnel centerline for bodies of revolution, and about 2.0 for typical airplane models.

For open jets the solid blockage may be taken as one-fourth the above values. Normally this results in a quantity that is negligible.

6.11. WAKE BLOCKAGE (THREE DIMENSIONS)

Maskell's Method

Although for many years wind tunnel engineers have been satisfied with wake-blockage corrections based on the single theory of simulating the wake by a source at the wing trailing edge (Thom, Ref. 6.5 and others) Maskell (Ref. 6.8) has reported the necessity of considering the momentum effects outside the wake when separated flow occurs. These effects are produced by lateral wall constraint on the wake, and result in lower wake pressure and hence lower model base pressures than would occur in free air.

There are three important contributions in Maskell's paper: The demonstration that wake blockage yields results similar to those on the same model in a higher-speed airstream (i.e., the wake does not vary significantly along or across the model); second, the natural tendency for the wake to tend toward axial symmetry (even for wings of $AR = 10$) permits a single correc-

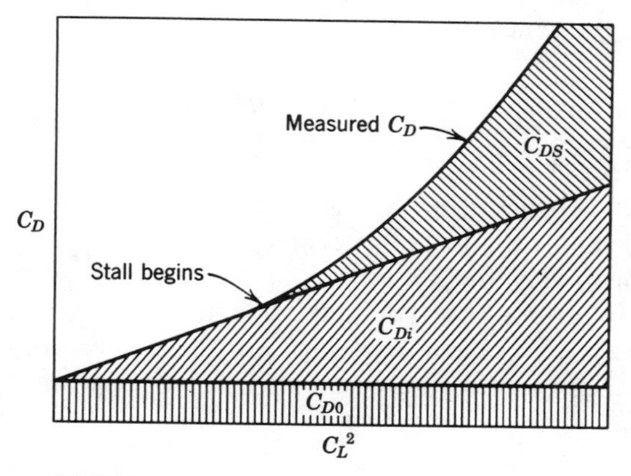

FIGURE 6.15. Drag analysis for a lifting wing.

tion for essentially all three-dimensional models (see Fig. 6.15). The complete wake blockage results for three-dimensional models are presented below.

The correction for wake blockage for *streamline flow* follows the logic of the two-dimensional case (Section 6.5) in that the wake is simulated by a source of strength $Q = D/\rho V$, which is matched for continuity by adding a downstream sink of the same strength. However for the three-dimensional case the image system consists of a doubly infinite source–sink system spaced a tunnel height apart vertically and a tunnel width apart horizontally. The axial velocity induced at the model by the image source system is again zero, and that due to the image sink system is

$$\Delta V = \frac{Q}{2Bh}$$

The incremental velocity is then

$$\epsilon_{\text{wb}} = \frac{\Delta V}{V_u} = \frac{S}{4C} C_{Du} \tag{6.31}$$

or

$$\frac{q_{\text{wb}}}{q_u} = 1 + \frac{S}{2C} C_{Du} \tag{6.32}$$

The increase of drag due to the pressure gradient may be subtracted by removing the wing wake pressure drag

$$\Delta C_{Dw} = \frac{K_1 \tau_1 \text{ (wing volume)}}{C^{\frac{3}{2}}} C_{Du} \tag{6.33}$$

and the body wake pressure drag

$$\Delta C_{DB} = \frac{K_3 \tau_1 \text{ (body volume)}}{C^{\frac{3}{2}}} C_{D0u} \tag{6.34}$$

The values of K_1, K_3, and τ_1 are again from Figs. 6.13 and 6.14. The values of Eqs. (6.33) and (6.34) are usually negligible.

The wake blockage for *separated flow* is as follows: Maskell through considerations mentioned above, added a term to account for the increased velocity outside the wake and its consequently lowered pressure. Dividing the total drag coefficient into a constant amount C_{D0}, one proportional to C_L^2, and one due to separated flow C_{Ds} (see Fig. 6.15) he obtains the total wake-blockage correction $\epsilon_{\text{wb}\,t}$ as

$$\epsilon_{\text{wb}\,t} = \frac{S}{4C} C_{D0} + \frac{5S}{4C}(C_{Du} - C_{Di} - C_{D0}) \tag{6.35}$$

$$\frac{q_c}{q_u} = 1 + \frac{S}{2C} C_{D0} + \frac{5S}{2C}(C_{Du} - C_{Di} - C_{D0}) \tag{6.36}$$

For angles below separated flow, the last term, which is equal to C_{Ds}, vanishes.

Reference 6.8 should be consulted for bluff models which have a substantial separated wake at $\alpha = 0$, for a correction based on base pressure then becomes necessary.

Figure 6.16 shows the variation of the "constant" from something over $\frac{5}{2}$ for high-aspect-ratio models on down to approximately 1.0 for two-dimensional models.

In Maskell's correction the basic problem is obtaining the values for C_{D0} and C_{Du}, two approaches will be outlined. The first approach is to plot C_{Lu}^2 versus C_{Du}. The minimum value of C_{Du} is picked as C_{D0} and the slope of the

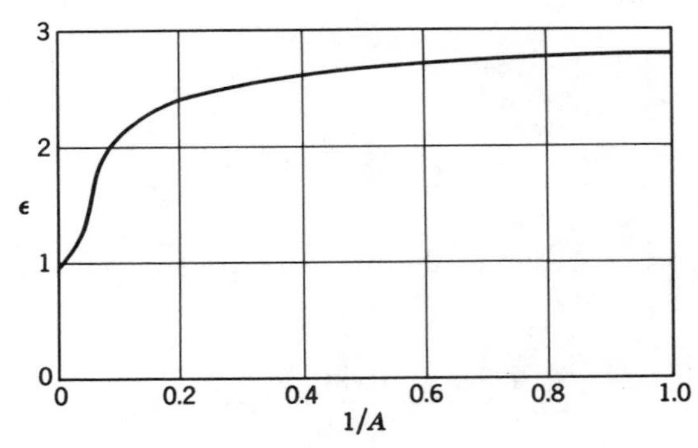

FIGURE 6.16. Variation of blockage factor with aspect ratio for nonlifting rectangular plates.

linear portion of the curve is used to calculate C_{Di}. For flap-down models the intercept of the linear portion of the C_{Lu}^2 versus C_{Du} when extended to the C_{Du} axis may have a negative value, this is accounted for in the following equations as ΔC_{Du} which is either zero or a negative number:

$$C_{Di} = C_{Lu}^2 \left[\frac{dC_{Du}}{dC_{Lu}^2}\right] + \Delta C_{Du} - C_{D0} \tag{6.37}$$

A second approach for C_{Di} and C_{D0} is based on Ref. 6.9 where Eq. (6.25) is

$$C_{Di} = \frac{C_L^2}{\pi AR} + \frac{K(\Delta C_L^2)}{\pi AR} \tag{6.38}$$

where C_L is the lift of the unflapped wing, ΔC_L is the lift increment due to flaps, and K is from Figs. 18, 19, and 20 in Ref. 6.9. It is a function of flap span, flap cutout, wing span, wing aspect ratio and the two-dimensional lift curve slope in radians. The values of K are based on the following assumptions:

1. Unswept wing with an elliptical load distribution or taper ratio of 2 : 1.
2. One set of flaps per semispan.
3. One cutout inboard of the flaps.
4. Ratio of local flap chord to local wing chord is constant along the wing span.
5. The effect of the flaps is considered to be the equivalent of a constant change in geometric incidence along the flap span resulting in a local lift increment due to the flap independent of the local wing incidence.

Figures 18–20 in Ref. 6.9 use a net flap span ratio that is defined for multiple flaps as

$$f_n = \sum \frac{\text{Flap spans}}{\text{Wing span}} \tag{6.39}$$

The overall flap span ratio is defined as

$$f_t = \frac{\text{Total flap span}}{\text{Wing span}} \tag{6.40}$$

The total flap span is from the outboard end of the outboard flap on the left wing to the same point on the right wing. The flap cut out ratio is

$$f_c = \frac{\text{Total flap cutout area}}{\text{Total wing area}} \tag{6.41}$$

These three ratios along with the wing aspect ratio and two-dimensional lift curve slope in radians are used with Figs. 18–20 to determine K. Then, two constants are defined:

$$A_1 = \frac{1}{\pi AR}, \qquad A_2 = \frac{K}{\pi AR} \tag{6.42a,b}$$

Thus, Eq. (25) (Ref. 6.9) becomes

$$C_{Di} = A_1 C_L^2 - A_2(\Delta C_L^2) \tag{6.43}$$

Note the following test values for C_D and C_L may or may not be corrected for tare and interference (Sections 6.30 and 4.17). C_{D0} is found by

$$C_{D0} = C_{D\,\alpha=0} - (A_1 + A_2)C_{L\,\alpha=0}^2 \tag{6.44}$$

The subscript $\alpha = 0$ means the values of lift and drag at $a = \psi = 0$ are used. It is assumed that there is no separated flow at these angles. The value of C_{D0} is defined as the $C_{D\,\alpha=0}$ minus the terms dependent on the wing lift plus the flap increment.

The induced drag is then

$$C_{Di} = A_1 C_{Lu}^2 + A_2(C_{L\,\alpha=0}^2) \tag{6.45}$$

This is the same lift increment as used in Eq. (25) from Ref. 6.9. Finally,

$$C_{DS} = C_{Du} - C_{D0} - C_{Di} \tag{6.46}$$

If C_{DS} is a negative number due to data scatter, it is set to zero.

This method allows Maskell's correction to be calculated in real time if desired. Care should be taken on models with leading edge flaps if they are separated at $\alpha = 0$.

There also is some controversy about the $\frac{5}{4}$ factor in the velocity correction equation. Maskell determined this factor from nonlifting flat plates perpendicular to the air stream. This then yields a separation bubble across the whole span of the plate. Wings of aspect ratio of 5 and greater do not initially stall across the whole span. Thus, Maskell's factor may be too large. Another approach to Maskell's correction is in Ref. 6.10.

Approximate Blockage Corrections

The total solid and wake blockage corrections may be summed according to

$$\epsilon_t = \epsilon_{sb} + \epsilon_{wb\,t} \tag{6.47}$$

When all is lost as far as finding blockage corrections for some unusual shape that needs to be tested in a tunnel, the authors suggest

$$\epsilon_t = \frac{1}{4}\frac{\text{Model frontal area}}{\text{Test section area}} \tag{6.48}$$

Blockage applies to everything in the test section, of course, and hence corrections to the free jet conditions must allow for the windshields and struts or other items necessarily in the test section during a test. If the image system method of evaluating tare and interference is used, the blockage contribution of the mounting system is automatically evaluated. That is, putting the image system in increases the model drag by the tare and interference *plus* the wake and solid blockage of the image windshields and support struts. When T and I are subtracted the windshield effects go with them.

On the other hand, when for some reason an image system is not to be used, ϵ_t should be taken as

$$\epsilon_t = \epsilon_{\text{model + struts}} + \epsilon_{\text{windshields}}$$

A maximum ratio of model frontal area to test section cross-sectional area of 7.5% should probably be used, unless errors of several percent can be accepted.

The windshield term will be a constant that can be evaluated with the help of Eqs. (6.28) and (6.31). For tests without yaw, the strut blockage may be included in the windshield term. An alternative method is to make a pitot-static survey with the windshields in, and use its values for "freestream," allowing for model blockage only in the work-up.

Blockage in open test sections is of opposite sign to and smaller than a closed tunnel.

Blockage Corrections Based on Measurements

Until the mid 1970s blockage corrections had been primarily based on simple potential flow solutions that yielded solid blockage based on geometric considerations. These corrections account only for the local flow acceleration around the model. In wind tunnel tests there is also a viscous wake for unseparated flow, and finally a separated wake that includes the viscous wake. The latter two blockages were more difficult to calculate. In 1957 Hensel (Ref. 6.11) developed a method based on measuring a wall pressure opposite the model and using doublets to model the flow. Maskell's method in 1965 (Ref. 6.8) is an empirical relation based on the measured drag, and a less empirical variant using base pressure. For rectangular plates Maskell demonstrated that the drag coefficient is related to the base pressure, which is a function of plate aspect ratio. The separaton of profile drag, both viscous and separated, from induced drag for complete models, or the selection of suitable base pressure, sometimes cause difficulty when attempting to use Maskell's method.

In 1974 Hackett and Boles (Ref. 6.12) showed that tunnel wall static pressures may be used to infer wake geometry and hence wake blockage. This was followed in 1975 by an extension involving a row of pressures along the center of the tunnel sidewall, which gave the axial distributions of both solid and wake blockage (Ref. 6.13 and 6.14). These "wall pressure signatures" were analyzed using chart look up methods which give a hands-on feel for the physics involved. In the interests of greater flexibility (shorter test sections, more complex "signatures") influence matrix methods were then introduced (Ref. 6.15), which also include the means for estimating tunnel induced upwash. The most recent application (Ref. 6.16) has considered tunnel interference for a jet in cross flow.

The "chart" method (Ref. 6.13) provides the best illustration of the approach. For a model on the tunnel centerline, pressures are measured along the sidewall from ahead of the model to a point downstream, yielding a pressure distribution similar to that shown in Fig. 6.17 with the velocity peak

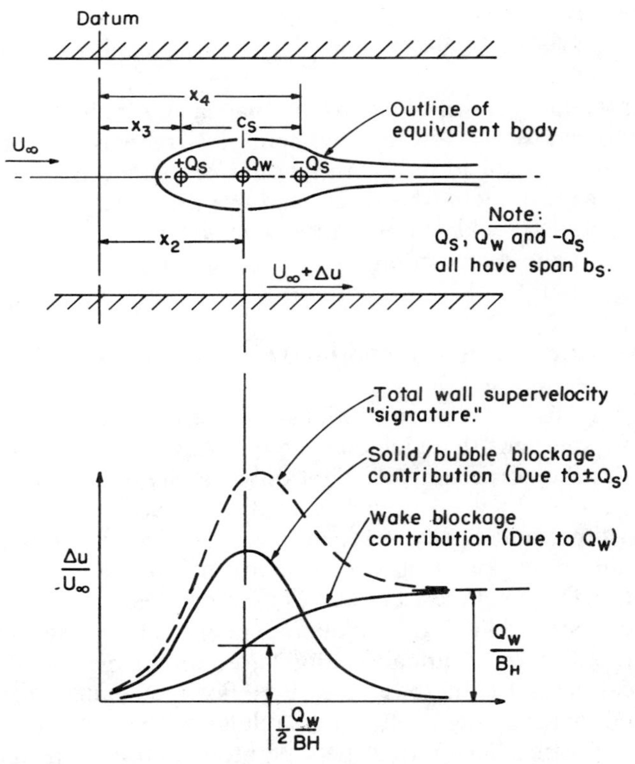

FIGURE 6.17. Nomenclature for Hackett's blockage correction.

just aft of the model. For analysis, the profile is divided into a part due to solid and bubble blockage (symmetric) and a part due to wake blockage (antisymmetric). The wake blockage source strength Q_w comes from the wake signature analysis. Initially the wake source is placed at a streamwise location X_2 coincident with the model midlength. The solid blockage parameters, Q_c, the source–sink strength and c_s, the spacing, are obtained from a chart look-up procedure (Refs. 6.13 and 6.14). When the wake blockage signature is measured in a sufficiently long test section, the velocity ratio goes from zero upstream of the model to a constant value downstream, determined by the wake size. To achieve this, the test-section length should be about 1.5 times its effective diameter with the model at 0.5 diameters from the nozzle. For shorter tunnels neither the upstream zero value nor the downstream constant value will be reached. However, the methods of Ref. 6.15 are able to handle this situation.

A least squares curve fit of a parabola is applied to the upper part of the symmetric curve. If the pressure peak does not match X_2 assumed initially, the wake is moved and another iteration is made. The use of a parabolic fit, and forcing the wake source to be midway between the solid source–sink pair, provides consistency of application. Once the wake source and the peak are aligned, parameters defining the symmetric, solid blockage curve are used with charts to obtain the source–sink spacing c_s and the strength Q_s. The three source strengths and their locations are used to build up an interference velocity $\Delta u/U$, or blockage "epsilon," by adding the three component parts on the centerline. Again, a chart is used for these calculations. Charts for a rectangular tunnel of $W/H = 1.43$ and a program listing for the method are given in Ref. 6.14.

As the previously described method was somewhat slow for "on-line," real-time calculations, an improved method was developed in Ref. 6.15. Beside the blockage correction from the source–sink model, an angle-of-attack correction using vortices was added. This new method also handles swept wings and jets in a cross flow (Ref. 6.16) and uses a skewed line singularity algorithm for sources, doublets, and horseshoe vortices. The method uses influence matrices at user-specified locations, for line-source elements and horseshoe vortices, and uses measured wall conditions to solve for their strengths. The source matrices are the sum of two other matrices corresponding to the direct influence of the line sources and a matching but opposite sign source situated far downstream to satisfy continuity. The sidewall pressures are still used for blockage corrections.

The ceiling and floor pressures at the same longitudinal distance in the tunnel are used to obtain the lift interference. The vortex-induced effects include vertical velocities at the sidewalls, which, in extreme high lift/large model cases, will affect wall pressures and hence appear to represent a blockage dependent on lift. For this reason, the lifting solution is done first and is available for an optional computation, which removes this spurious effect. The corrected sidewall data then reflect the true blockage.

The effect on the measured pressures of model offset from the tunnel center, sweep, and angle of attack are discussed in Ref. 6.15. To minimize computer-storage and run-time requirements the equations for singularity strength are solved for over-specified boundary conditions using a least-squares algorithm. This has the additional benefit of smoothing the curve fit through the tunnel wall data. These procedures are described in Ref. 6.15, as are the equations for the source and vortex strengths. The program is described along with its listing, and examples are given in the Appendices.

The matrix method of calculating the blockage correction is capable of being run on a computer in real time. For powered tests it is desirable to run at constant values of jet momentum coefficient, or tip speed ratio for rotors. The tunnel dynamic pressure with blockage corrections can be determined "on line" and adjusted to obtain constant normalized values.

In Ref. 6.14 the blockage corrections, together with Glauert's angle-of-attack corrections, were applied to four floor-mounted wings. These have wing area to tunnel cross section area ratios from 1.67% to 16.7%. The data agreement was quite good for lift and fairly good for induced drag. This same report also shows good correlation on sphere drag above and below the critical Reynolds number. The measured pressures in the tunnel were predicted and compared to measured results for Reynolds numbers above the critical value and good agreement was achieved past the major diameter. Similar correlation of measured pressures corrected for blockage was achieved for three geometrically similar idealized automobile models in the 16.25×23.25 ft Lockheed tunnel.

The ability to calculate blockage corrections and thereby determine corrected dynamic pressure in an on-line mode should be a boon to powered testing of V/STOL models as it eliminates the cross plotting to obtain constant tip speed ratio and jet momentum coefficient.

When testing powered lift models in the transition speed range, the minimum speed is limited by flow breakdown (see Section 6.33). This occurs when the powered wake impinges on the tunnel floor near the model.

6.12. STREAMLINE CURVATURE (THREE DIMENSIONS)

The corrections for streamline curvature for three-dimensional testing follow the same philosophy as those for the two-dimensional case (Section 6.6) in that they are concerned with the variation of the boundary-induced upwash along the chord. Once again the variation turns out to be essentially a linear increase in angle so that the streamline curvature effect may be treated as the loading on a circular arc airfoil. Similarly, the loading is treated as a flat-plate effect based on the flow angle change between the quarter and half chord, and an elliptic load based on the flow angle change between the half and three-quarter chord. But for the three-dimensional case the image system is vastly different from the simple system for two dimensions.

The three-dimensional image system is shown in Fig. 6.3. Basically it consists of the real wing with its bound vortex *CD* and trailing vortices $C\infty$ and $D\infty$. The vertical boundaries are simulated by the infinite system of horseshoe vortices and the horizontal boundaries by the infinite lateral system. Linking the two systems is the infinite diagonal system.

The effect of the doubly infinite image system at the lifting line of the real wing is the main boundary upwash effect, and it may be found in the familiar δ values for any particular condition. Here as mentioned previously we are interested in the *change* of upwash along the chord and, in some cases, along the span as well.

The amount of correction needed is most easily handled as a "τ_2" effect (see Section 6.21), where the "tail length" is, as in the two-dimensional discussion in Section 6.6, one-quarter of the wing chord length.

The τ_2 factor represents the increase of boundary-induced upwash at a point *P* behind the wing quarter chord in terms of the amount *at* the quarter chord. The total angle effect is then

$$\Delta\alpha_{\text{total}} = \Delta\alpha + \tau_2 \, \Delta\alpha \tag{6.49}$$

where $\Delta\alpha$ is the additive correction required for upwash at the quarter chord and τ_2 is the streamline curvature effect on the angle.

Values of τ_2 may be found in Figs. 6.49–6.54, using $c/4$ as the tail length needed to determine τ_2.

Another form of the same correction may be derived by assuming that the τ_2 curves are linear for the short "tail length" of the wing streamline curvature corrections. We then have

$$\begin{aligned} \Delta\alpha_{\text{sc}} &= \tau_2 \, \delta(S/C)C_{LW} \\ &= \frac{c}{4B} \frac{d\tau_2}{d(l_t/B)} \, \delta \, \frac{S}{C} \, C_{LW}(57.3) \end{aligned}$$

For a particular tunnel, both B and $d\tau_2/[d(l_t/B)]$, will be known, so that

$$\Delta\alpha_{\text{sc}} = kc(\Delta\alpha)(57.3) \text{ deg}$$

and once k is determined, no charts are needed to find $\Delta\alpha_{\text{sc}}$ for various models.

The additive lift correction is

$$\Delta C_{L\,\text{sc}} = -\Delta\alpha_{\text{sc}} \cdot a$$

where a = wing lift curve slope.

The additive correction to the moment coefficient is

$$\Delta C_{m\,\text{sc}} = -0.25 \, \Delta C_{L\,\text{sc}} \tag{6.50}$$

It should be noted that many tunnel engineers prefer to apply the correction entirely to the angle rather than to the angle and the lift. To make the correction to angle only, τ should be determined by using $c/2$ as a tail length instead of $c/4$. The moment correction will then be

$$\Delta C_{m\,sc} = +0.125\ \Delta\alpha_{sc\,(2)} \cdot a \tag{6.51}$$

6.13. GENERAL DOWNWASH CORRECTIONS

Very early in the century experimenters using open-throat wind tunnels found their tunnels giving very pessimistic results. The measured minimum drag and rate of change of drag with lift were too large, and the slope of the lift curve was too small. The minimum drag effect was largely due to the very low Reynolds numbers then found in the low-speed tunnels, but the other two effects were due to the tunnel boundaries. The discovery of a way to represent the walls mathematically led to a calculation of their effect. We may now present the theory and numerical values of the correction factors needed when a three-dimensional tunnel is used.

Consider the free-air streamlines caused by a pair of vortices such as are made by a uniformly loaded wing (Fig. 6.18). These streamlines extend to infinity in free air, but when the wing is enclosed in a round duct, they become contained, the wall itself becoming a streamline through which no

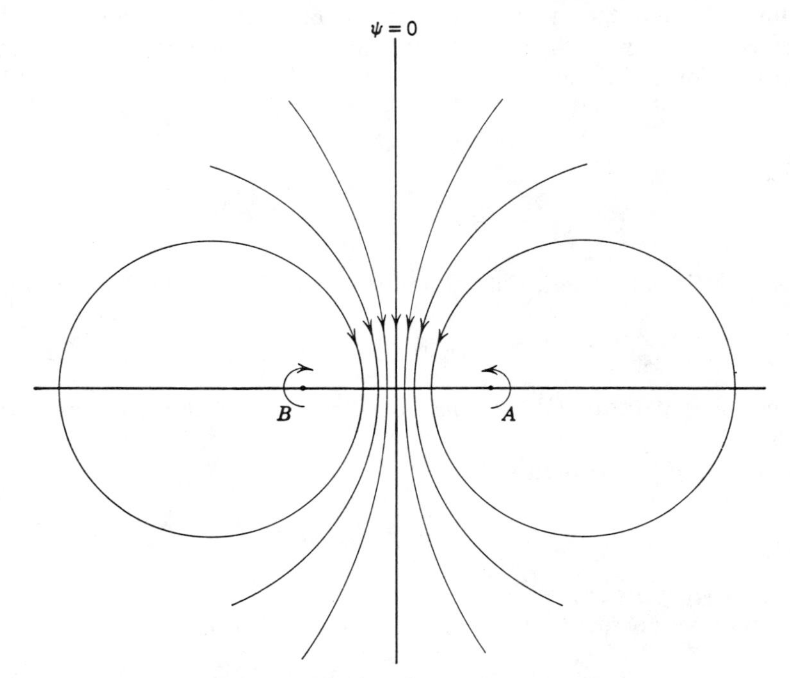

FIGURE 6.18. Field of bound vortices.

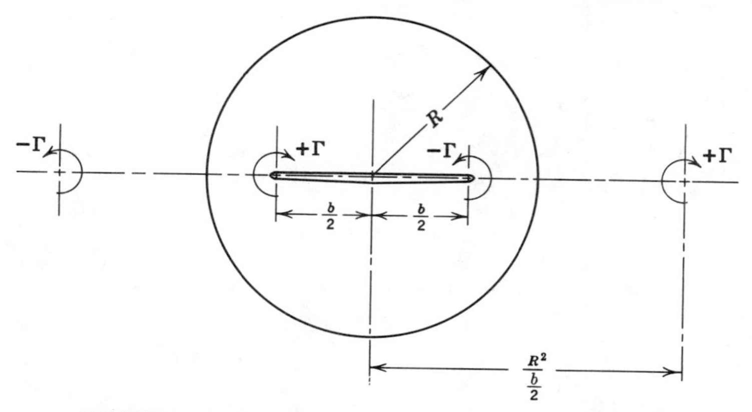

FIGURE 6.19. Location of added vortices, closed round jet.

fluid can pass. As in the two-dimensional case, the problem becomes one of finding the mathematical device that will simulate the walls by making a streamline that coincides with the walls. If we let the wing be small relative to the tunnel size, the problem becomes the simplest of all boundary problems; a streamline with the same radius as the tunnel test section is created by a pair of vortices placed out a distance $x = 2R^2/b$ on each side of the tunnel (Fig. 6.19). They must have the same strength as the wing vortices.

These two image vortices violate Helmholtz's vortex theorem, that is, a vortex cannot end in the fluid but must go the fluid boundary or form a closed loop. But like Prandtl's momentum theory for the induced angle of attack and drag of a finite wing they give the correct answers.

The streamlines due to the added vortices are shown in Fig. 6.20. It will take but a moment for the student to trace Fig. 6.18 on Fig. 6.20 and to see for himself or herself how the $\psi = 0$ streamline coincides with the tunnel wall (Fig. 6.21). Writing the stream function for the four trailing vortices, summing them, and setting the sum to zero (Fig. 6.21) will show that the spacing of the images must be at the distance shown in Fig. 6.19.

Another way to look at the effect of the added vortices is to consider their velocity field at the wing, as shown in Fig. 6.22. The upflow tends to offset the downflow caused by the wing trailing vortices, and the wing then has too little induced angle and too little induced drag. The exact amount may be found as follows:

The lift of a uniformly loaded wing may be written

$$L = \frac{\rho}{2} SV^2 C_L = \rho V \Gamma b$$

so that the circulation

$$\Gamma = \frac{SVC_L}{2b}$$

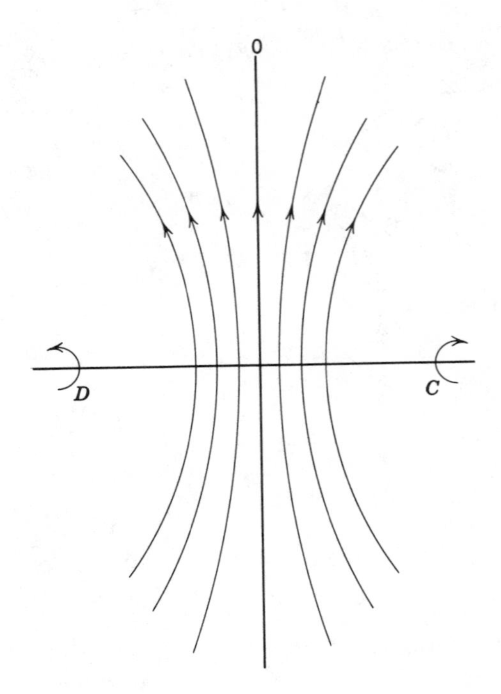

FIGURE 6.20. Flow field of added vortices.

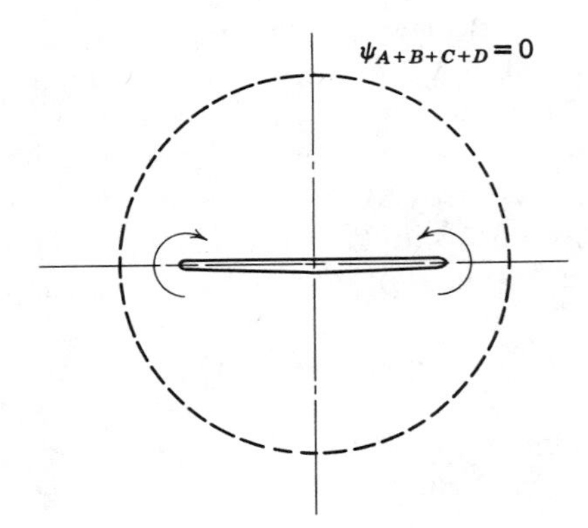

FIGURE 6.21. Total flow field of both bound and added vortices; $\psi = 0$.

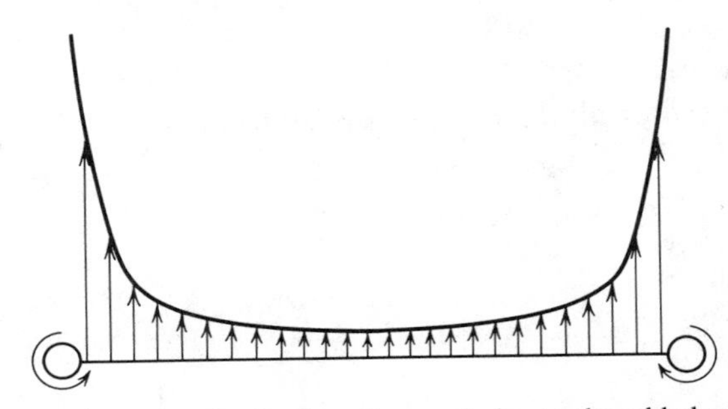

FIGURE 6.22. The distribution of upwash due to the added vortices.

The upwash at a distance r from a semi-infinite vortex is

$$w = \frac{\Gamma}{4\pi r}$$

and for two vortices at a distance $R^2/(b/2)$ this becomes

$$w = \frac{\Gamma b}{8\pi R^2}$$

Substituting for Γ we get

$$w = \frac{SV/\pi R^2}{8} C_L$$

and the induced angle due to the boundaries (let the tunnel cross-sectional area be C) is

$$\Delta\alpha_i = \frac{w}{V} = \frac{S/C}{8} C_L$$

The induced drag increment due to the boundaries is

$$\begin{aligned} \Delta C_{Di} &= \Delta\alpha_i C_L \\ &= \frac{S/C}{8} C_L^2 \end{aligned}$$

Both these effects reduce the free-air induced angle and induced drag attributable to a given C_L, making the wing appear to have a larger aspect ratio than it really has. The true values become (as in the development in Section 6.1)

$$\alpha = \alpha_u + \delta \frac{S}{C} C_{LW}(57.3) \tag{6.52}$$

and

$$C_D = C_{Du} + \delta \frac{S}{C} C_{LW}^2 \tag{6.53}$$

where $\delta = 0.125$ for a round closed test section when the wing is small and has uniform loading. The subscript W is added to emphasize that corrections are based on wing lift only.

But wings are seldom small and *never* have uniform loading. We shall have to reexamine these assumptions and see whether they induce serious errors. An examination of Fig. 6.22 will give an insight into the statement that the wing span should be less than 0.8 of the tunnel width. See Section 6.14.

The factor δ, it develops, is a function of the span load distribution, the ratio of model span to tunnel width, the shape of the test section, and whether or not the wing is on the tunnel centerline. The factor δ may be found for almost all conditions somewhere in this chapter, and Eqs. (6.52) and (6.53) are general; once δ is found it may be used to find the boundary effects. For most tunnels, however, only the δ's for uniform loading have been provided. This seems odd until one realizes that the shed vortices rapidly roll up into a single pair of vortices which exactly duplicate the trailing vortex pattern of uniform loading. They then have a vortex span b_v (which is given for a large number of wings in Fig. 6.23), and it is proper to use the uniform loading correction. However, since b_v is developed somewhat downstream, it is suggested that it is more reasonable to take an effective vortex span*

$$b_e = \frac{b + b_v}{2} \tag{6.54}$$

for use at the wing. The values of δ for elliptic loading are rarely used.

Thus, in summary, to find δ for a given wing, find b_e from Eq. (6.54) and use the proper δ for uniform loading. For wings not covered in Fig. 6.23, the approximation

$$b_e = 0.9b$$

will not result in a serious error.

If elliptic loading corrections are to be used, the geometric span b should be used in computing k = span/tunnel width.

The theory for open test sections will not be outlined other than to mention that the condition for a free boundary is that no pressures can be supported ($\phi = 0$, where ϕ is the velocity potential). For open jets, δ normally has a negative sign. Values of δ for open jets are given in the following sections.

The special effects of very large models are covered in the next two sections

This problem is presumably taken care of in the initial design of the model for a given tunnel. But if the model initially has the maximum permissible span, and then during its life the airplane is stretched, the wing loading increases to a point that more area is needed. A simple solution is to add an extension to the wing tip. This both increases the area and the aspect ratio, but it can be disastrous for the wind tunnel model.

* Values from Eq. (6.54) agree excellently with those suggested by Swanson and Toll in Ref. 6.17.

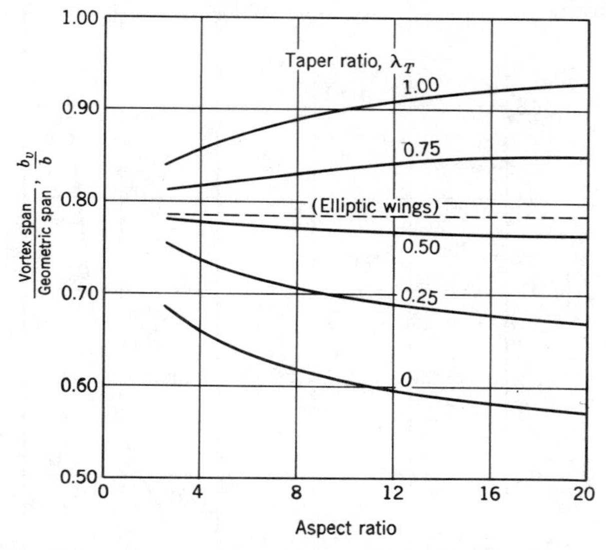

FIGURE 6.23. Values of the vortex span for elliptic, rectangular, and tapered wings.

6.14. LIFT DISTRIBUTION INTERFERENCE—ROUND JETS

The variation of spanwise distribution due to the walls of a closed throat wind tunnel is small unless the wing has a large span. If this condition exists, the data become discouraging, tip stall starting earlier and being more severe than it actually would be in free air.

Lift distribution interference in a round closed tunnel is discussed by Stewart (Ref. 6.18), who finds that ratios of span to tunnel width greater than 0.8 will indicate early tip stall. An interesting numerical example shows that, for a wing of AR = 7, span/tunnel width = 0.9, and $C_L = 1.2$, an effective wash-in amounting to 1.44° is caused by the tunnel walls.

The designer of wing tunnel models cannot correct for this in the model design since the effect of the walls varies with C_L. The amount of twist induced by a round closed tunnel on wings of elliptic planform is shown by Stewart to be

$$\frac{\Delta\alpha_i}{\alpha_i} = \frac{4R^2}{b^2}\left[\left(1 - \frac{b^4}{16R^4}\right)^{-\frac{1}{2}} - 1 - \frac{b^4}{32R^4}\right] \tag{6.55}$$

where $\Delta\alpha_i$ = induced wash-in of wing due to wind tunnel wall interference, α_i = induced angle of attack = $C_l/\pi AR$, r = wind tunnel radius. and b = wing span. A plot of Eq. (6.55) is shown in Fig. 6.24.

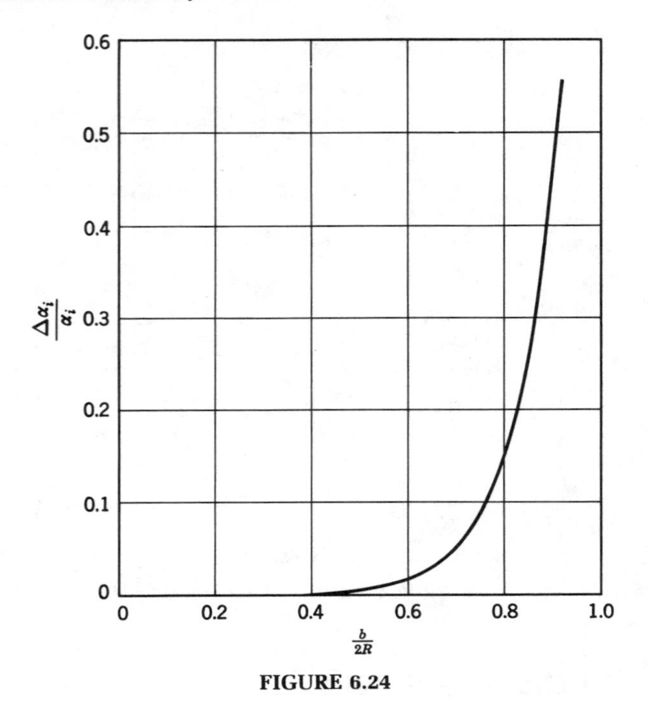

FIGURE 6.24

6.15. LIFT DISTRIBUTION INTERFERENCE—ELLIPTIC JETS

Gavin and Hensel (Ref. 6.19) have discussed the effect of the tunnel walls on the spanwise distribution of lift for closed elliptic jets with wings of aspect ratio = 8.0. Though the example discussed is very limited, further calculations using their method are possible. Their calculations may be summarized as follows:

1. At high lift coefficients when the wing tips lie outside the focal points of the elliptic jet, the variation of the induced angle of attack along the span is no longer negligible. This amounts to apparent wash-in, which becomes severe as the wing tip approaches the stall. As a result, when the wing span approaches the tunnel major diameter, determinations of stall characteristics in the tunnel are conservative; that is, the wing tips will stall at higher angles in free air.

2. Other things being constant, tunnel-wall interference is less for lift distributions in which the lift is concentrated toward the centerline. That is, for untwisted wings, those with high taper ratios have tunnel-induced upwash of smaller magnitude than wings with low taper ratios.

3. Tunnel-wall interference is less for wings of high aspect ratio, other conditions being held constant.

4. For wings with normal lift distributions, the mean upwash factor is a minimum when the wing tips are approximately at the tunnel foci.

6.16. DOWNWASH CORRECTIONS FOR CIRCULAR JETS

The corrections for uniform loading in a circular jet have been completed by Kondo (Ref. 6.20), and those for elliptic loading by Glauert (Ref. 6.21) following a method proposed by Rosenhead (Ref. 6.22). They are both based on the ratio of span to tunnel diameter $k = b/2R$, actual values being presented in Fig. 6.25. Glauert's data have been corrected to more modern units.

Owing to model length or mounting, it is sometimes necessary to place the model with its wing not on the jet centerline. This places the trailing vortices closer to one wall than to the other, altering the flow pattern and hence the proper value of δ. This condition has been examined by Silverstein (Ref. 6.23), who finds the values of δ with a displaced wing of uniform loading in a round jet to be as shown in Fig. 6.26. The nomenclature is described in Fig. 6.27.

6.17. DOWNWASH CORRECTIONS FOR RECTANGULAR JETS

Van Schliestett (Ref. 6.24) has discussed basic rectangular jet corrections for very small wings, correcting a mathematical error that appears in *TR* 461 (Ref. 6.25). His results are given in Fig. 6.28.

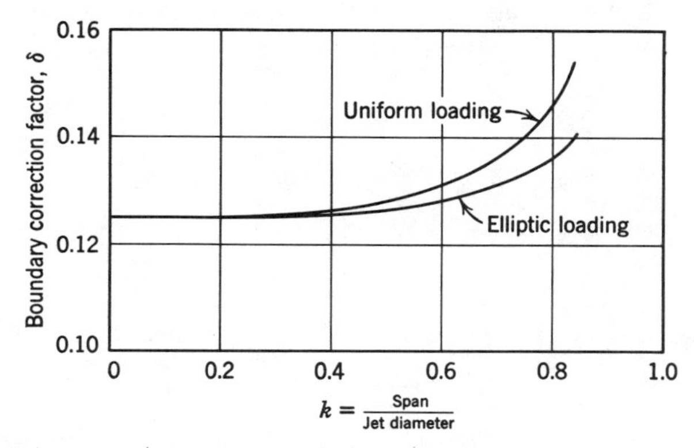

FIGURE 6.25. Values of δ for a wing with elliptic loading and for one with uniform loading in a closed round jet. For an open round jet the sign of δ is negative.

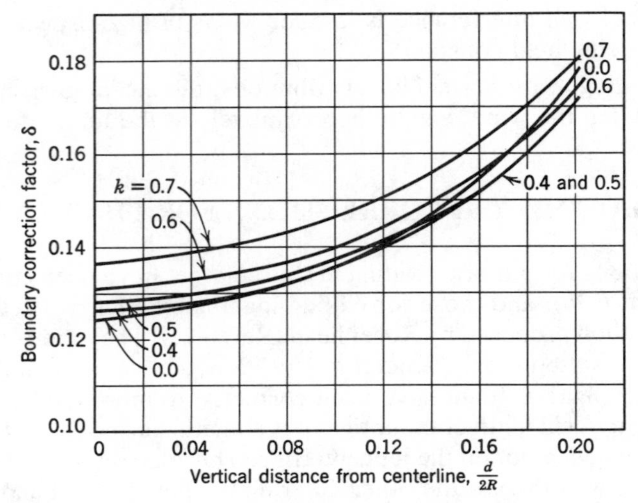

FIGURE 6.26. Values of δ when a wing with uniform loading is displaced above or below the centerline of a closed round jet. δ is negative for the open jet.

The boundary corrections for wings of moderate span compared to the tunnel width have been worked out for uniform loading by Terazawa (Ref. 6.26) and for elliptic loading by Glauert (Ref. 6.27). Figures 6.29–6.31 give the values for δ.

$$\lambda = \frac{\text{Tunnel height}}{\text{Tunnel width}}$$

Values of δ for the square and duplex tunnel when the wing of uniform span loading is above or below the centerline are found in Figs. 6.32–6.35.

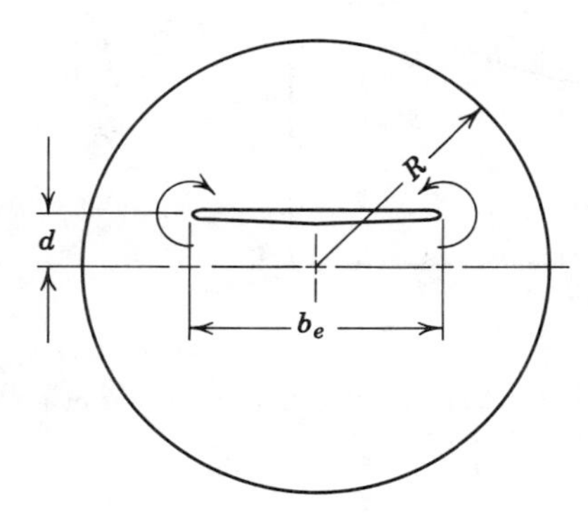

FIGURE 6.27

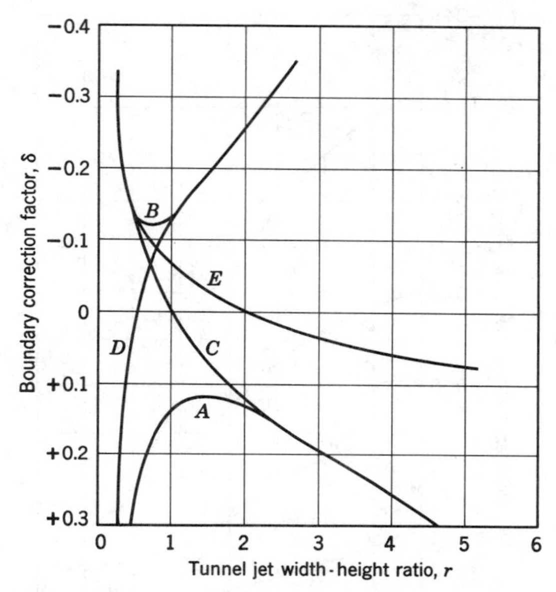

FIGURE 6.28. Values of δ for open and closed rectangular jets, very small wings only. A, closed tunnel. B, free jet. C, jet with horizontal boundaries only. D, jet with vertical boundaries only. E, jet with one horizontal boundary.

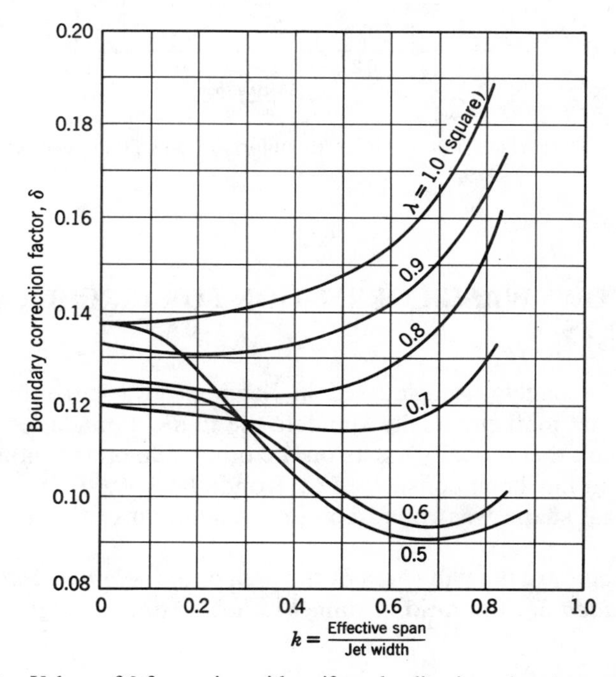

FIGURE 6.29. Values of δ for a wing with uniform loading in a closed rectangular tunnel.

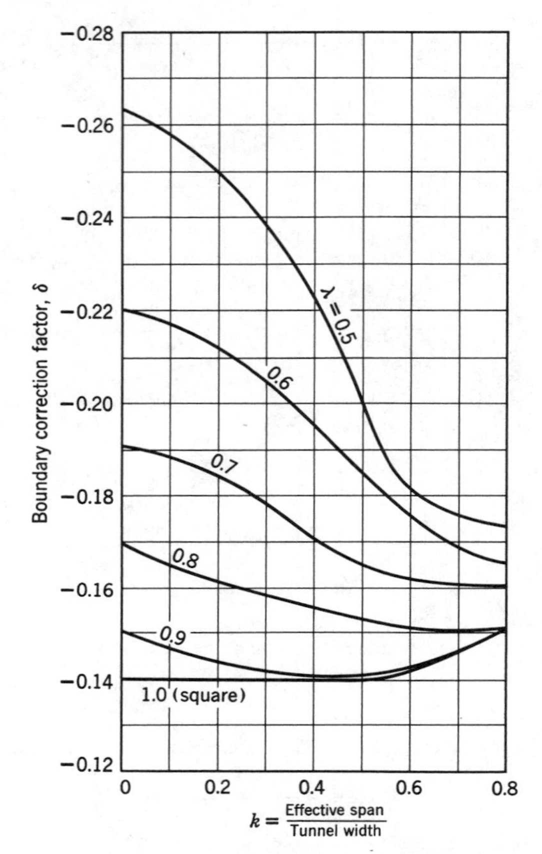

FIGURE 6.30. Values of δ for a wing with uniform loading in an open rectangular jet.

6.18. DOWNWASH CORRECTION FOR CIRCULAR-ARC JETS

The testing of panels as discussed in Sections 5.7–5.10 requires special corrections that mathematically simulate the tunnel boundaries and the image wing which theoretically exists on the other side on the mounting plate. This condition has been considered by Kondo (Ref. 6.20) for a test section whose original shape was round before the addition of the mounting plate (Fig. 6.36).

The variables are the wing area S, the area of the *original circle before the plate was added* S_0, the ratio of tunnel radius to tunnel height

$$\lambda = h/B \qquad (6.56)$$

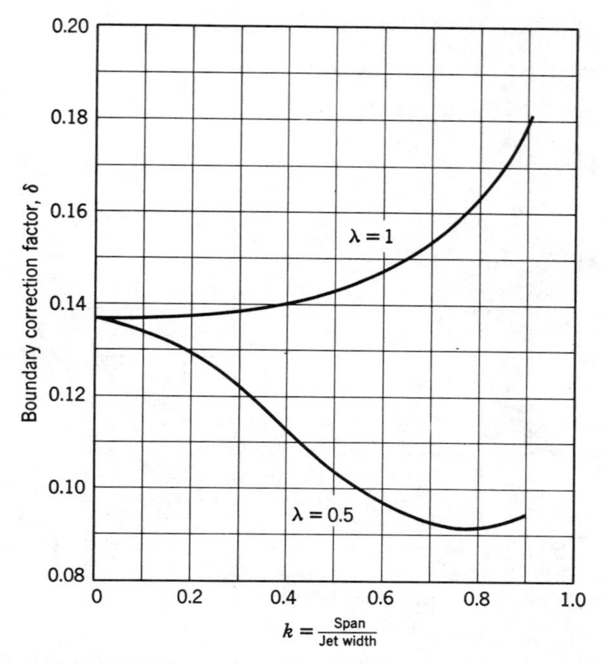

FIGURE 6.31. Values of δ for a wing with elliptic loading in a closed rectangular jet.

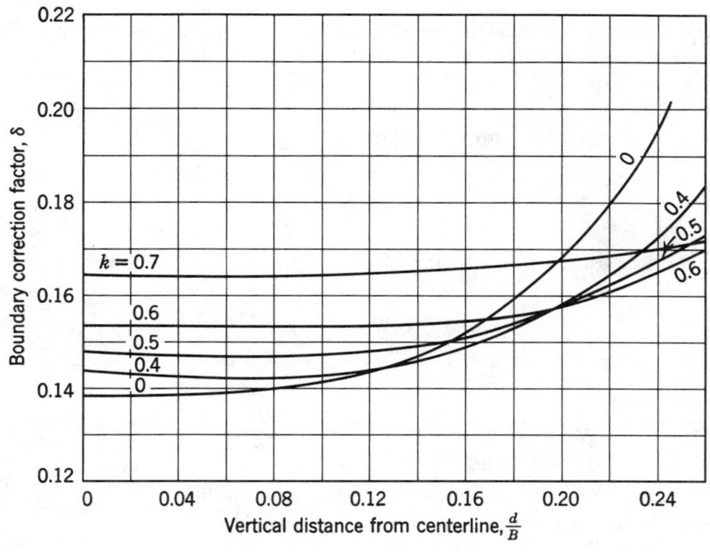

FIGURE 6.32. Values of δ when a wing with uniform loading is displaced above or below the centerline of a closed square jet.

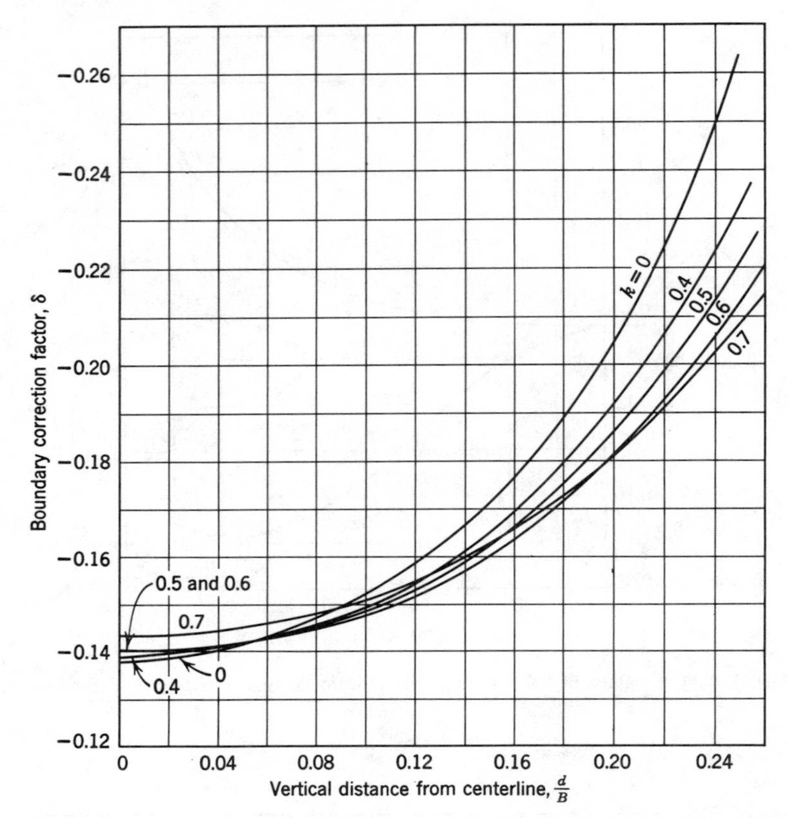

FIGURE 6.33. Values of δ when a wing with uniform loading is displaced above or below the centerline of an open square jet.

and the ratio of span to tunnel height

$$k = b_e/B \tag{6.57}$$

The additive corrections as usual take the form

$$\Delta\alpha = \delta(S/C)C_{LW}(57.3)$$
$$\Delta C_D = \delta(S/C)C_{LW}^2$$

but a word of caution is necessary. In Ref. 6.20 some confusion exists in the definition of "wing area" and "tunnel area." The wing area to be used is the *actual wing area including the image area,* and the tunnel area is the area of the original circle *not including the image circle.*

Values of δ for various values of k and λ may be found in Fig. 6.37 for closed circular-arc jets and in Fig. 6.38 for open circular-arc jets.

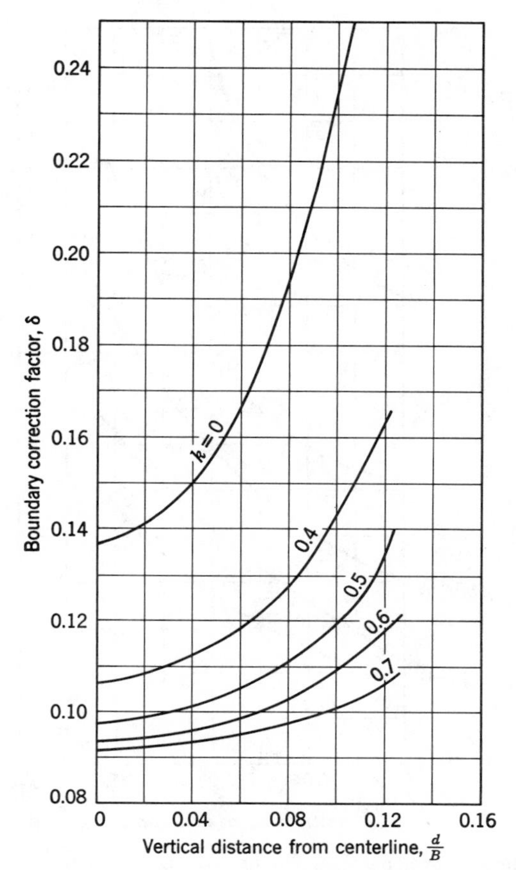

FIGURE 6.34. Values of δ when a wing with uniform loading is displaced above or below the centerline of a closed 2 : 1 rectangular jet.

It will be noted that these corrections are not strictly applicable to aileron tests, since in practice the "image" wing would have the control surface deflected oppositely. See Section 6.28. Further corrections for circular-arc jets are given in Ref. 6.28.

6.19. DOWNWASH CORRECTIONS FOR ELLIPTIC JETS

The corrections for wings with uniform loading in an elliptic jet have been completed by Sanuki (Ref. 6.29) and those for elliptic loading by Rosenhead (Ref. 6.30).

Sanuki bases his values for δ (uniform loading) on the ratio of the minor to the major axis of the jet λ, and the ratio of the span to the major axis k (Fig. 6.39).

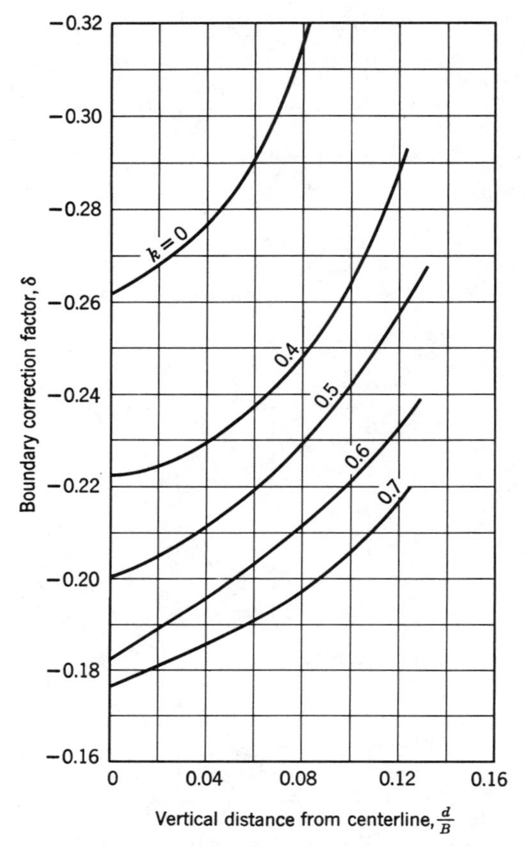

FIGURE 6.35. Values of δ when a wing with uniform loading is displaced above or below the centerline of an open 2 : 1 rectangular jet.

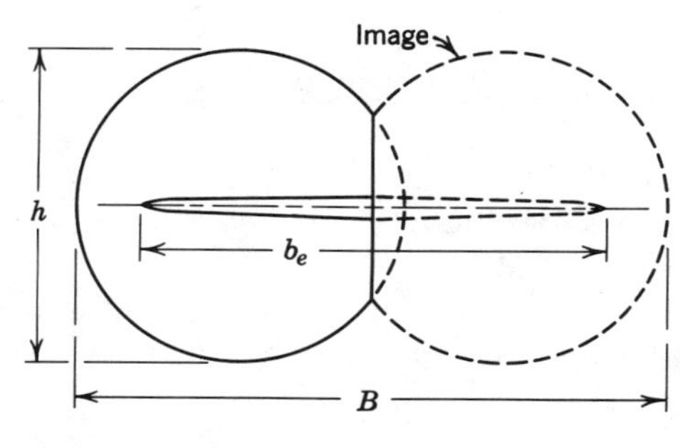

FIGURE 6.36. A model b_e in a tunnel whose boundaries are circular arcs.

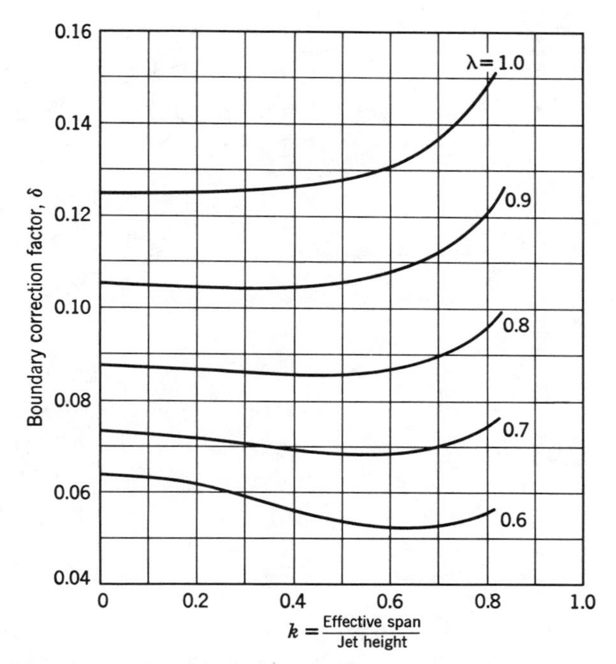

FIGURE 6.37. Values of δ for a wing with uniform loading in a closed circular-arc wind tunnel.

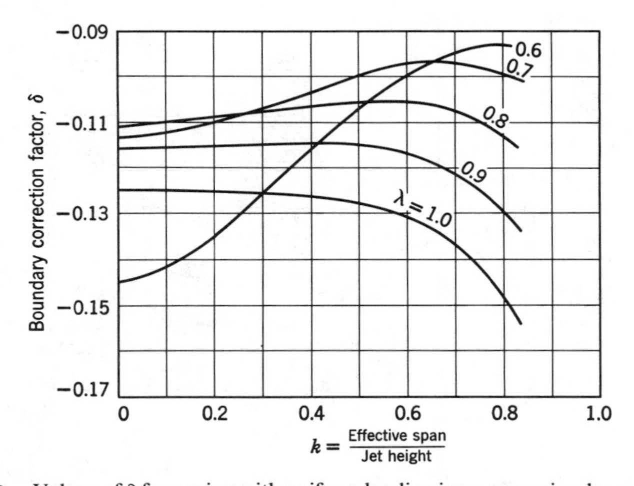

FIGURE 6.38. Values of δ for a wing with uniform loading in an open circular-arc wind tunnel.

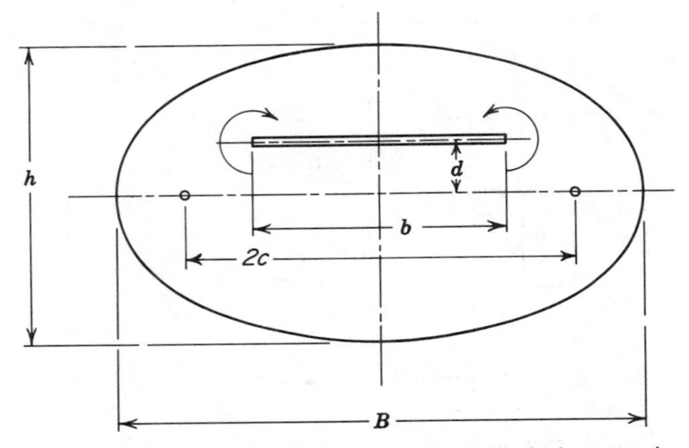

FIGURE 6.39. Nomenclature for wing in an elliptical test section.

$$\lambda = h/B \tag{6.58}$$

$$k = b_e/B \tag{6.59}$$

Values of δ are shown in Figs. 6.40 and 6.41.

The values for the wing not on the tunnel centerline (uniform loading) may be found in Figs. 6.42 and 6.43; nomenclature is given in Fig. 6.39.

Rosenhead bases his values for δ (elliptic loading) on the ratio of the axis containing the wing to the other axis of the jet λ and on the ratio of the span to the focal length b/c. For presentation here, however, the latter has been reconverted to the ratio of span to tunnel width $k = b_e/B$ (Fig. 6.39). The values of δ are shown in Figs. 6.44 and 6.45.

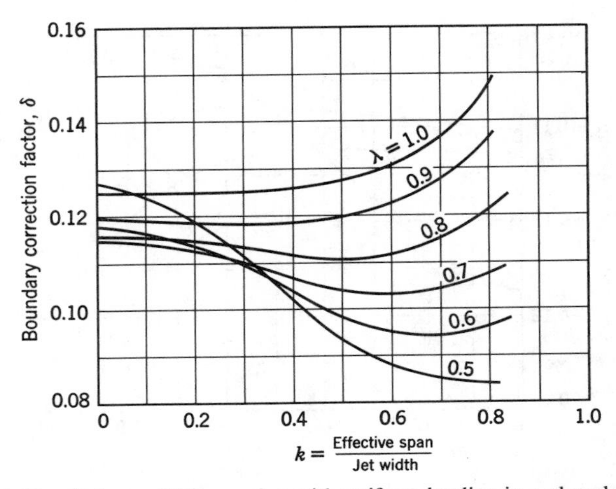

FIGURE 6.40. Values of δ for a wing with uniform loading in a closed elliptic jet.

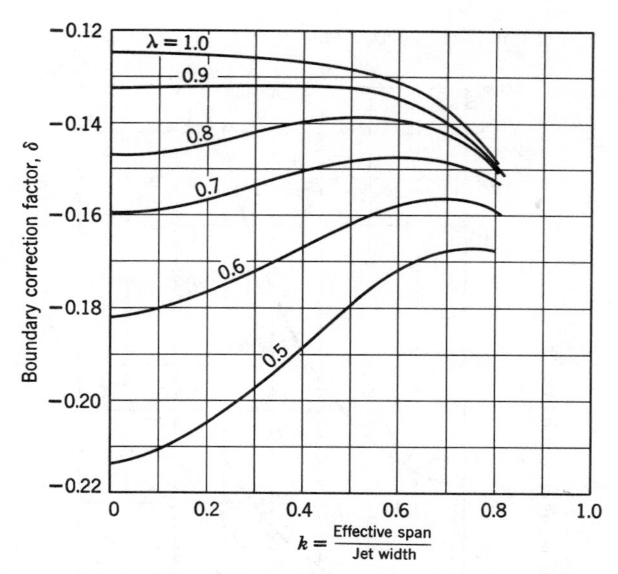

FIGURE 6.41. Values of δ for a wing with uniform loading in an open elliptic jet.

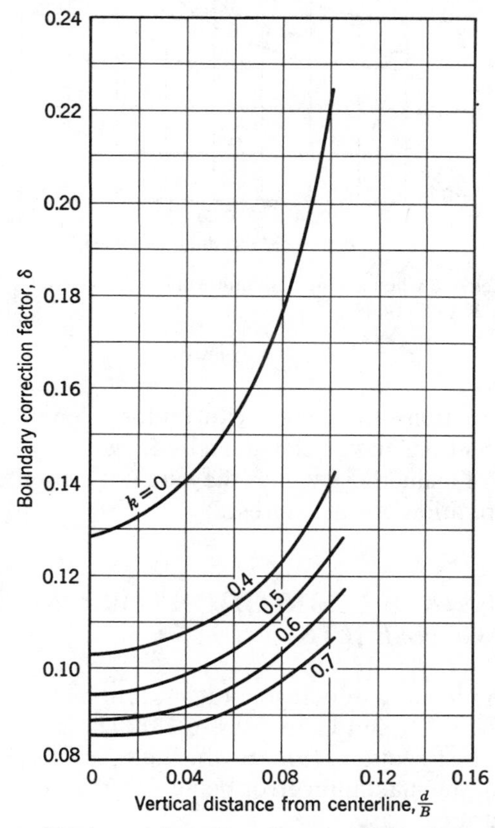

FIGURE 6.42. Values of δ when a wing with uniform loading is displaced from the centerline of a closed 2 : 1 elliptic jet.

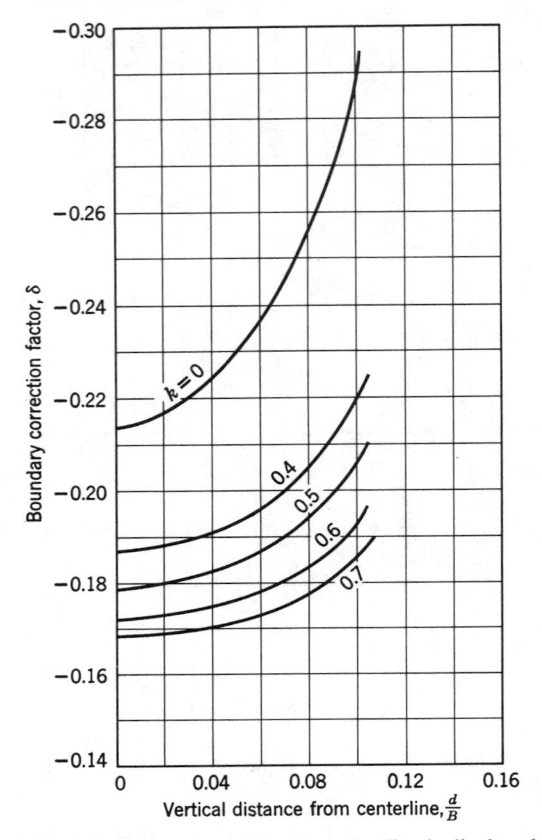

FIGURE 6.43. Values of δ when a wing with uniform loading is displaced above or below the centerline of an open 2 : 1 elliptic jet.

Additional corrections for wings with uniform loading in $1:\sqrt{2}$ partly open elliptic test sections have been given by Riegels in Ref. 6.31. In view of the improbability of using these values they are not presented, although their existence and derivation are of interest.

6.20. DOWNWASH CORRECTION FOR CLOSED OCTAGONAL JETS

Wings with elliptic loading in octagonal test sections have been considered by Batchelor (Ref. 6.32) and Gent (Ref. 6.33). The conclusion is that, for regular octagonal test sections (Fig. 6.46), the corrections for circular sections may be used, the maximum error being 1.5% in δ or well under 0.2% in C_D for the most critical case.

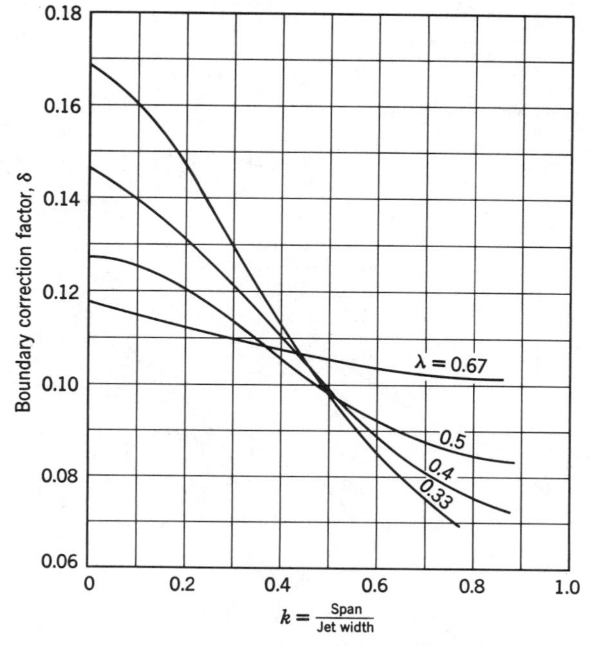

FIGURE 6.44. Values of δ for a wing with elliptic loading in a closed elliptic jet.

The octagonal test section formed by tempering the corners of a rectangular jet has been discussed only for the case where a 7 ft by 9 ft rectangular jet is reduced by 45° flat fillets whose vertical height reduced the amount of side wall exposed by one half (Fig. 6.47). The effect of these fillets is to make the basic rectangular jet more nearly approach the elliptic jet. The wind tunnel boundary factor is hence reduced.

The correction factors for both the 7 ft by 9 ft rectangular and the octagonal test section are shown in Fig. 6.48. As may be surmised, the corrections of the octagonal jet are essentially those of an elliptic jet of the same height-width ratio.

6.21. DOWNWASH CORRECTIONS FOR THE FLOW BEHIND THE WING

The method of simulating the boundaries by an image system in a plane taken through the wing quarter chord perpendicular to the axis of symmetry of the airplane has been covered in Section 6.13. However, the amount of velocity induced by a vortex increases rapidly as one moves from the end of the vortex, so that, from the three-dimensional picture (Fig. 6.3), the amount of upwash at the tail of a model in a tunnel is very much more than that at the

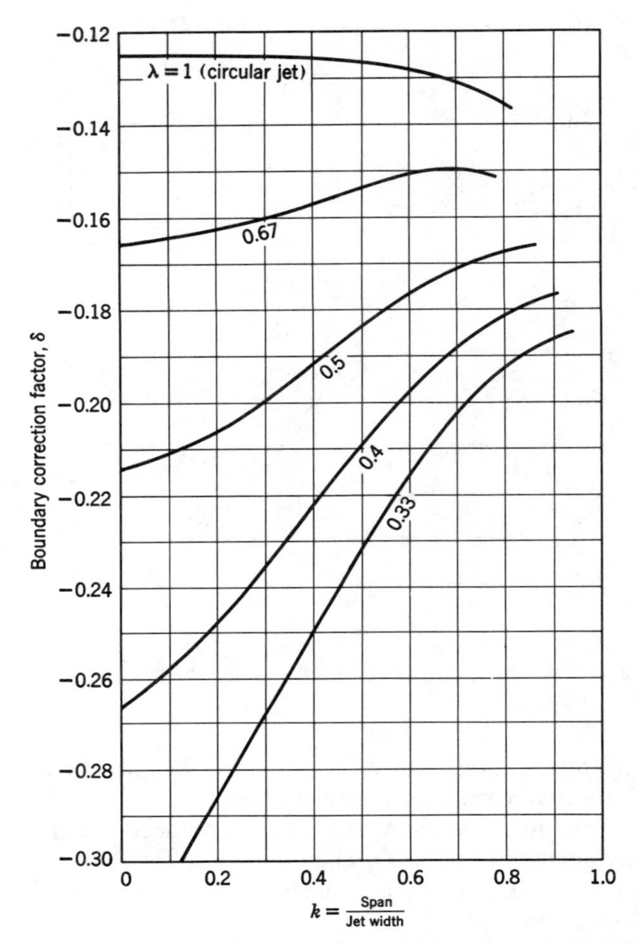

FIGURE 6.45. Values of δ for a wing with elliptic loading in an open elliptic jet.

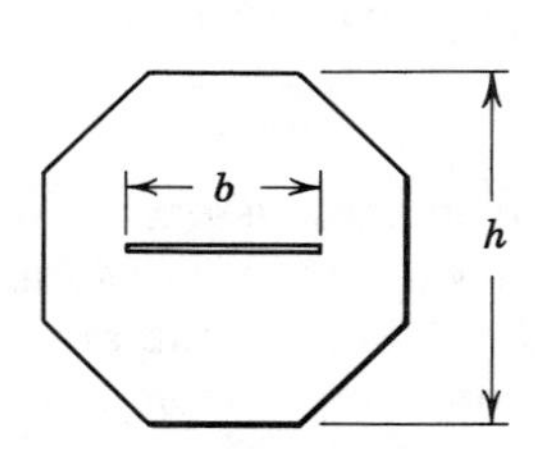

FIGURE 6.46. Wing in an octagonal jet.

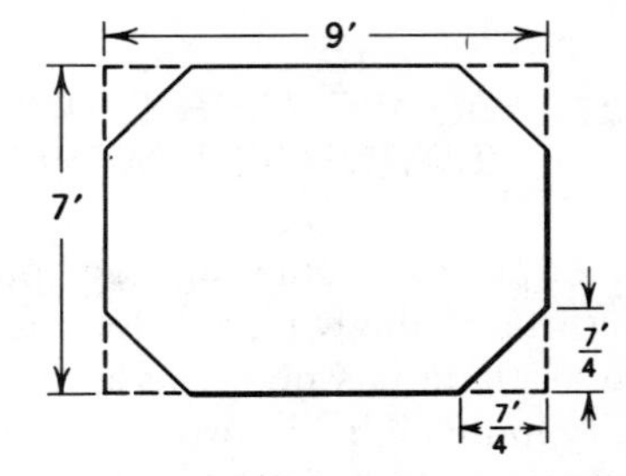

FIGURE 6.47. Tempering corners to form an o[illegible] test section.

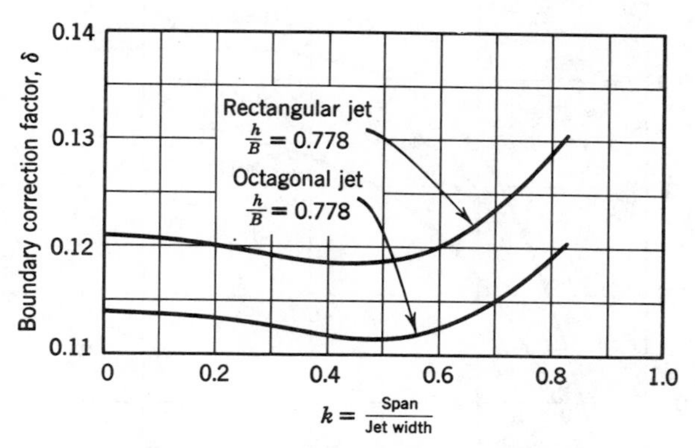

FIGURE 6.48. Values of δ for wings with elliptic loading in octagonal and rectangular test sections with tempered corners.

wing. Thus, for instance, at a time when the walls are reducing the wing angle of attack by 2°, they could conceivably be reducing the tail angle of attack by 3°. This large discrepancy, proportional to the lift coefficient, makes complete models in a closed test section appear very much more stable than they would in free air, while in an open jet the opposite effect is true.

This problem has been discussed by Lotz (Ref. 6.34), and the boundary-induced upwash velocity at a distance l_t behind the quarter-chord line has been presented as

$$w_k = \delta \frac{S}{C} C_{LW}(1 + \tau_2)V \tag{6.60}$$

where w_k = upwash velocity in the plane of symmetry at distance l_t behind the quarter chord (w_k does not vary greatly along the span), C = jet cross-section area, V = tunnel velocity, and τ_2 = downwash correction factor.

Values of τ_2 for a number of tunnels are given in Figs. 6.49 to 6.54, and values of δ for the relevant model may be found in Sections 6.16 through 6.20.

Some doubt exists about the validity of the downwash corrections for open-throat wind tunnels as regards the values behind the lifting line since they were derived for an infinitely long test section and not for one about one diameter long. This finite length does change δ and τ_2, but, for wing corrections and corrections for streamline curvature, the change is not serious. For stability corrections for complete models in open test sections it would be in order to consult Ref. 6.35 if the tail length is more than 0.4 B, where B is the tunnel width.

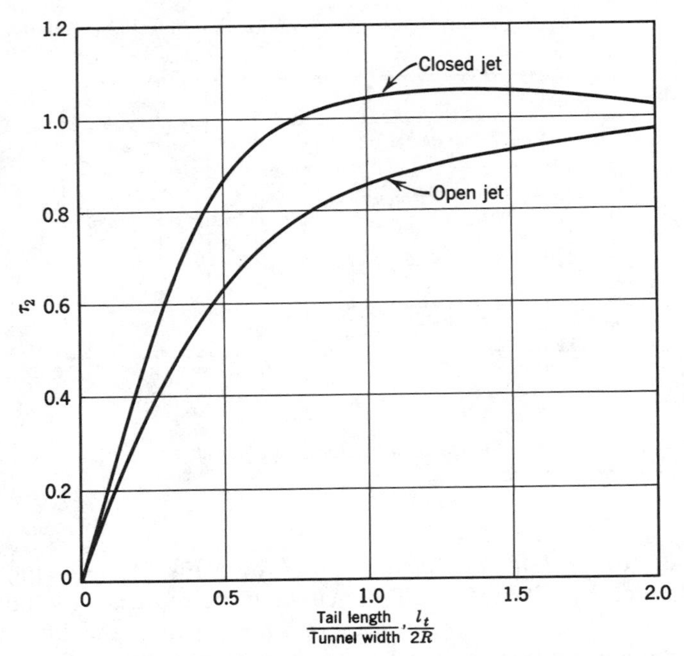

FIGURE 6.49. Values of τ_2 for open and closed circular jets.

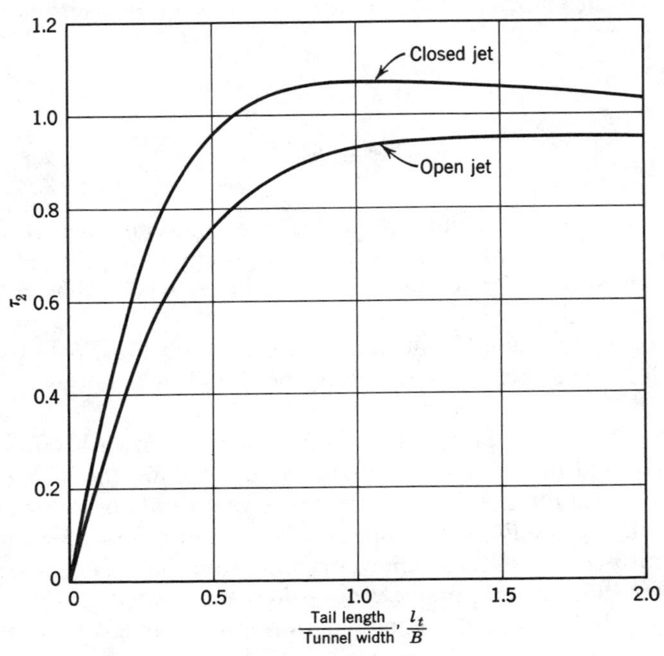

FIGURE 6.50. Values of τ_2 for open and closed elliptic jets.

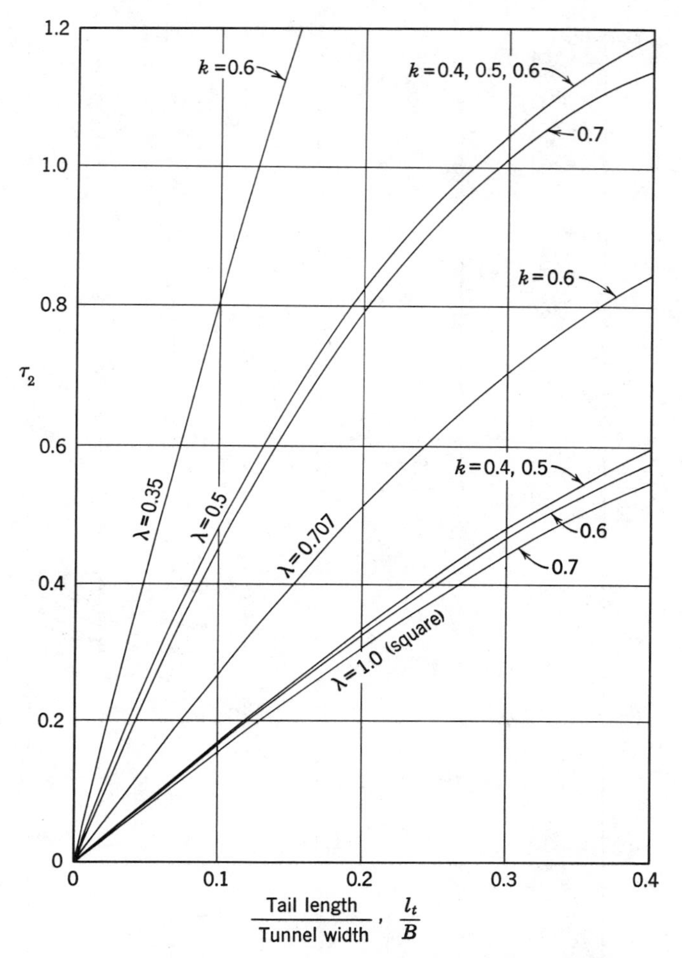

FIGURE 6.51. Values of τ_2 for several closed rectangular wind tunnels, wing on tunnel centerline, and horizontal tail on wing centerline. Values for $\lambda = 0.35$ are by extrapolation.

A parallel correction may be worked out for the static stability as follows: The moment about the center of gravity due to an aft tail upload is

$$m_{\mathrm{CG}_t} = -l_t L_t \tag{6.61}$$

where the subscript t refers to the horizontal tail, and l_t is the distance from the CG to the $\frac{1}{4}$ MAC of the tail. In coefficient form

$$C_{m\mathrm{CG}_t} = -\left(\frac{S_t l_t}{S\ \mathrm{MAC}}\right)\left(\frac{q_t}{q}\right) C_{L_t} = -\overline{V}\eta_t a_t \alpha_t \tag{6.62}$$

$$\left(\frac{dC_{m\mathrm{CG}}}{d\alpha}\right)_t = -a_t \overline{V}\eta_t \tag{6.63}$$

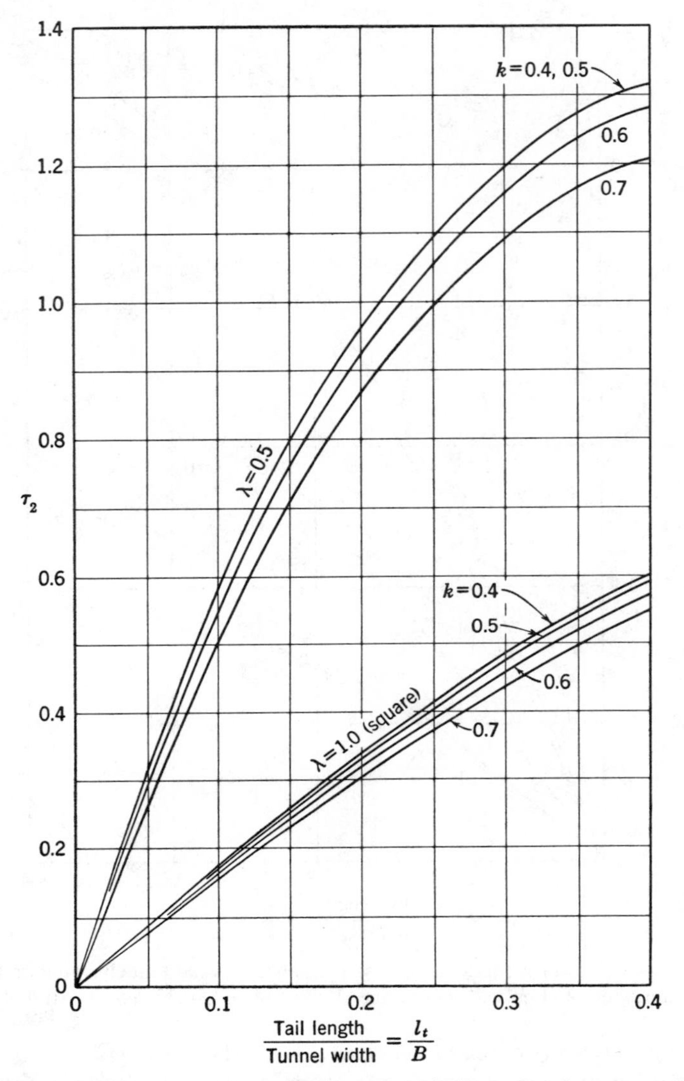

FIGURE 6.52. Values of τ_2 for two closed rectangular wind tunnels, wing on tunnel centerline, but horizontal tail 0.1 b_e above or below wing centerline.

By making two tail-on runs at different horizontal tail incidence angles ($i_t = \alpha_t$) this term can be evaluated. Then

$$\Delta C_{m\text{CG}_t} = \left(\frac{dC_{mCG}}{d\alpha_t}\right)\Delta\alpha_t \tag{6.64}$$

From Eq. (6.60) the additional correction to the tail angle of attack is

$$\Delta\alpha_t = \frac{w_k}{V} = \delta\,\tau_2\left(\frac{S}{C}\right)C_{LW}(57.3) \tag{6.65}$$

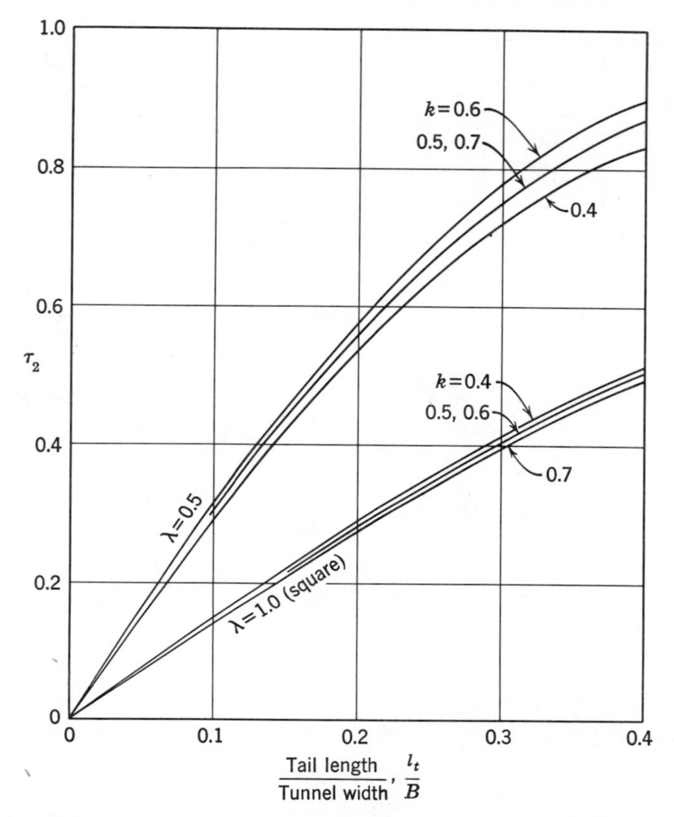

FIGURE 6.53. Values of τ_2 for two open throat rectangular wind tunnels, wing on tunnel centerline, horizontal tail on wing centerline.

hence

$$\Delta C_{m\text{CG}_t} = \left(\frac{dC_{m\text{CG}}}{d\alpha}\right)_t \delta\, \tau_2 \left(\frac{S}{C}\right) C_{LW}\,(57.3) \tag{6.66}$$

$$C_{m\text{CG}} = C_{m\text{CG}_u} - \Delta C_{m\text{CG}_t} \tag{6.67}$$

The subscript W is added to emphasize that the correction is based on the wing lift as are the corrections to angle of attack and the drag coefficient equations (6.4, 6.5 and 6.52, 6.53).

To properly evaluate the correction to pitching moment [Eq. (6.63)] requires a minimum of three runs: two tail-on runs at different tail incidence angles and a tail-off run. Basically what is being measured is the q at the tail. In many cases, for aircraft with both leading and trailing edge flaps a plot of $a_t\overline{V}\eta_t$ versus α will show an increasing value at low α's, a constant value in mid range α's followed by a decrease at higher α's. The low values at low α's is caused by the typical leading edge flap stall at low α's, while the decrease at

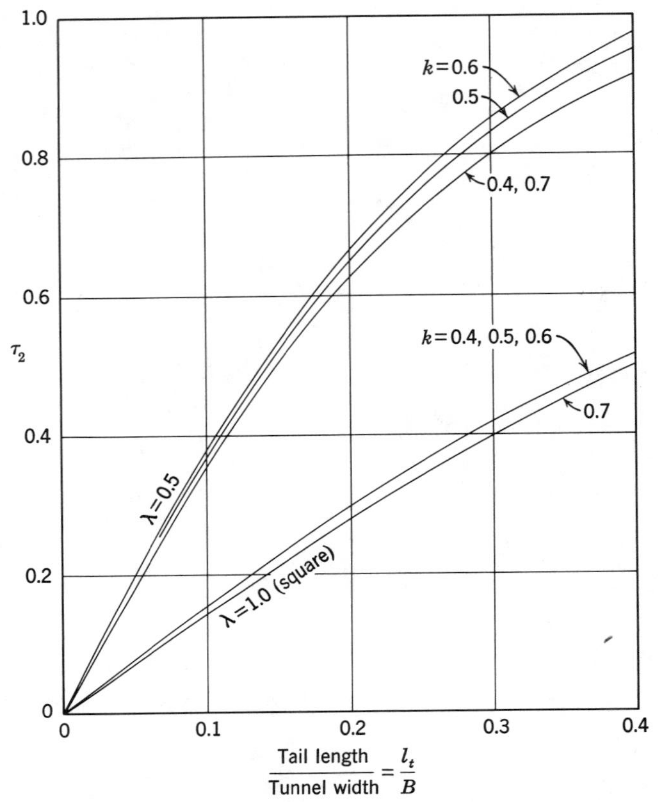

FIGURE 6.54. Values of τ_2 for two open throat rectangular wind tunnels, wing on tunnel centerline, but horizontal tail 0.1 b_e above or below wing centerline.

high α's near $C_{L\,\max}$ is the trailing edge flaps stalling. This effect is a function of the vertical location of the horizontal tail and the downwash of the wing. One note of caution: when evaluating $a_t\overline{V}\eta_t$ for models with remotely pitchable tails these runs should be made at moderate tail angles to avoid a stalled tail. Because these runs are required to correctly reduce the data, models should be designed to allow the tail incidence angle to be changed easily and its angle accurately determined.

The third run yields the basic information that is required to apply wall corrections. This is the wing lift coefficient, and should be evaluated any time a model change is made to the wing. This is the basis for the suggestions by wind tunnel engineers that tests to determine optimum flap settings be done tail-off. The problem with this is that flaps produce large negative pitching moments, and these in turn can require large downloads by the tail to balance the airplane which may negate the gain in $C_{L\,\max}$ due to the flaps. There is no easy solution to this problem except to require the model designer to design the model so that the time required to put the tail on and off is minimal. Care should be taken with changes to the inboard portion of the wing because any change in downwash can effect the tail both through the $d\epsilon/d\alpha$ term (Figs. 5.35 and 5.36) in the stability equation and the q at the tail.

The changes in pitching moment curve slope due to wall corrections can be quite large but their validity has been amply demonstrated by flight tests.

If it is impossible to evaluate $(q_t/q)(dC_{Lt}/d\alpha_t)$, the relation

$$\frac{q_t}{q}\frac{dC_{Lt}}{d\alpha_t} = \frac{0.1AR_t}{AR_t + 2}(0.8) \tag{6.68}$$

may be used. This is an approximate lift curve slope formula reduced by the factor 0.8. It is common to hear the factor 0.8 spoken of as a loss in dynamic pressure. Though there is some loss in average dynamic pressure over the horizontal tail produced by the fuselage boundary layer, it is far less than 20%, and the correction is more properly thought of as the result of the blanketing of the horizontal tail by the fuselage and its boundary layer, and the resultant loss in lift across the fuselage, than as a loss in average dynamic pressure.

6.22. COMMENTS ON DOWNWASH CORRECTIONS FOR OTHER THAN RECTANGULAR TUNNELS

Over the years solid wall tunnels have been built with a large variety of shapes other than rectangular. These shapes have included round (Sections 6.14, 6.16, and 6.21), circular arcs (Section 6.18), elliptic (Section 6.19), regular octagonal (Section 6.20), regular hexagonal, flat top and bottom with semicircular ends, rectangular with corner fillets, etc. The representation of these shapes by images outside of the walls have never been exactly correct. The curves that yield the wall correction factors from previous editions have been retained in this edition because the errors for moderate-size models and lifts are not too large and are further reduced by the wing-to-tunnel cross-sectional-area ratio, which is a small number. However, with the advent of high-speed digital computers the wall correction factors for these tunnels can be determined by vortex lattices in the walls as discussed in Joppa's corrections (Section 6.24), without the vortex wake relocation, if desired. The same technique can be expanded to handle swept surfaces also. Joppa's corrections will also handle rectangular tunnels.

6.23. V/STOL DOWNWASH CORRECTIONS—HEYSON'S METHOD

Glauert's classical wall corrections for wings assumed that all of the wing's lift is due to circulation and that the vortex wake trails straight aft in the plane of the lifting line and the freestream, which yields a vertical interference at the model. This assumption is not valid for V/STOL or powered lift systems, where the downwash at low forward speeds will approach the 90° required in hover. Heyson at NASA Langley began his work on the wall correction problem with rotors in Ref. 6.36.

Heyson in Ref. 6.37 represented the wake by a semiinfinite string of point doublets whose axes are tilted relative to the model's lift and drag. This wake proceeds linearly to the tunnel floor and then aft along the floor. Then images and superposition are used to find the effect at the model. Heyson's method yields both vertical and horizontal interferences rather than the vertical interference only given by Glauert. Both methods agree in interference factors for a wake trailing straight aft, when Heyson's $\delta_{w,L}$ is multiplied by $-\frac{1}{4}$ to account for different definitions of δ.

As outlined in Ref. 6.37, the four interference factors are calculated at selected points in the tunnel for a point or vanishingly small model. This same report demonstrates how doublet strings may be distributed along a wing or a rotor and shows the results for a tunnel with a width to height ratio of 2.

Heyson in Ref. 6.38 extended the use of superposition of Ref. 6.37 using digital computers to obtain interference factors for swept and nonswept wings, tails aft of wings, single rotors with tail, and tandem rotors. This work gives numerical results of the affect of many variables on the correction factors. The Fortran programs are given in Ref. 6.39.

Heyson further extended his work on Ref. 6.38 to a general theory of wall interference in closed rectangular test sections plus ground effects in Ref. 6.40. In this work large wake deflections are addressed, in both vertical and lateral directions. This work includes a program to determine interference factors at a point (vanishingly small model), and a discussion of the program and flow charts for the program. Owing to both lateral and vertical wake deflections, there are nine interference velocities. As lateral directional testing is concerned with moments, the gradient of the interference velocities are needed, which yields 27 velocity gradients. Rather then develop actual correction formulas for all possible models, the paper discusses tunnel-wall effects as a problem in similitude.

The following discussions of Heyson's corrections will be limited to the plane of symmetry of the aircraft and will follow Refs. 6.37 and 6.38. These corrections use a linear wake that is deflected downward. The angle of the wake deflection is a function of the lifting systems' lift and induced drag. The four wall interference factors are a function of the wakes's deflection. Hence a new set of factors is needed at each test point. In the classical corrections, the wall correction factor is constant for a given tunnel and model. Heyson's works have copious numbers of figures, which could be used to calculate the wall effects. However, the task of determining the wake skew angle, reading eight interference factors for the wing and tail as a minimum, calculating the interference velocities, and applying them to the data would be time consuming with a large chance for error. Thus the use of Heyson's methods implicitly assumes that a digital computer will be used.

To calculate the interference factors at any point in the tunnel, say the centerline of the wing, first requires the wake skew angle χ. This is the angle from the vertical to the wake and is the complement of the momentum downwash angle. The skew angle can be obtained from charts in Refs. 6.37 and 6.41, or calculated as outlined in Ref. 6.37.

The interference factors are based on χ effective, not the momentum value of χ. In forward flight the wake of a lifting system changes form and angular deflection as it moves aft of the model. The wakes for wings, rotors, and jets eventually evolve into a pair of trailing vortices. As these vortices roll up they do not move downward as rapidly as the central portion of the wake; almost all of the momentum transfer induced by the lift is in the central portion of the wake. This region is constrained to be eventually within the trailing vortex pair after roll up.

The actual downward deflection of the wake vorticity in the rolled up wake is approximately one-half that predicted by momentum. This was first pointed out by Heyson in Ref. 6.42. This paper established the criteria of an effective skew angle that is used to determine the interference factors. To handle the hover case (although testing in a tunnel may be questionable) the following definitions for downward deflection angle or wake skew angle are from Ref. 6.43:

$$\tan \theta_e = \frac{4}{\pi^2} \tan \theta \tag{6.69a}$$

or

$$\tan \chi_e = \frac{\pi^2}{4} \tan \chi \tag{6.69b}$$

where $\theta = 90° - \chi$. Note:

1. Effective angles are used only for obtaining interference factors.
2. Induced velocities are obtained directly from momentum theory.

The second statement is required to avoid an imbalance between the forces and the induced velocities engendered by the forces.

The theory in Refs. 6.37 and 6.38 and programs in Ref. 6.39 yield four interference factors that are used to determine the interference velocities at the chosen point:

$$\Delta w_L = \delta_{w,L} \frac{A_m}{A_T} w_o \tag{6.70a}$$

$$\Delta u_L = \delta_{u,L} \frac{A_m}{A_T} w_o \tag{6.70b}$$

$$\Delta w_D = \delta_{w,D} \frac{A_m}{A_T} u_o \tag{6.70c}$$

$$\Delta u_D = \delta_{u,D} \frac{A_m}{A_T} u_o \tag{6.70d}$$

where the δ terms are similar to the wall correction factor δ in Glauert's work. A_m and A_T are the momentum area of the lifting system and tunnel cross-sectional area. w_o and u_o are the momentum theory value of the vertical and longitudinal velocities at the lifting system due to the lifting system only. Δw_L and Δw_D are the boundary-induced vertical interference velocities (positive upward) resulting from L, lift, and D_i, induced drag. Δu_L and Δu_D are the boundary-induced longitudinal interference velocities (positive to the rear) from lift and induced drag.

The interference factors, $\delta_{w,L}$, $\delta_{u,L}$, $\delta_{w,D}$, and $\delta_{u,D}$ are functions of the tunnel geometry, height of the model in the tunnel, span loading, angle of attack, and wing sweep of the model. See Ref. 6.38 for superposition and uniform and elliptical span loads. As Ref. 6.37 derives these factors for a vanishingly small model, we will follow this report and apply the corrections.

The four interference velocities are combined to yield the total interference at the model:

$$\Delta w = \Delta w_L + \Delta w_D \tag{6.71}$$

$$\Delta u = \Delta u_L + \Delta u_D \tag{6.72}$$

The corrections are applied to the data as follows: the subscript c refers to values corrected for interference. Figure 6.55 is a sketch that illustrates the corrections of angles, velocities, and forces.

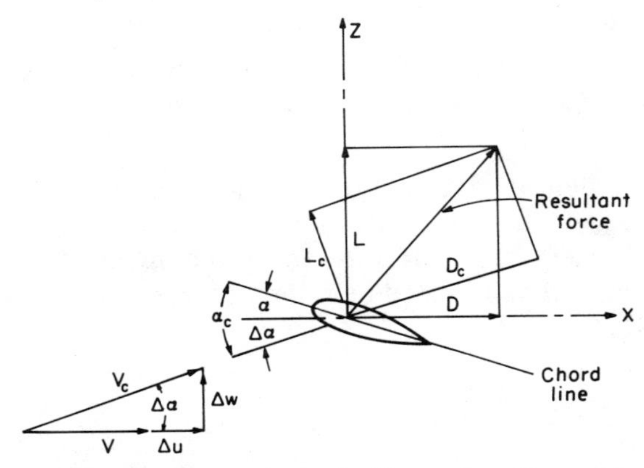

FIGURE 6.55. Nomenclature for Heyson's wall corrections.

$$\alpha_c = \alpha + \Delta\alpha \tag{6.73}$$

$$\Delta\alpha = \tan^{-1}\frac{\Delta w}{V + \Delta u} = \tan^{-1}\frac{\Delta w/V}{1 + \Delta u/V} \tag{6.74}$$

Note that Heyson does not use the small-angle assumption of classical wall corrections.

$$V_c = \sqrt{(V + \Delta u)^2 + \Delta w^2} \tag{6.75}$$

or, in terms of q or dynamic pressure:

$$\frac{q_c}{q} = \left(1 + \frac{\Delta u}{V}\right)^2 + \left(\frac{\Delta w}{V}\right)^2 \tag{6.76}$$

as lift and drag are defined as perpendicular and parallel to the relative wind or the corrected velocity.

$$L_c = L\cos\Delta\alpha - D\sin\Delta\alpha \tag{6.77}$$

$$D_c = L\sin\Delta\alpha + D\cos\Delta\alpha \tag{6.78}$$

and finally,

$$C_{L_c} = \frac{L_c}{q_c S}, \qquad C_{D_c} = \frac{D_c}{q_c S} \tag{6.79}$$

For the interference distribution over swept and nonswept wings, rotors, etc., and interference at the tail, see Ref. 6.38. The tail corrections, as with classical wall corrections, are based on the difference between the wall-induced angle of attack at the wing, and wall-induced angle of attack at the tail. This is the equivalent to a change in tail incidence and, of course, varies with the forces on the wing or lifting system. In Ref. 6.44 Heyson discusses two methods of correcting the tail pitching moment. The first accounts for the difference in pitching moment caused by the tail in the tunnel and the moments if the tail had the same wall interfrence as the wing. The second takes the difference in wall-induced angles of attack at the wing and tail and rotates the tail incidence by this amount.

In Ref. 6.38 there is a detailed discussion of the effect of model size (wing and rotor), wing sweep, and the location of a swept wing's pivot on the interference factor $\delta_{w,L}$. It should be realized that although the factor $\delta_{w,L}$ will vary, the induced velocity Δw_L is also influenced by the model and tunnel momentum areas [Eq. (6.70a)].

References 6.38 and 6.43 make extensive comparisons between Heyson's theory and the conventional corrections after Glauert. The correlation between the two methods is quite good at zero induced drag. At high lift coefficients (low skew angles) Refs. 6.37 and 6.38 indicate vertical interference factors greater than conventional theory. For an effective skew angle between 75° and 90° the magnitude of $\delta_{w,L}$ remains about constant at values equivalent to the conventional corrections. An effective skew angle of 75° is equivalent to a momentum skew of 57° or a momentum downwash angle of about 33°.

If a swept wing's aerodynamic center and pivot point coincide at the center of the tunnel, then the average $\delta_{w,L}$ shows little or no change up to an angle of attack of 20° for effective skew angles greater than 65° for a 45° swept wing. If the pivot is moved to the apex of the lifting line, there is little effect on $\delta_{w,L}$ at effective skew angles above 75° (momentum downwash = 33°) for a wing with 45° sweep.

Heyson's corrections would be useful for the occasional odd model, once the program in Refs. 6.37 and 6.38 are adapted to a tunnel's computer, with the assumption that the computer is of adequate size to carry out the calculations in a reasonable time.

One more point must be considered when testing models at low skew angles or large momentum downwash angles. As pointed out in Ref. 6.43, flow breakdown caused by an interaction between the model wake and the tunnel will result in a tunnel flow that does not reasonably approximate the uniform flow of level flight. This limit occurs approximately when X_f/b = 1.25 where $X_f = h \tan \chi$, h is the height above the floor, and χ is the momentum skew angle (see Section 6.33).

6.24. DOWNWASH CORRECTIONS: VORTEX–LATTICE (JOPPA'S METHOD)

Another approach to wall corrections for wings developing high lift was proposed by Joppa in Refs. 6.45–6.47. Of these, Ref. 6.47 is the most valuable as it contains the computer programs. Joppa's work, like Heyson's, has only become practicable with the advent of high-speed digital computers. Rather than allowing the wake or vortex to trail straight aft as in classical wall corrections, or assuming a deflected linear wake, Joppa took the following approach.

The wing vortex wake in the tunnel follows a different trajectory in the tunnel than in free air. Joppa calculated the wake trajectory and its flow field in free air by assuming a simple lift system represented by a horseshoe vortex. This lifting system then has its trajectory and flow field calculated in a wind tunnel at the same remote velocity and circulation strength. The differences are then interpreted in terms of wall interference.

The free-air trajectory is found using vortex segments that are about 0.10 of the vortex span. The wing is taken as having a uniform load with a bound vortex on the $c/4$ and a vortex span of $(\pi/4)b$. This is the usual representation of an elliptical load. The first trailing segment lies in the wing from the bound vortex to the trailing edge. The angle of the first downstream trailing vortex is determined by adding the induced angle of attack and the effective angle of attack at the plane of symmetry. The direction of each downstream segment in turn is calculated by the sum of velocities due to all other elements at its upstream end. This determines the coordinates of the downstream end; the following downstream segment is then translated to that point, keeping it attached. Then the process is repeated and the vortex trajectory is moved into place by sweeping along its length by successive iterations. As the vortex curves down it also curves inward at first, reducing the vortex span. The final trajectory is determined by first calculating the downward deflections and then a second pass for the horizontal or inward deflection. This double iteration is necessary to avoid an instability in the path after a few iterations. The path converges in three or four double passes before the instability develops. It is interesting to note that a curved vortex line is itself physically unstable.

The tunnel walls could have been represented by curved image vortex systems as in the conventional method, but this method only works for rectangular tunnels. To make the method applicable to any tunnel shape the tunnel walls are represented by vortex lattices. The strength of each element of the lattice is determined by simultaneously requiring the normal component of velocity to vanish at the center of each lattice. This satisfies the boundary condition of no flow through the tunnel walls at the control point. By this method the geometry of the image flow field does not change during each iteration and the large matrix of coefficients need only to be inverted once. For each vortex ring an equation at its control point is written where the strength of the ring Γ_i is the unknown. To keep the number of equations to a tractable level some judgment must be used as described in Ref. 6.47.

When the equations for Γ_i are solved and Γ_i is known, the wall-induced velocity at any point in the test section can be calculated. The interference is expressed as an angle, which is the arc tangent of the vertical-induced velocity and tunnel velocity.

For a round tunnel Prandtl used two images (Fig. 6.19) to represent the wall. This gives the upwash at the wing lifting line. But as the effect of the lifting line was not taken into account, it will not yield the downstream variation in wall-induced upwash. In fact, the use of two semiinfinite vortices to represent the walls violates Helmholtz's vortex theorems. However, at the wing it does give the correct answer. Joppa's method can represent a round tunnel by approximating it as a multisided polygon. In Ref. 6.47 Joppa compares his work with a nondeflected vortex wake with approximate results from Refs. 6.23 and 6.34. Joppa's value of δ lies very close to half way between the approximate methods.

The solution for the wake trajectory in the tunnel is an iterative combination of the free-air-trajectory solution and vortex lattice wall solution. The wake solution method for free air is used with the velocities from the wall for an undeflected wake added to those of the wing in free air. When the wake trajectory is found, a second solution for the Γ_i's in the wall is found for this wake. Then the cycle is repeated. As there is no strong interaction in the two systems for normal models and tunnels, only two to three cycles are required. When the solution is complete for the same tunnel velocity and wing circulation, they yield the following:

1. Complete velocity field in free air.
2. Complete velocity field in tunnel.
3. The separate contribution to item 2 of the wing and tunnel walls.

The interference velocities are the difference between the velocity in free air and the tunnel at a given point. In Ref. 6.47 Joppa only presented results for the vertical velocity components because he limited the work to moderate wake deflections that yield small longitudinal components. The vertical interference velocity is felt as an angle-of-attack change. Thus

$$\Delta\alpha = \tan^{-1}(V_y/V_x)_T - \tan^{-1}(V_y/V_x)_{\text{F.A.}} \tag{6.80}$$

where subscript y is vertical and x is horizontal, with T the tunnel and F.A. the free air. As the calculations are made for the same circulation and freestream velocity values in the tunnel and free air, then,

$$\Delta\alpha = \delta \frac{2\Gamma b}{CV} \tag{6.81}$$

where C is the tunnel cross-section area. And thus

$$\delta = \left(\frac{CV}{2\Gamma b}\right)[\tan^{-1}(V_y/V_x)_T - \tan^{-1}(V_y/V_x)_{\text{F.A.}}] \tag{6.82}$$

The results in Ref. 6.47 are based on circulation lift only and are for a wing of low aspect ratio to achieve high values of C_L/AR and relatively large wake deflection angles.

The effect of the tunnel's relocation of the vortex wake on the pitching moment correction is quite sensitive to the location of the tail relative to the wing.

The energy wake is also relocated in the tunnel, and this affects the dynamic pressure at the tail. The dynamic pressure of the tail may be less than freestream owing to wakes from wing flaps or greater than free-air stream due to propulsion devices. Thus, as described in Section 6.21, the values of $dC_m/d\alpha_t$ should be measured.

If both the vortex wake and the energy wake shift the same amount in the tunnel when testing in the STOL flight range, Joppa proposes that the tail on the model be made vertically movable. Then, prior to the test the vortex wake shift for each value of wing circulation is calculated. Then stability testing would be done at several tail heights. This would produce a family of pitching moment curves, each one of which would be valid for a given wing lift coefficient. The final moment curve would be a composite curve from these curves. This would reduce the energy wake shift correction to zero due to the relocation of the tail. The only correction then required would be due to wall-induced effects.

6.25. DOWNWASH CORRECTIONS FOR SWEPT WINGS AND NONUNIFORM LIFT DISTRIBUTIONS

Corrections for swept wings can be determined by both Joppa's method (Section 6.24) for any shape tunnel, and Heyson's for rectangular tunnels (Section 6.23). When the aerodynamic center of the wing and the pivot point coincide and are near the tunnel centerline for a swept wing, the change in the vertical wall correction factor from a nonswept wing for moderate downwash angles is very small (Section 6.23; Refs. 6.38 and 6.43).

For arbitrary and nonsymmetric lift distributions, Joppa's method can be adapted. For rectangular tunnels in these cases, Heyson's method appears to be easier to use. Heyson in Ref. 6.40 covers a vortex wake that is deflected both vertically and laterally, and this should cover almost any case.

It should be recognized that adapting the methods of both Joppa, for arbitrary and nonsymmetric lift, and Heyson, for the doubly skewed wake, will require an extensive amount of programming with the attendant expense. Both Joppa's method (if the correction is made a function of lift) and Heyson's (where the corrections are functions of lift and induced drag) can have an impact upon the time required to reduce any test condition, depending on the computer used. With the classical corrections where the correction factor is a constant for a given tunnel–model combination, the wall corrections are of the form of a constant times the wing or tail-off lift coefficient. Both Heyson and Joppa, on the other hand, would require calculations of the wall correction factors based on the wing lift only for each test point. These values would be used for tail-on runs, as are conventional corrections, using a table look-up for the same model attitude and power setting.

For a research test the increased time to reduce the data may not be critical. However, for a developmental test where the results of one run are often used to determine the next run, a long time to reduce data may be critical, because in these types of tests the drive has been toward real-time data.

The vortex lattice method of Joppa can also be adapted to slotted tunnels (see Section 6.27).

6.26. DOWNWASH CORRECTIONS FOR POWER-ON TESTS

The slipstream due to a propeller will increase the lift over those portions of the wing that is immersed in the slipstream. Depending on airplane geometry, the propeller may also change the induced angle of attack of the wing. For other than some V/STOL aircraft these effects are assigned to power effects and the power-off values of δ and τ_2 are used. However, $a_t \overline{V} \eta_t$ must be evaluated if the tail is in the slipstream. This term may be a function of T_C (see Section 5.12). V/STOL corrections are covered in Sections 6.23 and 6.33 when the propellers are tilted to provide direct lift.

For an aircraft where the whole span of the wing is immersed in the propeller slipstream there are two possible approaches. The first would be to mathematically model the wing propellers and their slipstream, then use the method of images (Section 6.1) to calculate the effect of the tunnel walls. This would be a time consuming and expensive process and is probably not justified for one or two wind tunnel tests. The second approach is to use the power-off C_L's and δ's for the wall corrections. This approximation is based on the model having very high C_L's based on freestream q. But the wing C_L is a function of α, not q. The high model lift is due to the high slipstream velocity passing over the wing. In the moment correction, $a_t \overline{V} \eta_t$ would be evaluated with power on as the term may be a function of T_C (see Section 5.12).

6.27. MINIMUM OR ZERO-CORRECTION TUNNELS

There are three approaches to achieve either zero or small wall corrections for large or V/STOL models with their high lifts. These are an active wall tunnel where blowing and suction through porous tunnel walls is used, an adaptive wall tunnel where the solid tunnel walls are deflected, and ventilated or slotted tunnels. In the first two schemes the purpose is to cause the tunnel walls to form a streamline at the walls that has the same shape at the walls as would occur in free air. Similar concepts are being studied for automotive testing with the purpose of reducing the blockage effect to allow the testing of larger models in a given size tunnel. The active wall tunnel will be discussed first.

After demonstrating that the tunnel walls could be represented by vortex lattices (Section 6.24) Joppa carried this work forward to the concept of a tunnel using active walls to reduce the wall effects for STOL vehicles in Ref. 6.48, using the basic assumption that the tunnel walls are in the far flow field and thus potential flow will give an accurate description of the model's flow. Since the detailed flow at the model is not required, the model can be represented by a simple horseshoe vortex. Then, near the walls a control surface is constructed. As solutions to potential flow problems are unique, and if at all points on the control surface the flow is identical to free air, then the model will be in free air. Since the control surface represents free air, it can be replaced with a stream tube, and therefore only the normal velocity to the control surface need be controlled and made equal to zero. The tunnel operates in the following manner:

With the model at some attitude and tunnel velocity the model lift is measured. Then, with the simple flow model, the flow conditions at the wall are calculated. This calculated flow is used to control flow into or out of the tunnel, which then changes the model lift. The new lift is measured and the process repeated until the walls (control surface) coincide with the free air at the tunnel walls for the measured lift. In an ideal case where injection is continuously distributed, both the mass flow and momentum across the control surface will be matched yielding a perfect simulation. For a practicable application the flow injection cannot be controlled to yield a continuous distribution. For a given wall porosity a large number of small jets were used to match the mass flow and momentum. But both mass flow and momentum could not be simultaneously matched.

To prove this concept tests were made with a two-dimensional wing. Free-air conditions were assumed to be met when the tunnel height was 12 times the wing chord. The active wall tunnel used a tunnel height of twice the chord. This was done by the use of inserts that extended $4\frac{1}{2}$ chord lengths ahead of and behind the model and contained 24 plenums to control the flow through the ceiling and floor. Tests were made with both 5% and 31% porosity, and using injection rates from matching mass flow to matching momentum.

Two-dimensional testing with large wing chords to tunnel heights results in the model acting like one with increased camber (Section 6.6). This increase in camber increases with angle of attack and results in a steeper lift curve slope that yields a higher lift coefficient for a given angle of attack. The active walls will reduce lift at a fixed angle of attack.

In his data, Joppa shows that matching mass with a large momentum mismatch overcorrects the data, and for momentum match with mass mismatch the data are undercorrected. This occurred for both porosities but with a much closer match to the free-air curve for 31% porosity when compared to the 5% porosity. These experiments by Joppa demonstrate that the active wall concept works in both theory and practice if both mass flow and momentum can be matched, and that only the flow normal to the tunnel walls need be controlled using the model's measured lift.

Similar work is being done in the transonic speed range (Ref. 6.49).

The second approach is the adaptive wall where the contour of the walls is made to conform to a free-air-stream tube. For a two-dimensional airfoil the tunnel has a flexible ceiling and floor. With the model installed the wall pressures are measured and compared with a calculated pressure distribution at the walls for an unlimited flow field external to the tunnel. The wall shape is altered until the two pressure fields agree within limits. If the agreement is perfect, then the tunnel flow is the same as free air. There are two approaches to calculating the unlimited flow field. One used by Ganzer in paper 4 of Ref. 6.50 uses wall pressures and wall contour to calculate the unlimited flow field. This method requires no knowledge of the model and its local flow field. The second method in Ref. 6.51 used a doublet to represent the model.

For two-dimensional wings both Ganzer and Goodyear in paper 7 of Ref. 6.50 have shown that the method works in general, although there are some difficulties when either shocks or large pressure gradients are present. Goodyear also demonstrated that the method works for the blockage of a nonlifting body (cylinder) up to a blockage ratio of 29.27%.

For the three-dimensional model, a more difficult problem, Ganzer is working on an octagonal shaped test section (Ref. 6.51), while Goodyear is keeping the side walls rigid and varying the ceiling and floor.

A small tunnel with variable contour side walls and ceiling and a fixed floor was used to investigate the blockage for automobile tests (Ref. 6.52). As the tunnel floor represents the road, it was kept fixed. There were three flexible strips in the walls and six in the ceiling. These strips slide on fixed wiper plates that extended into the tunnel when the strips deflected away from the model. Three similar representative models of automobiles were used to produce blockage ratios of 10%, 20%, and 30%, and tests were made at 0° and 10° yaw. Static pressures were measured on the model and the walls and ceilings. The model pressures for the three blockage ratios agreed for the contoured walls and their levels were different than straight walls for 10% and 20% blockage. The wipers did not appear to have any effect on the data, but this may not be true for a lifting model with a deflected wake.

Sears in Ref. 6.53 has proposed an adaptive wall tunnel that prevents the wake of a V/STOL model from impinging on the tunnel walls.

Slotted Tunnels

In transonic testing slotted or porous test section walls with a surrounding plenum have been used to prevent the test section from choking at high Mach numbers. There also have been several low-speed tunnels built for V/STOL testing that have slotted walls. This results in mixed tunnel boundary, part solid and part open. The mixed wall boundary conditions have made it very difficult to determine the proper wall corrections for this type of test section. Work is being done on two- and three-dimensions for both blockage and deflected wakes.

It might be well to review the potential theory methods of determining wall and blockage factors for corrections. These are:

1. The image method.
2. Vortex lattice method.
3. Wall perturbation method.
4. Methods using measured wall pressures.

For slotted or porous wall tunnels the image method will not work owing to the wall boundary conditions. The image system has been used for tunnels that have either an open or closed test section. Vortex lattice methods are used for closed test sections. For slotted tunnels, a vortex lattice is used for the solid portion of the walls and source panels for the open portion. Inflow or outflow from the plenum determines the sign. Linear small-perturbation theory was used by Ganzer for the adaptive wall tunnel. The wall static pressures are used in the adaptive wall methods also, as well as for blockage corrections with fixed solid walls (see Section 6.11).

Based on the reversal of sign of the wall correction factor between open and closed test sections, it has been reasoned that a ventilated test section could reduce both the magnitude and inhomogeneity of the wall corrections for large models and high lift coefficients.

There is a large amount of work using different approaches for low-speed and transonic tunnels in attempting to calculate wall corrections for slotted tunnels. The basic problem is in predicting how the tunnel air on the suction side of the model flows into and out of the plenum through longitudinal slots or porous walls. But at the time this edition was written there is no completely accepted solution for correcting force and moment coefficients for lifting models in this type of tunnel. It may well be that the problem is to match both the mass flow and momentum through the walls simultaneously as noted by Joppa.

The use of slotted walls and ceiling with a solid floor for automobiles has also been investigated in Ref. 6.54. Again, the purpose is to increase the allowable blockage ratio. In this work the model was at zero yaw and its location within a test section, equal to twice the automobile length, had to be empirically determined. The reentry of plenum air into the diffuser had to be carefully controlled. The slotted tunnel was able to test up to a 21.4% blockage ratio. It should be noted that solid body blockage is much simpler to handle than a high lift model.

6.28. DOWNWASH CORRECTIONS FOR REFLECTION PLANE MODELS

The main purpose of testing reflection plane models is to get the largest model size and hence the largest model Reynolds number. In turn, the large model size may require special attention to wall corrections. Another effect is that the reflection plane reflects, and under some conditions (aileron down, or vertical tail, for instance) an undesirable reflection for which special allowances must be made is obtained.

Reflection plane tests are conveniently divided into four classes:

1. Small symmetrical models (less than $0.6h$) such as halves of flapped wings, or horizontal tails.

2. Small unsymmetrical models such as aileron panels.
3. Small vertical tail models where a reflection is not desired.
4. Large reflection plane models of all kinds.

The first three may be reasonably treated in the space available, but detailed studies of large models will require consulting the work by Swanson and Toll (Ref. 6.17) or Heyson's work in Refs. 6.37–6.40 can be used. Heyson's work in Ref. 6.40 is capable of handling a skewed wake, both vertically and laterally and thus could handle a horizontal and vertical tail with both rudder and elevator deflections.

Small Symmetrical Models

The data for a small reflection plane model which is half of a symmetrical model may be corrected by treating the upwash and blockage as though the entire model were in a tunnel of double the width. (See Fig. 6.56.) The values of δ for such a setup are in Fig. 6.29, and those for a circular-arc tunnel are in Fig. 6.37. One normally gets a slightly lower lift curve slope and slightly higher induced drag than in the complete-model, complete-tunnel case, owing to some vortex shedding in the root boundary layer. When panel area and MAC are used the final data are directly applicable to the airplane if the split is along the plane of symmetry with an added amount of span to allow for the boundary layer displacement thickness. Sections 5.7 through 5.10 should also be consulted.

Small Unsymmetrical Models

When the model is unsymmetrical (aileron deflected, for instance), additional troubles accrue since the reflection will be symmetrical. Thus in this case the tunnel data include a small carry-over from the reflection and will show from one-tenth to one-fourth more increment of lift, drag, and pitching moment, yawing moment, and rolling moment than they should. Since tunnel data will be high for ailerons because of the failure to simulate aileron cable stretch and wing twist, some engineers plot up the span loadings with

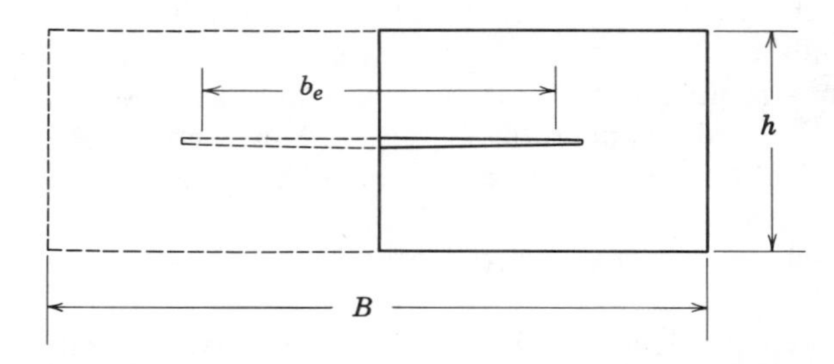

FIGURE 6.56. Nomenclature for a reflection plane model.

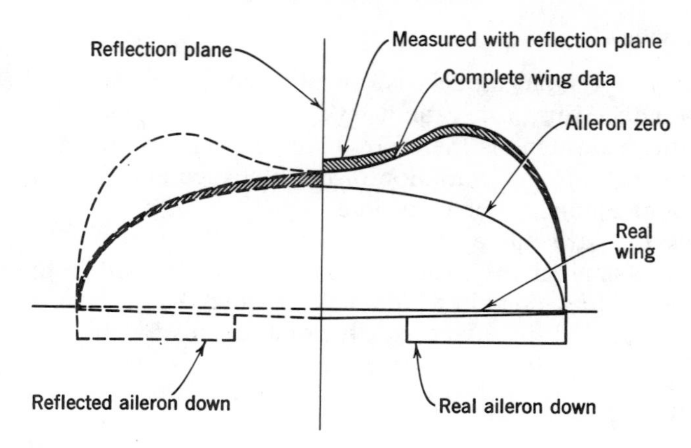

FIGURE 6.57. Effect of reflection plane on panel with aileron down.

ailerons zero and deflected and reduce their measured data by the proper carry-over increment.

The only time this effect can be misleading in comparing different ailerons on the same basic panel is when one aileron has more span than another. As seen in Fig. 6.57 the reflection effects increase with span and the aileron of greatest span yields the data most erroneously high.

Difficulties arising from the doubled increment of lift, drag, and moment have been covered in Section 5.8, which must be consulted for proper interpretation of the results.

Small Vertical Tail Models

The small vertical tail models present a number of additional difficulties since the degree of endplating given by the fuselage and horizontal tail is very difficult to predict. One approach is to consider the vertical tail a completely reflected symmetrical semispan wing and, after determination of the slope of the vertical tail lift curve from the complete model tests, to reduce the panel test data to conform. Normally the vertical tail drag is of so little interest that an increase of panel drag to allow for the reduced aspect ratio is not required.

Another approach, and perhaps the best one, is to determine the slope of the lift curve of the vertical tail from complete model tests and use the hinge moment data from panel tests at the same lift coefficient. This neglects the difference of span loading for panel and actual vertical tail installation. Still another alternative is to build a complete tail assembly model. Here the model is large enough for high Reynolds number, and actual endplating is well simulated. The effects of sidewash must be obtained from the complete model tests and incorporated into the data. The tail assembly model should have a fuselage stub nose at least one MAC ahead of the tail quarter chord.

Large Models

The analysis for large models is complicated fundamentally because the boundary-induced upwash cannot be considered constant along either the chord or the span. One is therefore justified in taking greater pains for the upwash corrections, but variation of the blockage along the model almost never is large enough to be considered.

Besides the streamline curvature effects, the variation of boundary-induced upwash along the span of the model tends to load up the tips. Some test section shapes do this to a smaller extent than others, or, if a split fuselage is included, its effect usually compensates for the tip loading, and standard wall corrections are satisfactory.

6.29. TESTING WITH A GROUND PLANE

When an aircraft operates within a semispan or less from the ground, its downwash pattern is altered and decreased by the ground (Ref. 6.55). This results in an increase in the lift curve slope, a decrease in induced drag, and an increase in the slope of the pitching moment curve. The latter results in an increase in elevator angle to balance the airplane; the hinge moments are usually unaffected. The flight condition at desired forward CG position and $C_{L\,\max}$ usually determines the elevator power. This condition is simulated in a wind tunnel by a ground board or ground plane. The ground plane spans the tunnel, and for an external balance it must be adjustable in height since it may have to be adjusted to different heights during a test. The supports under the board must have minimum blockage and in larger tunnels will have to support personnel working on the model. For a sting support, a sting that can change the model height at constant angle of attack is very convenient.

Turner in paper 25 of Ref. 6.42 determined a limit for testing with a conventional ground board. The limit was for models with full-span high lift configurations and checked data for models with aspect ratios from 6 to 10, using tilt wings, jet flaps, and double slotted flaps. The limit is a function of the ratio of the height above the ground plane to model span versus C_L. When $C_L > 20h/b$, a moving belt ground plane should be used. As with all such limits the line represented by the equation represents a gray area where the data should be questioned. The limit in this case is similar to flow breakdown for V/STOL models (Section 6.33). The downwash from the model tries to move forward in the ground plane's boundary layer and a vortex is formed on the ground plane that distorts the tunnel flow. The rule of thumb for V/STOL models with distributed lift is almost exactly the same. The effect on the tunnel data that are most easily detected is a decrease in the lift curve slope when testing the fixed ground planes.

Some tunnels are equipped with a moving belt ground plane. These usually require sting balances. The boundary layer is usually sucked off at the leading edge of the belt. Thus, if the belt moves at the test speed, there is no boundary layer on the belt as with a fixed ground plane. The speed of a moving belt appears to be limited by belt slippage around the drive roller. These units are expensive, and it is difficult to make the belt track. The sting must have adjustable, but positive, limits to the model α range and the sting lower height. Nothing is more embarassing than to have the sting put the model on the belt. This can rapidly sand off the bottom of the model or landing gear. The heat generated can weld the belt to its platen. The adjustable or positive limits are required for any ground plane tests to protect the balance, either internal or external. The combination of the model fouled on the ground plane and the power of most pitch actuators will quickly exceed the strength of the balance.

A second approach to removing the boundary layer in fixed ground planes can be achieved either by suction or by blowing through slots to replace the momentum lost in the boundary layer. The thickness of the boundary can also be controlled to some extent by a flap on the trailing edge. The large General Motors automotive tunnel, where cars are tested on the floor, uses suction at the leading edge of the test section to remove the boundary layer. This air then enters the tunnel behind the test section to maintain continuity.

For aircraft testing the ground plane should extend, as a minimum, to the nose of the model and aft of the tail.

Often both the upflow at the wing and the dynamic pressure will be a function of ground board height (Section 3.11). The data are usually corrected for blockage and, if required, buoyancy. The tunnel wall or boundary corrections usually are not applied since the model is very close to the floor. If boundary corrections are required, they only include the two walls and the ceiling. Interestingly enough, if all four boundaries are accounted for, you again get results for free air.

6.30. SUMMARY OF THREE-DIMENSIONAL CORRECTIONS FOR A CLOSED TEST SECTION

In this section the complete data reduction process for a wind tunnel test of a complete airplane will be summarized. This will include portions of Chapters 3, 4, and 6. The process can be done by hand (pocket calculator) or by digital computers. When a digital computer is used, it will be apparent that to attempt to reduce the data in a close to real-time mode requires a relatively large capacity computer.

1. *Keeping Track of the Data.* A typical pitch run can have 20 or more α's at which data are taken. In the data reduction there are either tables or curves that must be used to obtain required values, thus there has to be a common variable to keep track of the data.

Quite often when data are acquired by a computer, the computer will assign a sequential number to acquired data. This is analogous to the old IBM card number. This is not a good variable to use to keep track of the data for the following reasons. During a test the values of α or ψ used may not be the same for all runs, nor is it reasonable to require that they be the same. The number of angles may be reduced to save running time, they may be increased to define a curve better, or some angles may be repeated when checking for bad points.

Based on the above discussion the only way to keep track of the model attitude when acquiring data is to use a measured value from the model. Thus, the measured α and ψ or β, or roll angle from the model positioning mechanism should be used.

2. *Assumptions.* It is assumed that the model angles and the balance forces and moments are in engineering units, that is, angles in degrees and forces and moments in pounds and inch-pounds. This means that the measured quantities have been converted from bits or voltages to engineering units through the applicable calibrations and unit conversions.

The discussions that follow will assume that an external balance is used to avoid taking into account all of the possible variations that exist for stings and internal balances. Since the discussion is general in nature, it can easily be adapted for other model support and balance systems.

3. *True Balance Loads.* The loads read out of the balance are corrected to the true loads applied to the balance by the balance calibration equations (Sections 4.14 and 4.15). These are the values required in all subsequent calculations. They will be designated by subscript u, that is, L_u, etc. The balance calibration is applied to all loads, both aerodynamic and tares.

4. *Applying Mechanical Calibrations to α and ψ.* If the measured values are not the actual model geometric angles, apply the calibrations from the indicated angle to the geometric angle. This most often occurs for α due to location of the trunnion and the pitch arm.

If corrections are required for balance and model support deflections, these are also applied. Generally these deflections are obtained by applying known loads to the balance support system and model, so the loads from 3 are used. True geometric angle will be designated by the subscript g, that is, α_g, ψ_g.

5. *Basic Corrections to Dynamic Pressure.* Most tunnels are run by an indicated dynamic pressure obtained from two measured pressures. The indicated dynamic pressure is then corrected for the tunnel dynamic pressure calibration (see Section 3.11). This dynamic pressure will be designated q_A.

If the data system measures the error in the operator's maintaining the desired dynamic pressure, this value will be added to q_A as a Δq_{op} value.

Thus, at any data point the dynamic pressure approaching the model is $q_A + \Delta q_{\mathrm{op}}$.

6. *Summary of Corrections.* At this stage of the data reduction process there is a table for each run of the true test values for model geometric angles, dynamic pressure, and balance loads. The following corrections will be due to the tunnel walls.

7. *Blockage Corrections.* *Solid blockage,* ε_{sb}. This is a function of the model volume and the tunnel size. As indicated in Section 6.10 it usually is in two parts: one for the wing and a second for the body, which are summed to yield ε_{sb}. If possible, these values should be determined prior to the start of the test.

Wake blockage, ε_{wb} (see Section 6.11). If the wake pressure drag blockage [Eqs. (6.33) and (6.34)] are to be used they should be determined prior to the tests. These values are quite small for most models and are often neglected.

Maskell's corrections for a separated wake are often applied and to an extent include the wake pressure drag blockage terms. Maskell's correction is in two parts [Eq. (6.35)]. The first is based on the parasite drag C_{D0} and the second on the separated wake. The first part includes parts of Eqs. (6.33) and (6.34) and both corrections should not be applied. For unseparated flow, $C_{Du} = C_{Di} + C_{D0}$, and for separated flow, $C_{Du} > C_{Di} + C_{D0}$. Owing to data scatter, using a computer to reduce data usually requires a statement that if $C_{Du} - C_{Di} - C_{D0}$ is negative, set it to zero. Hackett's blockage must be determined by the wall pressures (see Section 6.11).

The final blockage correction is

$$\varepsilon_T = \varepsilon_{sb} + \varepsilon_{wb} \tag{6.83}$$

$$q_C = q_A(1 + \varepsilon_T)^2 \tag{6.84}$$

$$V_C = V_A(1 + \varepsilon_T) \tag{6.85}$$

It should also be noted that the blockage corrections are required to produce the correct dynamic pressure that is used to calculate all coefficients including pressure and hinge moment, etc. Thus, when blockage corrections are used, the force data must be acquired and reduced before the pressure coefficients, hinge moments, etc., are reduced.

The dynamic pressure is now fully corrected unless Heyson's wall corrections are used. In Heyson's corrections there is an additional correction to dynamic pressure due to the large wake delections (see Section 6.23).

8. *Support Tares and Interferences.* In tests where only incremental values are of primary interest the support tare and interference values are either not applied or a general set that do not take into account the interference of the supports on the model are applied. As the general tares are taken from another model, and as the mounting struts for an external balance enter the wing, these tares are best applied as a function of C_{LW} rather than α. This is because of possible differences in α_{L0} and the reference plane used to measure α. The tares for the appropriate q_I or q_A should be used. This is

done to save test time and costs incurred in evaluating all required support tares and interferences for each model and its various configurations.

If the tares are taken with the model being tested, the appropriate tare for the model configuration at the correct α_I or α_g and q_I or q_A are used (Section 4.17).

The tares are subtracted from the balance force data of Section 3.

9. *Weight Tares*. The weight tares are a result of the model center of gravity not being on the balance moment center. Thus when the model is pitched, there will be a pitching moment versus α_I due to the weight. When the model is yawed, there will be both a pitching and rolling moment due to weight. Similar tares will arise with a sting and internal balance. The tare values are subtracted from the balance data of Section 3. The weight tares are a function of model configuration, thus the correct tare must be applied to the correct run.

If additional tares are required, such as internal nacelle forces, they also must be applied at this point (Section 5.13). There is no order of preference in the order of subtracting support, weight tares, and other tares.

At this point the data have the correct geometric angles, correct dynamic pressure, and correct aerodynamic forces on the balance (corrected for support, weight, and other tares).

10. *Moment Transfers*. Now the balance forces and moments are transferred to the desired center of gravity location on the aircraft. These calculations only affect the three moments: pitch, yaw, and roll. The equations required are a function of the type of balance, which determines the relationship between the balance moment center, the model trunnion, and the location of the desired center of gravity. Thus, it is difficult to write the actual equations for all cases. As an example, if the balance moment center and the model trunnion coincide, then the following information is required. The three distances from the moment center to the center of gravity along the three orthogonal tunnel centerlines passing through the balance moment center act on each combination of model α and ψ. The three lengths will change as the model is pitched and yawed since the desired CG moves relative to the balance moment center. Because the equations are a function of geometry, they must be worked out for each tunnel and model. The pitching moment is affected by lift and drag, the yawing moment by drag and side force, and the rolling moment by lift and side force.

The customer or the model builder must furnish the distances between the trunnion and the desired center of gravity locations. The tunnel group knows the relationship between the trunnion and balance moment center.

Extreme care must be taken to make sure that all tare values (support, weight, other) have been applied before the balance data are transferred from the balance moment center to the desired center of gravity. Many of the these tares involve lift, drag and side force that are used in the moment transfer.

11. *Uncorrected Coefficients.* The table of data from section 9 is now reduced to coefficient form using the corrected dynamic pressure q_C. The dynamic pressure due to wake blockage and operator error (see Sections 5 and 7) will vary with α_g or ψ_g. The value associated with the angles under consideration is used. The lift coefficient as well as yawing moment, rolling moment, and side force have no further corrections, and thus are final corrected values. The drag, pitching moment, and angle of attack need further corrections.

12. *Wall Corrections.* The wall correction theories of Glauert, Heyson, and Joppa are based on the forces generated by the model lifting system. The spanwise lift distribution can be either taken as uniform, or can be distributed in an elliptical or other load for the wing. Which spanwise lift load is used depends on how the lifting surface was mathematically represented. In general, Glauert and Joppa use a uniform load with a reduction in vortex span and the same circulation at the centerline as an elliptical load. Heyson's latter work distributes the point doublets along the span and thus allows easily for either a uniform or elliptical load for wings, and uniform or triangular for rotors or any other desired loading. If desired, the load can be distributed along the span in Joppa's and Glauert's methods also.

For an airplane model this means that the wall corrections are based on using wing lift only. Thus, runs should be made with the wing only for wing and wing plus flap combinations. For some models this is not possible because the model was not designed to be tested with the wing alone. The reasoning behind this is that the fuselage is always there and thus it should be considered as a unit with the wing. Also, the amount of time required to assemble/disassemble the model for wing-alone runs may be prohibitive. In these cases runs are made with the horizontal tail off (tail-off runs) to obtain the required data for wall corrections.

The following discussion will use the conventional, or Glauert-type corrections, where the assumed vortex wake trails straight aft of the wing. In this case the correction factor at the wing is a constant. The plots in Chapter 6 are for this case. If either Heyson's or Joppa's method are used, the wall correction factors would be calculated based on the tail-off data for each angle of attack or C_L. See Sections 6.23 and 6.24.

Angle of Attack Correction

$$\alpha_C = \alpha_g + \Delta\alpha_{\text{up}} + \Delta\alpha_w \tag{6.86}$$

where α_C is the corrected angle of attack, α_g the geometric angle of attack (see Section 4), $\Delta\alpha_{\text{up}}$ is the tunnel upflow as determined in Section 4.16 by either probes or a calibration wing or as a part of evaluating the tare and interference, where

$$\Delta\alpha_w = \delta S/C\ C_{LW}(57.3) \tag{6.87}$$

δ is the wall correction factor and is a function of the model span to tunnel width ratio [see Eq. (6.55)], span load, location in the tunnel, and tunnel shape. C_{LW} is the wing or tail-off lift coefficient for the model wing configuration used in the run.

Drag Coefficient

$$C_{DC} = C_{Du} + \Delta C_{D\,\mathrm{up}} + \Delta C_{Dw} \tag{6.88}$$

C_{Du} is from Section 11.

$$C_{D\,\mathrm{up}} = C_{LW} \Delta \alpha_{\mathrm{up}} \tag{6.89}$$

$\Delta \alpha_{\mathrm{up}}$ is in radians.

$$\Delta C_{Dw} = \delta S/C\; C_{LW}^2 \tag{6.90}$$

Pitching Moment Coefficient

$$C_{m\mathrm{CG}C} = C_{m\mathrm{CG}_u} - \Delta C_{m\mathrm{CG}_t} \tag{6.67}$$

$$\Delta C_{m\mathrm{CG}_t} = \left(\frac{\partial C_{m\mathrm{CG}}}{\partial \delta_s}\right) \delta\, \tau_2 \left(\frac{S}{C}\right) C_{LW} (57.3) \tag{6.66}$$

$\partial C_{m,\mathrm{CG}}/\partial \delta_s$ is the rate of change of the pitching moment coefficient with stabilizer deflection, which is proportional to the change in angle of attack for the tail. It is $-a_t \overline{V} \eta_t$ [see Eq. (6.63)]. It is obtained by making at least two, but preferably three, tail-on runs at different stabilizer angles. Care must be taken in evaluating $\partial C_{m,\mathrm{CG}}/\partial \delta_s$ to make sure that the tail lift is in the linear portion of its C_L-α curve, that is, the tail is not stalled. Measuring $\partial C_{m,\mathrm{CG}}/\partial \delta_s$ avoids any error in assuming values for a_t and η_t. For a model with leading and trailing edge flaps the value of $\partial C_{m,\mathrm{CG}}/\partial \delta_s$ may vary. At low α's, when the leading edge is separated, it will have a lower value than the nonstalled region and again, at high α's, when the trailing edge flaps stall, the value will again decrease. In some cases the aerodynamic engineer prefers to use the variable value of $\partial C_{m,\mathrm{CG}}/\partial \delta_s$. τ_2 is the additional correction factor for differences in upwash from the tunnel walls at the tail and wing (see Section 6.21). Again, this term is a function of model geometry and size, tunnel geometry and size and location of the tail relative to the wing. The important parameters are tail length to tunnel width, and vertical location of the tail relative to the wing. If desired, the value of τ_2 can be varied to account for the vertical motion of the tail relative to the wing as the model is pitched.

The three wall corrections are usually worked up in the form of $K_1 C_{LW}$ for α, $K_2 C_{LW}^2$ for drag, and $K_3 C_{LW}$ for pitching moment. These constants, from Eqs. (6.87), (6.90), and (6.66), are stored in the computer versus the geometric angle of attack. There will be a set of these corrections for all combinations of the wing, flaps, spoilers, etc., that are tested. The correct set of wall corrections must be specified for each run.

13. *Summary of Data Corrections.* At this point the data have been corrected for angle calibrations, deflections, dynamic pressure calibration, operator error and blockage, forces and moments for balance calibration, support and interference tares, weight tares, and finally the tunnel boundaries. The data are at the desired center of gravity position or positions on the wind axes for an external balance.

14. *Transfer to Other Axes.* There are three axes systems that are used; these are wind, body, and stability axes. The axes can pass through either the balance moment center or the desired model center of gravity. The axes are defined as follows:

Wind Axes. The lift, drag, and crosswind force are parallel to the test-section centerlines. This axis system is used to define the lift as perpendicular to the remote velocity, the drag as parallel to the velocity, and the crosswind force as perpendicular to the plane of lift and drag. The moments are about the three axes.

Body Axes. This axis system is attached to and moves with the model. The lift is replaced with normal force, drag with axial, or chordwise, force and crosswind with side force.

Stability Axes. In this axis system the vertical axis, or lift, is perpendicular to the remote velocity, and the drag and side force yaw with the model (the yaw axis is not pitched with the model).

Most external balances measure about the wind axis and most internal balances measure about the body axis. Thus it becomes necessary to transfer from one axis system to another, depending upon how the data are intended to be analyzed. Since the data have already been transferred to the desired center of gravity (Section 10), the following equations can be used to transfer the data to the desired axes.

The wind axes to stability axes equations are:

$$\begin{aligned}
C_{LSA} &= C_L \\
C_{DSA} &= C_D \cos\psi - C_Y \sin\psi \\
C_{mSA} &= C_m \cos\psi + b/c\, C_l \sin\psi \\
C_{lSA} &= C_l \cos\psi + (c/b) C_m \sin\psi \\
C_{nSA} &= C_n \\
C_{YSA} &= C_Y \cos\psi + C_D \sin\psi
\end{aligned} \tag{6.91}$$

Subscript "SA" designates stability axis.

The wind axes to body axes equations are:

$$\begin{aligned}
C_{NB} &= C_L \cos\alpha + C_D \cos\psi \sin\alpha + C_Y \sin\psi \sin\alpha \\
C_{AB} &= C_D \cos\psi \cos\alpha - C_L \sin\alpha - C_Y \sin\psi \cos\alpha \\
C_{mB} &= C_m \cos\psi - (b/c)C_l \sin\psi \\
C_{lB} &= C_l \cos\psi \cos\alpha + (c/b)C_m \sin\psi \cos\alpha - C_n \sin\alpha \\
C_{nB} &= C_n \cos\alpha + C_l \cos\psi \sin\alpha + (c/b)C_m \sin\psi \sin\alpha \\
C_{YB} &= C_Y \cos\psi + C_D \sin\psi
\end{aligned} \tag{6.92}$$

The α is the geometric angle of attack and is not corrected for the tunnel upflow and walls. The subscripts are B = body axis, N = normal force, A = axial force, b = wing span, and c = mean aerodynamic chord.

The body axes to wind axes equations are:

$$\begin{aligned}
C_{LW} &= -C_A \sin\alpha + C_N \cos\alpha \\
C_{DW} &= C_A \cos\alpha \cos\psi + C_Y \sin\psi - C_N \sin\alpha \cos\psi \\
C_{mW} &= C_m \cos\psi + (b/c)C_l \cos\alpha \sin\psi + (b/c)C_n \sin\alpha \sin\psi \\
C_{lW} &= C_l \cos\alpha \cos\psi - (c/b)C_m \sin\psi + C_n \sin\alpha \sin\psi \\
C_{nW} &= C_n \cos\alpha - C_l \sin\alpha \\
C_{YW} &= C_A \cos\alpha \sin\psi + C_Y \cos\psi - C_N \sin\alpha \sin\psi
\end{aligned} \tag{6.93}$$

The subscript W = wind axes.

15. *Final Comments.* Although the calculations in the data reduction process are simple, a single run of 20 or 30 angles requires a large number of calculations. The number of calculations will be even larger if Heyson's or Joppa's wall correction theories are used, because they require the wall interference factors to be calculated at each test point. If it is desired to reduce the data in a very short period of time, a computer of medium size with lots of cache memory and disk storage will be required.

Furthermore, the support tare and interference values, weight tares, other tares such a internal nacelle drag, and wall corrections are all functions of the model configuration on a given test run. Thus, before a run is reduced the proper set of these corrections must be selected and the computer instructed to select the correct set. Under pressure to get the data out quickly, mistakes can be made. Therefore, at the end of the test these items must be checked on a run by run basis and any error corrected. Because of this, some tunnel operators will only release preliminary data during a test. This check on the data reduction process should be made before any computer readable set of data is released. If the data for support tares and other tares are taken during the test, considerable time will be required to work up and check these tares. Thus the data reduction of the final data will also be delayed.

The data reduction process has many constants that are functions of model size, such as wing area and moment transfer dimensions. Therefore, early in the test both a pitch and yaw run should be reduced using a pocket calculator (check point) to make sure that all the constants are installed correctly in the main computer program. The check point must be done independently of the computer program and use the original sources of model and tunnel values. If the two results do not agree, the error must be found and corrected. It is amazing how many constants can have numbers transposed or be entered in the wrong field.

It also should be noted that wall corrections usually are not applied to tests with ground planes because you desire the effect of the ground or lower boundary. Usually the dynamic pressure calibration and the upflow change when the ground plane is installed.

Although the data reduction process is a straightforward process, the chance for errors is large, so extreme care must be taken in carrying out the process, and the computer used must be capable of number crunching. A good data reduction person must be a good bookkeeper.

Example 6.3. Summary of Three-Dimensional Corrections (Closed Test Section). The model and tunnel dimensions are as follows:

The tunnel is 8 ft × 12 ft with four fillets 1.5 ft high. Thus the actual cross-section area is 91.5 sq. ft. The model dimensions are:

Wing: area = 7.1997 sq. ft, MAC = 12.8933 in., span = 88.182 in., AR = 7.50, 0.25 MAC is at model station (MS) = 49.980, waterline (WL) = 10.379, taper ratio = 0.30.

Horizontal Tail: area = 2.149 sq. ft, MAC = 9.3346 in., span = 35.184 in., AR = 4.0, 0.25 MAC is at MS = 90.240, WL = 16.902 at $\alpha = 0°$, $\delta_s = 0°$.

Model Trunnion: MS = 53.480, WL = 10.494.

The body has an overall length of 90.96 in., a maximum width of 8.88 in., and depth of 9.48 in. The maximum circumference is 29.5 in. for an effective diameter of 9.39 in. The wing volume is 0.6291 ft^3 and the body volume is 2.472 ft^3. The angle of attack is measured from the chord line of the wing MAC.

The data in Table 6.1 have been corrected for balance load interactions (3), and the mechanical calibrations have been applied to the angle of attack, α. There were no corrections applied for balance and model deflections.

Support tares and interference (8) will not be applied as the test was run for comparative data.

Blockage corrections (7). These are in two parts: a solid blockage due to the model volume and a wake blockage due to the wakes shed from the model. The solid blockage is from Eq. (6.28) for the wing and Eq. (6.29) for the body (Section 6.10).

TABLE 6.1

α_i	α_g	L_u	D_u	M_u	$-M_w$	$M_u - M_w$	M_{ac}	C_{LC}	C_{Du}	$C_{m_{ac_u}}$
					Tail Off					
2	2.48	356.0	41.18	−366.2	−28.4	−394.6	−1642.8	1.39914	0.16184	−0.50076
4	4.94	421.8	47.53	−80.2	−56.8	−137.0	−1622.2	1.65775	0.18680	−0.49428
6	7.41	482.4	54.39	183.4	−85.3	+98.1	−1601.8	1.89592	0.21376	−0.48793
					Tail On, $\delta_s = 0°$					
2	2.48	339.2	41.16	245.9	−36.5	209.4	−980.0	1.33258	0.16170	−0.29873
4	4.94	411.9	47.41	292.6	−70.0	222.6	−1228.0	1.16818	0.18625	−0.37417
6	7.41	481.6	54.89	346.2	−107.7	244.5	−1436.3	1.89200	0.21564	−0.43752
					Tail On, $\delta_s = -5°$					
2	2.48	31.81	41.39	1020.6	−36.5	984.1	−131.4	1.24934	0.16256	−0.04005
4	4.94	389.8	47.07	1064.1	−70.0	994.1	−379.0	1.53094	0.18487	−0.11548
6	7.41	456.6	53.64	1109.4	−101.7	1007.1	−602.1	1.79329	0.21067	−0.18341

TABLE 6.2

Configuration	C_{DP_e}	$1/\pi ARe$	$\frac{S}{4C} C_{D0}$	$5S/4C\ (C_{Du} - C_{D0} - C_{Di})$	ε_{wb_t}
Tail off	0.10068	0.031411	0.001980	0.00	0.001980
$\delta_s = 0°$	0.11108	0.028890	0.002185	0.00	0.002185
$\delta_s = -5°$	0.11820	0.028785	0.002325	0.00	0.002325

The wing thickness ratio on the average is 14% and the airfoil is similar to a 65 series, thus K_1 from Fig. 6.13 is 1.006. The model span to tunnel breadth is 0.61, for the tunnel $B/H = 1.50$, thus τ_1 is 0.88. The wing volume is 0.629 ft^3 and the tunnel cross sectional area is 91.5 sq. ft.

Thus

$$\varepsilon_{sbW} = \frac{(1.006)(0.88)(0.629)}{91.5^{1.5}} = 0.0006363$$

For the body, with an effective diameter of 9.39 in. and the length of 90.96 in., d/l is 0.1032. From Fig. 6.15 K_3 is 0.91. The volume of the body is 2.472 ft^3.

Therefore

$$\varepsilon_{sbB} = \frac{(0.91)(0.88)(2.472)}{91.5^{1.5}} = 0.002236$$

Then the total solid blockage, $\varepsilon_{sb_t} = 0.002872$.

Maskell's wake blockage (Section 6.11) will be applied by calculating the equivalent parasite drag and $1/\pi ARe$ (Section 5.5) for each of the three runs. The results are given Table 6.2, where $q_C = q_A(1 + \varepsilon_t)^2 = 35.00(1 + \varepsilon_t)^2$, the corrected dynamic pressure for each model configuration are in Table 6.3 (q_A from tunnel calibration).

Weight tares (9). The weight tares to pitching moment versus the angle of attack are listed and applied in Table 6.1. These tare values are corrected for balance interactions.

TABLE 6.3

Configuration	ε_{sb_t}	ε_{wb_t}	$(1 + \varepsilon_t)$	q_C
Tail off	0.002872	0.001980	1.004852	35.3405
$\delta_s = 0°$	0.002872	0.002185	1.005057	35.3549
$\delta_s = -5°$	0.002872	0.002325	1.005197	35.3647

Moment transfers (10). The moments will be transferred from the balance moment center that coincides with the trunnion to the 25% chord of the mean aerodynamic chord using the following dimensions taken from the model drawings at $\alpha = 0°$.

	Trunnion	25% MAC (wing)
Model station	53.480	49.980
Model waterline	10.494	10.379

The 25% MAC is 0.115 in. below and 3.50 in. forward of the trunnion. The slant distance from the trunnion to the center of gravity is 3.502 in. (this distance will be called a). The angle γ is positive clockwise from the tunnel centerline to the desired center of gravity. Thus,

$$\gamma = 360 - \tan^{-1}\frac{0.115}{3.50} = 360° - 1.88°, \quad \text{or } 358.12°$$

The horizontal transfer distance is $s = -a\cos(\gamma + \alpha)$ and the vertical transfer distance is $t = a\sin(\gamma + \alpha)$.

Applying the moment transfers to the lift and drag values gives the moment about the wing's aerodynamic center of the mean aerodynamic chord, MAC, in Table 6.1, $M_{ac} = M_u - M_w - sL_u - tD_u$ (see Section 5.5).

Uncorrected coefficients (11). Using the dynamic pressure corrected for blockage, wing area, and mean aerodynamic chord, the coefficients uncorrected for tunnel upflow and wall corrections are obtained. It should be noted that the lift coefficient is fully corrected (Table 6.1).

Wall corrections (12). Using Eq. (6.54) to find the effective vortex span for the wing aspect ratio of 7.5 and taper ratio of 0.3, $b_v/b = 0.72$ from Fig. 6.23, and thus $b_e = 0.86(b) = 75.84$, and $b_e/B = 0.53 = k$. From Fig. 6.29, $\delta = 0.108$.

$$\Delta\alpha_w = \delta\left(\frac{S}{C}\right)(57.3)C_{LW} = 0.4869C_{LW} \tag{6.87}$$

$$\Delta C_{Dw} = \delta\left(\frac{S}{C}\right)C_{LW}^2 = 0.008499C_{LW}^2 \tag{6.90}$$

The values of $\Delta\alpha_w$ and ΔC_{Dw} are based on the wing or tail-off lift coefficient and are listed in Table 6.4 verses the indicated angle of attack. In the same table the corrections due to tunnel upflow are also given.

The corrections to tail-on pitching moments is

$$\Delta C_{m\text{CG}_t} = \frac{\partial C_m}{\partial \delta_s}\left(\frac{S}{C}\right)\delta\,\tau_2(57.3)C_{LW} \tag{6.66}$$

TABLE 6.4

α_i	$\Delta\alpha_w$	ΔC_{Dw}	$\Delta\alpha_{\text{up}}$	$\Delta C_{D\,\text{up}}$	$\Delta\alpha_T$	ΔC_{DT}	ΔC_m
2	0.68	0.01664	−0.012	−0.00029	0.67	0.01635	−0.02488
4	0.81	0.02336	−0.012	−0.00035	0.80	0.02301	−0.02947
6	0.92	0.03055	−0.012	−0.00040	0.91	0.03015	−0.03371

The average value of $\partial C_m/\partial\delta_s$ from the data is −0.05143.

The ac of the tail is 6.52 in. above the wing ac at $\alpha = 0°$, thus $h_t/b_e = 0.086$. The tail moment arm (between ac's) is 40.26, therefore $l_t/B = 0.280$.

An examination of Figs. 6.29 and 6.31 shows a large difference in the shape of the curves at $\lambda = 1.0$ and $\lambda = 0.5$. Figure 6.52 for the tail above and below the wing only has $\lambda = 0.5$ and 1.0. Linear interpolations will lead to error. Using Fig. 6.51 for the wing and tail on the tunnel centerline, with data for $\lambda = 1.0$, 0.207, and 0.50, an increment can be obtained from $\lambda = 1.0$ to 0.67. This increment is then applied to the data of Fig. 6.52 from $\lambda = 1.0$ to yield a τ_2 of 0.71, as

$$\Delta C_{m\text{CG}_t} = \frac{\partial C_m}{\partial \delta_s}\left(\frac{S}{C}\right)\delta\,\tau_2(57.3)C_{LW} \tag{6.66}$$

$$\Delta C_{m\text{CG}_t} = -\,0.05143\left(\frac{7.1997}{91.5}\right)(0.108)(0.71)(57.3C_{LW})$$

$$\Delta C_{m\text{CG}_t} = -0.01778C_{LW}$$

Using charts for a $\lambda = 0.67$ tunnel, the value of $\delta = 0.108$ and $\tau_2 = 0.75$. Thus the approximations were fairly close. Note that most wind tunnels have calculated both δ and τ_2 for various model sizes (values are in Table 6.4).

If a measured value is not available for $\partial C_m/\partial\delta_s$, it can be estimated as follows:

$$\frac{\partial C_m}{\partial \delta_s} = -a_t\overline{V}\eta_t \tag{6.63}$$

From Eq. (6.68): $a_t\eta_t = 0.0533$. This assumes the two-dimensional lift curve slope is 0.100 per degree.

$$\overline{V} = \frac{l_t S_t}{S_W\,\text{MAC}} = \frac{(40.260)(2.149)}{(7.1997)(12.8933)} = 0.93229$$

$$\frac{\partial C_m}{\partial \delta_s} = -0.04969$$

TABLE 6.5

α_i	α_C	C_{LC}	C_{DC}	C_{mac_C}
		Tail Off		
2	3.15	1.39914	0.17819	−0.50076
4	5.74	1.65775	0.20981	−0.49428
6	8.32	1.89592	0.24391	−0.48793
		Tail On, $\delta_s = 0°$		
2	3.15	1.33258	0.17805	−0.27385
4	5.74	1.61818	0.20926	−0.34700
6	8.32	1.89200	0.24579	−0.40381
		Tail On, $\delta_s = -5°$		
2	3.15	1.24934	0.17891	−0.01517
4	5.74	1.53094	0.20788	−0.08600
6	8.32	1.79329	0.24082	−0.14970

This is within 3.7% of the measured value. It is, however, preferable to measure $\partial C_m/\partial \delta_s$, thus requiring no assumptions on a_t and q_t.

The use of total model lift (wing plus tail) rather than the wing lift will lead to the following values for $\alpha_I = 2.0°$, $\delta_s = -5.0°$

$$\Delta\alpha = 0.62, \quad \Delta C_D = 0.01395, \quad \Delta C_m = -0.02300$$

This is a decrease in α of 0.05°, C_D of 0.00240 and an increase in C_m of 0.00188.

In percent the errors are $\alpha = -1.6\%$, $C_D = -1.3\%$ and C_{mac} of -15.8%. The C_{mac} value is large because C_{mac} is a small number. However, the error is a function of the tail lift and would be greater for larger negative stabilizer angles or negative elevator angles. Because the wall corrections are functions of the wing circulation, it is not rational to use the total model lift.

At this point, if desired, the data can be transferred to the body or stability axis (8). In this example the pitching moment was calculated about the aerodynamic center of the wing's mean aerodynamic chord. However, in most wind tunnel tests the moments are also taken about forward and aft center of gravity positions that correspond to the desired limits for center of gravity travel. Table 6.5 gives the final corrected coefficients.

6.31. SUMMARY OF THREE-DIMENSIONAL CORRECTIONS FOR AN OPEN TEST SECTION

The airstream in an open test section is free to expand; therefore, the wake and solid blockage effects are small, but not zero. As suggested in Section

6.10 the solid blockage may be taken as one-fourth that of a closed tunnel. An examination of figures in this chapter for boundary correction factors (δ at the wing and τ_2 at the tail) shows that the numerical values are different for open and closed tunnels of similar geometry. The boundary correction factor at the wing (δ) is negative for open tunnels and positive for closed tunnels. Other than the blockage correction and the sign change of the boundary correction factor, the same methods can be used as for a closed tunnel.

One word of caution: many open test-section tunnels with external balances have a splitter plate to shield the balance from the airstream. In this case the tunnel may have one solid boundary, depending on the splitter plate location. This type of mixed boundary is discussed in Refs. 6.37 and 6.40.

6.32. BOUNDARY CORRECTION FOR PROPELLER TESTS

Glauert (Ref. 6.2) has examined the problem of testing propellers in a wind tunnel and suggests that the propeller diameter be kept small relative to the jet diameter and that an open-throat tunnel be employed. Under these conditions no boundary corrections are needed.

Unfortunately, for various practical reasons it is not always possible to adhere to the above stipulations. An approach to the wall corrections for propeller tests in a closed throat tunnel may be made as follows (Fig. 6.58):

In a closed jet the propeller slipstream under conditions of positive thrust will have a velocity u greater than the velocity in the jet without the propeller V. Since the same volume of air that passes section x ahead of the propeller must pass section y behind it, it follows that the velocity w outside the slipstream will be less than V. In free air, w would, of course, equal V. The lower-velocity outside air has a higher static pressure, and it follows that the slipstream also has too high a static pressure. This reacts back to the propeller so that it develops the thrust that might be expected at a lower speed V'. The test should therefore be run at a speed above V in order to develop the proper forces for V.

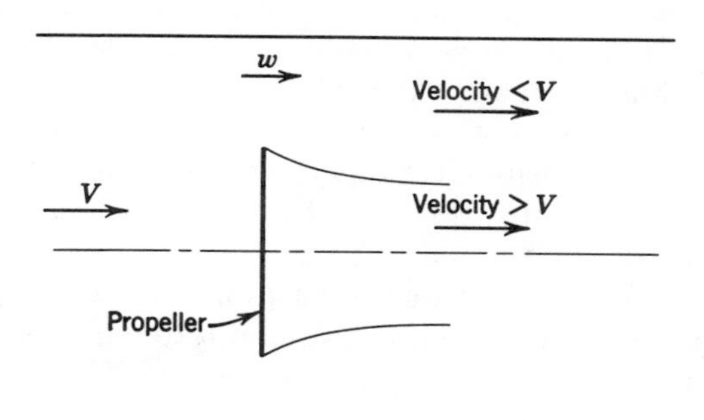

FIGURE 6.58. Nomenclature for propeller in a closed throat wind tunnel.

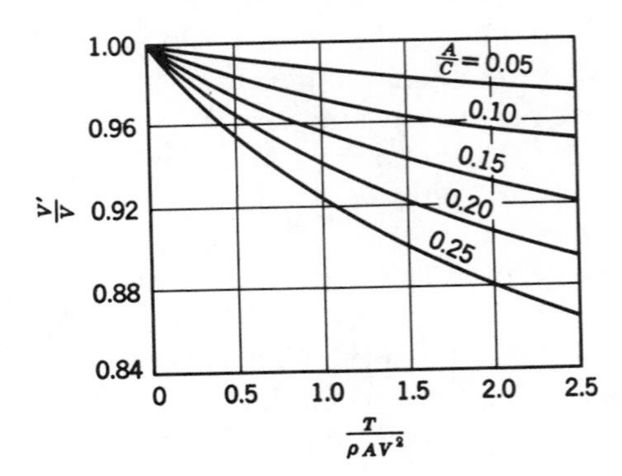

FIGURE 6.59

The amount of correction for this "continuity" effect may be found from

$$\frac{V'}{V} = 1 - \frac{\tau_4 \alpha_1}{2\sqrt{1 + 2\tau_4}} \tag{6.94}$$

where $\tau_4 = T/\rho A V^2$, $\alpha_1 = A/C$, A = propeller disk area, C = jet cross-section area, and T = thrust.

Values of V'/V may be obtained from Fig. 6.59. It is apparent that Eq. (6.94) will not work for negative thrust, since the correction becomes infinite at $\tau_4 = -0.5$ ($C_D = 1.0$ or $T_c = -0.392$). Although we are not aware of any particular studies of the problem, it is probable that the wake-blockage corrections of Section 6.11 would be satisfactory.

6.33. TUNNEL BOUNDARY EFFECTS ON V/STOL MODELS

Flow Breakdown

The basic concept in all V/STOL aircraft is to create lift by using power to produce a downward-directed momentum. Momentum can be produced in two ways. First with a large mass, small velocity, small Δv, such as rotors; second with a small mass, large velocity, large Δv, such as jets. The wake produced by the first is soft and can be deflected by the tunnel flow, while the second is stiff (it goes where you point it) and will have minimal deflection by the tunnel flow in the transition region. The powered lift system can either be distributed across the spans, that is, rotor or spanwise jet flap, or it can be point, that is, powered lift jets or lift fans.

As a V/STOL model goes through transition from level flight to hover, the downwash angle (measured from freestream to the centerline of the wake, plus downward) will change from a small angle to 90° at hover. As the wake angle changes so will the forward velocity from some relatively large value to zero at hover. When the transition flight region from forward flight to hover is simulated in a wind tunnel, the model and tunnel combination will encounter a phenomenon called flow breakdown, which is a limit on the minimum forward speed (Ref. 6.56).

Consider a model in a tunnel at a given lift. At high tunnel speed the tunnel flow is relatively undisturbed by the model and its relatively small downwash angle. As the tunnel speed is reduced the downwash angle increases and the model wake begins to interact with the tunnel boundary layer below and behind the model. A further decrease in tunnel speed and the wake penetrates the tunnel's floor boundary layer, a stagnation point near the center of the wake appears on the floor, and portions of the wake move laterally across the floor. Initially this is an unstable region and a portion of the wake near its leading edge will intermittently snap through, moving forward against the tunnel flow. When the wake snaps through, a parabolic-shaped vortex appears ahead of the rotor curving aft toward the tunnel walls. This has been called incipient stagnation. A further reduction in tunnel speed and the parabolic vortex becomes stable and relatively small, and well aft of the model (Fig. 6.60). This flow has two well-defined stagnations: one near the center of the wake and the second forward of the wake centered near the core of the vortex. As the tunnel speed is further reduced, the forward stagnation points move further forward and the vortex become larger. At some point, as the speed is reduced, the flow is not representative of free-air flow, and the data cannot be corrected. This point is called flow breakdown (Ref. 6.56). It should be noted that the vortex at low forward speeds moves across the floor up the tunnel wall and across the ceiling as it moves aft (Fig. 3.28).

By the use of plywood and plexiglass boxes open at the ends in the tunnel flow direction, different size tunnels can be simulated for a single-size model. Each of the inserts must have its flow calibrated for dynamic pressure and upflow (Refs. 6.57 and 6.58). As the same model is used in the main test section and the inserts at the same conditions, then both the Reynolds number and Mach number are matched. Using this technique for a tunnel width to height ratio of 1.5, a rotor with three disk loadings, run at the same rpm and tip speed ratios, gave the following results. The flow breakdown occurred at different tip speed ratios (tunnel speed), but the momentum downwash angle for a given rotor insert combination was constant. Thus flow breakdown is only a function of the momentum downwash angle for a given model area to tunnel cross-section-area ratio. The smaller the model to tunnel area ratio the larger the allowable downwash. Downwash for a given model to tunnel area ratio varies with the test-section width to height ratio. Corner fillets and curved end walls reduce the allowable downwash for a given W/H. Furthermore, the allowable downwash at the location of the tail is less than at the rotor or wing. As suggested in Refs. 6.43 and 6.59 the flow breakdown limit for a model with distributed lift can be estimated in terms of X_f/b, where X_f is the distance aft of the model at which a theoretically straight wake impinges on the floor, and b is the full model span, or rotor diameter.

$$S_f/b = \frac{h \tan \chi}{b}$$

Here χ is the wake skew angle and $\chi = 90° - \Theta_n$, where Θ_n is the momentum downwash angle (Fig. 6.60) and the values of Θ_n can be obtained from Refs. 6.37 and 6.41.

The flow breakdown limit is more severe at the tail of a model than at a wing. In Fig. 6.61 from Ref. 6.60 data are shown from a tail with a 1 ft span one diameter aft of a 2-ft-diameter propeller used to simulate a rotor. The tail was isolated from the propeller, had its own internal balance, and could be pitched independently of the propeller. The data, as expected, show an increase in the zero lift angle of attack with decreasing propeller tip speed ratio (propeller run at fixed rpm). As the tip speed ratio is reduced, the flow angle aft of the propeller increases very rapidly and then flattens out as shown by the 8 × 12 ft data, where there is no flow breakdown and the data are close to free-air data. In the inserts the flow angle follows the same trend to a maximum value and then decreases due to tunnel-wall effects. As flow breakdown is approached, the flow angle suddenly increases again due to the vortexlike flow below the model.

To illustrate the effect of wall corrections on pitching moment data in the form of the zero lift angle of the tail for the 4 × 6 ft insert and the 8 × 12 ft test section of Fig. 6.61, the data with Heyson's corrections (Section 6.23) are shown in Fig. 6.62. The data for the propeller in the 4 × 6 ft insert was corrected to free air, and these were used as test points for data in the 8 × 12 ft test section. It should be noted that the wall corrections change both the tip speed ratio and the tip plane path or shaft angle. The data in Fig. 6.62 show the delta in tail angle of attack (8 × 12 ft − 4 × 6 ft) with no wall corrections, data corrected for a fixed tail location. These data arose because the tail was attached to the propeller fairing, hence it was at a fixed vertical location. Because the rotor was run at −3° shaft angle, the tail was below the tip plane path. Correcting the 4 × 6 ft insert data rotated shaft angle and thus the tail should have been rotated to maintain its relation to the tip plane path.

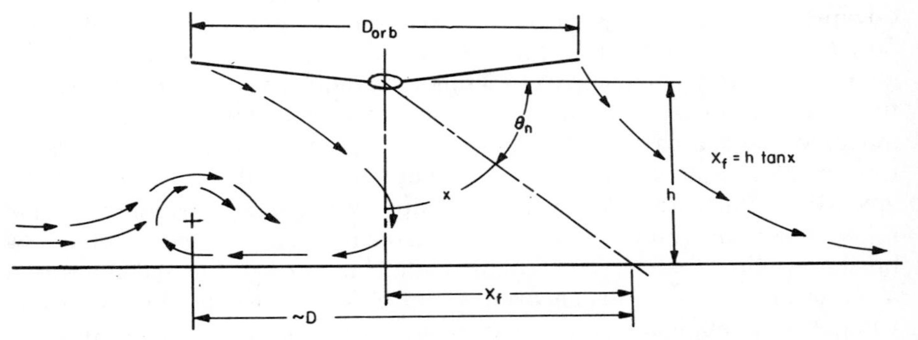

FIGURE 6.60. Nomenclature sketch of rotor wake in flow breakdown showing floor vortex.

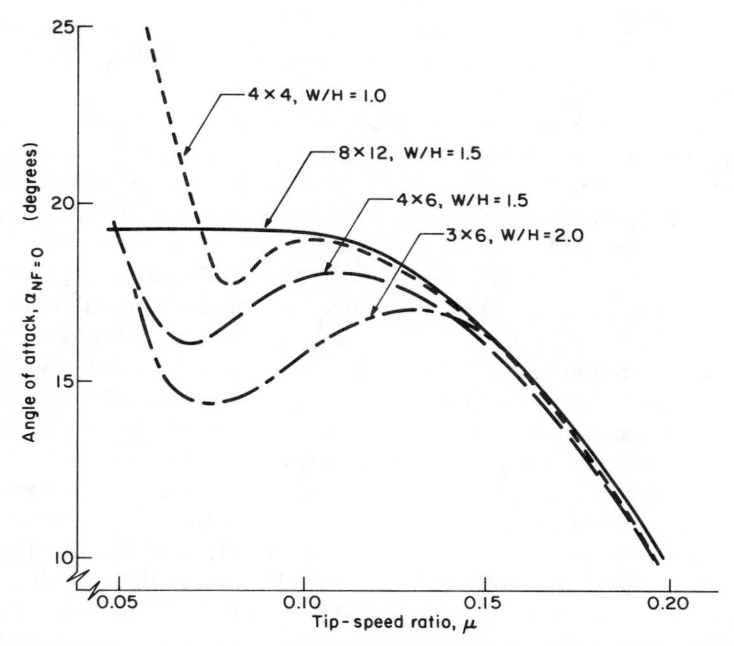

FIGURE 6.61. Effect of flow breakdown on tail mounted behind a rotor in various size tunnels.

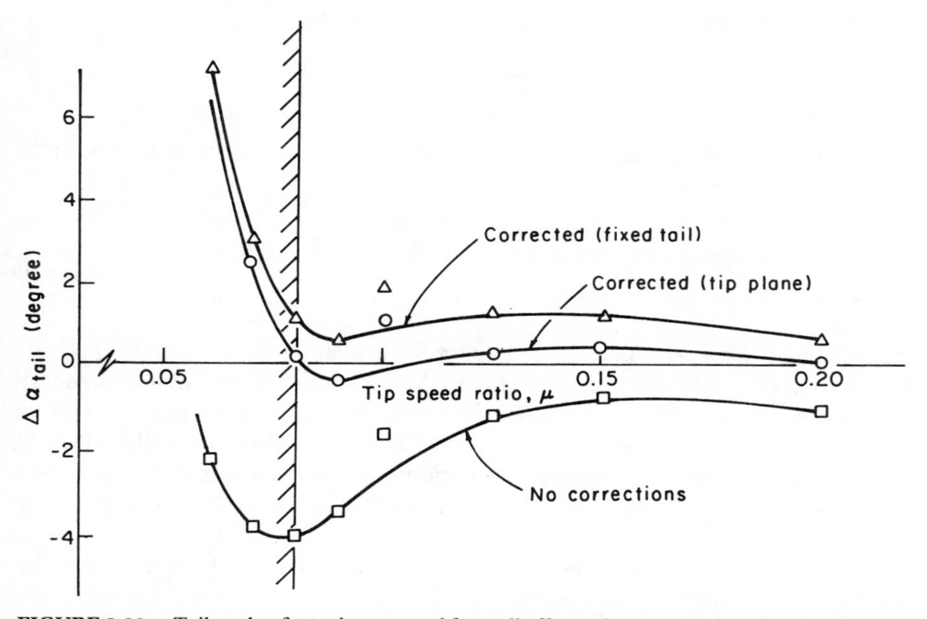

FIGURE 6.62. Tail angle of attack corrected for wall effects above and below flow breakdown.

This mislocation of the tail was corrected by the methods of Refs. 6.61 and 6.62 to find the change in flow angle at the tail due to vertical displacement yielding the curve labeled corrected tip plane.

The curve labeled corrected tip plane demonstrates that Heyson's theory will correct the tail angle of attack, and hence pitching moment, quite well as long as data are not taken below the flow breakdown point (shaded vertical line). Below the limit the wall corrections overcorrect the data by large amounts. At a tip speed ratio of 0.06, the overcorrection is 7° out of 22° for the 8 × 12 ft tunnel, a 32% error. This error increases very rapidly, which implies that the test limit must be known and approached with caution.

The application of Heyson's corrections to the angle of attack, lift, and drag of both rigid and hinged rotors will correct the data only above the flow breakdown limit. The rigid rotor (propeller) was corrected by assuming a uniform load distribution (Ref. 6.60). The hinged rotor (Ref. 6.63) was corrected by using (a) a uniform load (9 doublets) and (b) a triangular load (20 doublets). Both loadings gave similar results. The data can be corrected only above the flow breakdown. As Heyson's corrections change both the angle of attack and tip speed ratio it is necessary to cross-plot data to obtain a constant tip speed ratio. References 6.38 and 6.43 give detailed discussions of wall corrections and estimation of test limits for V/STOL vehicles using a distributed lift system.

Raising the rotor in the tunnel does not increase the test limit, in fact the lowest limit was with the model on the centerline (Ref. 6.64).

The location of the vortexlike core that occurs when operating in a flow breakdown mode can be detected by six methods:

1. Static pressure variation on the tunnel floor.
2. Smoke.
3. Tuft wands to determine streamlines and hence vortex core or tufts on the walls and floor.
4. Vortex meters.
5. Variation of total pressure above floor.
6. Laser.

The path of the vortexlike floor was traced by methods 1, 2, and 5 and the results agree quite well in Ref. 6.63. An example of the static pressure method is also shown in Fig. 6.63. This is a useful method since it could be installed in the tunnel floor permanently. The core of the vortexlike flow is located at the minimum pressure coefficient. As the tip speed ratio decreases the downwash angle increases and the vortexlike core moves forward in the tunnel.

The presence and locations of the floor vortex can also be determined by tufts on the tunnel floor and walls. This method does not locate the core, but it is a reliable indication of flow breakdown.

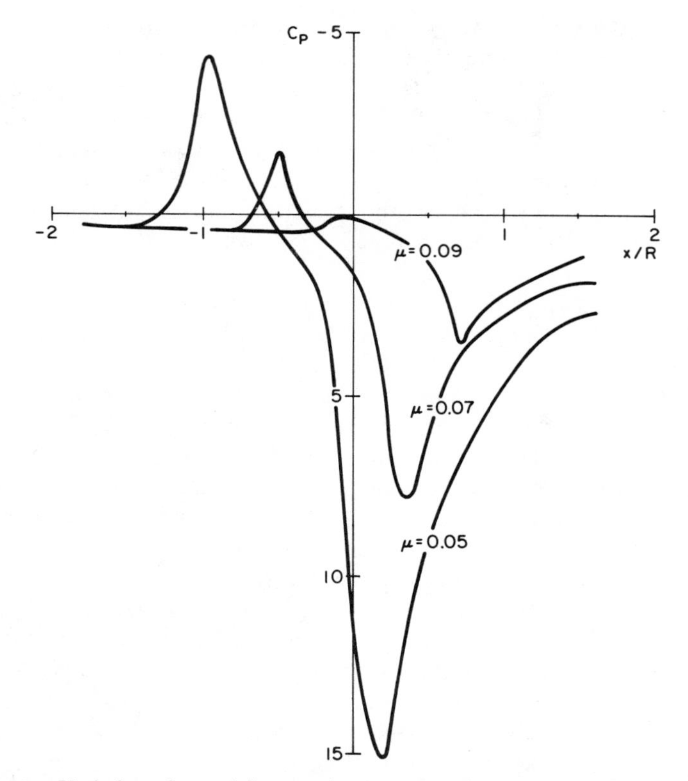

FIGURE 6.63. Variation of tunnel floor static pressure under a rotor. $\mu = 0.09$ is above flow breakdown, while $\mu = 0.07$ and 0.05 are below breakdown. Positive C_P's are wake center stagnation; negative C_P's flow breakdown vortex core.

The problem of flow breakdown is much more complicated when a discrete or point-type powered lift system, as produced by jets or lift fans, is used. The first problem with these configurations is the wide range of locations at which the engine can be located on the aircraft.

Tyler and Williamson in Refs. 6.65 and 6.66 studied flow breakdown for lift jets in the NRC V/STOL propulsion tunnel. Since they were using lift jets, they defined incipient stagnation as a conservative but safe minimum tunnel speed. This is the highest tunnel velocity for the formation of the floor vortex. By defining a thrust coefficient as

$$C_T = \frac{F}{q_\infty h^2}$$

where h is the height of the nozzle exit above the floor. The limit was established based on incipient stagnation as $C_T = 0.62$ for both a single jet and a pair arranged in tandem over a tilt range of $-5°$ to $+20°$. When two jets were side by side the limit was $C_T = 0.90$. The nozzle spacing for the two jets was 4.3 times their jet diameter. They concluded that a tandem pair were the same as a single jet of the same total nozzle area centered between the pair.

Studies were made of local flow in the plane of the nozzle exit. It was concluded that the tunnel velocity could be reduced so very little below incipient stagnation that incipient stagnation should be taken as the lower test limit. It should be noted that the thrust per unit area for lift jets and lift fans is much larger by orders of magnitude than a rotor or other distributed power lift systems.

A small amount of data for a representative nonswept wing model with two nonmetric lift jet engines forward of the wing is reported in Ref. 6.67. The model had a 3 ft span and was tested in an insert, the large tunnel, and over a ground plane at the same location as the insert floor. Changes in the aerodynamic characteristics of the model were measured. These data are only applicable to the configuration that was tested. The data showed that the flow breakdown occurs at higher ratios of freestream to jet velocity ratios in the insert due to the side walls than for the ground plane. The flow breakdown point was within 5% of Tyler and Williamson's values.

There have been several proposed methods of extending the onset of flow breakdown. These include moving belt ground plane (Ref. 6.42) or as an alternative, floor blowing (Ref. 6.15) and opening the tunnel floor extensively (Ref. 6.68). The first two can delay the onset of flow breakdown, but once it occurs they cannot remove it.

In tests on a 46% span knee-blown jet-flap model, Hackett et al. (Ref. 6.15) found that it is possible to use floor blowing to reduce the size of this floor vortex to a small region near the wake impingement. Static pressure measurements along the floor centerline were used to detect the vortex location: floor blowing was increased so as to remove the suction region. Laser velocimeter measurements confirmed that the floor vortex was reduced in size and moved aft. The angle of attack at the model increased substantially. Inspection of wall pressures, and subsequent analysis, showed a reduction in tunnel blockage.

The model gave good agreement with data taken in a larger tunnel at model C_μ's of 4.0 (C_L about 11.0) with and without floor blowing. However, at higher C_μ's the data did not agree with no blowing and only agreed at low α's with blowing. This indicates that flow breakdown was still present, as did the laser data. The wall pressure signatures used to determine blockage (Section 6.11) showed a second pressure peak aft of the main peak and of comparable magnitude at model C_μ's above 4.0, with and without blowing. Since the second peak was further aft in the tunnel, it did not have a major effect on data at the wing, but it may have an effect on pitching moment generated by a tail. This second peak is probably the result of the flow breakdown vortex moving up the tunnel walls (Ref. 6.56).

The removal of portions of the tunnel floor can also extend the test range. Using a rotor in Ref. 6.68 the allowable downwash before flow breakdown was extended from 45° to 75°. When removing the floor of a tunnel to allow the sharply deflected wake to escape the tunnel, there is the concern of the loss of tunnel air. In this test series there were no indications of the wake reentering the tunnel through the floor. The mass of air lost by venting was recovered through a breather ring aft of the test section. Pitch and yaw flow angles measured at the start of the test section were more uniform with the tunnel vented at low advance ratios. When tests are made with a vented tunnel, the effect of venting on the tunnel flow should be determined, as well as tracing the exiting air to ensure that a portion does not return through the vent.

6.34. BOUNDARY CORRECTIONS FOR CONTROL SURFACE HINGE MOMENTS

Little information has been released on boundary corrections for control surface hinge moments. It is apparent that the small size of such surfaces on complete models makes hinge-moment corrections quite unnecessary. Larger models, such as those used for reflection plane tests, have their hinge moments increased by solid jet boundaries in a manner similar to the increase of pitching moment. The increase of hinge moment due to the walls is of the order of 8% for a 30% flap on a large reflection plane model.

Note: hinge moment data must be reduced with the dynamic pressure corrected for both solid and wake blockage. The same is true for pressure coefficients.

REFERENCES

6.1. H. Julian Allen and Walter G. Vincenti, Wall Interference in a Two-Dimensional-Flow Wind Tunnel with Consideration of the Effect of Compressibility, *TR* 782, 1944.

6.2. H. Glauert, Wind Tunnel Interference on Wings, Bodies, and Airscrews, *R&M* 1566, 1933.

6.3. Ira H. Abbott, Albert E. Von Doenhoff, and Louis S. Strivers, Jr., Summary of Airfoil Data, *TR* 824, 1948.

6.4. Alan Pope, *Basic Wing and Airfoil Theory,* McGraw-Hill, New York, 1951.

6.5. A. Thom, Blockage Corrections in a High Speed Wind Tunnel, *R&M* 2033, 1943.

6.6. E. B. Klunker and Keith C. Harder, On the Second-Order Tunnel Wall Constriction Correction in Two-Dimensional Compressible Flow, *TN* 2350, 1951.

6.7. John G. Herriot, Blockage Corrections for Three Dimensional-Flow Closed-Throat Wind Tunnels with Consideration of the Effect of Compressibility, *TR* 995, 1950.

6.8. E. C. Maskell, A Theory of the Blockage Effects on Bluff Bodies and Stalled Wings in a Closed Wind Tunnel, ARC *R&M* 3400, 1965.

6.9. A. D. Young, The Induced Drag of Flapped Elliptic Wings with Cut Out and with Flaps that Extend the Local Chord, ARC *R&M* 2544, 1942.

6.10. J. C. Vassaire, Correction de Blocage dans les Essais en Soufflerie, Effets de deColle-ments, AGARD *CP*-102, Paper 9, 1972.

6.11. Rudolph W. Hensel, Rectangular Wind Tunnel Blocking Corrections Using the Velocity Ratio Method, *TN* 2372, 1951.

6.12. J. E. Hackett and R. A. Boles, Highlift Testing in Closed Wind Tunnels, *J. Aircraft,* **12,** August 1976 (see also AIAA Paper 74-641, July 1974).

6.13. J. E. Hackett and D. J. Wilsden, Determination of Low Speed Wake Blockage Correc-tions via Tunnel Wall Static Pressure Measurements, AGARD *CP* 174, Paper 23, Octo-ber 1975.

6.14. J. E. Hackett, D. J. Wilsden, and D. E. Lilley, Estimation of Tunnel Blockage from Wall Pressure Signatures: A Review and Data Correlation, NASA *CR* 152,241, March 1979.

6.15. J. E. Hackett, S. Sampath, and C. G. Phillips, Determination of Wind Tunnel Con-straints by a Unified Wall Pressure Signature Method. Part I: Applications to Winged Configurations, NASA *CR* 166,186, June 1981.

6.16. J. E. Hackett, S. Sampath, C. G. Phillips, and R. B. White, Determination of Wind Tunnel Constraint Effects by a Unified Wall Pressure Signature Method. Part II: Appli-cation to Jet-in-Crossflow Cases, NASA *CR* 166,187, November 1981.

6.17. Robert S. Swanson and Thomas A. Toll, Jet Boundary Corrections for Reflection Plane Models in Rectangular Wind Tunnels, *TR* 770, 1943.

6.18. H. J. Stewart, The Effect of Wind-Tunnel-Wall Interference on the Stalling Characteris-tics of Wings, *JAS,* September 1941.

6.19. J. R. Gavin and R. W. Hensel, Elliptic Tunnel Wall Corrections, *JAS,* December 1942.

6.20. K. Kondo, The Wall Interference of Wind Tunnels with Boundaries of Circular Arcs, *ARI, TIU,* 126, 1935.

6.21. H. Glauert, The Interference of the Characteristics of an Airfoil in a Wind Tunnel of Circular Section, *R&M* 1453, 1931.

6.22. L. Rosenhead, Uniform and Elliptic Loading in Circular and Rectangular Tunnels, *PRS,* Series A, Vol. 129, 1930, p. 135.

6.23. A. Silverstein, Wind Tunnel Interference with Particular Reference to Off-Center Posi-tions of the Wing and to the Downwash at the Tail, *TR* 547, 1935.

6.24. G. Van Schliestett, Experimental Verification of Theodorsen's Theoretical Jet-Bound-ary Correction Factors, *TN* 506, 1934.

6.25. T. Theodorsen, Interference on an Airfoil of Finite Span in an Open Rectangular Wind Tunnel, *TR* 461, 1931.

6.26. K. Terazawa, On the Interference of Wind Tunnel Walls on the Aerodynamic Character-istics of a Wing, *ARI, TIU,* 44, 1932.

6.27. H. Glauert, The Interference on the Characteristics of an Airfoil in a Wind Tunnel of Rectangular Section, *R&M* 1459, 1932.

6.28. James C. Sivells and Rachel M. Salmi, Jet Boundary Corrections for Complete and Semispan Swept Wings in Closed Circular Wind Tunnels, *TN* 2454, 1951.

6.29. M. Sanuki and I. Tani, The Wall Interference of a Wind Tunnel of Elliptic Cross-Section, *Proc. Physical Mathematical Soc. Japan,* Vol. 14, 1932.

6.30. L. Rosenhead, The Airfoil in a Wind Tunnel of Elliptic Cross-Section, *PRS,* Series A, Vol. 140, 1933, p. 579.

6.31. F. Riegels, Correction Factors for Wind Tunnels of Elliptic Section with Partly Open and Partly Closed Test Section, *TM* 1310, 1951.

6.32. G. K. Batchelor, Interference in a Wind Tunnel of Octagonal Section, *ACA* 1, January 1944.

6.33. Betty L. Gent, Interference in a Wind Tunnel of Regular Octagonal Section, *ACA* 2, January 1944.

6.34. I. Lotz, Correction of Downwash in Wind Tunnels of Circular and Elliptic Sections, *TM* 801, 1936.

6.35. Abe Silverstein and S. Katzoff, Experimental Investigation of Wind-Tunnel Interference on the Downwash behind an Airfoil, *TR* 609, 1937.

6.36. H. H. Heyson, Jet Boundary Corrections for Lifting Rotors Centered in a Rectangular Wind Tunnel, NASA *TR* R-71, 1960.

6.37. Harry H. Heyson, Linearized Theory of Wind Tunnel Jet Boundaries Corrections and Ground Effect for VTOL/STOL Aircraft, NASA *TR* R-124, 1962.

6.38. H. H. Heyson, Use of Superposition in Digital Computers to Obtain Wind-Tunnel Interference Factors for Arbitrary Configurations, with Particular Reference to V/STOL Models, NASA *TR* R-302, 1969.

6.39. H. H. Heyson, Fortran Programs for Calculating Wind-Tunnel Boundary Interference, NASA *TM* X-1740, 1969.

6.40. H. H. Heyson, General Theory of Wall Interference for Static Stability Tests in Closed Rectangular Test Sections and in Ground Effect, NASA *TR* R-364, 1971.

6.41. H. H. Heyson, Nomographic Solution of the Momentum Equation for VTOL-STOL Aircraft, NASA *TN* D-814, 1961.

6.42. H. H. Heyson and K. J. Grunwald, Wind Tunnel Boundary Interference for V/STOL Testing, Paper 24, Conference on V/STOL and STOL Aircraft, NASA *SP* 116, 1966, pp. 409–434.

6.43. H. H. Heyson, Rapid Estimation of Wind-Tunnel Corrections with Application to Wind-Tunnels and Model Design, NASA *TD* D-6416, 1971.

6.44. H. H. Heyson, Equations for the Application of Wind-Tunnel Wall Corrections to Pitching Moments Caused by the Tail of an Aircraft Model, NASA *TN* D-3738, 1966.

6.45. R. G. Joppa, A Method of Calculating Wind Tunnel Interference Factors for Tunnels of Arbitrary Cross-Section. *Report* 67-1, University of Washington, Department of Aeronautical Engineering, 1967 (see also NASA *CR* 845, 1967).

6.46. R. G. Joppa, Wall Interference Effects in Wind-Tunnel Testing of STOL Aircraft, *J. Aircraft,* **6,** 209–214, May–June 1969.

6.47. R. G. Joppa, Wind Tunnel Interference Factors for High-Lift Wings in Closed Wind Tunnels, NASA *CR* 2191, 1973.

6.48. S. Bernstein and R. G. Joppa, Development of Minimum-Correction Wind Tunnels, *J. Aircraft,* **13,** 243–247, April 1976.

6.49. R. J. Vidal and J. C. Erickson, Jr., Experiments of Supercritical Flows in a Self Correcting Wind Tunnel, Paper 78-788, 10th Aerodynamic Testing Conference, 1978, pp. 136–141.

6.50. Wind Tunnel Corrections for High Angle of Attack Models, AGARD Report 692, Munich, February 1981.

6.51. U. Ganzer and Y. Igeta, Transonic Tests in a Wind Tunnel with Adapted Walls, ICAS-82-5.4.5, *Proceedings,* 13th Congress of the International Council of the Aeronautical Sciences, Vol. 1, 1982, pp. 752–760.

6.52. J. D. Whitfield, J. L. Jacocks, W. E. Dietz, and S. R. Pate, Demonstration of the Adaptive Wall Concept Applied to an Automotive Wind Tunnel, Paper 82-0584, AIAA 12th Aerodynamic Testing Conference, 1982.

6.53. W. R. Sears, Wind-Tunnel Testing of V/STOL Configurations at High Lift, ICAS-82-5.4.1, *Proceedings,* 13th Congress of the International Council of the Aeronautical Sciences, Vol. 1, 1982, pp. 720–730.

6.54. S. Raimondo and P. J. F. Clark, Slotted Wall Test Section for Automotive Aerodynamic Test Facilities, Paper 82-0585, AIAA 12th Aerodynamic Testing Conference, 1982, pp. 101–109.

6.55. I. G. Recant, Wind Tunnel Investigation of Ground Effect, *TN* 705, 1939.

6.56. W. H. Rae, Jr., Limits on Minimum-Speed V/STOL Wind-Tunnel Tests, *J. Aircraft,* **4,** 249–254, May–June 1967 (see also AIAA Paper 66-736).

6.57. V. M. Ganzer and W. H. Rae, Jr., An Experimental Investigation of the Effect of Wind Tunnel Walls on the Aerodynamic Performance of a Helicopter Rotor, NASA *TN* D-415, 1960.

6.58. J. L. Lee, An Experimental Investigation of the Use of Test Section Inserts as a Device to Verify Theoretical Wall Corrections for a Lifting Rotor Centered in a Closed Rectangular Test Section, Masters Thesis, University of Washington, Dept. of Aeronautics and Astronautics, 1964.

6.59. H. H. Heyson, Wind-Tunnel Wall Effects at Extreme Force Coefficients, International Congress of Subsonic Aerodynamicists, New York Academy of Sciences, 1967.

6.60. W. H. Rae, Jr. and S. Shindo, Comments on V/STOL Wind Tunnel Data at Low Forward Speeds, *Proceedings,* 3rd CAL/AVLABS Symposium, Aerodynamics of Rotary Wing and V/STOL Aircraft, Vol. II, 1969.

6.61. H. H. Heyson and S. Katzoff, Induced Velocities Near a Lifting Rotor with Non-Uniform Disk Loading, NACA *Report* 1319, 1956.

6.62. J. W. Jewell, Jr. and H. H. Heyson, Charts of the Induced Velocities Near a Lifting Rotor, NASA *TM* 4-15-596, 1959.

6.63. W. H. Rae, Jr. and S. Shindo, An Experimental Investigation of Wind Tunnel Wall Corrections and Test Limits for V/STOL Vehicles, Final Report DA-ARO-31-124-G809, U.S. Army Research Office–Durham, 1973.

6.64. S. Shindo and W. H. Rae, Jr., Low-Speed Test Limit of V/STOL Model Located Vertically Off-Center, *J. Aircraft,* **15,** 253–254, April 1978.

6.65. R. A. Tyler and R. G. Williamson, Wind-Tunnel Testing of V/STOL Engine Models—Some Observed Flow Interaction and Tunnel Effects, AGARD *CP* 91-71, December 1971.

6.66. R. A. Tyler and R. G. Williamson, Experience with the NRC 10 ft. × 20 ft. V/STOL Propulsion Tunnel—Some Practical Aspects of V/STOL Engine Model Testing, *Canadian Aeronautics and Space Journal,* 18, September 1972, p. 191.

6.67. S. Shindo and W. H. Rae, Jr., Recent Research on V/STOL Test Limits at the University of Washington Aeronautical Laboratory, NASA *CR* 3237, 1980.

6.68. R. E. Hansford, The Removal of Wind Tunnel Panels to Prevent Flow Breakdown at Low Speeds, *Aeronautical J.* **75,** 475–479, November 1975.

CHAPTER 7

The Use of Wind Tunnel Data

One of the top airplane designers in Great Britain has been credited with the statement that he "could go on designing airplanes all day long if he had not also to build them and make them fly," and his point is surely well taken. Data may *easily* "be taken all day long"—as long as they are not used to design airplanes.

Indeed, the very subject of extrapolating wind tunnel data to full scale will probably elicit a grim smile from aeronautical engineers who see this page. The aerodynamicist disparages the wind tunnel engineer; the wind tunnel engineer thinks the aerodynamicist wants too much; and if any poor soul is assigned the combination of jobs, well, one is reminded of the classic experiment of crossing a hound dog and a rabbit wherein the offspring ran itself to death.

Probably the nearest approach to the truth lies in the fact that wind tunnels are very rarely called upon to test exact models of items that may be flown. Though this offers a magnificent "out" to the wind tunnel engineer, it is not meant that way. Reynolds number effects on small items are too great even if they could be accurately constructed; hence the small excresences are left off the models. In many cases the aerodynamicist who plans on adding these items selects the lowest possible drag values with the net result that he underestimates their interference and overestimates the performance of the airplane. The cure for this situation is to consolidate these items and minimize their effect. Room for improvement can surely exist when examples can be cited of airplanes that have no less than twenty-two separate air intakes and over thirty removable inspection panels. Of course, the effects of small protrusions can be tested in the full-scale tunnel on the actual airplane itself, provided the wingspan is not too great.

Unfortunately undesirably little correlation between flight test and wind tunnel data is available. This lack is attributed to the dual reasons that after flight test there is rarely time to back up and correlate with the wind tunnel, and that, even when it is done, the success or failure of the methods used is generally a company secret. The literature reveals only a handful of papers on correlating wind tunnel and flight results.

Any flight test and wind tunnel correlation always suffers from a great number of unknowns. The tunnel data suffer from inexact or unknown Reynolds number extrapolation, possible uncertainties in corrections to the data such as tare and interference and wall effects, errors in duplicating the power on effects with fixed-pitch propellers, simulation of flow around or through jet engine nacelles, omission of manufacturing irregularities and small excrescences, and insufficient deflections of the model under load. Some models are built to simulate a 1-g load. The flight test data suffer from the pilot techniques, accelerations due to gusts, errors in average center of gravity locations, determination of true airspeed, and unknowns of propeller efficiencies and other power-plant effects. Considering the impressive room for disagreements, the generally good agreement found is remarkable.

We will take up, in turn, each of the important aerodynamic quantities usually measured in the tunnel and say what we can about their use. First, however, we must consider the boundary layer, for the understanding of scale effects is essentially the understanding of the boundary layer.

7.1. THE BOUNDARY LAYER

Owing to viscosity of the air, the air very near the wing is slowed gradually from some local velocity a short distance out to zero right at the wing. The region in which this velocity change takes place is called the *boundary layer,* and the velocity gradients in the boundary layer very largely determine whether the drag of a body is x, or $10x$. A boundary layer in which the velocities vary approximately linearly from the surface is called *laminar**; one whose velocities vary approximately exponentially from the surface is called *turbulent*. Their drag values based on *wetted area* (approximately double *wing area*) are given by

$$C_{D\,\text{laminar}} = 2.656/\sqrt{\text{RN}} \tag{7.1}$$

* Both laminar and turbulent flow may be demonstrated simply with a cigarette held very still in still air. The rising smoke column will be smooth (laminar) for about 10 in. and then will turn turbulent. Talking or any other tiny disturbance will also make the laminar flow become turbulent.

$$C_{D\,\text{turbulent}} = 0.910/(\log_{10} \text{RN})^{2.58} \tag{7.2}$$

and plotted in Fig. 7.1 along with a drag curve of a 23012 airfoil.

The boundary-layer thickness, defined as the distance from the surface to the point where the velocity in the boundary layer is 0.99 times the velocity just outside the boundary layer, is given by

$$\delta_{\text{laminar}} = 5.2\sqrt{l^2/\text{RN}} \tag{7.3}$$

$$\delta_{\text{turbulent}} = 0.37\,l/(\text{RN})^{\frac{1}{5}} \tag{7.4}$$

where l = distance from body leading edge and RN = Reynolds number based on l and freestream velocity.

Several important phenomena are known about the boundary layer. First, both its drag and its thickness are related to the Reynolds number. Second, laminar flow, having far less drag, has less energy with which to surmount roughness or corners and it hence separates from a surface much more easily than does turbulent flow. Third, the maintenance of a laminar boundary layer becomes more difficult as the Reynolds number (its length) increases. Fourth, laminar flow is encouraged by a pressure gradient falling in the direction of flow.

Modern computer-aided airfoil design produces favorable pressure gradients over a large portion of the chord. The extent of laminar flow can be increased by very smooth surfaces (composite materials), and by either removing the boundary layer by suction or energizing it by surface blowing. In the light of these actions we may examine how a flow can be changed widely under conditions of changing Reynolds number.

Assume that the wing shown in Fig. 7.2 is in a stream of such turbulence that laminar flow will change to turbulent at a Reynolds number of 1,000,000, and further assume a model size and velocity such that the Reynolds number of the entire flow length shown in 7.2*a* is 1,000,000. We note two items: first,

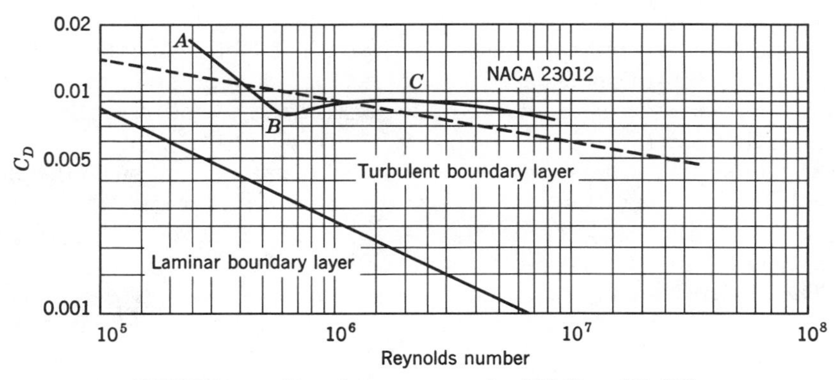

FIGURE 7.1. Plot of $C_{D0\,\min}$ versus log RN (from *TR* 586).

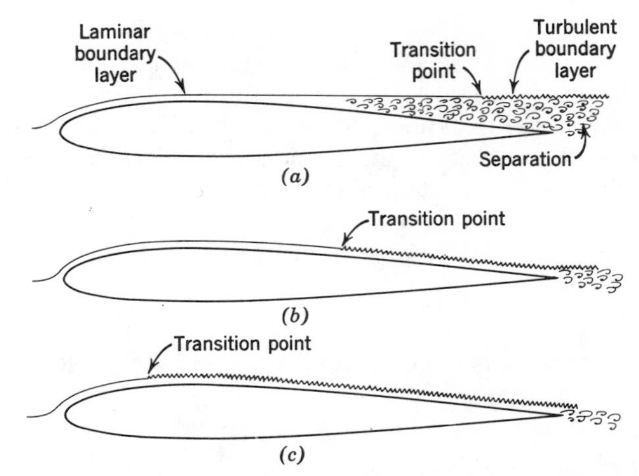

FIGURE 7.2. Effect of increasing Reynolds number on boundary layer flow.

the laminar flow is unable to negotiate the curve of the airfoil and excessive separation is evident; and, second, transition takes place *before* 1,000,000 since the flow downstream of the maximum thickness has a rising pressure gradient that discourages laminar boundary layer. Here then is a case where we have too much laminar flow, and the resultant drag is excessively high, corresponding to a point *A* in Fig. 7.1 and the way the boundary layer behaves on a 23012 airfoil at a Reynolds number of 300,000.

Returning to Fig. 7.2*b*, which corresponds to a higher Reynolds number, we see that the transition point has moved forward according to item 3, and now we have the maximum laminar flow and minimum drag. This corresponds to point *B* in Fig. 7.1, and a 23012 airfoil at a Reynolds number of 650,000.

The still higher Reynolds number illustrated in Fig. 7.2*c* fails to show a decrease of drag, even though both laminar and turbulent drag decrease with increasing Reynolds number, since there has been a great increase in the region of turbulent flow. This is point *C* in Fig. 7.1. (RN = 1,200,000.)

Further increase in Reynolds number yields a reduction in drag coefficient although the transition has now reached the minimum pressure point and its further motion is resisted by the falling pressure gradient from the leading edge to that point.

Since the pressure pattern of every airfoil is unique, and since the same may be said of every airplane design, it is apparent that tests made in the Reynolds number range where laminar separation is developed will be exceedingly difficult to interpret for full scale.

The effects can be profound on essentially all qualities of interest—forces, stability moments, and hinge moments. Obviously, it behooves the tunnel engineer either to provide wind tunnel Reynolds numbers equal to flight (a procedure rarely possible), or to somehow make the model boundary layer duplicate that of the full-scale aircraft. This is accomplished by a trip strip.

7.2. THE TRIP STRIP

A trip strip is an artificial roughness added to the model to fix the location of transition from a laminar to turbulent boundary layer on the model. The trip strip can also prevent a separation of the laminar boundary layer near the leading edge. If, however, the trip strip is either too high above the surface or has too great an amount of roughness, it can affect the model drag and maximum lift. Several methods of making trip strips are:

1. *Grit*. The traditional trip strip based on NACA/NASA reports is a finite width strip of grit. When properly applied, the grit gives a three-dimensional trip that simulates the way full scale transition occurs.

The two commercially available materials that are used are carborundum and ballotine micro beads or balls. They both are available in graded sizes with nominal diameters (see Table 7.1). The width of the trip strip is usually 0.125–0.250 in. Masking tape is used to lay out the trip strips and then one

TABLE 7.1. Commercial Carborundum Grit Numbers and Corresponding Particle "Diameters"

Grit Number	Nominal Grit Size (in.)
10	0.0937
12	0.0787
14	0.0661
16	0.0555
20	0.0469
24	0.0331
30	0.0280
36	0.0232
46	0.0165
54	0.0138
60	0.0117
70	0.0098
80	0.0083
90	0.0070
100	0.0059
120	0.0049
150	0.0041
180	0.0035
220	0.0029

paints or sprays on shellac, lacquer, artist's clear acrylic, or superhold hair spray. The hair spray has the advantage of being soluble in water and thus it can be removed without harm to the model finish. After spraying or painting, the grit is dusted or blown on the wet adhesive. The grit is difficult to apply to vertical and lower surfaces, but this can be done by bending a card into a V and with skill blowing the grit onto the surface. It is difficult to obtain a repeatable, uniform, relatively sparse distribution of the grit. If the grit is dense-packed, the trip will approach a two-dimensional trip. Also, during the course of the test the grit breaks away from the adhesive. Both of these problems can lead to difficulty in repeating data during, and from test to test. The data repeatability is most severe in drag. Sample boards and photographs of grit density can be helpful in attaining some measure of repeatability.

2. *Two-Dimensional Tape.* These consist of 0.125-in. printed circuit drafting tape or chart tape. The trip strip is built up with multiple layers of the tape. The surface must be clean and oil free to obtain good adherence of the first layer of tape. If layers of multiple color tape are used, it is easy to detect a layer that has blown off the model. The tape requires the height of the trip strip to vary in discrete steps due to the tape thickness. The two-dimensional tape acts on the boundary layer in a different manner than the three-dimensional grit and thus its simulation is not as good as the grit.

3. *Thread or String.* This is similar to the two-dimensional tape but the thread or string is glued down in the same manner as the grit by spraying the surface with an adhesive and stretching and pressing the thread onto the surface. This method is not often used.

4. *Three-Dimensional Pinked Tape.* Tape is cut in two using dressmaker's pinking shears. This leaves two pieces of tape with triangular edges. The tape is applied to the model with the 90° points pointing forward. Layers of tape can be built up to give various heights, and the points staggered in the span direction to give a more representative transition than the two-dimensional tape of 2.

5. *Triangles.* Tape and pinking shears are used to make small triangles that are applied to the model. Multiple layers can be used to increase the thickness. This method is slow to apply.

6. *Epoxy Dots.* A vinyl tape with backing is run through a modified computer-paper-tape machine. This produces holes of 0.050 in. diameter, 0.100 in. center to center. The tape less backing is then applied to the model surface. Then an epoxy compound is forced into the holes. When the epoxy sets a series of dots above the surface form the trip strip. The tape can be obtained in different thicknesses, or multiple layers can be used. See Figs. 7.3–7.5 for a model with natural transition and with fixed transition using the dot-type trip strip.

7. *Other.* The exposed portion of streamlined mounting struts may be "staked" by the use of center punches to yield a permanent transition strip. Round struts have been knurled to yield the same results.

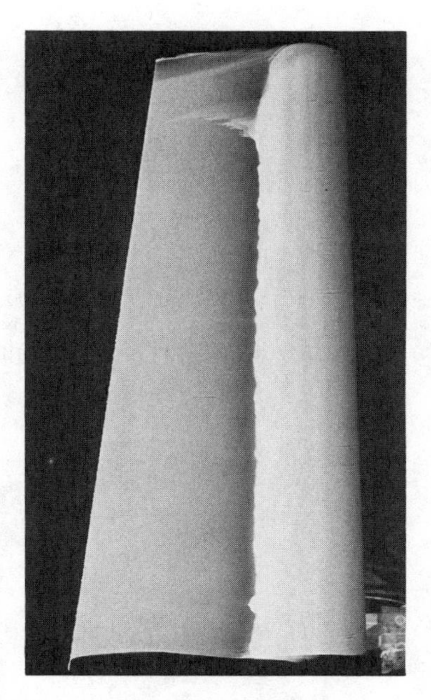

FIGURE 7.3. Oil flow visualization showing natural transition at approximately 40% chord. $C_L = 0.28$, RN = 1,260,000 (based on average chord). (Photograph courtesy of University of Washington.)

Location

The trip strip is intended to simulate the boundary layer transition on the full-scale aircraft. The following is a general guideline for location:

1. *Lifting Surfaces.* Lifting surfaces would include wings, horizontal and vertical tails, and auxiliary fins such as winglets, etc. The trip strip is applied to both sides of the surface. For the NACA four- and five-digit airfoils and conventional wing construction, the full-scale transition will occur at approximately 10% of the chord at cruise lift coefficients. For the newer laminar flow airfoils and smooth composite skin, the full-scale transition can be as far aft as 60% of the chord. If the transition is fixed near the leading edge for composite skin laminar airfoils, there will be large changes in the lift, drag, and pitching moment. For general-aviation-type aircraft with production quality airfoils, Ref. 7.1 presents a summary of measured transition in flight and some comparison with tunnel tests, plus extensive references.

2. *Fuselage, Nacelles.* For fuselages that have the maximum thickness well forward, the trip strip is often located where the local diameter is one-half the maximum diameter. For a low drag body, one designed for

FIGURE 7.4. Naphthalene–fluorene flow visualization showing natural transition at approximately 40% chord. $C_L = 0.28$, RN = 1,260,000 (based on average chord). (Photograph courtesy of University of Washington.)

extensive laminar flow, the trip strip is put at 20–30% of the body length. Care should be taken to ensure that laminar flow is not reestablished aft of the trip strip. Where the windshield, inlets, or blisters protrude from the main contour, an additional trip strip is often used to prevent the reestablishment of laminar flow. For flow through nacelles, the trip strip inside the nacelle is about 5% aft of the hi-lite, and usually flow visualization is used to ensure that the boundary layer is tripped. The trip strip on the outside is also about 5–10% aft of the hi-lite.

In application of trip strips where there is a possibility of reestablishment of laminar flow, flow visualization such as sublimation or oil flow should be used to ensure that there is no laminar flow aft of the trip strip (Section 3.10). See Figures 7.3, 7.4, and 7.5.

FIGURE 7.5. Fixed transition at 8% chord using 0.007-in.-high epoxy disks. $C_L = 0.28$, RN = 1,260,000 (based on average chord). (Photograph courtesy of University of Washington.)

Height of the Trip Strip

If the test is to be run at various dynamic pressures, the height of the trip strip will be determined by the lowest dynamic pressure. If the height is determined by the highest q, then there is a risk that the boundary layer will not be tripped at the low q.

The height of the trip strip can be determined by

$$h = \frac{12K}{R} \tag{7.5}$$

where K is a constant (actually a Reynolds number) based on grit roughness. Its value is 600 for Reynolds numbers greater than 100,000 based on freestream and distance from leading edge to trip strip. This is the minimum Reynolds number for transition based on roughness. If the Reynolds number based on distance to the trip strip is less than 100,000, the value of K increases to about 1000 at a Reynolds number of about 0 (see Fig. 7 of Ref. 7.2, and Ref. 7.3); R is the Reynolds number per foot based on freestream velocity.

The use of Eq. (7.5) yields trip strip heights that can be used for stability and control tests but does not yield enough information for performance tests, that is, drag due to the trip strip. When drag data are required, a trip strip height buildup should be run.

Results of a study made to determine the required grit size are presented in Fig. 7.6. With increases in grit size to about 0.003 in., the drag coefficient increases quite rapidly. This is interpreted as an indication that a completely turbulent boundary layer has not been established downstream of the transition strip. Above a grit size of 0.004 in., the rise of drag with grit size is considerably smaller and is constant. This indicates that transition is complete and that the increasing drag is a pressure drag on the transition strip. In this case, the grit size that should be used for testing is indicated to be between 0.003 and 0.004 in. To correct the measured wind tunnel drag for the effect of the trip strip the data for fully established transition is extrapolated back to zero trip strip height as indicated by the dash line in Fig. 7.6. The drag correction is then the delta between the model drag with trip strip and the zero-height-extrapolated drag value. This delta then corrects for the pressure drag caused by the trip strip, while the trip strip itself ensures that the laminar to turbulent transition is at the proper location. If the tunnel test is run at various q's and the drag correction is required at all of them, then the trip strip height should be determined for each q. This implies that the trip strip height will be changed with each q, thus it becomes of paramount importance that the trip strip application is repeatable, and easy to apply,

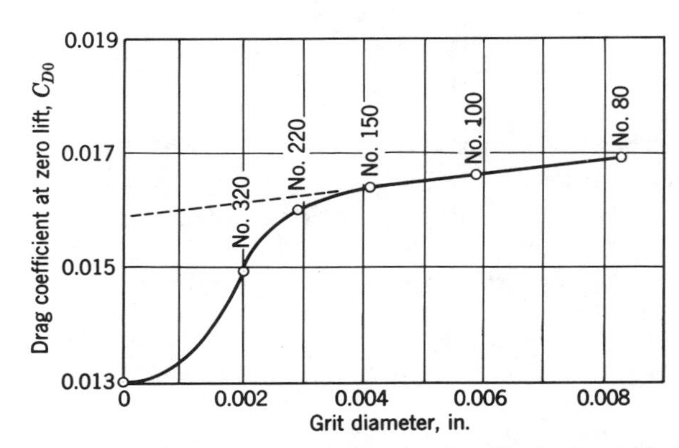

FIGURE 7.6. Results of a transition study using a series of grit sizes. (Redrawn from data courtesy of Calspan Corp.)

which will influence the type of trip strip that is used. As noted in item 1, grit may be hard to repeat, therefore two-dimensional tape (2) or pinked tape (4) and (5) or epoxy dots (6) may be preferred.

The height of the trip strip should also be determined over the range of C_L's from cruise through $C_{L\max}$. Since the trip strip height is determined for a given test Reynolds number or test q, there will be the following effect on $C_{L\max}$. Below the q used to determine the trip strip height the $C_{L\max}$ will increase with q or Reynolds number as expected. Above the q used to determine the height of the trip strip, the $C_{L\max}$ may decrease with increasing q or Reynolds number. If this happens, the trip is extended above the edge of the boundary layer and, if located far enough forward on the surface, the trip strip has become a stall strip. This is almost always the case when the trip strip is at 5–10% of the chord, as required by the old NACA four and five digit series airfoils. The trip strip acting as stall strip is usually not a problem with laminar flow airfoils with the trip at 30% chord or greater. It may be the best policy to make the runs for the variation of $C_{L\max}$ with Reynolds number without trip strips. If leading edge laminar separation is suspected, it can be checked by flow visualization.

Transition or trip strips are not needed on all models. If the model shape is such that it precludes the formation of laminar flow, then trip strips are not required. This is often the case with many architectural models of buildings, bridges, open lattice towers, trees, shingled roofs, etc. If there is a question about laminar flow and transition, the affect of trip strips can be measured, but care must be taken to ensure that the trip strip does not cause premature separation. Again, flow visualization will be useful as a positive check.

Figure 7.7 shows the drag coefficients corresponding to different amounts of laminar and turbulent boundary layers at various Reynolds numbers. Zero form drag is assumed. In this chart are shown the theoretical minimum wing drag of 100% laminar flow and the decrease in drag coefficient due to extension of the laminar layer. For example, extension of the laminar layer from 20 to 60% at RN = 2,000,000 reduces the drag coefficient from 0.0073 to 0.0048. In fact it is just this extension of laminar flow that reduces the profile drag of the laminar flow airfoils to values much less than the "conventional" ones. Some general observations concerning scale effect on minimum drag for these sections are in order because of their greater extent of laminar flow. Over the lower range of Reynolds numbers (up to RN $= 9 \times 10^6$) a gradual decrease in minimum drag occurs. This reaction is attributed to the thinning of the boundary layer. At the higher Reynolds numbers the drag increases steadily at least up to RN $= 25 \times 10^6$ because of the forward movement of transition. Of course, again airfoil geometry (camber, thickness, thickness form) influences the general behavior, but the scale effect is due primarily to the relative strengths of the two interacting boundary layers. Comparison tests of two different airfoils of different families can be most deceiving if the testing is done at low Reynolds numbers. For example the 65_3-418 airfoil displays much larger profile drag than the 0012 section at low Reynolds

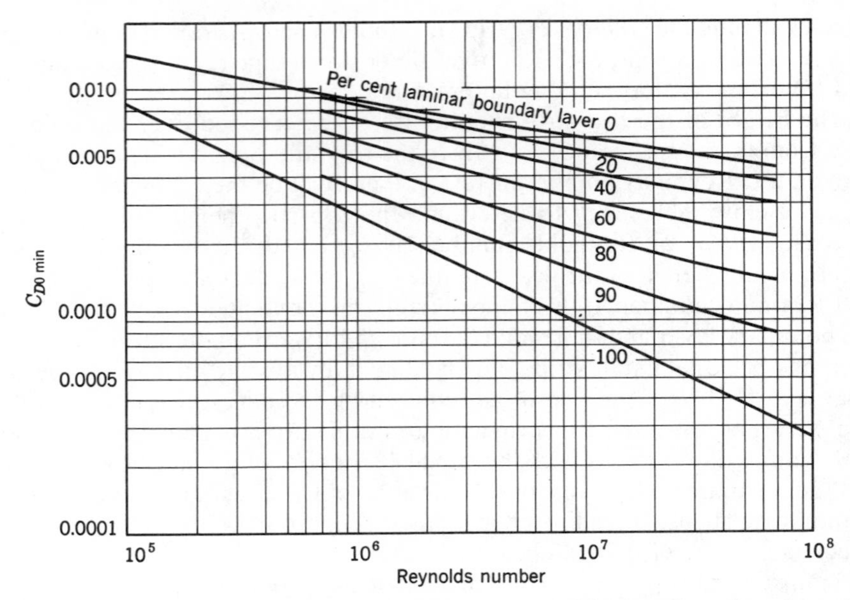

FIGURE 7.7. $C_{D0\,min}$ versus RN for various percentages of laminar flow. Zero form drag assumed.

numbers, whereas at full-scale conditions just the opposite is true. The 6 series section has extensive laminar separation in the low Reynolds number range; the 0012 has far less separation region and hence has lower drag.

In extrapolating drag coefficients, it is necessary to make due allowance if the Reynolds number of the tunnel data is the "effective" Reynolds number. This procedure is necessary because the part of the drag associated with skin friction decreases with increasing Reynolds number. Thus for a given effective Reynolds number the friction coefficients are larger than at a numerically equal test Reynolds number. The difference between measured drag and the actual drag at the equivalent free-air Reynolds number may be read from the turbulent drag curve of Fig. 7.1. An example is given.

Example 7.1. The drag of a wing is measured at a test Reynolds number of 3,000,000 and a turbulence factor of 2.0. The measured drag coefficient is 0.0082. Find the equivalent free-air drag coefficient.

1. As $RN_e = TF \times RN$

$$RN_e = 2.0 \times 3{,}000{,}000 = 6{,}000{,}000$$

2. From the turbulent drag curve of Fig. 7.1, $C_D = 0.0073$ at RN = 3,000,000 and 0.0066 at RN = 6,000,000.

$$\Delta C_D = 0.0073 - 0.0066 = 0.0007$$

3. The measured drag is therefore too high by 0.0007, and $C_D = 0.0082 - 0.0007 = 0.0075$.

7.3. SCALE EFFECT ON DRAG

Although one might argue that the difficulties of measuring drag are less important than stability and control (because drag affects performance, but not safety), the simple truth is that whether or not a plane is ever built depends very much on the drag data obtained in the wind tunnel, and how well one can make potential customers believe it.

The first conclusion (from Fig. 7.1) is that, even though comparison tests between objects may be made with fair accuracy, a test Reynolds number of 1,500,000 to 2,500,000 would be needed if extrapolation is intended. An ameliorating condition is that, if a low Reynolds number separation exists and is cured by some change, the probability that it will arise at a higher Reynolds number is extremely small.

If an effort is made to maintain true profile on the full-scale airplane, tunnel wind drag extrapolated to the net Reynolds number (by the aid of Fig. 7.7) should be only slightly optimistic owing to the drag of the flap and aileron cutouts, control surface inspection doors, and pitot tubes, antennas, other excrescences, etc., that are not represented on the model.

The effects of the minimum drag of the complete airplane are more difficult to handle. There are truly an immense number of small items on the full-scale aircraft that cannot be represented at model scale. Whether they undercompensate or overcompensate for the drag reduction expected from higher Reynolds number is anyone's guess. Frequently the engineer assumes that the two exactly compensate, but he would hate to guarantee better than ± 0.0010 in C_D.

Similarly, the rate of change of drag with lift, usually considered as the change in span efficiency factor e, should not change for straight wings.* For instance (Fig. 7.8), it has been observed that, when C_L^2 for a given airplane is plotted against the total drag coefficient C_{DT}, the graph is nearly a straight line. Furthermore, since we may write

$$e = \frac{1}{(dC_D/dC_L^2)\pi AR} \tag{7.6}$$

it becomes apparent that the slope of the line dC_D/dC_L^2 may be used to find e (Section 5.5). Fortunately the slope of this line is practically independent of Reynolds number, and a wind tunnel test may hence be used to determine full-scale e. A plot of the 23012 airfoil at two Reynolds numbers is given in Fig. 7.8.

* e for swept wings is not as tractable as for straight wings. Sometimes the tip stall is reduced by a higher Reynolds number in a manner that increases e, and other times e is reduced.

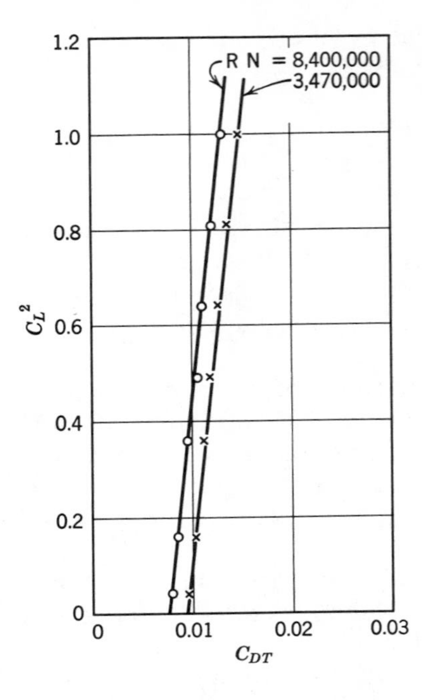

FIGURE 7.8. Plot of C_L^2 versus C_{DT} for an NACA 23012 at two Reynolds numbers.

As we have noted, the determination of the amount the C_L^2 versus C_{DT} curve is moved over (i.e., the scale effect on $C_{D0\,\text{min(aircraft)}}$) with increasing Reynolds number is not so simple; in fact, no direct rule is known. If similar tests have been completed in the past and flight tests made, perhaps the comparison may yield the ΔC_D necessary. For an entirely new aircraft, the minimum drag may be measured at several velocities, and a plot of $C_{D0\,\text{min}}$ versus log RN may be made. The straight line that usually results from such a plot may be extrapolated to find the approximate full-scale $C_{D0\,\text{min}}$ (Fig. 7.9).

Another method, particularly for components, consists of converting the tunnel test point to a skin friction coefficient by the equation

$$C_f = C_{D0} \times (S/\text{wetted area})$$

and spotting this point on a plot of C_f versus RN. The point is then extrapolated along the turbulent skin friction curve to the full scale Reynolds number. The resulting skin friction coefficient is then converted to a ΔC_{D0}.

A third method of extrapolating total aircraft to full-scale Reynolds number is as follows (see Fig. 7.10):

1. Plot C_D versus C_L, tunnel data.
2. Subtract $C_{Di} = C_L^2/\pi AR$ from the C_D plot to obtain $C_{D0\text{(tunnel)}}$.
3. Estimate $C_{L\,\text{max}}$ (full scale) from Fig. 7.11, Ref. 7.4, or other sources, and extend C_{D0} until it is horizontal at $C_{L\,\text{max}}$. The increased curvature of the

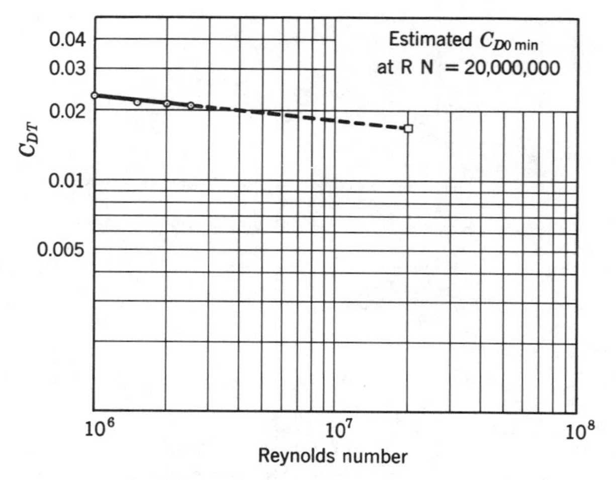

FIGURE 7.9. Plot of C_{DT} versus log RN.

C_{D0} curve should be moved to an increased C_L in a manner similar to that described in Section 7.4.

4. Decrease $C_{D0\,\text{min (tunnel)}}$ by the C_D change in *wing* drag from tunnel Reynolds number to full scale. See Fig. 7.1. (This is the controversial step. Some engineers make no change to tunnel $C_{D0\,\text{min}}$ because manufacturing irregularities on the actual aircraft may increase the drag as much as increased Reynolds number decreases it.)

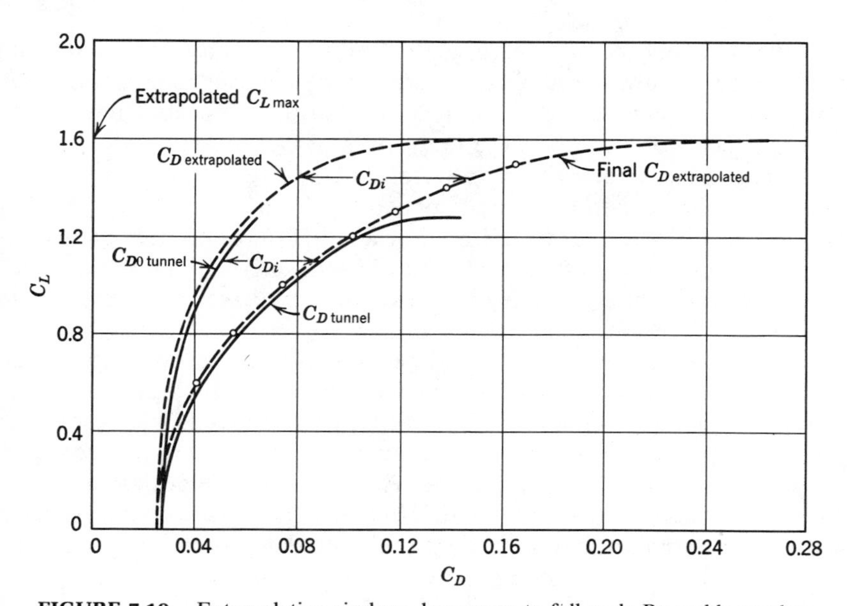

FIGURE 7.10. Extrapolating airplane drag curve to full-scale Reynolds number.

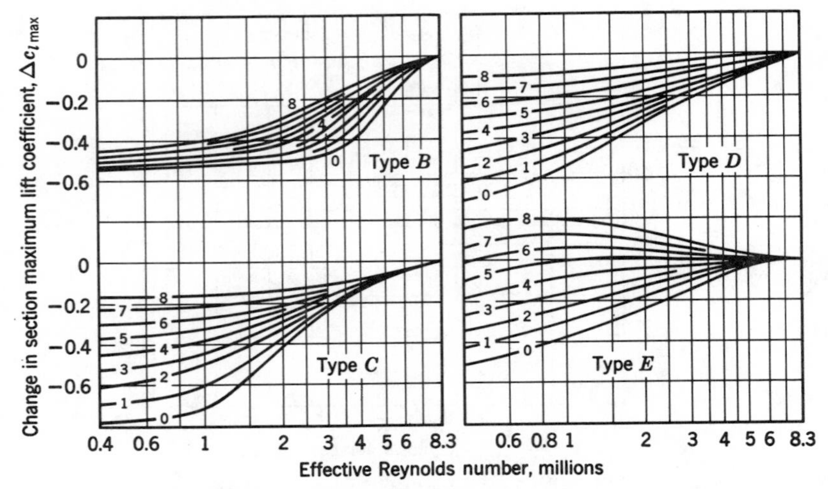

FIGURE 7.11. Effect of Reynolds number on $C_{L\,max}$.

5. Add C_{Di} back in to get the final extrapolated drag curve. In this step use values of C_L up to $C_{L\,max}$ (full scale).

Special attention should be paid to the extrapolation of C_D at $C_{L\,max}$. In many cases the engineer neglects the drag increase that accompanies the increase of $C_{L\,max}$ with Reynolds number, and predicted glide angles near $C_{L\,max}$ are then considerably above the attained values.

The methods outlined above for getting full-scale values from tunnel data are successful only when applied by experienced aerodynamicists.

In closing we may state that, though the low-drag "bucket" found in the tunnel may be a bit optimistic, the shape of the drag curve up until the $C_{L\,max}$ effects predominate seems to follow closely the drag obtained in flight.

If comparisons are to be made between tunnel and flight test, the tunnel data must be corrected for trim drag. For a tail-aft aircraft there is usually a download on the tail to trim the airplane. This negative tail lift increases the drag of the horizontal tail due to its lift, and increases both the lift and induced drag of the wing.

The prediction of aircraft drag and performance has always been an empirical process. The preliminary design prediction methodology accounts for the drag of existing designs from flight tests, plus results of wind tunnel and flight research. As the aircraft design moves to specific aircraft configurations, these are tested in wind tunnels. At the end of the design process an optimum aircraft shape has evolved. This final shape is not necessarily an aerodynamic optimum, but an aircraft optimum that has involved trade-offs between weight, function, complexity, costs, and performance.

7.4. SCALE EFFECTS ON THE LIFT CURVE

The effect of Reynolds number on the lift curve is indeed profound, and often quite unpredictable. We will first discuss the work of Jacobs (Ref. 7.4) on the NACA forward thickness airfoils, and then treat the newer profiles.

The Forward Thickness Airfoils

In *TR* 586 (Ref. 7.4), Jacobs indicates that variations in lift curve slope caused by increasing Reynolds number are very small, but in general the lift curve will be more linear, the slope will increase slightly,* and the stall will become more abrupt. (See Fig. 7.12). Lift curves already linear at the lower Reynolds numbers will be extended at higher ones. It follows that $C_{L\,\max}$ and the angle at which it occurs are increased. The amount of the increase of both angle and $C_{L\,\max}$ is of paramount value to the engineer.

* At very low values of the Reynolds number, about 150,000, the lift curve again steepens, and $dC_L/d\alpha$ may then exceed 2π/radian (see Section 8.4).

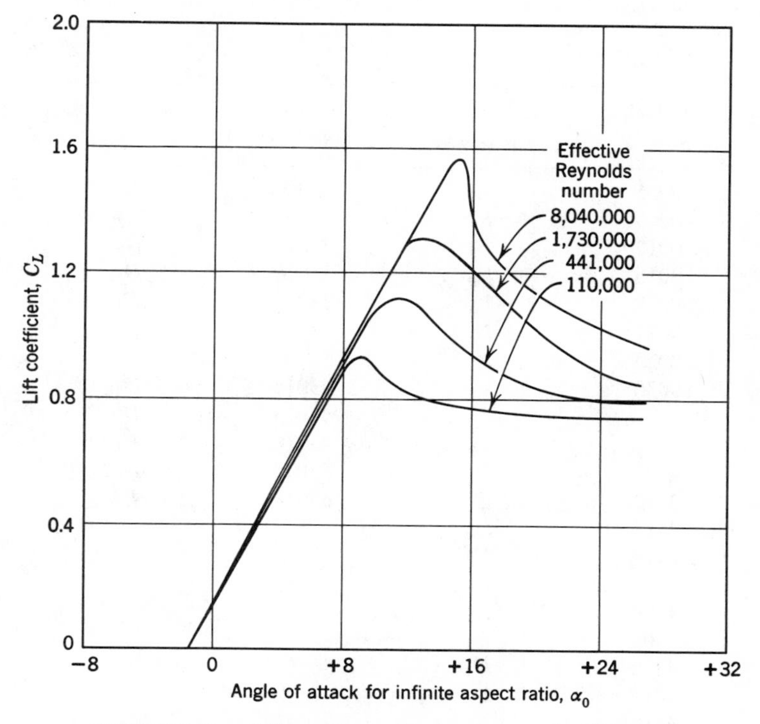

FIGURE 7.12. Effect of Reynolds number on the lift curve.

The method outlined in Ref. 7.4 makes it possible to determine the $C_{L\,\max}$ at Reynolds numbers below 8,300,000 for a large group of airfoils and enables the engineer to estimate possible Reynolds number effects on new airfoils.

The method is to read the $C_{L\,\max}$ at $\mathrm{RN}_e = 8{,}300{,}000$ and the stall type from Table 7.2. Then the increment (usually negative) is selected from Fig. 7.11 and added to the high Reynolds number $C_{L\,\max}$ to get $C_{L\,\max}$ at the desired Reynolds number. Unfortunately this seemingly simple method is of lessened value in most practical cases for two reasons: first it concerns *section* $c_{l\,\max}$ values when wing $C_{L\,\max}$ values are usually needed; and, second, the engineer will probably not find the desired airfoil in Table 7.2.

Many analytical methods exist based on using vortices either on a lifting line or as vortex lattices to predict the span load distribution of the wing and hence its downwash. The local Reynolds number is known, and the location of the first stalled region and wing $C_{L\,\max}$ can be estimated. The success of these methods depends on knowing the effect of Reynolds number on the wing airfoil sections.

The way around the problem is largely empirical. Many engineers have had sufficient experience correlating tunnel data with flight tests so that they feel qualified to estimate $C_{L\,\max}$ due to Reynolds number. Most of their

TABLE 7.2

Airfoil NACA	Scale Effects on $C_{L\,\max}$	Airfoil NACA	Scale Effects on $C_{L\,\max}$
0006	A	23006	A
0009	B_0	23009	C_2
0012	C_0	23012	D_2
0015	D_0	23015	D_2
0018	E_0	23018	E_2
0021	E_1	23021	E_2
0025	E_2		
0030	—	43012	D_4
		43015	D^4
2212	C_3	43018	E_4
2409	B_2	63012	D_6
2412	C_2	63018	E_7
2415	D_2		
2418	E_2		
4406	A_3		
4409	B_4		
4412	C_4		
4415	D_4		
4418	E_4		
4421	E_5		

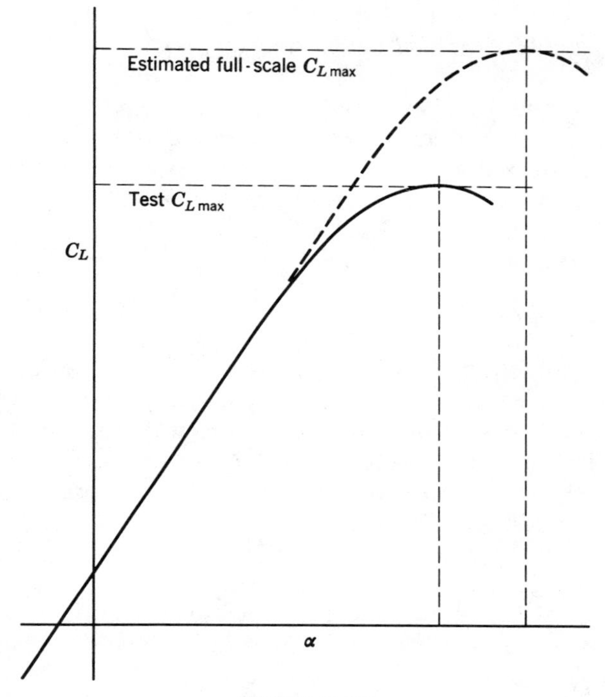

FIGURE 7.13

estimations run around $\Delta C_{L\ max} = 0.15$ for the range from a tunnel test at $RN_e = 1{,}500{,}000$ to full-scale Reynolds number $= 6{,}000{,}000$. For large jet-transport aircraft with leading edge and multielement trailing edge flaps, the extrapolation to much higher RN in the range of 60,000,000 at the wing root for a B-747-type aircraft is much more difficult. Such an extrapolation requires an experienced aerodynamicist and flight test data. The method is as follows (Fig. 7.13):

1. The linear part of the lift curve from tunnel data is extended with the same slope.
2. Through the value of $C_{L\,max}$ (full scale) as estimated, a horizontal line is drawn.
3. The curved portion of the test lift curve is then raised until it has the proper value of $C_{L\,max}$ and shifted laterally until it joins the linear part of the constructed full-scale lift curve.

The net result is a full-scale lift curve having the proper value of zero lift, slope, and $C_{L\,max}$, but probably having an angle of maximum lift that is too great and a stall that is too gentle. These two deficiencies are not serious, however, and the engineer has at least something with which he can work.

Since the speed of the airplane is reduced for landing, it is sometimes possible to obtain tests at landing Reynolds number in a tunnel of moderate capacity.

Maximum lift coefficients measured in different wind tunnels agree much better when based on "effective" Reynolds numbers (see Section 3.15) than when based on the test Reynolds number. Increased Reynolds number obtained by added turbulence are satisfactory for maximum lift measurements.

In Ref. 7.5, agreement on $C_{L\,\max}$ within ±0.1 was found when data from 1,500,000 were extrapolated to 26,000,000 by Jacobs' method.

Low-Drag Airfoils

The effect of scale on the lift characteristics of the 6 series of laminar flow airfoils has not been as thoroughly investigated as for the other airfoils previously discussed. Fortunately some data on Reynolds number effects are available in Ref. 7.6 for a number of 6 series sections up to $RN = 25 \times 10^6$. Although the effects vary with thickness form, thickness, and camber, some general remarks can be made. The angle of zero lift and the lift curve slope are virtually unaffected by scale, but the effects on maximum lift follow one of two general trends depending on the airfoil thickness ratio. For thickness ratios of 12% or less there is little effect over the lower Reynolds number range (up to $RN = 6 \times 10^6$). Increasing Reynolds number produces a rapid increase in $c_{l\,\max}$ to a more or less constant value, which then decreases slowly on up to $RN = 25 \times 10^6$. Turbulent separation beginning at the trailing edge seems to be responsible for this.

For the thick sections the trend is toward a continual increase in $c_{l\,\max}$ with increasing RN. The large-scale effect exhibited by an 18 percent thick airfoil seems to be related to the rapidly changing condition of the boundary layer at the leading edge. Any new airfoil tested at fairly low Reynolds number will present a difficult task of estimating its behavior at much greater Reynolds numbers.

7.5. SCALE EFFECTS ON FLAP CHARACTERISTICS

We are usually justified in expecting a little more from a flap full scale than is found in a tunnel at low Reynolds number, provided that the basic airfoil does not suffer extreme effects itself. In a number of fairly typical examples flight turned up about 0.2 more flap lift coefficient increment than did the tunnel. Figure 7.14 illustrates a tunnel range in which the flap increment was unaffected by scale.

With the current computational ability to design airfoil sections via computer with the desired pressure distributions for cruise flight and then use leading and trailing edge flaps for high lift, the problem of extrapolating wind tunnel data for RN effects is quite difficult. See Fig. 7.15 for the effect of scale on the F-111A flaps down.

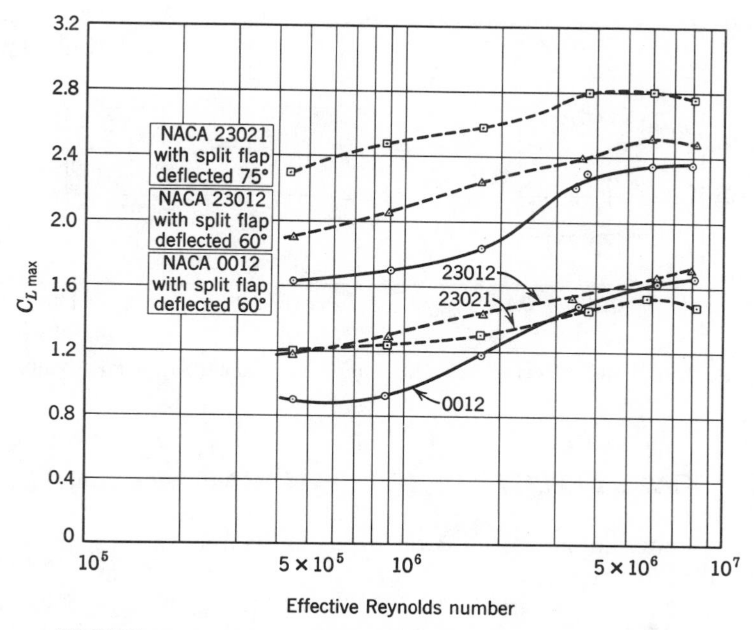

FIGURE 7.14. Effect of Reynolds number on $C_{L\,max}$, flaps down.

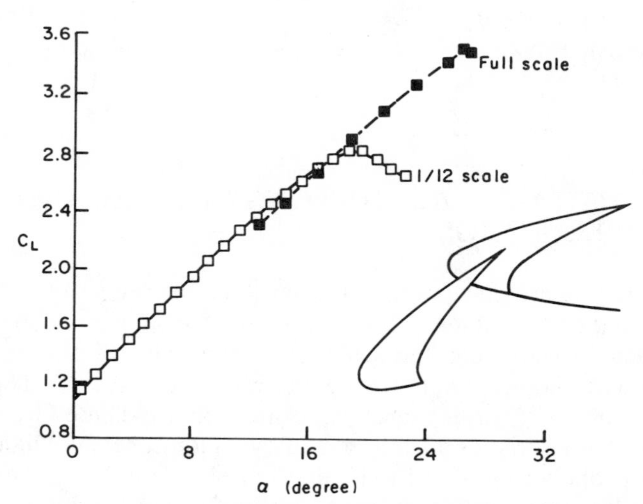

FIGURE 7.15. Effect of scale on the lift characteristics of the F-111 aircraft. (Courtesy of NASA Ames.)

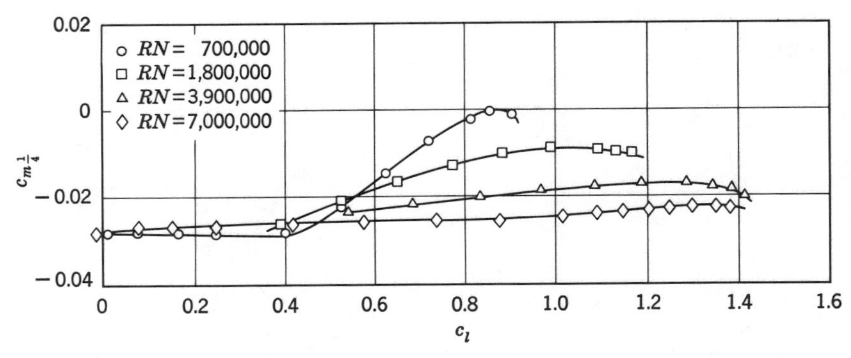

FIGURE 7.16. Effect of Reynolds number on a 66,2-215 (a = 0.6) airfoil.

7.6. SCALE EFFECTS ON THE PITCHING MOMENT CURVE

The static longitudinal stability of airplanes having airfoils with forward thickness (such as the 23012, etc.) seems to change little from typical tunnel Reynolds numbers of around 1,000,000 up to flight values of around 20,000,000. Usually the tunnel-flight discrepancy is in the direction of slightly more stability of the airplane than was predicted in the tunnel.

But the extrapolation of pitching moments when the airfoil thickness is well rearward (such as in the 65 and 66 series airfoils) is far more difficult to handle. The pitching moment variation of the 66,2-215 (a = 0.6) airfoil is shown in Fig. 7.16 from tests made in the British compressed-air tunnel and reported in Ref. 7.5. Obviously, the extrapolation of even the data made at 1,800,000 would be a very difficult job. Nor were trip strips of much value here, and satisfactory tests were accomplished only by going to high Reynolds numbers.

7.7. EFFECT OF SCALE ON LONGITUDINAL STABILITY AND CONTROL

In general the longitudinal characteristics will not be seriously different from those indicated by tunnel tests. Frequently one finds the model neutral point a little farther forward than full scale when power is off, and in pretty fair agreement with power on for propeller-powered aircraft. The elevator needed to trim rarely varies more than 2° from that indicated by the tunnel. Some other characteristics as determined by wind tunnel and flight tests of a two-engine propeller-driven aircraft may be of interest, even though they cannot be taken as typical; they are given in Table 7.3.

TABLE 7.3

Parameter[a]	Wind Tunnel	Flight Test
$dC_h/d\alpha$ ($\delta_E = 0°$)	−0.0014	−0.0011
$dC_h/d\delta$	−0.0044	−0.0044
$d\epsilon/d\alpha$	0.45	0.41

[a] Ch = hinge moment coefficient, δ = control deflection.

7.8. EFFECT OF SCALE ON DIRECTIONAL STABILITY AND CONTROL

The directional stability in flight in a number of instances has been seriously less than indicated in the wind tunnel, necessitating a redesign of the vertical tail. The reasons for this difference are not fully understood. In a number of airplanes the rudder was not powerful enough to yaw the aircraft into rudderlock on the model, but did so in flight. Catastrophies have been avoided only because rudderlock flight tests are always made quite gradually and with full expectancy of trouble. The comparative values in Table 7.4 are from wind tunnel and flight tests of a two-engine aircraft, propeller driven.

7.9. EFFECT OF SCALE ON LATERAL STABILITY AND CONTROL

The agreement between flight and tunnel tests on the lateral parameters seems to be generally satisfactory. The disagreement between tunnel and flight for aileron power is expected from considerations of cable stretch and wing flexibility. The values listed in Table 7.5 are from wind tunnel and flight tests of a two-engine propeller driven aircraft. Though they are normal values they cannot be taken as typical for all airplanes of course.

TABLE 7.4

Parameter	Wind Tunnel	Flight Test
$dC_n/d\delta_R$	−0.00126	−0.00085
$d\psi/d\delta_R$	−0.787	−1.39
$dC_h/d\psi$	−0.0010	−0.0005
$dC_h/d\delta_R$	−0.0037	−0.0027
$dC_n/d\psi$ (free)	−0.00125	−0.00045

TABLE 7.5

Slope	Wind Tunnel	Flight Test
$dC_l/d\delta_a$	0.00175	0.00144
$C_{l\,\max}/\delta_a$	0.023/18°	0.022/20°
$dC_t/d\psi$	0.00128	0.00076
dC_l/dC_n	−0.688	−1.69

7.10. CORRELATION BETWEEN WIND TUNNEL AND FLIGHT

This is an area where most of the data are kept by the companies as a closely guarded secret.

When comparisons are made between tunnel and flight there are uncertainties in the data obtained by each method. In flight test the data are taken at full-scale Reynolds numbers and the aircraft's roughness and excrescences are properly simulated. The full-scale drag effects of control surface and flap gap leakage, air conditioning, cabin pressurization, heat exchanger, thrust reversers, engine-nozzle modifications, etc., can best be evaluated by flight test. The limitations in flight test are the inability of doing an airplane build-up. It is difficult to fly an aircraft tail on and tail off to obtain the horizontal tail effect. In flight test the total drag of the complete aircraft is obtained from which only general conclusions can be drawn. Quite often there is not sufficient data to assess the effect of variations in Reynolds number. Thus Reynolds number effect can be concealed in the parasite drag or drag rise due to Mach number. This may require the analysis to use an assumed effect of Reynolds number. For analysis it is desirable to separate the drag polar into the following parts:

1. $C_{DP\,\min}$ due to friction, pressure, interferences, excresences, roughness.
2. ΔC_{DP} the variation of C_{DP} with C_L due to friction, pressure, nonelliptic span load, vortex, and so on.
3. C_{Di} induced drag variation with C_L^2 and aspect ratio, due to span load or vortex lift.
4. ΔC_{Dm} the drag due to compressible waves and shock wave induced separation.

If this separation of the various parts of the total drag is to be obtained, a large number of flight test data points are required and very complete wind tunnel data are needed as a guide to the analysis.

In analysis of flight test data it is not correct to assign all of the drag variation with lift to the wing, as the drag of the fuselage and other parts of the aircraft also vary with lift. Trim drag from the lift and drag of the horizontal tail must be accounted for when trying to determine the drag of the wing. The negative lift of the tail increases the wing lift required for level flight and therefore the induced drag of the wing. The tail itself also has a parasite and induced drag that must be taken into account. The tail operates in the wing downwash so its lift is usually not perpendicular to the flight path. Drag data from flight tests are only as accurate as the installed thrust (either jets or propellers) can be measured or predicted.

An additional problem with flight test data is that it often is acquired while the aircraft is accelerating. For example, $C_{L\,max}$ is obtained with the aircraft either decelerating at about 1 to 3 knots/second, and/or sinking. This makes the $C_{L\,max}$ from flight test higher than the true $C_{L\,max}$ by 0.2 or so. From this discussion it can be seen that it is difficult to both obtain and analyze data taken from flight tests.

The use of wind tunnel tests to predict airplane drag has the following advantages. A specific aircraft configuration can be studied one component part at a time over the lift coefficient and Mach number range. This allows trim drag, interference, nonelliptic span load, pitch up on swept wings, etc., to be studied in detail. As deltas in many cases are not affected by Reynolds number, new designs or aerodynamic features can be compared against model standards whose full-scale values are known.

There can be problems with wind tunnel data when corrections such as support and interference tares, blockage corrections, wall corrections, etc., are either improperly or not applied.

Two-dimensional pressure drag can be extrapolated to full-scale Reynolds number using the skin friction laws of Squire and Young (Refs. 7.7, 7.8) when the experimental pressure distribution of the airfoil is used to compute the local velocities. See also Refs. 7.9 and 7.10.

Depending on the size of the program there may only be a small number of tunnel entries (in some cases only one) during the design stage. When this occurs, there is a strong chance that the tunnel results have been used to modify the design and that the model and aircraft are no longer similar, making comparison difficult, if not impossible.

With V/STOL vehicles it is difficult to obtain steady-state data, especially in transition between vertical and horizontal flight. In the wind tunnel the data are often steady state, assuming that it is taken outside of the flow breakdown region (Section 6.33). This requires extreme care in attempting to match tunnel and flight data.

In Ref. 7.11 a comparison is made using both flight test data and wind tunnel test data in the NASA Ames 40 × 80 ft tunnel of a Learjet Model 23 using a 0.15 scale model and the second airplane built. Based on mean aerodynamic chord the 0.15 model was tested at RN $= 1.4 \times 10^6$. The airplane in the 40 × 80 ft was tested at RN $= 4.1 \times 10^6$ and 8.4×10^6. The flight test data were taken at RN $= 8.6 \times 10^6$. The data correlation between the 40 × 80 ft tunnel and flight test was good. The variation in $C_{L\,max}$ with RN was similar to NACA series 6 airfoils. The $\Delta C_{L\,max}$ from 0.15 scale to full scale was 0.32, both for flaps up and 40° flaps. Other than an anomally in α_{L0} between the 0.15 and full scale, the lift curve was similar to Fig. 7.12. The 0.15 scale model had a higher drag level and the polar was rotated. There were also slight changes in C_{m0} flaps up tail on, but not for tail off, and slight changes in the slope of C_m versus C_L.

Additional information can be found in Refs. 7.5 and 7.12 through 7.18.

REFERENCES

7.1. B. J. Holmes and C. J. Obara, Observations and Implications of Natural Laminar Flow on Practical Airplane Surfaces, ICAS-82-5.1.1, *Proceedings,* 13th Congress of International Council of the Aeronautical Sciences, Vol. 1, 1982, pp. 168–181.

7.2. A. L. Braslow, R. M. Hicks, and R. V. Harris, Jr., Use of Grit-Type Bounday-Layer-Transition Trips on Wind-Tunnel Models, NASA *TN* D-3579, 1966.

7.3. A. L. Braslow and E. C. Knox, Simplified Method for Determination of Critical Height of Distributed Roughness Particles for Boundary-Layer Transition at Mach Numbers from 0 to 5, NASA *TN* 4363, September 1958.

7.4. Eastman N. Jacobs, The Variation of Airfoil Section Characteristics with Reynolds Number, *TR* 586, 1937.

7.5. R. Hills, Use of Wind Tunnel Model Data in Aerodynamic Design, *JRAS,* January 1951.

7.6. Laurence K. Loftin, Jr., and William J. Bursnall, The Effects of Variations in Reynolds Number between 3.0×10^6 and 25.0×10^6 upon the Aerodynamic Characteristics of a Number of NACA 6-Series Airfoil Sections, NACA *TN* 1773, December 1948 (also reissued as *TR* 964).

7.7. H. Schlichting, *Boundary Layer Theory,* McGraw-Hill, New York, 1960, pp. 620–625.

7.8. H. B. Squire and A. D. Young, The Calculation of the Profile Drag of Aerofoils, ARC *R&M* 1838, 1938.

7.9. B. M. Jones, Flight Experiments on the Boundary Layer, *Journal of the Aeronautical Sciences,* **5,** 81–101, January 1938.

7.10. J. Bicknell, Determination of the Profile Drag of an Airplane Wing in Flight at High Reynolds Numbers, NACA *TR* 667, 1939.

7.11. R. D. Neal, Correlation of Small-Scale and Full-Scale Wind Tunnel Data with Flight Test Data on the Lear Jet Model 23, Paper 700237, SAE National Business Aircraft Meetings, 1970.

7.12. Clark B. Millikan, J. E. Smith, and R. W. Bell, High Speed Testing in the Southern California Cooperative Wind Tunnel, *JAS,* February 1948.

7.13. Marion T. Hockman and Robert E. Eisiminger, The Correlation of Wind-Tunnel and Flight Test Stability and Control Data for an SB2C-1 Airplane, *JAS,* January 1948.

7.14. J. E. Linden and F. J. Obrimski, Some Procedure for Use in Performance Prediction of Proposed Aircraft Designs, SAE *Preprint* 650800, October 1965.

7.15. L. W. McKinney and D. D. Baals (eds.), Wind-Tunnel/Flight Correlation—1981, NASA *CP* 2225, 1982.

7.16. H. L. Stalford, High-Alpha Aerodynamic Model Identification of T-2C Aircraft Using EBM Method, *J. Aircraft,* **18,** 801–809, October, 1981.

7.17. T. M. Moul and L. W. Taylor Jr., Determination of an Angle-of-Attack Sensor Correction for a Light Airplane, *J. Aircraft,* **18,** 838–843, October 1981.

7.18. R. F. Stengel and W. B. Nixon, Stalling Characteristics of a General Aviation Aircraft, *J. Aircraft,* **19,** 425—437, June 1982.

CHAPTER **8**

Small Wind Tunnels

In order to avoid the impression that useful wind tunnels must have a large jet and a speed of 100 mph or more, it seems pertinent to discuss some uses of smaller tunnels. A 30-in. tunnel was used by Van Schliestett in the program presented in *TN* 506 (Ref. 6.24), and a still smaller tunnel was used by Merriam and Spaulding (Ref. 3.1) in their outstanding calibrations of pitot-static tubes. Other examples could be given of successful programs carried out with the most inexpensive equipment.

The fundamental advantage of a small wind tunnel is traceable to the economics of tunnel operation. Small tunnels cost less to build and less to run. Though economy in operation is frequently neglected in tunnel proposals, it should not be, especially when it is realized that the electricity cost alone of some tunnels exceeds $2000 an hour!

A further advantage of a small tunnel is the small size of the models and the consequent saving in construction time. Small size may be a disadvantage, it is true; but those who have built or had built a 5–8 ft model are well aware of the time and cost of such models.

The most successful tests made in a small tunnel are, obviously: (1) those unaffected by Reynolds number and (2) those where any change due to Reynolds number is inconsequential.

8.1. TESTS UNAFFECTED BY REYNOLDS NUMBER

The tests most completely free from Reynolds number effects are those embracing pressure readings. Such experiments as static-pressure surveys and the aforementioned pitot-static calibration are in this group. The value of pressure distributions around airfoils is well known and has been given renewed attention by a greatly increased interest in boundary layer flow and in airfoils in general. Individual companies have been tending to design their own airfoils. The small tunnel can play an important part in this work. The many criteria (see Sections 4.18 and 4.19) that are determinable from pressure surveys are within reach of nearly all wind tunnels.

Many experiments concerning wind tunnel corrections are suitable for the small tunnel. These, too, are unaffected by Reynolds number.

A further use of small tunnels is in the study of flow patterns. These studies can be made with tufts or china clay, lamp black, flourescent oil, etc. (see Section 3.10). The study can be accomplished by either sketching the flow pattern which requires both artistic skill and time, or they can be photographed. Studies of this type can give insight into separation, both on wings and bodies, or other shapes such as automobiles, etc.

The progress of the stall over a wing may be unchanged by Reynolds number, although the entire stall is unusually delayed on the full-scale airplane.

Two tests for military airplanes that could be performed in a small tunnel include jettison tests of the drop-type external fuel tanks and tests of the flow from fuel dump valves. Both these actions frequently develop unforseen complications.

It should be noted that the expression "unaffected by changes in Reynolds number" must not be taken too broadly. By this is customarily meant that *reasonable* changes in Reynolds number produce little effect, and in turn this implies that the range under consideration will be free from movement of the transition point, i.e., above $RN_e = 2,500,000$, or transition points artifically fixed (see Section 7.2). The absence of change in e and $\alpha_{Z.L.}$ is limited by these stipulations. Pressure distributions at low angle of attack and the slope of the lift curve seem almost unaffected on down to RN = 200,000. There is a slowly growing body of data at Reynolds numbers below 200,000. It appears that there are many interesting problems that occur below RN = 200,000 and these are discussed in Section 8.4.

8.2. WHEN REYNOLDS NUMBER MAY BE IGNORED

Tests wherein any variations of results due to Reynolds number are inconsequential embrace both qualitative tests and tests wherein the taking of data is secondary, that is, where tunnel testing is being used for instruction purposes.

Qualitative tests are those that are expected to lead either to more testing or to abandonment of the project. They include the testing of radical ideas with a searching attitude for something promising.

8.3. THE SMALL WIND TUNNEL FOR INSTRUCTION

Almost no type of testing is performed in a large tunnel that cannot be duplicated in a small tunnel, the possible exception being tests of powered models. It might be possible by the use of small dc can motors (used for computer tape drives and model airplanes with spans up to 6.0 ft) and model railroad gear trains to simulate propeller-powered aircraft. For instructional purposes plastic model kits can often be reinforced and used. If aircraft models have the rivet and panel details removed, they can serve as an inexpensive model for flow studies. A small tunnel is invaluable for instructional purposes.

Many schools have a small wind tunnel, not unlike that shown in Fig. 8.1, for the use of undergraduates. The jet size is from 14 to 30 in. square and the dimensions are such that a space 14 ft by 30 ft is sufficient for the tunnel and motor. Twenty to twenty-five horsepower will provide 100 mph in the test section.

Walls for the test section may be made so that they may be wholly or partially removed, thus making it possible to perform tests with open or closed jet and to study asymmetrical boundaries. Open test sections cannot be used for nonreturn tunnels without a plenum around the test section (Section 1.2).

Many of these smaller tunnels do not use a six-component balance. The necessity for completing a test in the usual laboratory period of 3–4 hr precludes as complete a test as might be desired.

FIGURE 8.1. A small wind tunnel. (Courtesy Zumwalt and Darby.)

Suitable experiments for instructional purposes are listed below; some of these, as indicated, provide an opportunity to introduce the student to some of the more sophisticated instrumentation that is frequently used. Most of these require about 3 h to complete. Few, if any, tunnels use an open jet anymore, so it is to be taken for granted that a closed test section will be employed.

Experiment 1. Jet Calibration

Tunnel condition: Balance out.

Apparatus: Pitot-static tube, yawhead, turbulence sphere, 2 micromanometers, meter stick.

Tests:

1. Read dynamic pressure at 2-in. stations across jet. Plot percent variation in dynamic pressure from centerline value against station.
2. Read angle of flow at 2-in. stations across jet. Plot flow inclination in degrees against station.
3. Determine the turbulence by either a sphere or hot wire.
4. Read static pressure at tunnel centerline from plane of jet to exit cone at 3-in. intervals. Plot static pressure against station.
5. Read dynamic pressure on tunnel centerline at the trunnion at various tunnel indicated dynamic pressures. Plot measured versus indicated dynamic pressure for tunnel q calibration.
6. If available, parallel the pitot dynamic pressure across a pressure transducer that has been calibrated. A suitable range for the transducer would be ± 0.5 psi or less. Connect output of transducer to a digital voltmeter. Plot indicated tunnel versus measured dynamic pressure for tunnel q calibration.

Experiment 2. Balance Alignment and Aspect Ratio

Tunnel condition: Balance in.

Apparatus: Two wings of similar profile and chord but different aspect ratio, 4 and 6.

Tests:

1. Install wing of $AR = 6$. Read L, D, M from below zero lift to stall.
2. Invert model and repeat.
3. Same for wing of $AR = 4$.
4. Plot all data, and make alignment and boundary corrections. (Final data include tare drag and interference, but with models of about 3-inch chord the evaluation of these effects is extremely difficult.) Note on plots $\alpha_{Z.L.}$, $dC_L/d\alpha$ $C_{L\ \max}$, $C_{D0\ \min}$, C_{m0}, ac.

Experiment 3. Tailsetting and Downwash

Tunnel Condition: Balance in.
Apparatus: Model with horizontal tail having variable incidence.
Tests:

1. Read L, D, M from zero lift to stall with tail off.
2. Repeat with tail on, elevator zero, and at least three tail incidences between $\pm 8°$.
3. Plot α against C_m, and downwash ϵ against α. Determine α_{tail} and tail incidence for $C_L = 0.2$.

Experiment 4. Static Longitudinal Stability

Tunnel condition: Balance in.
Apparatus: Model with removable tail and movable elevators.
Tests:

1. Run model from zero lift to stall reading L, D, M. Tail off.
2. Repeat with tail on and elevators 0°, −5°, −10°, −15°.
3. Plot C_m against C_L for each elevator setting and C_m against δ_e. State dC_m/dC_L. Also plot C_L against α.

If a sting-type balance is available, this is a good experiment to use it on. Balance gage output would be fed to whatever type system is used with the sting balance. A separate experiment would be the calibration of the sting type balance.

Experiment 5. Profile Drag by Momentum Theory

Tunnel condition: Closed tunnel, balance out.
Apparatus: 12-in. chord airfoil, wake survey rake and multiple tube manometer, or pitot or pitot plus static that can traverse wake, scanivalve or pressure transducer plus associated electronics can be used in place of the manometer. Back-lighted manometers can be photographed and have the advantage of showing students the shape of the wake, plus ensure that the entire wake is obtained.
Tests:

1. Read wake $0.7c$ behind airfoil trailing edge with wake survey rake from = −3° to 6°. Plot c_{d0} against α.
2. Read wake at $\alpha = 0$ and at two or three dynamic pressures. Plot c_{d0} against Reynolds number.

Experiment 6. Pressure Distribution

Tunnel condition: Closed jet, balance out.

Apparatus: Pressure wing (should be same wing as Experiment 5), multiple tube manometer or scanivalve pressure transducer and associated electronics. Manometer will show pressure distribution as data is acquired.

Tests: With tunnel set at desired dynamic pressure read pressures for several angles of attack for zero lift through stall. If run in conjunction with Experiment 5, plot c_l versus c_d. If run as separate experiment, plot c_n, c_c, and c_{mac} versus α.

Experiment 7. Dynamic Stability

Tunnel condition: Open jet, balance out.

Apparatus: Flying wind model, dynamic stability rig.

Tests: At 40, 60, 80 mph disturb model and time oscillations. Plot period against velocity. Note that this is good only for the short-period constant-speed mode.

Experiment 8. The Boundary Layer

Tunnel condition: Open jet, balance out.

Apparatus: 12–15-in. chord NACA 0012 wing (used because of large amount of data on this airfoil), boundary layer mouse plus manometer or scanivalve and associated electronics, or traversible hot wire or thin film gage, or very small pitot.

Tests: Place mouse, pitot, etc., at 5, 10, 15, 20, 25, 30, and 35% chord, and read dynamic pressures at 0.03, 0.06, 0.09, 0.12 in. from surface. Determine transition region by plotting velocity profiles and velocity at constant height. Flow visualization can be used to check transition point.

8.4. LOW REYNOLDS NUMBER TESTING

The small tunnels operate at low Reynolds numbers, and it is fitting to have a good understanding of flow at these values in order to avoid the pitfalls into which many engineers have fallen. Indeed, a remarkable correlation exists at any time between the current capabilities of wind tunnels and the type of airfoils designers select. At Reynolds numbers of about 50,000 a thin wing with 4–6% camber appears best, and was used in early aircraft such as the Breguet, and many World War I fighters. At Reynolds numbers around 150,000 the Clark Y performs properly, as do other sections with perhaps 4% camber and 12% thickness. They in turn were used on the *Spirit of St. Louis* and many other airplanes of the period 1925–1935. At 4,000,000 to 8,000,000 the symmetrical sections of slightly higher thickness show up well, and we find those on many aircraft of 1935–1940. Later, of course, the tunnels with lower turbulence became available, and they in turn influenced design greatly from 1940 to 1950 until high subsonic effects began to crowd out other problems.

The point of the above discussion is to draw attention to the fact that most "modern" airfoils will yield embarrassingly poor results at low Reynolds numbers, and teachers trying either 0015 or 65 series wings at RN = 150,000 will find themselves with extremely wiggly lift curves, and drag curves for symmetrical wings showing less drag at 5° angle of attack than at zero—an "impossible" state of affairs.

Mueller in Refs. 8.1, 8.2, and 8.3 gives results of extensive studies of flow over two-dimensional airfoils at Reynolds numbers as low as 40,000, based on chord (Figs. 8.2 and 8.3). At RN = 40,000 the lift curve for a NACA 66_3-018 from negative to positive stall is in three distinct pieces. Two parts are near the stall and at $\alpha = \pm 8°$, there is a linear region with very low slope (Fig. 8.2). In Ref. 8.1 smoke photographs taken at $\alpha = 0°$ show a laminar separation on both top and bottom of the airfoil at 65% chord with periodic vortex shedding. At $\alpha = +6°$ the flow (Fig. 8.4) is attached to the lower surface but separates at 10–15% chord on the upper surface. At $\alpha = 8°$ there is a laminar separation bubble on the upper surface that acts similarly to a trip strip inducing transition with turbulent reattachment and a large increase

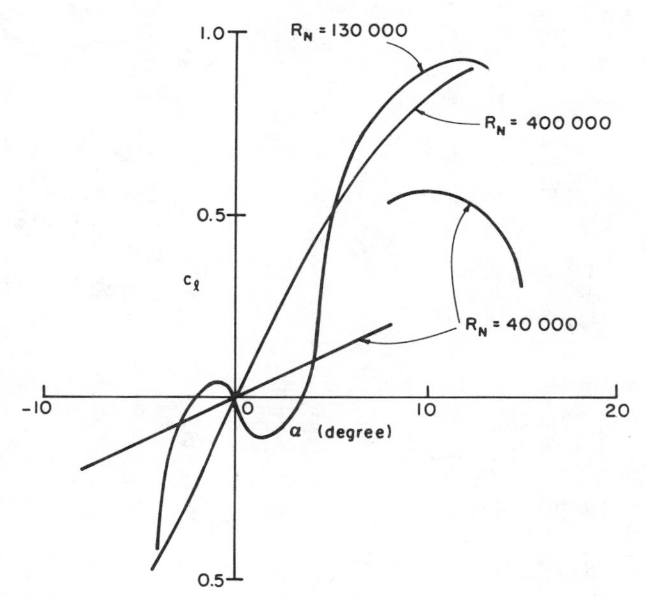

FIGURE 8.2. Lift curves for a smooth NACA 66_3-018 airfoil at three low Reynolds numbers. (Adapted from *AIAA Journal,* Vol. 20, No. 4, April 1982, pp. 459–460. Copyright American Institute of Aeronautics and Astronautics.)

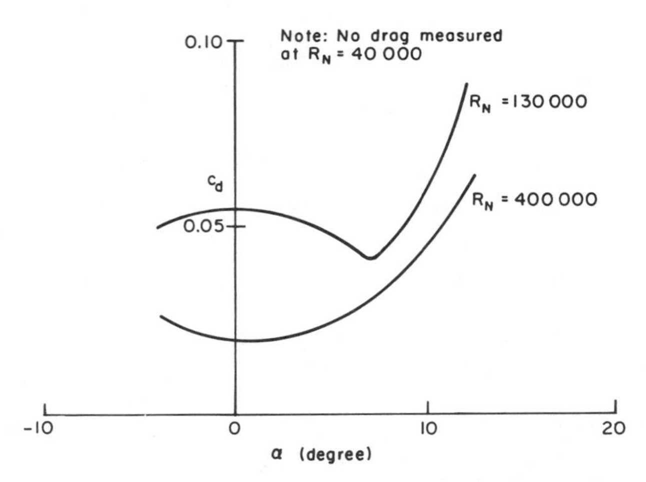

FIGURE 8.3. Drag curves for a smooth NACA 66_3-018 airfoil at two low Reynolds numbers. (Adapted from *AIAA Journal,* Vol. 20, No. 4, April 1982, pp. 460, 462. Copyright American Institute of Aeronautics and Astronautics.)

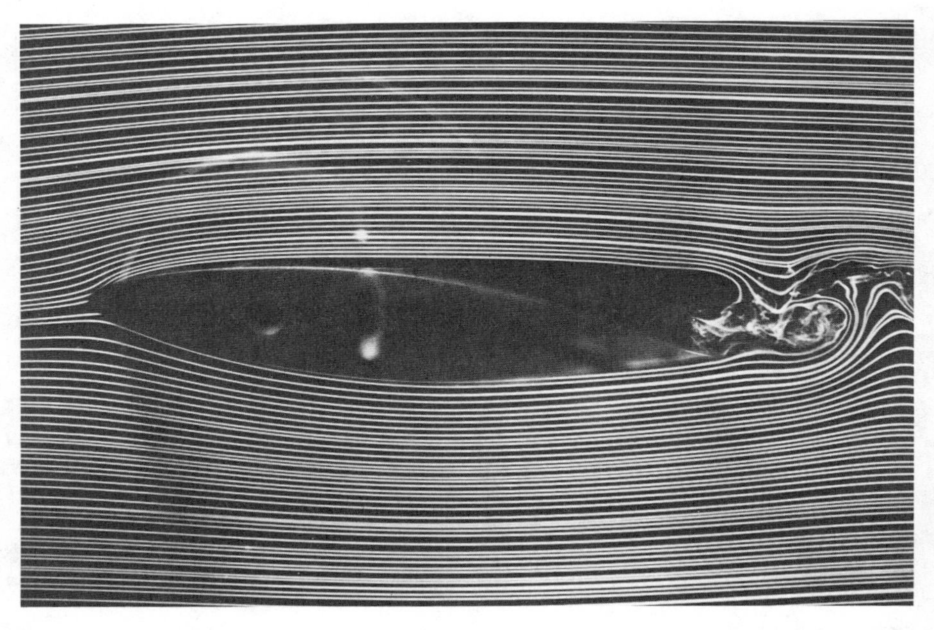

FIGURE 8.4. Smoke-wire flow visualization for a smooth NACA 66_3-018 airfoil at 6° angle of attack, RN = 40,000. (Reprinted from *AIAA Journal,* Vol. 20, No. 4, April 1982, p. 459. Copyright American Institute of Aeronautics and Astronautics.)

in lift. The lift increase is limited, however, by a trailing edge separation. This trailing edge separation moves forward with α causing the typical $c_{l\,max}$ shape of the lift curve. At RN = 40,000 drag was not measured due to the small forces.

At RN = 130,000 the airfoil shows a complete reversal of the lift curve slope at $\alpha = 0°$, and the drag is lower at 6° than at 0°. The addition of a trip strip near the leading edge gave more normal curves for lift and drag. At a Reynolds number of 400,000 the lift curve and drag curve were typical of low Reynolds number performance. In Ref. 8.2 lift and drag data for the same NACA 66_3-108 airfoil are given for additional Reynolds numbers. In Ref. 8.3 smoke-wire flow visualization at RN = 55,000 for the NACA 66_3-018 airfoil gives an insight into the surface flow at low Reynolds numbers. This work of Mueller gives an illuminating perspective into the statement at the start of the section regarding the pitfalls that can occur on an airfoil at low Reynolds numbers.

The spanwise variation of profile drag as discussed in Section 4.18 is considered by Mueller in Ref. 8.2. He suggests that the three-dimensional flow in the boundary layer at RN = 55,000 as shown in Ref. 8.3 may cause an error in either pitot or hot-wire measurements and that this accounts for the spanwise variation in profile drag. The momentum is based on the change in velocity parallel to the tunnel centerline and the shed vortices and periodic variation in the wake could cause the error. It has been long accepted that the momentum method is questionable for airfoils where separation is present.

At the present time there is a growing interest in acquiring data at low Reynolds numbers. For use on high-performance, high-aspect-ratio sailplanes in Ref. 8.4 Althaus and Wortmann have published data on many airfoils at Reynolds numbers from 1×10^6 to 3×10^6 with some data at 0.28 and 0.50×10^6. Other users of low Reynolds number airfoil data would possibly be general-aviation, remotely piloted vehicles, fan blades, wind turbines, model airplanes, etc. The data in Ref. 8.1 through 8.4 have been taken in Eiffel-type tunnels with very large contractions with many screens, prior to the contraction, which leads to very low values of turbulence. Mueller's tunnel has 12 antiturbulence screens followed by a 24 : 1 contraction. The tunnel has a turbulent intensity of less than 0.1%. Research of this nature is the case where the power cost of screens is acceptable and laminar flow is required.

In Ref. 8.3 Mueller uses the method of Ref. 8.5 to determine the uncertainties in the data. This method uses a careful specification of the uncertainties associated with primary experimental measurements such as pressure, temperature, and so on, and the accuracy of the instruments used.

The uncertainty of c_l is W_c and is a function of the uncertainty of the measurements. For a force coefficient measured by a balance, then

$$\text{Force} = F = BE = c_f(\tfrac{1}{2}\rho_\infty V_\infty^2)S \tag{8.1}$$

where BE is the output voltage E times a calibration constant to put it into engineering units. The c_f is any aerodynamic coefficient, such as c_l, c_d, and so on.

$$c_l = \frac{BE}{q_\infty S} \tag{8.2}$$

and the force uncertainty is

$$W_{c_f} = \left[\left(\frac{\partial c_f}{\partial B} W_B\right)^2 + \left(\frac{\partial c_f}{\partial E} W_E\right)^2 + \left(\frac{\partial c_f}{\partial S} W_S\right)^2 + \left(\frac{\partial c_f}{\partial V_\infty} W_{V_\infty}\right)^2 + \left(\frac{\partial c_f}{\partial \rho_\infty} W_{\rho_\infty}\right)^2\right]^{\frac{1}{2}} \tag{8.3}$$

This may be simplified by dividing by c_f:

$$\frac{W_{c_f}}{c_f} = \left[\left(\frac{W_B}{B}\right)^2 + \left(\frac{W_E}{E}\right)^2 + \left(\frac{W_S}{S}\right)^2 + \left(\frac{\partial W_{V_\infty}}{V_\infty}\right)^2 + \left(\frac{W_{\rho_\infty}}{\rho_\infty}\right)^2\right]^{\frac{1}{2}} \tag{8.4}$$

Now assuming that the tunnel test section is vented to the atmosphere, then by the equation of state

$$\rho_\infty = \frac{P_{\text{atm}}}{RT_{\text{atm}}} \tag{8.5}$$

thus

$$\frac{W_{\rho_\infty}}{\rho_\infty} = \left[\left(\frac{W_{P_{\text{atm}}}}{P_{\text{atm}}}\right)^2 + \left(\frac{W_{T_{\text{atm}}}}{T_{\text{atm}}}\right)^2\right]^{\frac{1}{2}} \tag{8.6}$$

Since $q_\infty = \frac{1}{2}\rho_\infty V_\infty^2$, then

$$\frac{W_{V_\infty}}{V_\infty} = \left[\left(\frac{W_{T_{\text{atm}}}}{2T_{\text{atm}}}\right)^2 + \left(\frac{W_{q_\infty}}{q_\infty}\right)^2 + \left(\frac{W_{P_{\text{atm}}}}{P_{\text{atm}}}\right)^2\right]^{\frac{1}{2}} \tag{8.7}$$

and

$$V_\infty = \left(\frac{2q_\infty}{\rho_\infty}\right)^{\frac{1}{2}} = \sqrt{\frac{2q_\infty RT_{\text{atm}}}{P_{\text{atm}}}} = K\sqrt{\frac{q_\infty T_{\text{atm}}}{P_{\text{atm}}}} \tag{8.8}$$

As a second example consider the determination of the lift coefficient by integration of pressures along the chord

$$c_l = \sum_{i=1}^{n} (C_{P_{l_i}} - C_{P_{u_i}}) \left(\frac{\Delta x}{c}\right)_i \tag{8.9}$$

In this case the uncertainties are the result of the length along the chord increment and the pressures. As $W_{\Delta P} = W_q =$ some value in %

$$W_{C_P} = \left[\left(\frac{\partial C_P}{\partial \Delta P} W_{\Delta P}\right)^2 + \left(\frac{\partial C_P}{\partial q_\infty} W_{\Delta q_\infty}\right)^2\right]^{\frac{1}{2}} \tag{8.10}$$

where

$$\frac{\partial C_P}{\partial \Delta_P} = \frac{1}{q_\infty} \quad \text{and} \quad \frac{\partial C_P}{\partial q_\infty} = \frac{1}{q_\infty^2} \tag{8.11}$$

The average uncertainty of the pressure coefficients are used to determine the uncertainty of the lift coefficient:

$$W_{c_l} = \left[\left(\frac{\partial c_l}{\partial C_{P_u}} \overline{W_{C_{Pu}}}\right)^2 + \left(\frac{\partial c_l}{\partial C_{P_l}} \overline{W_{C_{Pl}}}\right)^2 + \left(\frac{\partial c_l}{\partial \Delta x/c} W_{\Delta x/c}\right)^2\right]^{\frac{1}{2}} \tag{8.12}$$

This is a straightforward method applicable to cases where single samples of the measurement are taken as is often the case in wind tunnel tests.

The above analysis for lift coefficient by the integration of pressures requires that the $P_i - P_\infty$ for the C_P and the tunnel q_∞ be measured simultaneously. The same is true for data with the force balance. The simultaneous measurement of tunnel q is required to avoid the problem of time-dependent fluctuations in tunnel speed, which are very difficult to control at the low speeds required for very low Reynolds numbers. Almost all wind tunnels give difficulty in holding airspeed precisely when run at low speeds owing to the inertia of both the drive system and the air.

To summarize the problems with acquiring accurate data at very low Reynolds numbers below 100,000 to 150,000: First, the tunnel must have very low turbulence to promote laminar boundary layers on the model. This is necessary since the data are dependent on the behavior of the laminar boundary. The low turbulence requires large contraction ratios and damping screens before the contraction.

Second, since the test speeds are low, there is almost always present time-dependent excursions in tunnel velocity about the average, which requires the data and tunnel q to be taken simultaneously. This generally requires a computer-controlled electronic data system.

Third is the problem of balance zero shifts or drift.

The use of electronic data systems usually requires amplification with response down to dc. The low-signal level due to small forces and pressures requires high gains. Thus the amplification must be of high quality (cost).

The drift with time must be closely monitored. The signal conditioner or power supplies must be of the low noise type. And finally, owing to low transducer output voltages, care must be taken with shielding, etc., to avoid a signal to noise ratio problem. One millivolt noise on a 1000 mV signal is quite different than a 1 mV noise on a 3 mV signal.

When balances are used, they will have a relatively low spring rate or stiffness to generate an adequate signal from strain gages from the low applied loads. To measure lift and drag Mueller used a strain gage balance with two flexures: one for small loads and a second, stiffer, flexure that was engaged for larger loads. As the material frequency is proportional to the square root of the spring stiffness over mass, the natural frequency of the balance plus model will be very low. This may require the judicious use of electronic filtering of the balance output signal or a low-frequency cutoff. There also is the possibility of the model–balance having a large enough amplitude vibration that can lead to either a broken flexure in the balance or an inability to prevent the balance–model fouling with adjacent parts that have very small clearances.

For other exploratory tests of aircraft models at low Reynolds numbers (below 150,000 based on chord) it may be possible to apply a grit strip or other types of trip strips to fix transition as on large models and avoid the laminar bubble at the leading edge of the lifting surface (Ref. 8.2, Section 7.2). The model may have a slightly higher drag and lower maximum lift. However, these results should be acceptable when used for trends and increments. Similar acceptable results should be obtained on automobiles and trucks as long as the desired results are not sensitive to the Reynolds number.

Also, small tunnels can, of course, be used for any test of exploratory nature when the model is not sensitive to laminar and turbulent boundary layers and the transition. It may well be that in many cases the problem of building the model, especially in the case of structures such as transmission towers, cranes, etc., will be the limiting factor.

REFERENCES

8.1. T. J. Mueller and S. M. Batill, Experimental Studies of Separation on a Two Dimensional Airfoil at Low Reynolds Numbers, *AIAA Journal,* **20,** 457–463, April 1982.

8.2. T. J. Mueller and B. J. Jansen Jr., Aerodynamic Measurements at Low Reynolds Numbers, Paper 82-0598, AIAA 12th Aerodynamic Testing Conference, 1982.

8.3. S. M. Batill and T. J. Mueller, Visualization of Transition in the Flow over an Airfoil Using the Smoke-Wire Technique, *AIAA Journal,* **19,** 340–345, March 1981.

8.4. D. Althaus and F. X. Wortmann, *Stuttgarter Profilkatalog I,* F. Vieweg & Sohn, Braunschweig, West Germany, 1981.

8.5. S. J. Kline and F. A. McClintock, Describing Uncertainties in Single-Sample Experiments, *Mechanical Engineering,* 3, 1953.

CHAPTER 9

Nonaeronautical Uses of the Wind Tunnel

A large and fertile field for extending the utility of wind tunnel testing in general is that concerning nonaeronautical problems: those produced by natural winds, or by traveling on the surface of the ground or water, or by air pollution due to lack of wind. These types of studies are divided into three classes:

1. *Wind engineering* is the treatment of the interactions of the wind in the atmospheric boundary layer with the structures and activities of humans. Since each year winds cause several hundred deaths, several thousand injuries, and several hundred million dollars in damages, wind engineering is an important area. It requires the effort being put on it, and more. Wind engineering includes solving the problems of proper design and construction of buildings, bridges, glass installation, towers, and chimneys, and the problems of air pollution, soil erosion, snow drifts, and much more.
2. *Surface vehicle testing* considers travel on the surface of the earth or water. Auto drag and cooling are major areas of work.
3. *Miscellaneous nonaeronautical problems* is a catch-all for tests such as aerodynamics of insects and birds, minimum drag positions for skiers and bicyclists, energy from the wind, and so forth.

Besides a need for studies of the above problems for engineering purposes, there is now a legal reason in that the Clean Air Act (Public Law 95-95) describes ''good engineering practice'' as including pollution effects, and the Environmental Protection Agency reserves the right to require fluid modeling or field tests for proposed installations.

In the following we will discuss the above types of wind tunnel studies, including the design features of the specialized tunnels required.

9.1. GENERAL NONAERONAUTICAL TESTING PROCEDURES

The person needing a wind tunnel test for his or her nonaeronautical problem almost certainly has no conception of the complexity of running a tunnel properly. He or she also will have little idea of the various ratios (Reynolds number, Strouhal number, etc.) that must be matched to provide data that can be extrapolated to full scale, nor of the boundary layers and other flow characteristics so familiar to aerodynamicists. Nor will he have heard of wall corrections, or evaluating tare and interference through the use of dummy systems (not often needed in non-aero tests).

The complexities of model design and construction will also be foreign to the nonaerodynamicist, who usually proposes a combined wind tunnel and display model, the latter far below necessary strength requirements.

Thus the tunnel engineer must be prepared to give a short course on wind tunnel capabilities and model requirements. It will help immensely if data are presented which have led to cost savings through avoidance of wind damage. An important consequence of the early discussions is both a gain in safety and a better understanding as to why wind tunnel models and tests are so expensive.

9.2. METEOROLOGICAL WIND TUNNELS

Meteorological wind tunnels are those designed to simulate testing in the natural boundary layer. Cermak in Refs. 9.1, 9.2, and 9.3 describes five requirements:

1. Proper scaling of buildings and topographic features.
2. Matching Reynolds numbers, $\rho Vw/\mu$.
3. Matching Rossby numbers, $V/L\Omega$.
4. Kinematic simulation of air flow, boundary layer velocity distribution, and turbulence.
5. Matching the zero pressure gradient found in the real world.

Reynolds number effects are usually small due to the sharp edges of most objects under study, but check runs should be made at varying speeds to make sure. If problems show up they may usually be cured through the use of trip strips (Section 7.2). Reynolds numbers for buildings are based on width, w.

The Rossby number is concerned with the effect of the rotation of the earth on its winds. It accounts for a change of wind direction of perhaps 5° in 600 ft. This is of little significance and would be hard to simulate if it were necessary.

The velocity distribution in the natural boundary layer should be simulated as completely as possible. For example, at a scale of 1 : 450 a 900-ft building will be 2 ft high. The boundary layer must be matched to at least 3 ft high, and preferably all the way to the test-section ceiling. The boundary layer velocity distribution and turbulence can be well duplicated by an installation of spires in the entrance cone (Figs. 9.1*a* and 9.1*b*) followed by a roughness run of 10–15 test-section heights often made with small cubes on the floor. The building or locale to be tested and its environs are placed on a turntable, and rotations then made such that the test area is subjected to winds from all directions. There is a small error in that the boundary layer and turbulence in the real world may be different according to the approach direction. Sometimes, especially when a building is on a lakeshore, a different wind structure is employed.

FIGURE 9.1*a*. The strakes, floor roughness, and upstream buildings needed to simulate the proper conditions for studying the wind pressures and pedestrian-level velocities for the Equitable Center West, New York. Scale is 1 to 400. (Courtesy Cermak-Peterka Associates.)

FIGURE 9.1*b*. Close-up of surprisingly high-drag strakes needed to produce atmospheric like boundary layers. (Courtesy Canadian National Research Council.)

The longitudinal pressure gradient normally found in a wind tunnel and exacerbated by the very thick boundary layer needed can be made negligible by providing an adjustable test-section roof that may be adjusted to provide the extra cross-sectional area needed.

While it is necessary to provide cooled or heated air and test section floor for some types of tests of pollution (see Section 9.8) the tunnel engineer gets a break for force, pressure, and dynamic tests, because when the wind blows hard, it is so well mixed that temperature gradients do not occur. For these tests simulating the boundary layer structure and turbulence are adequate. Ground conditions influence the boundary layer [see Eq. (9.1)] and so no one boundary layer works for all tests. Nor does the atmosphere have only one temperature gradient; both daily and seasonal changes occur.

Two types of ''atmospheric'' wind tunnels have evolved, both with long tests sections:

1. *Meteorological* wind tunnels with test sections perhaps 15 test-section heights long having the capability of both cooling and heating the air and test section floor.
2. *Environmental* wind tunnels that use winds only for studies. These normally have test sections about 10 test-section heights long.

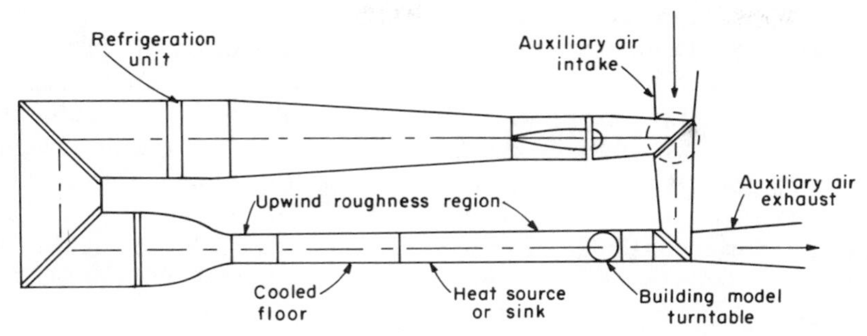

FIGURE 9.2. Layout of meteorological wind tunnel at Wind Engineering Laboratory at Colorado State University, Ft. Collins, Colorado.

Besides the spires and cubes on the floor mentioned before,* it is necessary to simulate the environs for about 1000–2000 ft full scale already extant about the building site. They change not only the wind loads, but may add significantly to the pollution problems. Model scales vary from 1 to 50 to 1 to 500 for buildings to 1 to 10,000 or so for topographic studies.

Tunnels of the atmospheric type are shown in Fig. 9.2 (from Ref. 9.3) and in Fig. 9.3. Both have the capability of temperature gradients for low-speed tests of inversions and general diffusion studies.

* Standard tunnels with short test sections may also have their flow altered to simulate a natural boundary layer for some atmospheric tests.

FIGURE 9.3. The atmospheric simulation tunnel at Calspan, Buffalo, New York.

There is an advantage in having a straight-through tunnel in this type of work (or a closed circuit tunnel which may be converted to open circuit) because tests are often made that could damage a closed circuit type. These include smoke, "snow material," erosion tests with sand, when model failure is part of the test (roofing material or gravel), or when a water trough or rain simulation puts moisture in the tunnel which could hurt plywood construction or rust balance components.

Air pollution tests sometimes require that the tunnel be run at very low velocities. A direct-current drive motor plus a variable-pitch propeller is the best arrangement for this, but an alternating-current motor with a variable-pitch propeller is cheaper and usually adequate. Adjusting tunnel speed by rpm is less necessary with thick boundary layer flow because the flow pattern is adjusted by the spires and roughness. Atmospheric test engineers like to hold flow speeds to within 1–2%. At the time of writing there are probably less than two dozen atmospheric tunnels in the world, although double that number is needed.

9.3. WIND ENGINEERING

Wind engineering combines the fields of meteorology, fluid dynamics, structural mechanics, and statistical analysis to minimize the unfavorable effects of the wind and maximize the favorable. It is an only recently recognized field. In general four separate areas are studied:

1. Wind forces on buildings and structures. These problems are concerned with forces, moments, deflections, local pressures and velocities.
2. Dynamics of structures. This area includes buffet, flutter, swaying, and breathing.
3. Local winds. These problems require measuring mean wind velocities, turbulence, and turbulence energy and scales.
4. Mass transport by winds. This covers soil erosion, pollution, blowing soil, efflux from smokestacks and diffusion. (See Fig. 9.4.)

9.4. STATIC TESTS OF BUILDINGS

Wind tunnel engineers may be called upon to help correct buildings already built and in trouble, or to guide the architect in a new design. "Trouble" can mean any of a long line of complaints: building swaying, losing glass or cladding, losing roofing, people getting knocked down by wind,* whistling noises, smoke or other fumes coming in ventilation systems, etc. These tests are less fun than those on proposed buildings because the architect is usually

* Local ground winds can be as much as four times those aloft.

FIGURE 9.4. Laying in roughness to a terrain study. (Courtesy Calspan Corporation.)

very defensive. On proposed designs it is less costly to make changes. A full building test program encompasses:

1. Preliminary smoke tests to search for possible trouble spots where pressure ports are needed.
2. Static wind loads, which may lead to dynamic tests later.
3. Ventilation intake studies with smoke being emitted by nearby factories, or efflux from the proposed building itself. (The high wind velocities near the top of a building result in a higher total pressure near the top, and wind blowing downward. This can lead to all sorts of unexpected intake problems.) (See Fig. 9.5.)
4. Local high velocity areas that might cause problems for people.
5. Detailed smoke studies after force tests have been made and the model painted a dull red to make the smoke show up better are used to convince the architect and his or her backers that the suggested changes are needed. Sometimes a location problem is identified so that simply moving the proposed building will solve the problem.*

* If Candlestick Park had been built one playing field to the north, much of its wind problems could have been avoided. (J. E. Cermak, *Science Year, World Book Annual,* Field Enterprise Education Corp., 1978.)

FIGURE 9.5. Smoke blowing downward on building model. (Courtesy Calspan Corporation.)

Pressure measurements are needed in order to ensure the integrity of cladding and windows, and this is difficult to obtain in very turbulent flow. Current practice is to use transducers that have a flat frequency response to say 150 Hz, and then to average samples taken at rates of 50–150/s for say 20s (see Section 3.8). Locations for pressure ports are from Item 1 above and from the tunnel engineer's experience.

A special case of wind loading occurs when one tall building is adjacent to another. Usually this produces shielding from winds in some directions, and more serious buffeting in others. Buffeting from adjacent cylindrical or hexagonal structures can be reduced by fixing transition through added roughness. (Ref. 9.28.)

If the actual boundary layer is known at the location of a proposed site, the same boundary layer should be simulated. If not, the maximum speed at 30 ft altitude is measured or estimated and the boundary layer is structured

TABLE 9.1. Coefficients for Boundary Layer Shape[a]

	h (ft)	α
Open country	280	0.15
Low rise buildings	1200	0.28
Urban	1700	0.4

[a] The value of h is a rough indication of boundary layer thickness.

according to

$$\frac{u}{u_{\text{ref}}} = \left(\frac{z}{z_{\text{ref}}}\right)^{\alpha} \tag{9.1}$$

where u = mean velocity at height z,
u_{ref} = mean velocity at reference height.

The boundary layer shape coefficient α varies according to the terrain (Table 9.1). Wind speed increases with height, while turbulence is greatest near the ground. Several boundary layer profiles are shown in Fig. 9.6.

The dynamic pressure used for reducing the forces and moments to coefficient form may be taken as an average value over the model, the value noted at the middle of the model, or freestream q. It must be defined or the data are useless. Sometimes the velocity at 30 ft is used as a reference velocity, or, alternately, the reference dynamic pressure may be the model-out dynamic pressure at the exact height of the pressure orifice.

Several atmospheric tunnels are described in Table 1.1.

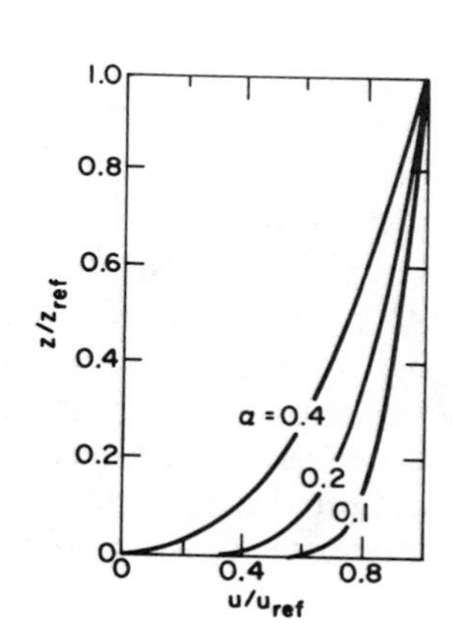

FIGURE 9.6. Boundary layer profile for several values of α.

9.5. DYNAMIC TESTING OF BUILDINGS

Buildings with more than a 6 to 1 ratio of height to width are apt to develop accelerations that are quite discomforting to the occupants. There are two approaches to solving this problem.

1. The building structure engineer knows the natural frequencies of his or her building; but not the buffet frequencies expected. Testing of a rigid model on a high response balance can obtain the needed information, and added stiffening can then be incorporated into the design.

2. An elastic model can be built and actual accelerations and displacements measured for the various wind speeds and wind directions, but a number of uncertainties remain. For example, the reference velocity is unknown, and there are some uncertainties as to the dynamic characteristics that should be built into the model. Reference 9.4 and the references therein consider the comparison between a model and a building that has already been built, noting a fair agreement with dynamic data from each, but acknowledging that the test program was helped by having a building already constructed from which coefficient ranges could be determined. The model ended up with 21 degrees of freedom, sensors that measured three displacements, and 15 accelerometers. Of interest was finding that torsional motion resulted in accelerations in some parts of the building being 40% higher than in others.

Some ordered tests have been made in this country and Belgium trying to define building designs apt to have dynamic problems, and more needs to be done. Most tunnel engineers can "eyeball" a design and come close to estimating any problems.

9.6. UNSTEADY AERODYNAMICS

The wind can produce structural oscillations in several ways, not all of which can be duplicated in the wind tunnel.

Simple Oscillations

All natural and man-made structures have a natural frequency at which they will oscillate unless critically damped. The natural frequencies of many items (trees, signposts, etc.) are close enough to those of wind puffs that it is not unusual for a failure to result from a second puff's catching a structure deflected from a previous puff. This sort of failure is not amenable to wind tunnel study and is only mentioned for completeness.

Aeolian Vibrations

Long, singly supported structures such as smokestacks and towers have a tendency to oscillate in a direction normal to the wind and at their natural frequency. The condition arises from the shed vortex street: the most recently shed vortex (*a* in Fig. 9.7) has the predominant effect, inducing an asymmetrical flow over the tower and a side force that bends the stack sideways in the direction of V_a. As the vortex moves downstream and its induced effect at the stack diminishes, the stack moves toward its undeflected position. At resonance, the next shed vortex induces a flow that encourages the movement to return, pass center, and continue in the new direction. This motion occurs on stacks, TV towers, and transmission lines. Typically the oscillations increase beyond the critical speed but reach a limit and decrease to zero when the exciting frequency gets far beyond the natural frequency. Such an oscillation is called Aeolian although sometimes the term is restricted to the case when the structure has a high natural frequency (10–100 cps) and displacements are small. The humming of power lines is Aeolian.

Galloping

A second type of motion arises when a body has a negative slope of the lift curve, and motion across a wind then produces a force in the direction of the wind. This is not unusual for bluff bodies whose in-wind side becomes unstalled with a small angle of attack (Fig. 9.8), and Bernoulli-type flow then is able to act. Galloping oscillations are usually violent and must be eliminated through design changes. For instance twisting stranded wires or adding spiral wrappings have been found effective in many cases.

Breathing

Sometimes large-diameter stacks and other structures will distort at some natural frequency such that the flow pattern is changed toward that fre-

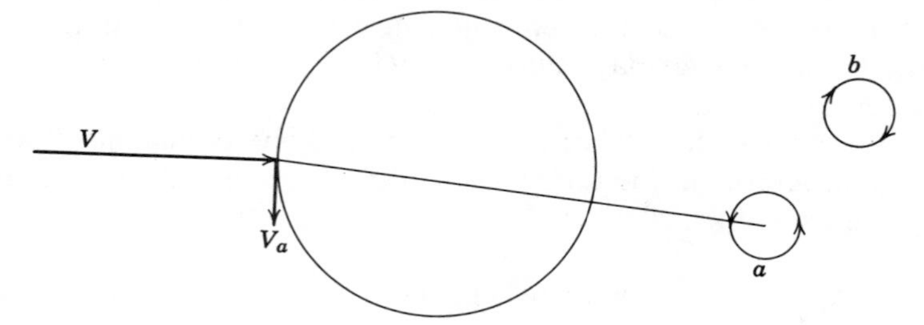

FIGURE 9.7. The downward component V_a from vortex *a* is greater than the upward component V_b (not shown) from the more distant vortex *b*.

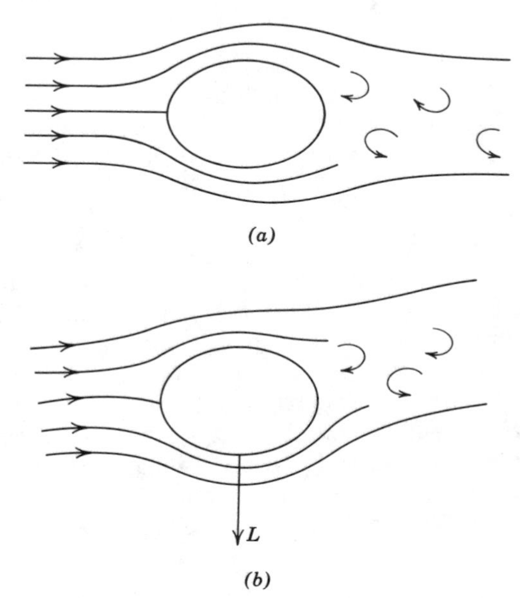

FIGURE 9.8. Downward lift, L, produced at positive angle of attack through unstalling the separated part of a blunt shape.

quency. This is different from motion of the body with very little distortion.

Transmission lines exhibit both Aeolian and galloping oscillations. The Aeolian singing is understood, usually occurs at from 10 to 100 cycles full scale, and is suppressed by dampers.

Galloping produced by wind loads is reasonably well understood, but unfortunately very large deflections (of say 100 diameters) are not uncommon and are hard to accommodate in a tunnel. Test have been made on spring-mounted sections employing both wire, stranded wire, and wires with simulated ice accretion. Here some galloping has been developed, but apparently only small deflections have occurred. Far more interesting are the electrical effects, which include observations of everything from substantial increases of galloping when the current is turned on to powered transmission lines which galloped for days during no wind conditions! Here is a fertile field for research.

We shall consider the above types of unsteady aerodynamic effects on a number of structures in the following sections, noting that oscillations may also be categorized as:

1. Aerodynamic instability, where the movement has only a single degree of freedom.
2. Flutter, more than a single degree of motion.
3. Buffet, where the motion arises from the wake of a nearby structure.

Ideally, to duplicate the full-scale motion the model should be similar to full scale in shape and the following characteristics should correspond:

1. Reynolds number, $\rho Vl/\mu$.
2. Structural damping, δ_s.
3. Stiffness, $S/\rho V^2$.
4. Density ratio, σ/ρ.

Fortunately, complete compliance with the above conditions (which would require essentially a full-scale model) is not necessary, as discussed in the following sections.

In the above

l = characteristic length, ft,
S = modulus of elasticity, lb/sq. ft,
V = velocity, fps,
δ_s = logarithmic decrement of oscillation,
ρ = air density, slug/ft^3,
σ = structural material density, slug/ft^3,
μ = air viscosity, lb-sec/sq. ft.

The sharp-edged construction of bridges precludes Reynolds number troubles, but stacks should have roughness added along a line from 30° to 60° from the incoming wind direction if the test is below a Reynolds number of about 400,000. The similarity rules become those for density ratio, σ/ρ; structural damping, δ_s; and for bridges, maintenance of equal V/NB, where V is velocity, in ft/s; N is cycles/s; and B is some typical dimension, usually bridge width, in feet. The spring constants of the model are adjusted to bring the V/NB ratio into equality with that of full scale, and the axis of rotation must also match full scale. Under the above conditions one can vary the tunnel speed and note whether resonance occurs during the correspondingly probable wind velocities. In a few cases bridge buffet has been introduced by the wake of a nearby bridge (Fig. 9.9). Oscillations of cable bundles in power lines have also been traced to wake buffet. When Aeolian or buffet oscillations occur, they are usually cured by the addition of saw-teeth or spoilers added in such a manner as to prevent the structure from emitting discrete vortices. Cures for stacks have included spoilers (called "strakes") wound helically around the stack (Fig. 9.10). These increase the wind loads, but tend to make them steady.

9.7. TESTING SUSPENSION BRIDGES

The effect of natural winds is important to the proper design of long or even intermediate bridge spans. Two wind instabilities must be studied: (1) vortex shedding, which causes limited vertical movement or torsional oscillations

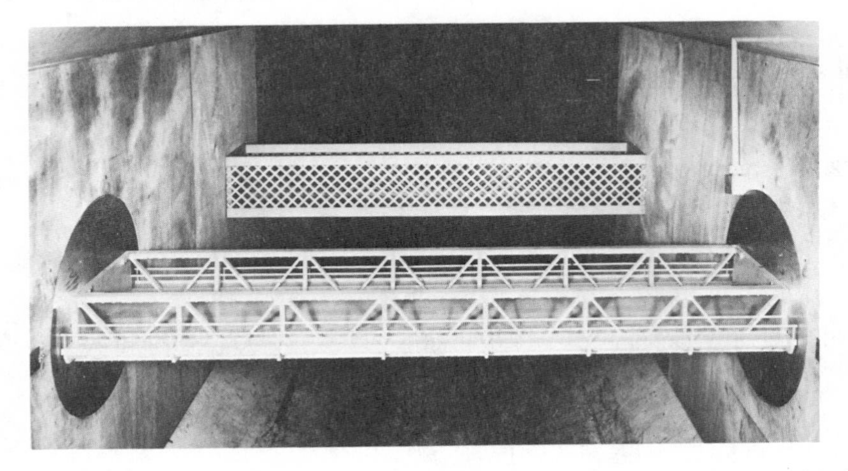

FIGURE 9.9. Wind tunnel experiment on the buffeting of bridges. (The sectional model of a suspension bridge in the foreground is in the wake of the sectional model of a massive railway bridge.) (Courtesy National Physical Laboratory.)

at generally low wind speeds; or (2) flutter instability, which can result in both vertical movement and torsional oscillation. Four types of tests have been tried:

1. Full models tested in smooth air.
2. Section models tested in smooth air (Ref. 9.6).
3. Section models mounted on taut wires in smooth air.
4. Full elastic models in turbulent air.

Section models would of course be desirable for ease of construction, especially since the ''final'' design is usually not available until well into the

FIGURE 9.10. A circular section cylinder model with three-start helical strakes mounted in the wind tunnel. (Courtesy National Physical Laboratory.)

study, and the construction time (and cost) of a full elastic model are onerous. Tested in smooth air, the section model does indicate changes that would reduce undesirable motions, and the section model data do agree fairly well with the full bridge model, both in smooth air, but they both give overly conservative data; the proposed design really will not be in as much trouble as indicated. Full models in turbulent air are the best. To be most useful, the turbulence must be scaled to the model size, and such real air turbulence is very large. Whereas a tunnel might have a u'/U of 0.5%, the real air might have a value of 10% or more. (u' is the root-mean-square variation in the horizontal wind velocity, and U is the mean horizontal wind measured over a long time.) Apparently the random turbulent air does not feed into the natural frequency oscillations. Scaling laws for bridge testing are covered in Ref. 9.7.

9.8. TESTING FOR POLLUTION

In these days of various types of pollutants being discharged into the air, it becomes important to know not only where they will go, but how far they must go before diffusion makes them harmless. The number and types of pollution problems are amazing. They include industrial smoke, throw-away gases from chemical processes, efflux from nuclear-power-plant holding tanks, using wind to carry silver iodide aloft to increase the probability of rain, the necessary distance from an LPG spill to obtain a noncombustion mixture, getting rid of hydrogen sulfide smell from a geothermal power plant, the dispersal of pesticides through trees, etc. Furthermore, it is often necessary to know when the above will suffer from inversions that hold the pollutants close to the ground.

A great contribution to testing for dispersal in a wind tunnel has been made by Skinner and Ludwig in Refs. 9.8 through 9.11. They argue that (1) the dominant feature of plume mixing is turbulent exchange, and that (2) fluid viscosity is not important. They also note that there is a minimum Reynolds number below which there will be a laminar sublayer in the tunnel* which cannot be tolerated, or conversely, when testing above that speed, a greater area can be examined by going to larger model scale ratios as long as the tunnel speed is increased proportionately. Their conclusion, provided the stack flow is "reasonably turbulent," is that by exaggerating the buoyancy of the exit momentum of the stack effluent and compensating with an increase of tunnel speed, the same dispersion is obtained as when modeling the ratio of stack exit density to ambient density. Thus hot stack gases are not necessary for this type of test, permitting larger ground areas to be simulated by a model of fixed diameter. Having the plume in the tunnel, its dispersion rate may be measured, or a program of changes may be studied to

* RN = 3.0 based on roughness height.

move its location away from certain areas. Another type of pollution test employs a pollutant distributed at various locations to see how the dispersal is affected by changing conditions.

The change in plume height caused by nearby buildings is shown in Figs. 9.11*a* and 9.11*b*. Figure 9.12 demonstrates a successful design for keeping stack pollutants away from the containers on a ship's deck. Figure 9.13 shows how smoke can blanket a whole factory under inversion conditions.

Other types of dispersion studies include those of Ref. 9.8 where the probability of recirculation of high-temperature air to a compressor installation, and the effect of the efflux on nearby buildings, was modeled at a scale of 1 to 600.

FIGURE 9.11. The influence of nearby buildings on smoke flow as duplicated in a wind tunnel. In both setups the efflux velocity is equal to the wind velocity and the stack height is 1.5 building height. In (*a*) the building is downstream of the stack; in (*b*), upstream. (Courtesy National Physical Laboratory.)

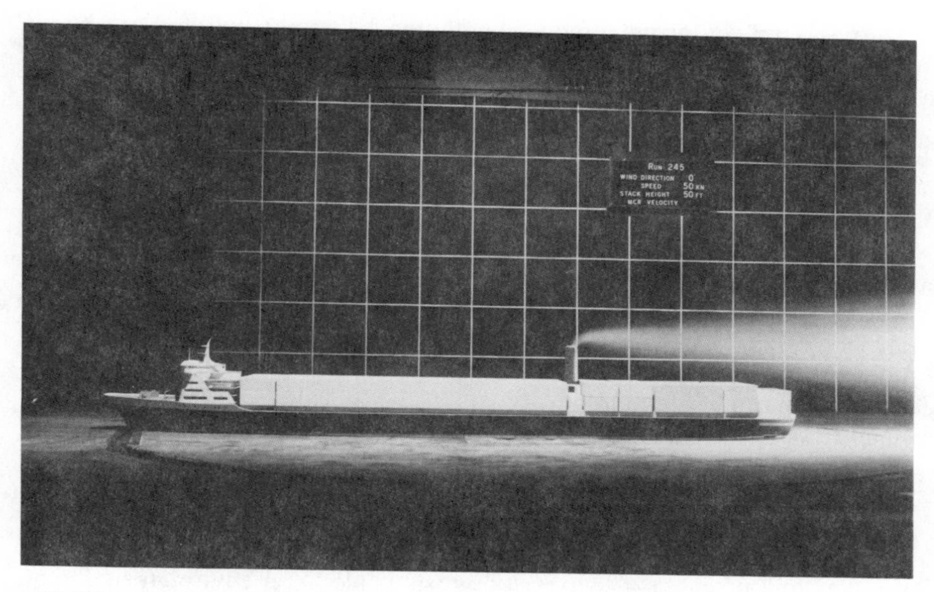

FIGURE 9.12. Measuring smoke emissions from a container ship's engines. (Courtesy Colorado State University.)

FIGURE 9.13. Smoke covering factory during an inversion test. (Courtesy Calspan Corporation.)

An example of terrain effects on pollution is covered in Ref. 9.9, wherein the incredible number of 49 SO_2 emitters are located in a valley. The model was scaled to 1 to 2400, and it is hoped that analyzing the data will enable future site locators to determine the effect their additional plant would have on local levels of SO_2.

9.9. TESTING WIND POWER DEVICES

Currently, and for the forseeable future, there is a lot of interest in developing devices that will either supply direct energy for pumping water for irrigation or to prevent pond freezing, or for the generation of electricity. Tests are usually run at low tunnel speeds, but the tunnel engineer should satisfy himself or herself of the model integrity (and that it has a brake) and that damage to his or her tunnel is not likely to occur. Tests will probably encompass runs under various power loadings and at different Reynolds numbers. It would be preferable to test windmills in the wind gradient that they will eventually see, but this is rarely done. The tunnel engineer should encourage the windmill promoter to have siting tests made in an environmental wind tunnel to get the best results in the field (Refs. 9.12, and 9.13.) Prior to siting studies, one may guess that the winds on a mountain top will run 80–100% higher than along a local plane. A long-term record cannot be established for the mountain location, but a ratio can be, and from this 50–100-year probabilities can be estimated.

At the moment, aside from several exotic but unproven designs, the types of windmills of most interest are the ordinary windmill, the Darrieus, and the Savonius, which is a bucket type. For estimating the maximum power coefficient one should use, respectively, 0.4, 0.4, and 0.3. The lower maximum power coefficient for the Savonius type is offset in practice by its lower manufacturing cost. Measurement of side force is not normally made, but it should be as all types develop small to substantial lateral "lift" force, as does an airplane propeller.

A test of a Darrieus rotor (without boundary layer simulation) is shown in Fig. 9.14 from Ref. 9.14. Presentation of results is shown in Fig. 9.15, where it is seen that Darrieus rotors need help to get started (unlike propellers or Savonius rotors). It is also seen that they are self-braking. Definition of the items of interest are as follows:

$$\text{Power coefficient} \quad C_P = \frac{Q\omega}{\frac{1}{2}\rho_0 V_0^3 A}$$

$$\text{Torque coefficient} \quad C_Q = \frac{Q}{\frac{1}{2}\rho_0 V_0^3 A} \tag{9.2}$$

$$\text{Tip speed ratio} \quad T_R = \frac{R\omega}{V_0}$$

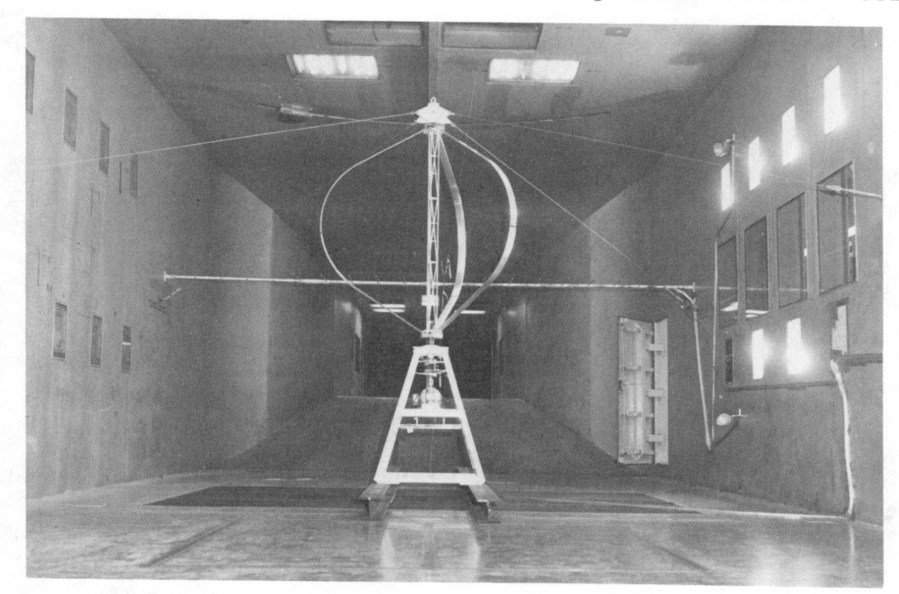

FIGURE 9.14. Testing Darrieus wind turbine in Vought wind tunnel. (Courtesy Sandia National Laboratories.)

where A = swept area of device
R = maximum radius
Q = torque
V_0 = freestream velocity
ρ_0 = freestream density
ω = rotational speed

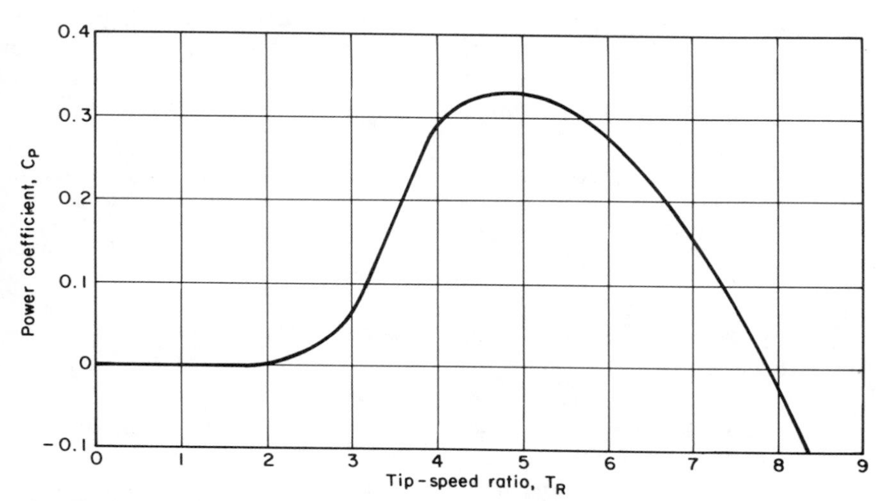

FIGURE 9.15. Power curve for Darrieus rotor. (Courtesy Sandia National Laboratories.)

Testing windmills in a wind tunnel requires the wake-blocking corrections of Section 6.11. The analysis in Section 5.19 demonstrates that a maximum of only 59.4% of the stream power is even theoretically available.

Large installations seem to suffer from making a disturbing buzzing sound, interrupting TV reception, and having instabilities, which lead to a short mechanical lifetime. They should never be put on a rooftop, even to get a high local velocity. Considering that natural winds have many times the total power needed for the entire country, wind tunnel engineers should do their best to help.

9.10. TESTING FOR EVAPORATION

The growing need for studying transpiration from plants and evaporation from open bodies of water has resulted in the construction of wind tunnels in which the moisture content of the air and its temperature may be controlled. Tunnels of this type are in Japan, at the University of Nottingham in England, at the Colorado State University in the United States, and elsewhere. They are all low-speed tunnels employing controlled air exchange.

Transpiration tests are full scale in that the weight of moisture removed from actual plants is studied. Electric lights or other heaters are used for temperature control.

Evaporation is of interest both to agronomists who are concerned with the loss of water from storage areas and channels and to process engineers who have drying problems. Both types of tests have been explored in wind tunnels.

Evaporation tests employ open bodies of water and the surroundings are changed to study the increase or decrease of evaporation. Evaporation, as one might guess, depends on the surface area, the relative dryness of the air, and the effective wind velocity. This last term is the "catch," since the effective velocity may vary from a fraction of 1% to perhaps a few per cent of the nominal velocity, depending on the type of boundary layer formed by the air over the water.

The use of an evaporation constant N has been proposed. It is given as

$$N = \frac{\dot{W} A x}{(C_i - C_o)\nu_e} \tag{9.3}$$

where $\dot{W}$ = weight of water evaporated per square foot per second,
A = area of water, sq. ft,
x = distance measured from upstream edge of water, ft,
C_i = saturation water concentration at the average temperature of the air over the surface, lb/sq. ft,
C_o = water vapor concentration of the ambient air, lb/sq. ft,
ν_e = molecular diffusivity coefficient for water into air, sq. ft/s.

The constant N has been found to be a function of Reynolds number, whether the boundary layer is laminar or turbulent, and for a circular body of water is

$$N = 0.256R^{0.87} \tag{9.4}$$

where $R = V_* x/\nu_e$,
V_* = mean apparent shear velocity at the downstream edge of the water surface, fps.

Again, V_* is only a fraction of the ambient velocity.

Since test velocities approximate those of real wind, and ν_e is the same for model and full scale, wind tunnel tests of evaporation suffer from scale effect. A series of speeds may be used and extrapolation then essayed.

9.11. TESTING FOR SNOWDRIFT PATTERNS

Many people in the southern part of the country do not realize the yearly cost of snow removal and many in the northern part do not realize the savings possible by wind tunnel studies of ameliorating changes (preferably before but sometimes after construction) which reduce local drifting about access areas.

Drifting snow can block doorways and roads and may even inactivate a facility completely. Fortunately, the problem can be studied in a tunnel with good correlation to full scale. Snow patterns that might take years (in the arctic) to accumulate may be duplicated in a few hours. Work of this nature has been done at Colorado State University, the Canadian National Research Council, and elsewhere. In general an environmental wind tunnel as described in Section 9.2 is a necessity in order that the boundary layer approximates real conditions. See Fig. 9.16.

The basic parameters of snowdrift and erosion problems are discussed below, but Refs. 9.15, 9.16, and 9.17 should be consulted when preparing for a test.

When a stream of wind flows over a bed of loose particles, those higher than the others produce more drag until finally they roll and bump in the wind direction. This process is called "saltation" and is responsible for most of the motion (and end deposits) that form snowdrifts or soil erosion. When the wind velocity exceeds about five times the threshold speed, the particles bump hard enough to bounce into the stream when they are then said to be "in suspension."* Modeling to match this phenomenon is by the following ratios:

1. Scale factor d/L, where d = diameter of simulated snow particle (in.) and L = length of a full-scale reference dimension (in.).

* Saltation and suspension can occur at much lower speeds in the presence of falling particles.

FIGURE 9.16. Snowdrift study for new Timberline Lodge (building on right) on Mt. Hood. Note how the tower at the left rear provides self-removal of snow from entrance with prevailing winds from left.

2. Coefficient of restitution e. This concerns the rebound distance/drop distance, and is 0.555 for ice.
3. Particle velocity V_p/V, where V_p = velocity of simulated snow particle (fps) and V = velocity of real snow particle (fps).
4. Fall velocity V_f/V, where V_f = free fall velocity of simulated snow particle, fps. Here we have one of the rare instances where we may "scale gravity"—at least to the extent that the fall velocity may be varied by changes in the particle density.
5. Particle Froude number V^2/gd, where g = acceleration of gravity, ft/s^2.

Selecting a model scale of $\frac{1}{10}$, we find that the test velocity becomes 0.316 full-scale velocity. The fall ratio may be maintained by using borax ($Na_2B_4O_7$), whose density providentially yields the right value and whose coefficient of restitution is 0.334—lower than ice, but possibly close to snow—and whose diameter may be controlled to be $\frac{1}{10}$ that of snowflakes.

In a tunnel saltation has been found (for the above simulated snow) to occur at 11 mph without snow falling, and at lower velocities when snow is falling.

Data for snow tests are obtained in the form of photographs and depth contours. One substantial contribution from tunnel tests is the technique of reducing drifting by erecting a building on piles with a free space beneath. This has been applied in arctic building. Strom in Ref. 9.16 has an authoritative discussion of the snow simulation problem.

9.12. TESTING FOR SOIL EROSION

Soil erosion caused by wind is of interest to the agronomist from the standpoint of losing topsoil. The road engineer, on the other hand, would like to see his or her roads stay clear. Another facet of soil erosion is the damage done to car windshields and paint by windborne particles.

The mechanism of natural soil pick-up is beginning to be understood. Submicron particles will not erode from a smooth surface in a wind of gale velocities, but mixed with 5- to 50-micron grains they become highly erodable. Fundamental studies of large-scale grain loads have been made using strain-gage-mounted grains.

The important area of wind tunnel testing for soil erosion is not to be entered lightly, and the authors suggest first consulting the soil test references in Ref. 9.18. Much of the correlation with full scale is encouraging.

Another type of erosion is the determination of the wind speed at which gravel begins to be eroded from the roof of a building, damaging the roof, nearby parked cars, and passersby. Reference 9.19 covers this type of test.

9.13. TESTING AUTOMOBILES AND TRUCKS

Testing autos and trucks gives the tunnel engineer a little more than average fun because he or she probably owns a car (and probably not an airplane) and will enjoy telling others how the mysterious device he or she employs helps improve the mileage of the car he or she or anybody else drives, and because it gives him or her an insight into future models.

The automotive engineer uses two types of wind tunnels: standard tunnels for testing scale models (mostly for drag) and full-scale automotive tunnels, wherein the car or small truck is tested with the engine running and putting out the proper power for the speed. It is interesting that there is no counterpart of the full-scale auto test in the aviation world. That would require an enormous tunnel with incredible heat removal capabilities. As a counterbalance, a wind tunnel test of an airplane model does not require the presence of a host of experts covering each of the many auto components which may be under test.

We will first discuss auto and small truck models in conventional tunnels. Then we will cover the requirements of a full-scale auto test tunnel, and finally testing under full-scale conditions.

9.14. TESTING SCALE MODELS OF CARS AND SMALL TRUCKS

Because typical car drag at 55 mph accounts for about 60% of the total power, the great preponderance of the scale model testing is concerned with the reduction of drag. Table 9.2 shows some average values.

TABLE 9.2. Typical Values of Drag Coefficient Based on Frontal Area for Several Types of Cars and Trucks

Configuration	Drag Coefficient Based on Frontal Area
Flat plate normal to wind	1.2
Truck and trailer	0.8
Station wagon	0.6
Good four door	0.4
Optimum	0.25

A fine example of a case where wind tunnel tests *were* used to great advantage was reported by Schlichting in Ref. 9.20. Here the fuel consumption of a severely squared-off bus was decreased substantially by paying attention to tunnel tests (Fig. 9.17).

Occasionally there are tests for down force produced by small wings for racing cars and, rarely, determining pitch divergence regions to avoid catastrophic soaring of record attempting race cars. The usual reasons apply for using wind tunnels instead of road tests:

1. It is difficult to make good measurements from a vibrating platform.
2. Lack of any control over environmental conditions.
3. The ease of making changes to small models.

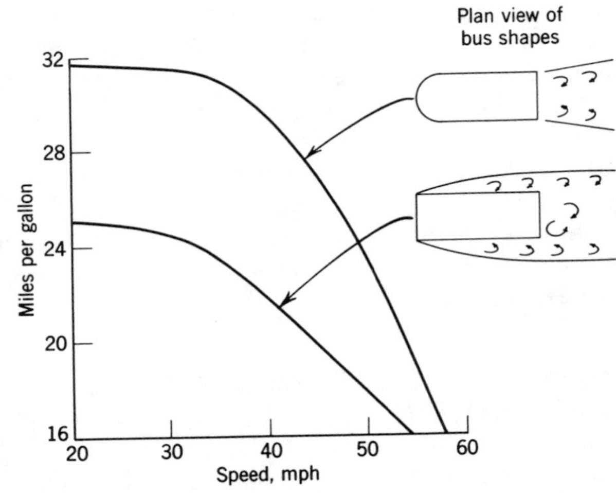

FIGURE 9.17. Increase of gas mileage obtained by rounding front of Volkswagen bus. (Redrawn from Ref. 9.20).

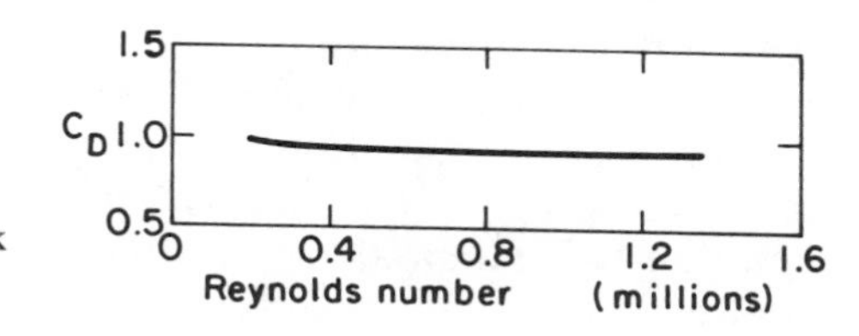

FIGURE 9.18. The independence of truck drag with Reynolds number.

The high drag of most auto models results in excessive wake blockage corrections unless the model is kept fairly small: frontal area ratios of 5% for closed test sections and 10% for open ones. Two types of very interesting work have been reported that would permit larger models. Reference 9.21 seems to show that area ratios of 16% or even 21% may be possible with a slotted test section and proper selection of the model test location. Reference 9.22 studies the approach of contouring the test section walls to match a streamline not far from the model. This procedure could permit even larger models than given above.

A second big problem with testing any surface vehicle is dealing with the boundary layer if the model is placed on the tunnel floor. Tunnel engineers prefer to keep the displacement boundary layer to less than 0.1 of the body clearance. This dictates using a ground board or a moving belt under the car, or removing the tunnel boundary layer with a suction slot ahead of the test section. All of these work, along with the complications of added set-up time for the ground boards and the fact that the belt is not yawed for yaw tests. The slot saves ground board installation time and loss of tunnel test-section area. It does require extra calibration runs because some of the tunnel air is now passed around the test section.

It is also necessary to keep the tunnel cool enough so that modeling clay will not sag or fall off.

Runs consist of first determining that the tunnel is capable of developing a high enough Reynolds number to get on the drag plateau (Fig. 9.18), usually above 1×10^6 based on model width. If this is the case, runs are made varying yaw up to about 15°. The yaw runs are important since sidewinds exist most of the time, and streamling has to include drag at small angles of yaw. The yaw runs are probably not very realistic since natural side winds have a very thick boundary layer. Wheels are usually not spun for model tests; spinning has more effect when the wheels are exposed. Presentation of car data follows that of truck/trailer data in Section 9.15.

9.15. SCALE-MODEL TESTING OF TRUCKS WITH TRAILERS

Few aerodynamicists who have been buffeted from the "bow wave" of an oncoming truck have not felt strongly that "they could do better" as regards the excessive drag that the bow wave represents. Accordingly many truck configurations have been tested in wind tunnels, usually $\frac{1}{10}$ to $\frac{3}{10}$ scale models, almost never with running wheels. The usual procedure is to mount the truck–trailer combination over a ground board that has been adjusted to have zero pressure gradient with the tunnel empty, and to use the ground-board-in calibration along with the regular tunnel piezometer rings in the contraction section. Either the six-component external balance or an internal sting balance may be used.

The flow is so bad over, around, and under a trailer–truck, and subject to such large changes when a configuration or tunnel flow change is made, that the comparison of data taken in different tunnels with full scale is not really good. It is clear that more work needs to be done to get reliable data for the effect of configuration changes. Very probably, as the truck designs show improved aerodynamics, the correlations will improve. (See Ref. 9.24.) As a matter of interest, the most pressure that can be developed over the flat front of a truck is something under $+1.0q$, where q is the dynamic pressure, and over the flat rear about $-0.5q$. Thus it is easier to reduce drag by working on the front than streamlining the rear, although both are desirable.

In general, test engineers do some preliminary smoke tests to get a feel as to just how bad the flow around the truck is. Then a Reynolds number run is made to ensure that the tests are in the flat area as far as drag at $\psi = 0°$ and, say, 7° is concerned. This often requires Reynolds numbers greater than 500,000 based on truck width. Then measurements are taken at a single speed up to ±15° yaw. Typical presentation of truck data is in Fig. 9.19. Coefficients are defined as follows:

A	projected frontal area of model
C_D	D/q_0A
C_L	L/q_0A
C_S	S/q_0A
C_{PM}	PM/q_0AW
C_{RM}	RM/q_0AW
C_{YM}	YM/q_0AW
D	aerodynamic drag force acting parallel to the groundplane and in the longitudinal vertical centerplane of the model, and positive aft
L	aerodynamic lift force acting normal to the ground and positive upward
q_0	freestream dynamic pressure, $\frac{1}{2}\rho U_0^2$
S	aerodynamic side force acting normal to the vertical centerplane of the model, and positive to the right
W	model width
PM	aerodynamically induced pitching moment; moment about a horizontal axis in the groundplane which is normal to the vertical centerplane of the model, and positive when it tends to raise the vehicle's nose

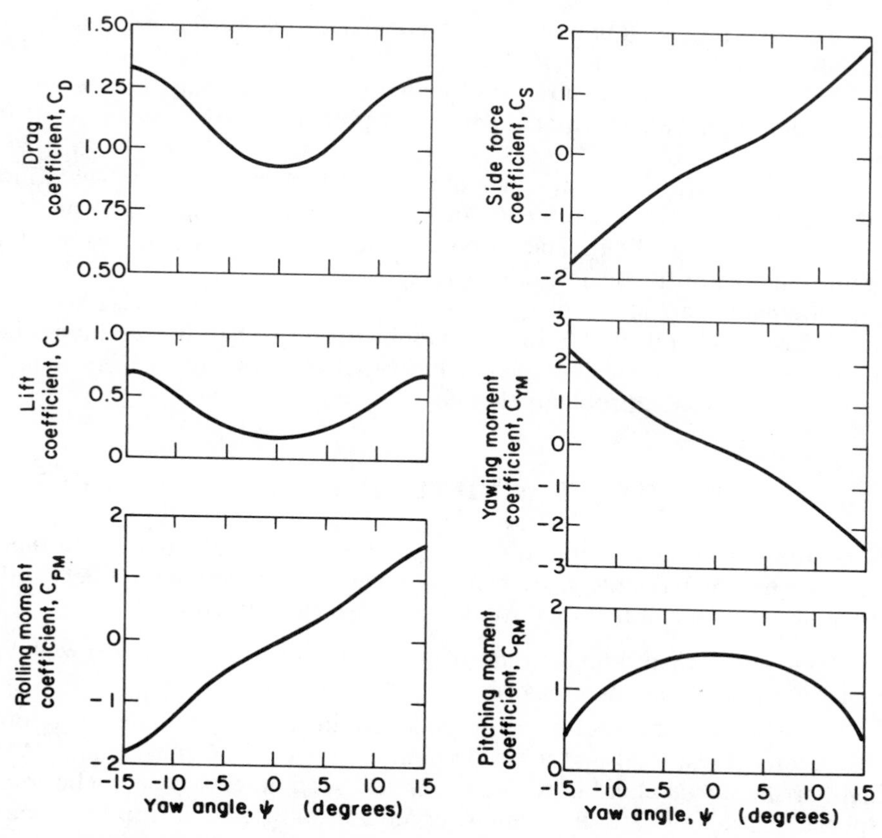

FIGURE 9.19. Presentation of typical truck/trailer data. (From Ref. 9.25.)

RM aerodynamically induced rolling moment; moment about a horizontal axis located in the groundplane and in the vertical centerplane of the model, and positive when it tends to lower the right side of the vehicle

YM aerodynamically induced yawing moment; moment about a vertical axis fixed in the vertical centerplane of the model, and positive when it tends to rotate the vehicle nose to the right

Of particular interest is the large increase of drag with yaw, especially when one realizes that winds almost never blow directly along a curving highway. Unreported studies seem to indicate that assuming an average cross wind of 7 mph is probably reasonable. A second problem, especially with trucks towing trailers, is the overturn moment, which can easily exceed the weight-restoring moment. This is the reason that one sees trucks parked along the highway during a windy day. This factor also argues for getting some roll and side force data in the 90° yaw position.

Two areas that need exploring are tandem and side-by-side driving. Side-by-side driving is requested by the state police during windy days out west, and it would be useful to know how much it helps. There may be substantial economies from tandem driving, provided proper controls can be worked out to avoid rear-ending.

Several truck/trailer companies are considering having their own wind tunnels, both for the convenience of rapid entry to testing and to protect proprietary improvement. The authors are unaware of any plans to build a tunnel capable of full-scale truck testing.

A companion area for trailer wind load and overturn moments is of course the mobile home industry. While the mobile homes or half-homes cannot be designed for minimum roll moments during delivery, the tow drivers should be well aware of the problems that could arise.

9.16. AUTOMOTIVE WIND TUNNEL DESIGN FEATURES

Wind tunnels designed for testing automobiles and small trucks with their engines running have the same general layout as conventional tunnels with the following special features usually provided (Ref. 9.23):

1. A lower than conventional high speed. Tests are usually run at 55 mph, and tunnel maximum speeds are around 120 mph.

2. As large a test section as possible within space and tunnel cost considerations (Figs. 9.20 and 9.21). Desirable would be 20 ft wide and 15 ft high, or slightly smaller but with an open test section, closed on the bottom. Special test sections features which permit unusually large models have been proposed (see Sections 9.13 and 9.14) but have not yet been applied.

FIGURE 9.20. General Motors automotive wind tunnel.

FIGURE 9.21. Full-scale car in General Motors wind tunnel.

3. A turntable provided with roller sets for all four wheels. These rollers are connected to dynamometers so that engines can be run to produce the same horsepower that would be called for in real life, making cooling tests possible. Rollers must be adjustable in tread width and dynamometers must be arranged to absorb power for either front or rear drive systems.

4. A substantial cooling system, usually consisting of a scoop to remove engine heated air, and a special exhaust removal system to keep the tunnel air free of contaminants. The exhaust system must not be a direct connection to the car's exhaust pipe or it would act like a supercharger.

5. A capability to run at very low speeds and yet remove the engine heat and exhaust, and in some cases to run backward at a low velocity to simulate being stopped at a light with a slight tailwind blowing.

6. A slot across the test section floor near the entrance cone to remove the boundary layer, or at least 50% of it. Tunnel calibrations must be made with and without the boundary layer removal system functioning, because it will pass some of the tunnel air around the test section.

7. A tunnel refrigeration system adequate to keep the tunnel cool enough that clay may be used for styling changes (Fig. 9.22).

8. A rain simulator so that both windshield wiper operation may be checked, and design changes made to keep the side windows clear of water. Freezing rain is also needed. Leaks may be discovered, but are considered bad manufacturing procedures rather than part of the tests.

9. Auxiliary hot water facilities such that cars can have cooling studies made without the engines running; hot water being pumped through the cooling system and its efficacy measured. This is done to make sure styling changes have not hurt cooling prospects in proposed designs for which engines are not yet ready.

FIGURE 9.22. Working on car model in General Motors wind tunnel.

While the auto manufacturers would like to be able to study car noise both inside and outside the body, wind tunnel noise runs about 80 decibels at 50 mph and so far does not permit it. (See Section 9.25.)

9.17. TESTING FULL-SCALE CARS

The capability of operating car engines in a full-scale wind tunnel and thus ensuring adequate cooling for a wide number of driving conditions demonstrates the worth and utility of automotive wind tunnels. General cooling conditions may be simulated for level highway, mountain grades, city traffic, and build-up resulting from highway heating followed by reduced air and coolant flow such as encountered when leaving a freeway and in short-period stopping of the engine in a filling station, or continued idle in off ramp traffic. In some of the above cases, the added load of a trailer can be simulated by programming the dynamometers. While the primary target may be the monitoring of the coolant temperature,* it may be necessary to measure temperatures in as many as 200 places. These include the transmission, thermal protection systems, passenger compartment, underbody systems,

* The coolant must be kept below 267°F, which is the boiling point for a 50–50 (by volume) mixture of water and ethylene glycol with a 16 psi limit.

and other components that are temperature sensitive. The trend toward having a smooth underside is exacerbating many cooling problems.

The best part of using a climatic wind tunnel for all this is that there is no wait for ambient conditions, nor are there long drives to find the grades needed.

A new problem is high catalytic converter temperatures after overrich engine conditions arising from a stuck choke or bad plugs. A test program for a new design might encompass the following:

1. Determining actual drag, particularly with slight yaw, and other force and moment values.
2. Determining cooling of the car under a number of operating conditions (which rarely can be duplicated for recheck in the outdoors).
3. Determining cooling under special conditions such as idling with a slight tailwind, cooling at a traffic light after a freeway run with air-conditioner on; cooling of transmission oil.
4. Cooling of interior with sun lamps (Fig. 9.23).
5. Braking power and cooling.
6. Windshield wiper operation under many weather conditions.
7. Finding consequences of mistakes such as power erection of a convertible top at 55 mph. Operation of disappearing headlights or wipers after icing or at high speed.
8. Cooling while towing trailer up long hill, low speed.

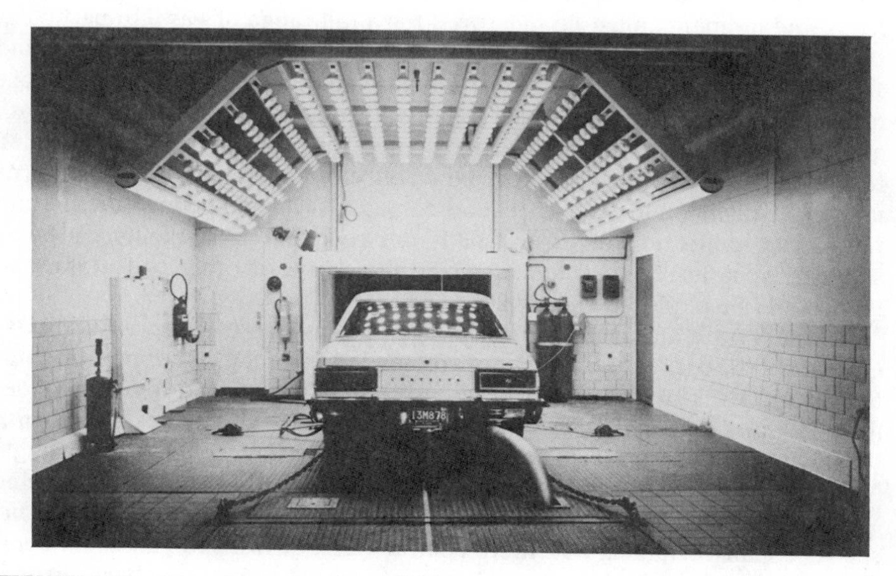

FIGURE 9.23. Set-up for testing air-conditioning system. (Courtesy Chrysler Motors Corporation.)

9. Effect of real car details on drag.
10. Rain deflection.
11. Exhaust gas recirculation.
12. Check of ventilation systems.

A cooling run would include warm-up, a selected speed and load with duration being either time (typically 30–45 min) or miles (such as a 20 mile grade). This could be followed by a 15–45 s stop and a heat soak either with or without engine idle for 10 min. Usually all this would be done with an ambient temperature of 110°F, except that for mountain grades a more reasonable 85°F is used.

Component cooling tests are also made with thermocouples on various engine components. Tests have shown that spinning the wheels when they are partially shrouded (as with a passenger car) is probably not necessary. Data presentation for the forces and moments usually follows the charts shown in Fig. 9.19. The cooling data may be plotted against time.

9.18. TESTING SOLAR COLLECTORS

The interest in solar energy has spawned a need for wind loads and moments on the various solar collectors, usually of parabolic cross section, of various aspect ratios, and arranged in various arrays all the way from being in sheltered ground installations to being on rooftop locations subjected to all sorts of local wind concentrations. Since winds come from all directions, loads and moments must be measured for a full range of yaw. In particular there is interest in drag (also called lateral load) and pitching moment and with the loads along the long axis. Yawing moments and roll moments are measured but are usually of lesser import. Besides needing loads for strength and preservation of the proper focal distance, the pitching moment is needed to size the drive motor which keeps the collector aimed at the sun, and turns it over at night or in hail to reduce damage to the reflecting surface and reduce the collection of dust. Loads are reduced substantially by being shielded by a nearly solid fence around the array, or other collectors, and this should be explored (Fig. 9.24).

The test program usually consists of force, moment, and pressure data. Obviously the forces and moments are needed from the standpoint of foundation and structural design. The pressure data are needed for holding deflection limits of the reflector itself. An atmospheric-type tunnel is used with the boundary layer modeled according to Eq. (9.1). The collector pitch angle is varied from $-180°$ to $+180°$ and yaw varied from 0° to 90°. Array spacing and distance above ground are additional variables. Runs at several airspeeds usually establish that there is little variation with Reynolds number. The tunnel speed may be set at 90 mph at 30 ft altitude, or higher speeds if the proposed site requires it.

FIGURE 9.24. Solar collectors being tested behind wind barriers. (Courtesy Sandia National Laboratories.)

$$C_{\text{force}} = \frac{\text{Force}}{qA}$$

$$C_m = \frac{\text{Moment}}{qAc}$$

where c = gap span
$q = \frac{1}{2}\rho V_{CL}^2$
$A = Lc$
V_{CL} = velocity at trough centerline

For estimating balance loads, one could expect maximums of $C_D = 2.0$ and $C_L = 2.0$. Data may be presented as plots of lateral force, lift force, and pitching moment coefficients against pitch angle for various heights above the ground, aspect ratios, and collector designs, and at fixed pitch angles for a range of yaw. Yaw normally does not produce large changes in forces and moments. Owing to the buffeting of such irregular nonstreamlined shapes, it will be necessary to take a large number of data points and compute average values. (See Ref. 9.26.)

9.19. TESTING RADAR ANTENNAS

The same type of tests described above may be made for radar antennas or other dish-type receivers. A major difference is that one would not expect to find them in arrays, and more than likely radars will be placed on mountain tops where signals may be received from 360°. As local winds are terrible, it is not unusual to find them protected by radar-transparent domes.

FIGURE 9.25. Testing a model of a radio telescope in one of the tunnels at the National Maritime Institute, England.

For wind tunnel tests of exposed antennas the model should be mounted in the tunnel on top of the same structure it will see in the field—trailer, antenna stand, or small house (see Fig. 9.25)—and measurements of drag, side force, and torque should be made every ten degrees or so from $-10°$ to $+190°$. If there is any question at all about structural integrity, pressure measurements should be taken near corners, particularly on the dome. Data may be put in coefficient form by using the dynamic pressure, maximum projected frontal area, and dish diameter or span. For essentially all designs the sharp edges of the dish preclude effects due to Reynolds number. Some antenna data are shown in Fig. 9.26. The drag shown includes that of the supporting structure and is therefore high in comparison to that for the dish only.

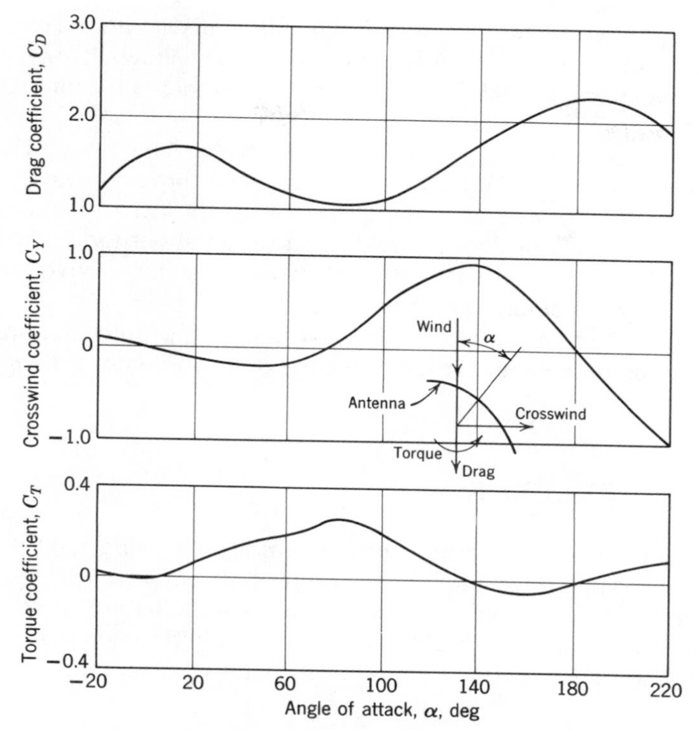

FIGURE 9.26. Presentation of data from elliptic dish-type radar antenna. C_D includes support drag.

9.20. TESTING SAILS AND SHIPS

Sails have been tested in wind tunnels in limited numbers over the years. Added impetus has recently been given to such tests by the construction of the Southampton University tunnel in England especially for such work. In general, sail tests embrace measurements of side force, drag, yaw, and roll on a model mounted on the tunnel floor in an atmospheric tunnel where the boundary layer is properly simulated.

Sail material should be varied during a test to see what effect, if any, arises. Material roughness and porosity will probably be out of scale, and several variations of each should be tried for comparison.

Ship tests have been somewhat more extensive. These usually embrace a floor or ground plane model cut off at the waterline, with measurements made of side force and drag only, although yaw and roll would be of interest. Again, a wind gradient should be provided as above.

Both sail and ship model tests should have their data corrected for wake and solid blockage. The models should be kept small enough so that at 90 deg yaw bow and stern remain no less than half a ship length from the tunnel walls. If a ground plane is used, the difficulties discussed in Section 6.29 should be evaluated. Extreme care to duplicate model detail, such as ships' railings, ventilators, mast detail, etc., are not warranted in studies to reduce aerodynamic drag.

Tests of speed boats are primarily to find a body shape that has minimum nose-up characteristics. Here the model is tested through a range of pitch angles about the stern, and the angle at which the aerodynamic moment overcomes the moment due to gravity and thrust about the stern is determined. The current wide, flat-bottomed speedboats can survive only a few degrees of nose up before they become unstable.

A new problem that arises with tankers carrying liquid natural gas is ascertaining that the vents needed as the gas boils off do not constitute a fire hazard.

9.21. WIND AGRONOMY

Pollination for many agricultural products is by the wind, and thus plant distribution and planting patterns must be properly employed. Trees near the edge of the fields may need to be cut to improve the natural winds. On the other hand, some plants do better without a lot of buffeting by the wind, and these need special attention. The area of wind engineering as applied to agriculture is only in its infancy.

9.22. TESTING WINDBREAKS

The efficacy of a windbreak has been mentioned in connection with reducing the loads on solar collectors. A second and important use is to reduce winter heating loads by reducing the convective cooling of a house, and a third is to increase the yield from some types of wheat that do not like being blown about. Instrumentation downstream of the windbreak should continue for at least six times its height, and data averaging employed to determine the windbreak efficiency.

9.23. TESTING INSECTS, BIRDS, AND PEOPLE

Over the years a number of wind tunnel tests have been made of natural fliers, alive, frozen, and simulated. Initially experimenters were seeking mysterious and incredibly efficient devices that nature's creatures were supposed to have. No such things have been found; nor are they needed to explain natural flight. The high landing angles of some birds have been duplicated with highly latticed wings, and bird power has been estimated to be in line with demonstrated performance. Differences which remain may be explained by the sculling action of feathers reported by some observers.

Live insects have been somewhat more cooperative than birds, and have flown in tunnels for close observation.

Since live models are expected to be quite small, no special wind tunnel techniques are needed. The authors suspect that they will produce more hilarity than useful data.

More recently, attention has been paid to people-drag of the type encountered by bike racers and skiers. The bicycles should be arranged so that the biker can pedal, and with a belt so that the front wheel rotates along with the back one. Tests have shown a change of drag with the number of wheel spokes. For skiers, the "model" is mounted in the tunnel and during a run at 55 mph (which is close to full scale "flight") assumes a series of positions endeavoring to learn which minimizes his/her drag. The fascinating part of these tests, and they are not without hilarity despite the seriousness of the end results, is that the model corrects his/her own drag by watching a drag indicator. Position changes are shown on frontal and side TV projectors, and coaching suggestions sent in as needed. A programmed computation in real time is also presented so the skier is shown how much each change helps. Record photographs are taken. Substantial improvements in clothing have resulted from such programs. The model set-up is shown in Fig. 9.27 (from Ref. 9.27).

9.24. TESTING UNDERWATER VEHICLES

There are many underwater devices whose characteristics may be determined in the wind tunnel. For those intended to operate entirely submerged there will be no free surface to worry about and the test need not be run at equal Froude numbers for model and full scale, although equal Reynolds numbers would be desirable. Submarine control surface design has been based largely on data obtained from panel models mounted on the tunnel floor. More recently, flow conditions have been better simulated by mounting the surfaces on at least a portion of the submarine hull, and the results have shown this sort of setup to be desirable. However one should make sure that a sufficient portion of the hull is simulated.

9.25. TESTING FOR NOISE

Studies of flow-generated noise from submarines, ships, other types of marine vehicles and appendages and their wake distributions have long been a problem in the military world, just as noise suppression for land vehicles has been in the civilian area. Realizing the advantages of holding the model and

FIGURE 9.27. Skier in wind tunnel. (Courtesy Calspan Corporation.)

the measuring instrumentation still and letting the fluid move, as well as the enormous cost of running full-scale tests, and that noise suppression is more easily handled with air than with water, the Naval Ship Research and Development Center designed and built a unique wind tunnel known as the Anechoic Flow Facility. While basically a single return wind tunnel with a closed test section upstream of an open one, its other features both as a wind tunnel and low noise facility have made an enormous contribution to the world of wind tunnels for others to copy.

A sketch of the tunnel is shown in Figure 9.28. The special features include:

1. The use of a wide angle diffuser to permit a contraction ratio of 10 to 1 without a long diffuser and return path with its high construction costs.

2. The use of two 100° turns and two 80° turns instead of the more customary four 90° turns. This permitted the length needed for the fan noise suppressors, again with a shortened passage.

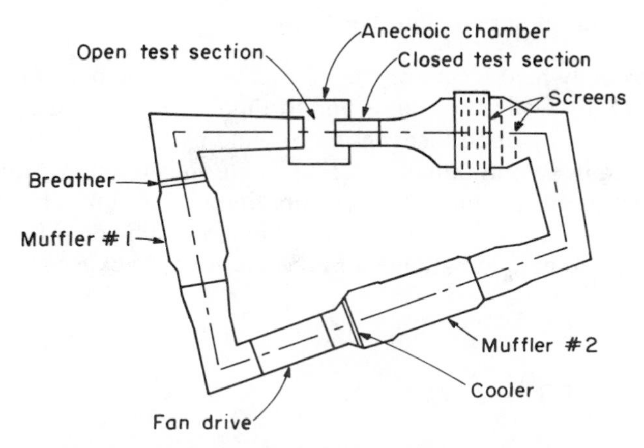

FIGURE 9.28. Outline of Anechoic Flow Facility.

FIGURE 9.29. Model being tested in Anechoic Flow Facility. (Courtesy David Taylor Navel Ship Research and Development Center.)

3. Extremely heavy concrete construction plus the use of noise suppression materials on walls and ceiling and turning vanes, and an anechoic chamber surrounding the open test section to yield by far the lowest noise levels ever achieved in a wind tunnel. Section isolation is practiced throughout, as well as isolation of the entire tunnel from the ground through several feet of crushed rock. Figure 9.29 shows a model in approach condition being tested for noise. The anechoic baffles can be seen in the background.

REFERENCES

9.1. J. E. Cermak, Wind Tunnel Testing of Structures, JEMD, *ASCE* **103,** 1977.

9.2. J. E. Cermak, Wind Tunnel Design for Physical Modeling of the Atmospheric Boundary Layer, *Proceedings ASCE* **107,** 1981.

9.3. E. J. Plate and J. E. Cermak, Micrometeorological Wind Tunnel Facility, Description and Characteristics, *CER* **63,** 1963.

9.4. W. A. Dalgliesh, K. R. Cooper, and J. T. Templin, Comparison of Model and Full-Scale Accelerations of the High Rise Building, NRC Presentation Paper, 1983.

9.5. W. A. Dalgliesch, J. T. Templin, and K. R. Cooper, Comparisons of Wind Tunnel and Full Scale Building Surface Pressures with Emphasis on Peaks, DBR Paper 961, NRC, 1979.

9.6. H. P. A. P. Irwin and G. D. Schuyler, Wind Effects on a Full Elastic Bridge Model, ASCE Preprint 3268, 1978.

9.7. R. L. Wardlaw, Sectional vs. Full Model Wind Tunnel Testing of Bridge Road Decks, DME/NAE Quarterly Bulletin No. 1978 (4), 1979.

9.8. G. R. Ludwig, Wind Tunnel Model Study of the Hot Exhaust Plume from the Compressor Research Facility at WPAFB, *TR* AFAPL TR 77-58, 1977.

9.9. G. R. Ludwig and G. T. Skinner, Wind Tunnel Modeling Study of the Dispersion of SO_2 in South Allegany County, EPA 903/9-75-019, 1976.

9.10. G. T. Skinner and G. R. Ludwig, Physical Modeling of Dispersion in the Atmospheric Boundary Layer, Calspan #201, 1978.

9.11. G. T. Skinner and G. R. Ludwig, Experimental Studies of Carbon Monoxide Dispersion from a Highway Model in the Atmospheric Simulation Facility, Calspan NA 5411-A-1, 1976.

9.12. R. N. Meroney, Sites for Wind Power Installations, Proceedings Workshop on Wind Energy Conversion Systems, Washington D.C., June 9–11, 1975 CEP 74 RNM 87.

9.13. R. N. Meroney, Prospecting for Wind: Windmills and Wind Characteristics, Reprint 3555, CSU CEP 78-79 RNM 1979.

9.14. B. F. Blackwell, R. E. Scheldahl, and L. V. Feltz, Wind Tunnel Performance for the Darrieus Wind Turbine with NACA 0012 Blades, Sandia Laboratories, UC-60, 1977.

9.15. D. Kabayashi, Studies of Snow Transport in Low Level Drifting Snow, Institute of Low Temperature Science, Sapporo, Japan, Report A-31, 1958.

9.16. G. Strom et al., Scale Model Studies of Snow Drifting, Research Report 73, U.S. Army Snow, Ice and Permafrost Research Establishment, 1962.

9.17. R. J. Kind, A Critical Examination of the Requirements for Model Simulation of Wind Induced Phenomena Such as Snow Drifting, Atmospheric Environment, Pergammon Press, New York, 1976, Vol. 10, p. 219.

9.18. J. E. Cermak, Note on the Determination of Snow Drift Locational Stability. *CER* 66-67, 1966.

9.19. R. J. Kind, Further Wind Tunnel Tests on Building Models to Measure Wind Speeds at which Gravel is Blown off Roofs, NRC LTR LA-189, 1977.

9.20. H. Schlichting, Aerodynamic Problems of Motor Cars, AGARD Report, 307, 1960.

9.21. S. Raimondo and P. J. F. Clark, Slotted Wall Test Section for Automotive Aerodynamic Test Facilities, AIAA paper 82-0585, 1982.

9.22. J. D. Whitfield, J. L. Jacobs, W. E. Dietz, and S. R. Pate, Demonstration of an Adaptive Wall Concept Applied to an Automotive Wind Tunnel, AIAA paper 82-0584, 1982.

9.23. K. B. Kelly, L. G. Provencher, and F. K. Schenkel, The General Motors Aerodynamics Laboratory, A Full Scale Automotive Wind Tunnel, *SAE* 820371, 1982.

9.24. K. R. Cooper et al., Correlation Experience with the SAE Wind Tunnel Test Procedure for Trucks and Buses, *SAE* 820375, 1982.

9.25. L. T. Duncan, Aerodynamic Evaluations of the 1980 F-Series Light and Medium Trucks, *SAE* 801405, 1980.

9.26. D. R. Randall, Donald McBride, and Roger Tate, Steady State Wind Loading on Parabolic Trough Solar Collectors, Sandia National Laboratories 79-2134, 1980.

9.27. M. S. Holden, Aerodynamics in Alpine and Nordic Skiing Competition, Calspan Report 1981.

9.28. J. A. Peterka and J. E. Cermak, Adverse Wind Load Induced by Adjacent Buildings, ASCE Preprint 2456, 1976.

9.29. K. R. Cooper and J. A. Watts, Wind Tunnel and Analytical Investigations Into the Aeroelastic Behavior of Bundled Conductors, *IEEE Trans. PAS* **94,** 1975.

APPENDIX

Numerical Constants and Conversion of Units

1. Speed of Sound.
   $a = 49.01\sqrt{°R} = 65.77\sqrt{°K}$, ft/s.
   °R = °Fahrenheit + 459.6.
   °K = °Centigrade + 273.0.
2. Standard Sea-Level Conditions.
   Pressure p_0 = 14.7 lb/in.^{2} = 29.92 in. mercury.
   Density ρ_0 = 0.002378 slug/ft^3.
   Viscosity $\mu_0 = 3.74 \times 10^{-7}$ lb-s/ft^2.
   Speed of sound a_0 = 761 mph = 1116 ft/s.
   Temperature t_0 = 59°F.

3. Standard Atmosphere. Temperature decreases 1°F for each 280 ft of altitude until 36,500 ft. From 36,500 to 82,000 ft temperature is constant at −69.7°F. Pressure decreases according to

$$p = (1.910 - 0.01315Z)^{5.256}$$

up to 36,500 ft, and according to

$$p = 6.94e^{(1.69-0.0478Z)}$$

from 36,500 to 82,000 ft. In both formulas above, Z is in thousands of feet, and p is in inches of mercury. Density decreases according to

$$\rho = \rho_0 \frac{p}{p_0} \frac{T_0}{T}$$

Viscosity varies with temperature according to

$$\mu = 2.27 \frac{(°\text{R})^{1.5}}{°\text{R} + 198.6} \times 10^{-8} \frac{\text{lb-s}}{\text{sq. ft}}$$

4. Conversion Factors

Multiply	by	to obtain
A. LENGTH		
Inches	2.54	centimeters
Feet	30.48	centimeters
	0.3048	meters
Miles	5280	feet
	1.609	kilometers
	0.8684	nautical miles
Centimeters	0.3937	inches
Meters	39.37	inches
	3.281	feet
	1.094	yards
Kilometers	3281	feet
	0.6214	miles
	1094	yards
B. AREA		
Square inches	6.452	square centimeters
Square feet	929.0	square centimeters
	144	square inches
Square centimeters	0.1550	square inches
Square meters	10.76	square feet
C. VOLUME		
Cubic feet	1728	cubic inches
	0.02832	cubic meters
	7.4805	U.S. gallons

Multiply	by	to obtain
Imperial gallons	0.1605	cubic feet
	4.546	liters
	277.4	cubic inches
U.S. gallons, liquid	0.1337	cubic feet
	231	cubic inches
	0.83267	imperial gallons
	4	U.S. quarts
Cubic meters	35.31	cubic feet
	1.308	cubic yards
	264.2	U.S. gallons

D. VELOCITY

Multiply	by	to obtain
Feet/minute	0.01667	feet/second
	0.01136	miles/hour
Feet/second	1.097	kilometers/hour
	0.5921	knots
	0.6818	miles/hour
Miles/hour	0.447	meters/second
	1.467	feet/second
	1.609	kilometers/hour
	0.8684	knots
Kilometers/hour	0.9113	feet/second
	0.5396	knots
	0.6214	miles/hour
	0.2778	meters/second
Meters/second	3.281	feet/second
	3.6	kilometers/hour
	2.237	miles/hour
Knots	1.152	miles/hour

E. WEIGHT

Multiply	by	to obtain
Ounces (avoirdupois)	0.0625	pounds (avoirdupois)
Pounds (avoirdupois)	16.0	ounces (avoirdupois)
Tons (short)	2000	pounds (avoirdupois)
	907.18	kilograms
	0.90718	tons (metric)
Tons (long)	2240	pounds (avoirdupois)
	1016	kilograms
Tons (metric)	1000	kilograms
	2205	pounds
	1.1025	tons (short)
Kilograms	2.2046	pounds

Multiply	by	to obtain
F. PRESSURE		
Pounds/square inch	51,710	microns
	0.06804	atmospheres
	2.036	inches of mercury
	703.1	kilograms/square meter
Pounds/square foot	0.1924	inches of water
	4.883	kilograms/square meter
Atmospheres	76.0	centimeters of mercury
	29.92	inches of mercury
	1.033	kilograms/square centimeters
	14.7	pounds/square inch
	2116	pounds/square foot
Inches of water	5.204	pounds/square foot
	25.40	kilograms/square meter
	0.07355	inches of mercury
Kilograms/square meter	0.2048	pounds/square foot
Microns (of mercury)	0.00001934	pounds/square inch

G. TEMPERATURE

To change Fahrenheit to Centigrade

1. Add 40.
2. Multiply by $\frac{5}{9}$.
3. Subtract 40.

To change Centigrade to Fahrenheit

1. Add 40.
2. Multiply by $\frac{9}{5}$.
3. Subtract 40.

Index